A Textbook on Modern Quantum Mechanics

A Textbook on Modern Quantum Mechanics

A C Sharma

CRC Press is an imprint of the
Taylor & Francis Group, an informa business

First edition published 2022
by CRC Press
6000 Broken Sound Parkway NW, Suite 300, Boca Raton, FL 33487-2742

and by CRC Press
2 Park Square, Milton Park, Abingdon, Oxon, OX14 4RN

CRC Press is an imprint of Taylor & Francis Group, LLC

ISBN: [978-0-367-72344-6] (hbk)
ISBN: [978-0-367-72347-7] (pbk)
ISBN: [978-1-003-15445-7] (ebk)

Typeset in Palatino
by KnowledgeWorks Global Ltd.

Access the Support Material: https://www.routledge.com/9780367723446

Contents

Preface

Quantum mechanics not only describes the microscopic world but is also essential to explain many macroscopic quantum systems such as semiconductors, superconductivity, and superfluidity. Concepts such as the particle properties of radiation, the wave properties of matter, quantized energy and momentum, and the idea that one can no longer know exactly where a single particle like an electron is at any instant of time are the foundations of modern sciences and are indispensable when explaining most of the experimental results in physical, chemical, material, and biological sciences. Consequently, advances in modern technologies are not possible without prior knowledge of modern quantum mechanics.

I taught multiple courses on quantum mechanics during my 33-year teaching career at The Maharaja Sayajirao University of Baroda, Vadodara (India) and Jiwaji University, Gwalior (India), and as subject matter expert to several other universities. The purpose of writing this book has been to design a textbook that

i. provides everything a student needs to know for succeeding at all levels of undergraduate and graduate studies,
ii. meets international standards with detailed and elegant mathematical treatment and the up-to-date interpretation/presentation style, and
iii. contains enough solved examples to illustrate the fundamental concepts.

Over the past decades, quantum mechanics has continually evolved, and the tools of presentation and mathematical treatment of the subject have gone through significant changes. Therefore, I felt the need to write a textbook on modern quantum mechanics that has a rejuvenated approach to the subject matter. As a teacher, I believe that testing the understanding of a topic by problem-solving is vital for learning the concepts.

The topics covered in this book are divided into 12 chapters. Several newly emerged topics from contemporary physics are incorporated either as text or as solved examples. There are over a hundred solved examples and an equal number of unsolved exercises in this book. Each chapter of this book consists of solved examples followed by unsolved exercises, with answers, to consolidate the readers' understanding. Every effort has been made to make this book reader friendly. Chapter 1 starts with the historical development of quantum mechanics and explains the key concepts behind the fundamental discoveries, which made important contributions to the foundation of modern quantum mechanics. The emphasis of this chapter is to bring out clearly the concepts behind Schrödinger's wave mechanics, Heisenberg's matrix mechanics, and Dirac's relativistic quantum mechanics. Development of the Schrödinger wave equation and its 1D applications are covered in Chapter 2. The solution of Schrödinger's wave equation for free particles, quantum wells, quantum barriers, and potential steps along with its relevance and importance to contemporary physics and chemistry are highlighted and demonstrated through examples from condensed matter physics. Chapter 3 deals with matrix and vector formulation of quantum mechanics. Review of important properties of matrices and vector space which are relevant to quantum mechanics, the change of basis and irrelevance of choice of basis of a vector space, the numerical approach to finding eigenvalue and the inverse of a matrix, the equivalence of the Heisenberg and Schrödinger approaches to quantum mechanics, the general uncertainty principle, the 1D harmonic oscillator, and its matrix formulation are some of the main attractions of this chapter. A modern approach to explain the transformations, conservation laws, and symmetries is presented in Chapter 4. Reflection (parity operator) and time reversal are discussed at length in the chapter. Chapter 5 covers orbital, total, and spin angular momenta and the Clebsch-Gordon coefficients. The matrix approach to deriving eigenvalues of orbital angular momentum and the Pauli theory of spin half systems are special features of this chapter. Chapter 6 discusses the solution of Schrödinger's equation for central potentials and 3D systems along with free particle motion and discretized energy states in finite volume. Chapter 7 deals with approximation methods: Perturbation theory for non-degenerate and degenerate systems, variational methods, and WKB approximations. The kth order corrections to energy and wave function for non-degenerate perturbation theory, two types of treatment for WKB approximation, and detailed derivation of connecting formulae, and application of the variational method to compute energy for excited states are some of the topics which are not generally found in textbooks on quantum mechanics, but are covered in this chapter. The quantum theory of scattering along with detailed derivations on partial wave analysis and integral equation are presented in Chapter 8. Chapter 9 details the quantum theory of many particle systems and includes explanations of space and spin parts of wave function, detailed derivations for direct and exchange integrals for the helium atom, and the Thomas-Fermi and density functional theories for systems of N-particles which are essential to teach at the graduate level but are not covered in most of the textbooks on quantum mechanics. Time-dependent perturbations and the semi-classical treatment of interaction of fields with matter are covered in Chapter 10 along with a discussion of the appropriate mathematical treatment of time-dependent potentials and exactly solvable time-dependent two-state systems. Chapter 11 deals with relativistic quantum mechanics which includes derivation of the Dirac equation, constants of motion, and a solution of the Dirac equation for central potentials. Essential contents on the quantization of fields and second quantization are presented in Chapter 12. A contemporary approach is used for presentation and discussions on field quantization and second quantization for systems of Bosons and Fermions.

Access the Support Material at https://www.routledge.com/9780367723446.

Acknowledgments

It is my pleasure to acknowledge the questions followed by intense discussions with my all students, on whom the matter of this book is well tested. I am indebted to two of my colleagues, Professor Mahesh Prakash, former head, School of Physics, Jiwaji University, Gwalior, and Professor J. P. Singh, former head, Physics Department, Faculty of Science, The Maharaja Sayajirao University of Baroda, Vadodara, for their critical reading of the first draft of this book and making several valuable suggestions. I would also like to acknowledge, with a sense of gratitude the valuable input and the assistance in drawing the figures from Dr. Digish K. Patel (one of my former graduate students), Assistant Professor, Ganpat University, Mehsana (Gujarat). Dr. Pushpendra Tripathi, Associate Professor, Aligarh Muslim University has also made many useful suggestions for the content of this book. With a deep sense of gratitude I thank my wife, Ragini Sharma, for her constant encouragement and wholehearted support throughout my professional career, especially when I was writing this book. Finally, I acknowledge with pleasure the encouragement and support received from our sons Sanshit and Mayank and our daughter-in-law Vrishti.

About the Author

Dr. A C Sharma is a retired Professor of Physics from the Faculty of Science, The Maharaja Sayajirao University of Baroda, Vadodara (Gujarat) India. He has been head of the physics department, dean of the faculty of science for two terms, chairman of the board of studies in physics, chairman of the faculty board for two terms, and coordinator of many research programs awarded to physics department and to science faculty from the Department of Science and Technology, and from the University Grants Commission, New Delhi. He has distinguished educational records, obtained his Ph.D. from Roorkee University (now IIT Roorkee) and did his post-doctoral work at Cavendish Laboratory, University of Cambridge (U. K.). Professor Sharma has served as scientists pool officer at IIT Delhi, lecturer at Jiwaji University Gwalior, and then reader and professor at Maharaja Sayajirao University of Baroda. He has also been a visiting scientist to Chalmers University of Technology, Sweden and Indian Institute of Science, Bangalore. He has 33 years' experience of teaching undergraduate, post-graduate, and graduate students and 40 years' research experience in different areas of theoretical condensed matter physics. Semiconductors, surfaces and interfaces, superconductivity and nanostructures have been the areas of his research interest. Many-particle theory and numerical computations have been his expertized techniques to perform the research work. He has published 125 research papers. Some of his research papers are published in journals like *Physical Review, Journal of Condensed Matter Physics*, and *Journal of Applied Physics and Solid State Communications*. Professor Sharma has successfully completed 8 major research projects, delivered 41 invited talks at national and international conferences, organized 6 conferences, and supervised 12 students for their Ph.D. degrees. He has been awarded several scholarships, visiting fellowships and medals. Professor Sharma is a member of the Indian Physics Association, and a Fellow of Gujarat Science Academy. He has worked as a subject expert to evaluate the graduate thesis and research proposals for several universities/institutions and referee to several search journals.

1

Introduction to Quantum Mechanics

During the late 1800s and early 1900s, it became clear that physics was due for major revision. Existing laws of physics at that time failed to explain adequately or even approximately a large number of phenomena and observations. The explanation of phenomena involving small systems like electrons and atoms and their interaction with electromagnetic fields faced serious problems when attempted with classical physics. The following clearly demonstrated the necessity for a departure from classical mechanics:

1. Classical physics laws suggest that an electron moving in an orbit of an atom must lose energy by emission of synchrotron radiation, and it must spiral gradually towards the nucleus of an atom, which was not observed experimentally.
2. The experiments that were performed for observing the interference of light, the photoelectric effect, and diffraction of electrons clearly demonstrated that under certain situations waves act like streams of particles, and streams of particles were found to act as if they were waves. The laws of classical physics were completely unable to explain this.
3. The laws of classical physics suggest that the energy density of an electromagnetic field in vacuum cannot be finite because of the divergence of energy carried by short wavelength modes, which was not observed in experiments, and the total energy density was found to be finite.

Quantum mechanics, which was developed between 1900 and 1930 along with the general and special theory of relativity, completely revolutionized the field of physics, and it is now one of the pillars of modern sciences. The new concepts such as the wave properties of a particle and particle properties of radiation, quantized energy and momentum, and the idea that the position of a particle like an electron cannot be known exactly at any one instant of time had been found necessary to explain all the new experimental evidence that was available at that time. Classical mechanics and electromagnetism were replaced by quantum mechanics when a system was of a size comparable with that of an atom. Today, it is well understood that quantum mechanics is not only needed to describe systems at very small scales, but it is also required to explain *macroscopic systems* such as semiconductors, superconductivity, and superfluids. At present, several fields of physics and chemistry, like condensed matter physics, atomic/molecular physics, computational chemistry, particle physics and nuclear physics, are better understood within the mathematical framework of quantum mechanics. Since, everything is divisible into quantum-mechanical particles, the laws of quantum mechanics must reduce to the laws of classical physics in the appropriate limit known as the *classical limit*. Quantum mechanics was developed in both a non-relativistic and relativistic manner. Some of the most accurate theories of physics had been worked out within the framework of relativistic quantum mechanics and *quantum field theory*. However, non-relativistic quantum mechanics is widely used, because of its simplicity for situations where relativistic effects are not so important.

The research that contributed to the foundations of quantum mechanics was as follows:

1.1 Blackbody Radiation and Planck's Hypothesis

Heated up objects such as a hot iron generally glow to give off light, which displays a range of colors lying between red to yellow. On further increases in temperature, more light and a different spectrum of light result. Though different spectra are exhibited by different substances, the majority of these are approximated by an ideal known as *blackbody radiation*. An idealized body that completely absorbs all the energy of the radiation that is falling upon it and attains an equilibrium temperature and then reemits that energy as quickly as it is absorbed is termed a blackbody.

The radiated energy has various wavelengths, and for every wavelength a maximum of energy occurs at some value that depends on the temperature of the body. The hotter body exhibits a shorter wavelength for maximum radiation. Calculations of the energy distribution for the radiation from a *black body* with the use of classical physics were unsuccessful. German theoretical physicist Max Planck suggested that the radiation energy is not emitted continuously but is emitted in discrete packets, called quanta. The energy E and frequency ν of a quanta are related by the relation $E = h\nu$. The quantity h, which is approximately equal to 6.62607×10^{-34} joule·second, is a universal constant. The energy of a spectrum was calculated and it was shown by Planck that the calculated energy spectrum agreed with observations made on the black body, for the entire wavelength range. It was assumed by Planck that the atoms in an oscillating state were the sources of radiation. The vibrational energy of each oscillator can take any number of discrete values, but no value in between two discrete values. He further assumed that when an oscillator changes its energy state from E_1 to a state of lower energy E_2 the discrete amount of energy $(E_1 - E_2)$ is equal to $h\nu$, which is determined from blackbody radiation data. *Planck's law* for E_λ, radiated energy per unit volume of a black body in the wavelength interval of λ to $\lambda + \Delta\lambda$, is given in terms of h, speed of light c, the Boltzmann constant k_B, and the absolute temperature T:

$$E_\lambda = \frac{8hc}{\lambda^5} \frac{1}{\left(e^{\frac{hc}{k_B T \lambda}} - 1 \right)} \tag{1.1}$$

The sun releases its internal energy as light from the surface and does not reflect any incoming light and therefore it acts very similarly to a blackbody. Fig. 1.1 given below demonstrates what the shape of the curve for blackbody radiation looked like for objects at different temperatures, including the sun. The wavelength of maximum power emitted by a blackbody at temperature T is given by Wien's displacement law; $\lambda_{max} T = 2898\ \mu m$ K.

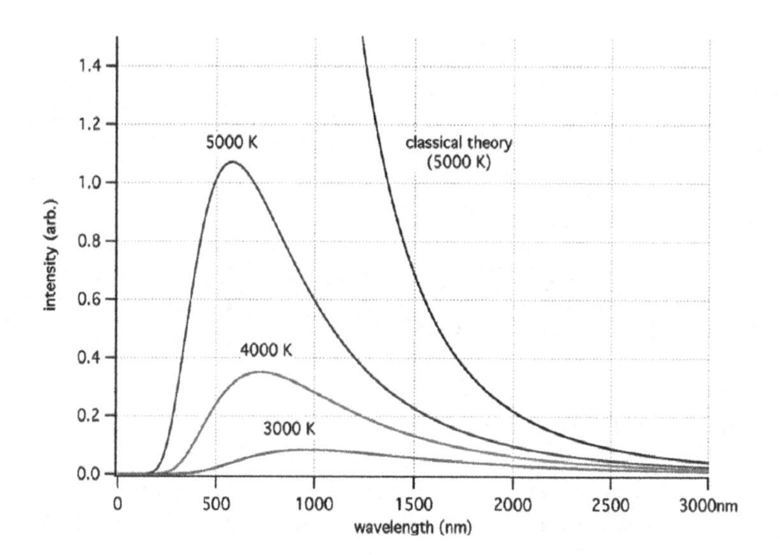

FIGURE 1.1
An illustrative diagram for blackbody radiation spectrum.

1.2 Photoelectric Effect

Emission of electrons when light hits a metal surface is known as the photoelectric effect. The emitted electrons are called *photoelectrons*. The theory of photoelectric effect explains the following experimental observations on emission of electrons from an illuminated metal surface.

1. There exists a *threshold frequency* for every metal surface, which is a minimum frequency of incident light below which no photoelectrons are emitted.
2. Electrons are emitted above the threshold frequency. The maximum kinetic energy of each emitted photoelectron depends on the frequency, not on the intensity of the incident light, unless the intensity is too high.
3. The rate of ejected photoelectrons is directly proportional to the intensity of the incident light, for a given metal and frequency of incident radiation. For fixed frequency, an increase in the intensity of the incident beam increases the magnitude of the photoelectric current, when stopping voltage remains the same.
4. A small time lag, less than 10^{-9} second, between the incidence of radiation and the emission of a photoelectron is observed.
5. Direction of distribution of emitted electrons peaks in the direction of polarization (the direction of the electric field) of the incident light, for linearly polarized light.

In 1905, Albert Einstein published a paper by borrowing Planck's hypothesis about the quantized energy from his blackbody research, and he assumed that the incident radiation on a metal surface should be thought as quanta of energy $h\nu$. The quanta is known as a photon. In photoemission, one such photon is absorbed by one electron. Some energy will be lost in moving towards the surface when the electron is at some distance into the material of the cathode. There will always be some electrostatic cost as the electron leaves the

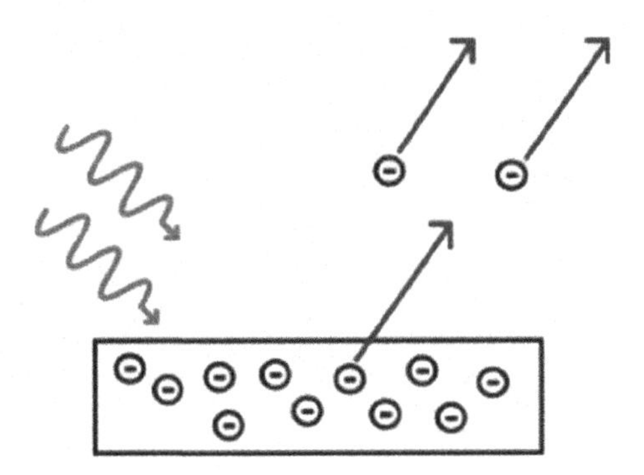

FIGURE 1.2
An illustrative diagram for ejection of photoelectrons.

surface, which is termed as work function ϕ. The electrons emitted from regions very close to the surface are the most energetic. These electrons leave the cathode with kinetic energy:

$$\frac{1}{2}mv^2 = h\nu - \phi \tag{1.2}$$

There is a minimum energy for a given metal, for which the quantum of energy is equal to ϕ. Light below that energy, no matter how bright, will not eject electrons.

Every photon with energy more than threshold energy excites only one electron. An increased intensity of light only increases the *number* of released electrons and not their kinetic energy. Since, a very small time is required for an atom to be heated to a critical temperature, release of the electron is nearly instantaneous upon absorption of the light.

1.3 Bohr's Atomic Model and the Hydrogen Atom

A model of an atom to explain how electrons can have stable orbits around the nucleus was proposed by Bohr in 1913. The size of the orbit determines the energy of an electron; it is lower for smaller orbits. When the electron jumps from a higher energy orbit to a lower energy orbit, radiation is released. Three postulates made by Bohr are:

1. The electron revolves in a stable orbit (stationary) around the nucleus without radiating any energy. These stationary orbits are attained at certain discrete distances from the nucleus. The electron is not allowed to have any energy in between the discrete energies of orbits.
2. The magnitude of angular momentum of a revolving electron in a stationary orbit is quantized and it attains the values which are integral multiples of $\hbar = h/2\pi$:

$$mvr = n\hbar \tag{1.3}$$

where n is an integer, which takes values $1,2,3,4,.....$ and is termed a principle quantum number. The smallest value of n is 1, which corresponds to an orbital radius equal to 0.0529 nm, known as the Bohr radius. Here, it is assumed that an electron of mass m and charge $-e$ moves with velocity v, in an orbit of radius r.

3. Electrons can jump from one allowed orbit to another by gaining (absorption) or losing (emission) energy from electromagnetic radiation of frequency ν determined by the energy difference of the levels according to $\Delta E = E_1 - E_2 = h\nu$.

For the hydrogen atom, an attractive coulomb force provides the centripetal acceleration $\frac{v^2}{r}$ to maintain orbital motion. The total force on the electron thus is:

$$\begin{aligned} \mathbf{F} &= \frac{e^2}{4\pi\varepsilon_0 r^2}\hat{\mathbf{r}} = m\frac{v^2}{r}\hat{\mathbf{r}} \\ \Rightarrow mv^2 &= \frac{e^2}{4\pi\varepsilon_0 r} \end{aligned} \tag{1.4}$$

where $\varepsilon_0 = 8.854 \times 10^{-12}$ F/m (Farad per meter) is the permittivity of free space. The total energy of the electron in a hydrogen atom is given by:

$$\begin{aligned} E &= \frac{1}{2}mv^2 - \frac{e^2}{4\pi\varepsilon_0 r} \\ &= \frac{e^2}{8\pi\varepsilon_0 r} - \frac{e^2}{4\pi\varepsilon_0 r} \\ &= -\frac{e^2}{8\pi\varepsilon_0 r} \end{aligned} \tag{1.5}$$

Also, Eqns. (1.3) and (1.4) can be combined to give the radius of the nth orbital of the hydrogen atom:

$$r_n = \left(\frac{4\pi\varepsilon_0\hbar^2}{me^2}\right)n^2 = a_0 n^2 \tag{1.6}$$

Here, $\frac{4\pi\varepsilon_0\hbar^2}{me^2} = a_0 = 0.0529$ nm is the radius of the first orbital of the hydrogen atom, termed the Bohr radius. On substituting Eqn. (1.6) into Eqn. (1.5), one gets energies of quantized levels:

$$E_n = -\frac{me^4}{32\pi^2\varepsilon_0^2\hbar^2 n^2} = -\frac{e^2}{8\pi\varepsilon_0 a_0 n^2} = -\frac{E_1}{n^2} \tag{1.7}$$

Here, $E_1 = -13.6$ eV is the ground state energy of the hydrogen atom.

1.4 Compton Scattering of Photons

During the early 20th century, research on interaction of X-rays with matter demonstrated that when X-rays of a known wavelength interact with atoms, the X-rays are scattered through an angle θ, and they emerge at a different wavelength that is related to θ. The wavelength of the scattered rays was found longer (corresponding to lower energy) than the initial wavelength, in multiple experiments, whereas classical electromagnetism predicted that the wavelengths of scattered and initial X-rays should be equal. The observed X-ray shift was attributed to the particle-like momentum of light quanta, in the research work published by Compton in 1923. The energy and frequency of light quanta are interdependent. Compton assumed that each scattered X-ray photon interacted with only one electron, and derived the mathematical relationship between the shift in wavelength and the scattering angle:

$$\lambda' - \lambda = \frac{\hbar}{m_0 c}(1 - \cos\theta) \tag{1.8}$$

where λ' and λ are the wavelengths after and before scattering; m_0 is the rest mass of the electron and θ is the scattering angle. The $\frac{\hbar}{m_0 c}$ is known as the Compton wavelength of the electron, which is equal to 2.43×10^{-12} m. The wavelength shift, $\Delta\lambda = \lambda' - \lambda$, can take any value from zero to (for $\theta = 180°$), twice the Compton wavelength of the electron (for $\theta = 0°$).

Compton also observed that no wavelength shift was experienced for some X-rays, despite being scattered through large angles. This happened because of the photon failing to eject an electron. The magnitude of the shift is related to the Compton wavelength of the entire atom (which could be 10,000 times smaller) not to the Compton wavelength of the electron. This is a different phenomenon known as coherent scattering off the entire atom, as the atom remains intact, gaining no internal excitation. In modern experiments, measurement of energies is more conventional than the measurement of wavelengths of the scattered photons. By writing $E = \frac{\hbar c}{\lambda}$ and $E' = \frac{\hbar c}{\lambda'}$ in the above relation we obtain:

$$E' = \frac{E}{1 + \left(\frac{E}{m_0 c^2}\right)(1 - \cos\theta)} \tag{1.9}$$

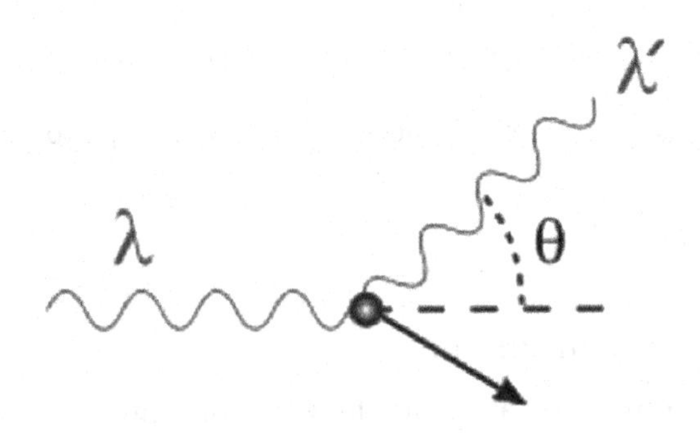

FIGURE 1.3
An illustrative diagram of Compton scattering.

1.5 De Broglie Hypothesis

In 1924, de Broglie proposed that just as light has both wave-like and particle-like properties, electrons too have both particle-like and wave-like properties. One of the most important contributions to the development of quantum mechanics comes from matter waves. Just like a beam of light, all matter exhibits wave-like behavior, and a beam of electrons can be diffracted. This is known as the de Broglie hypothesis, and the matter waves are referred as de Broglie waves.

The de Broglie wavelength λ, which is associated with a particle, relates to the momentum of particle $\mathbf{p}$ through the relation $\lambda = h/|\mathbf{p}|$. George Paget Thomson's thin metal diffraction experiment and the Davisson-Germer experiment independently demonstrated the *wave-like behavior* for electrons. It had also been confirmed for other elementary particles, neutral atoms and even molecules. All matter exhibits properties of both particles and waves, and hence the de Broglie relationship hold for all types of matter.

1.6 Pauli Exclusion Principle

Pauli discovered the exclusion principle in 1925, which states that each electron must be in its own distinct quantum state. In other words, no two electrons in an atom or no two identical Fermions in more generalized terms, can simultaneously occupy the same quantum state, defined by a set of quantum numbers (n,l,m,m_s), where n,l,m, and m_s are termed the principal quantum number, azimuthal quantum number, magnetic quantum number and spin quantum number, respectively. If two electrons occupy the same orbital, then the values of n, l, and m are the same, and therefore their m_s must be different, which means that the electrons must have opposite half-integer spin projections of $\frac{1}{2}$ and $-\frac{1}{2}$. Particles with whole number spin (Bosons) are not subject to the Pauli exclusion principle. The exclusion principle was initially formulated by Wolfgang Pauli for electrons and later on it was extended to all Fermions. The Pauli exclusion principle and the requirement of wave function to be antisymmetric with respect to exchange are equivalent.

1.7 Schrödinger Wave Equation

During the early 20th century, physicists began to acknowledge that matter possesses wave-like behavior. However, there existed no known equations that obeyed matter waves. In 1926, Austrian physicist Erwin Schrödinger, in continuation to de Broglie's hypothesis, formulated a second-order differential equation to explain the wave nature of the matter and particle associated with the wave. The Schrödinger equation assumes that a particle behaves as a wave and yields a solution in terms of the function known as the *wave function* ψ and the energy E of the particle under consideration. Once the wave function is known, then everything about the particle can be deduced from the wave function. This makes ψ a most important quantity, which itself does not have any physical significance; however, the absolute square of it, i.e. $|\psi|^2$ gives the probability of finding the particle in a region of space at an instant of time. The E of the particle, depending upon the potential energy V, and boundary conditions (constraints on the particle), can be continuous or quantized. One of the most remarkable features of quantum mechanics *is* the *quantization of energy of the particle.*

The Schrödinger equation has been very popular, despite its initial criticism by the scientific community because of its limitations in regard to relativistic particles. The Schrödinger equation is used in two forms, one consisting of time, termed as the time-dependent equation, and the other in which the time factor is eliminated, and hence named the time-independent equation.

The Schrödinger equation, like Newton equations, cannot be derived but can be verified from experimental results. A simple way to get the Schrödinger equation could be as follows:

Consider the motion of a free particle along the x-axis with momentum p. According to de Broglie's hypothesis, a wave of $\lambda = \frac{h}{p}$ is associated with the particle and hence we can assume a wave function $\psi(x,t)$ given by:

$$\psi(x,t) = Ae^{i(kx-\omega t)} \tag{1.10}$$

The hypotheses of Planck and de Broglie suggest that $E = h\nu = \hbar\omega$ and $p = \frac{h}{\lambda} = \hbar k$. Therefore:

$$\psi(x,t) = Ae^{\frac{i}{\hbar}(px-Et)} \tag{1.11}$$

On partial differentiation of Eqn. (1.11) with respect to x and t, we obtain:

$$\frac{\partial\psi(x,t)}{\partial x} = \frac{ip}{\hbar}\psi(x,t) \Rightarrow \left[\frac{\partial}{\partial x} - \frac{ip}{\hbar}\right]\psi(x,t) = 0 \tag{1.12}$$

$$\frac{\partial\psi(x,t)}{\partial t} = -\frac{iE}{\hbar}\psi(x,t) \Rightarrow \left[\frac{\partial}{\partial t} + \frac{iE}{\hbar}\right]\psi(x,t) = 0 \tag{1.13}$$

Equations (1.12) and (1.13) suggest that $p = -i\hbar\frac{\partial}{\partial x}$ and $E = i\hbar\frac{\partial}{\partial t}$. For a non-relativistic free particle $E = \frac{p^2}{2m}$ and hence we get:

$$-\frac{\hbar^2}{2m}\frac{\partial^2\psi(x,t)}{\partial x^2} = i\hbar\frac{\partial\psi(x,t)}{\partial t} \tag{1.14}$$

which is the Schrödinger equation for a free particle moving along the x-axis. It is to be noticed here:

i. For a free particle moving along an arbitrary direction, Eqn. (1.14) can be generalized to:

$$\begin{aligned} -\frac{\hbar^2}{2m}\left[\frac{\partial^2}{\partial x^2} + \frac{\partial^2}{\partial y^2} + \frac{\partial^2}{\partial z^2}\right]\psi(\mathbf{r},t) = i\hbar\frac{\partial\psi(\mathbf{r},t)}{\partial t} \\ \Rightarrow -\frac{\hbar^2}{2m}\nabla^2\psi(\mathbf{r},t) = i\hbar\frac{\partial\psi(\mathbf{r},t)}{\partial t} \end{aligned} \tag{1.15}$$

ii. For a particle moving under the influence of a field characterized by potential energy, $V(\mathbf{r})$, the total energy of the particle is $E = \frac{p^2}{2m} + V(\mathbf{r})$ and therefore Eqn. (1.15) becomes:

$$\left[-\frac{\hbar^2}{2m}\nabla^2 + V(\mathbf{r})\right]\psi(r,t) = i\hbar\frac{\partial\psi(r,t)}{\partial t} \tag{1.16}$$

iii. It is important to note that Eqn. (1.14) has been obtained by assuming the wave function given by Eqn. (1.10). However, several solutions like Eqn. (1.10) and their linear combination will also be the solutions of Eqn. (1.14). Therefore:

$$\psi(x,t) = \int A(k)e^{i(kx-\omega t)}\,dk \tag{1.17}$$

will also be the solution of Eqn. (1.14). Equation (1.17) represents a group of waves having different k-values. The group of waves is known as a wave packet. Each k corresponds to a wave and in principle, k can take any value in between $-\infty$ and ∞. However, for practical cases it varies over a certain range. The $\frac{d\omega}{dk}$ and $\frac{\omega}{k}$ have dimensions of velocity. With the use of $E = h\nu$ and $\lambda = \frac{h}{p}$, we can write $\frac{d\omega}{dk} = \frac{\hbar k}{m} = v$, which is the velocity for a freely moving particle. Therefore, $\frac{d\omega}{dk}$ is termed group velocity and it is represented by v_g, while $\frac{\omega}{k}$, which is the velocity of an individual wave, is known as phase velocity.

1.8 Born Interpretation of Wave Function

As stated above, it became clear that matter must be considered to have wave-like properties to explain experimental data. In 1926, Max Born formulated a physical law of quantum mechanics, which gives the probability of getting the given results from a measurement on the quantum system. The Born law, which is one of the key principles of quantum mechanics, states that the probability density of finding a particle at a given point of time is proportional to the square of the magnitude of the wave function of the particle at that point. Thus, the wave function itself has no physical significance, but the square of its absolute magnitude has significance when evaluated at a point.

The probability of finding a particle between a and b at time t is given by:

$$\int_a^b |\psi(x,t)|^2\,dx \tag{1.18}$$

where $|\psi(x,t)|^2 = \psi(x,t)^*\ \psi(x,t)$ is a complex square. At any given time:

$$\int |\psi(x,t)|^2\,dx = 1 \tag{1.19}$$

which would remain true at any other point of time, because the probability of finding a particle over a space cannot change with time. We need (i) the integral to be time-independent, because otherwise a probabilistic interpretation wouldn't be possible, and (ii) the probability of finding the particle over the entire space to be 1.

1.9 Heisenberg Uncertainty Principle

German physicist Werner Heisenberg introduced first in 1927 the uncertainty principle, which states that the more precisely the position of a particle is determined, the less precisely its momentum can be known, and vice versa. The formal inequality relating to standard deviations in one dimensional position Δx and momentum Δp was derived by Earle Hesse Kennard and by Hermann Wey in 1928:

$$\Delta x \Delta p \geq \frac{\hbar}{2} \tag{1.20}$$

Two alternative explanations for the uncertainty principle are offered in quantum mechanics: (i) the wave mechanics picture and (ii) the matrix mechanics picture. The matrix mechanics approach formulates the uncertainty principle in a more generalized manner. In the wave mechanics description of quantum mechanics, the uncertainty relation in terms of position and momentum arises because the expressions of the wave function in the two corresponding orthonormal bases in Hilbert space are the Fourier transform of each other. A nonzero function and its Fourier transform both cannot be sharply localized, because position and momentum are conjugate variables. A similar relation arises between the variances of all pairs of Fourier conjugates. All pairs of non-commuting self-adjoint operators representing observables are subject to similar uncertainty relations in matrix formulations of quantum mechanics. One of the most fundamental concepts in quantum theory lies in the idea that $xp \neq px$, which is not followed by pure numbers but by the operators and matrices. Our understanding of physical properties being represented by numbers is altered on being represented by operators and matrices. We know that matrices have the same non-commutativity as operators, and they too can be the mathematical formalism of quantum theory. In fact, Heisenberg's original approach to quantum theory, which is called matrix mechanics, was to find matrices x and p, such that $xp - px = i\hbar$.

An eigenstate of an observable represents the state of the wave function for a certain measurement of value (the eigenvalue). For example, if a measurement of an observable A is performed, then the system is in an eigenstate of that observable. However, the eigenstate of the observable A need not be an eigenstate of another observable B. In such a case, it does not have a uniquely associated measurement for it, as the system is not in an eigenstate of that observable.

1.10 Davisson and Germer Wave Properties of Electrons

To test the de Broglie's hypothesis that matter behaved like waves, Clinton Davisson and Lester Germer fired slow moving electrons at a crystalline nickel target, in 1927 at Bell Labs. In the experiment, they measured and determined the angular dependence of the

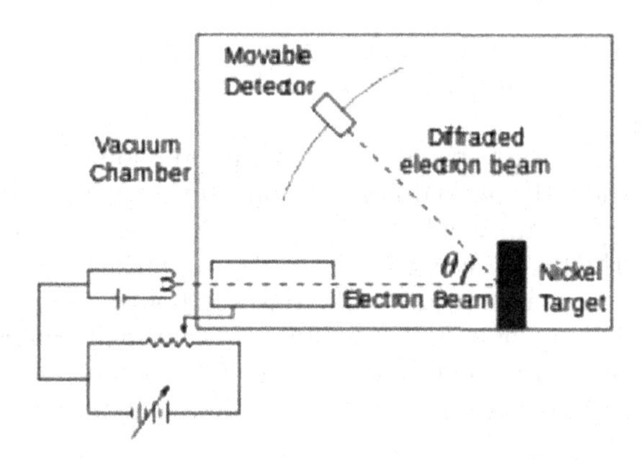

FIGURE 1.4
Illustrative set up of Davisson-Germer experiment.

reflected electron intensity to verify the diffraction pattern that was predicted by Bragg for X-rays. The same effect by firing electrons through metal films to produce a diffraction pattern was independently demonstrated by George Paget Thomson at the same time. The de Broglie hypothesis that matter has wave-like behavior was confirmed by the Davisson-Germer experiment. The Compton effect and the Davisson-Germer experiment established the wave-particle duality hypothesis, which is a fundamental step in quantum theory.

The Davisson-Germer experiment consisted of (i) firing an electron beam from a heated tungsten filament working as an electron gun, (ii) an electrostatic particle accelerator perpendicular to the surface of the nickel crystal, and (iii) measurement of how the number of reflected electrons varied with respect to variation in angle between the detector and the nickel surface. The electrons released from a heated tungsten filament were accelerated through an electric potential difference towards the nickel crystal to give them kinetic energy. The experiment was conducted in a vacuum chamber to avoid collisions of the electrons with other atoms on their way towards the surface. A Faraday cup electron detector was used to measure the number of electrons that were scattered at different angles. The detector accepted only elastically scattered electrons.

1.11 Bohr-Sommerfeld Quantization Condition

One of the main tools of the old quantum theory is the Bohr-Sommerfeld quantization condition. The condition allows one to select out only certain states, no other states, of a classical system as allowed states in which the system can exist. The basic idea of Bohr-Sommerfeld is a quantization condition that means that the motion in an atomic system is quantized, or discretized:

$$\oint p_i \, dq_i = n_i h \tag{1.21}$$

where q_i are generalized coordinates and the p_i are corresponding canonical conjugate momenta of the system. The quantum numbers n_i are *integers* and the integral is taken over one period of the motion at constant energy, described by the Hamiltonian. The integral is equal to an area in phase space, a quantity known as action, which is quantized in terms of Planck's constant.

The quantization condition, given by Eqn. (1.21) is often known as the *Wilson-Sommerfeld rule*, proposed independently by William Wilson and Arnold Sommerfeld. The Bohr theory was extended by Sommerfeld, and it turned out to be a powerful tool of atomic research at that time. The theory was adopted and further developed by physicists. Bohr originally used only one quantum number, while the Bohr-Sommerfeld theory is more generalized and describes the atom in terms of two quantum numbers.

For a circular orbit, angular momentum is mvr, which is a constant of motion and hence for a circular orbit Eqn. (1.21) reduces to:

$$\oint mvr\, d\theta = nh \quad \text{or} \quad 2\pi mvr = nh \tag{1.22}$$

which is the Bohr quantization rule given by Eqn. (1.3).

1.12 Correspondence Principle

The correspondence principle states that the behavior of systems described by quantum mechanical theory should reproduce that which is given by classical physics, in the limit of large quantum numbers. The conditions under which results from quantum mechanics agree with those from classical physics are called the correspondence limit. Quantum calculations must agree with classical calculations for large orbits and for large energies. In a more generalized manner, the *correspondence principle* states that a new scientific theory must reproduce the earlier existing scientific theory in appropriate circumstances, which means that a new theory explains all the phenomena under appropriate limits for which the preceding theory was valid. Two examples of this are the following: (i) Einstein's theory of special relativity follows the correspondence principle, because it reduces to classical mechanics when velocities are small as compared to the speed of light. (ii) When the number of particles is large, statistical mechanics reproduces thermodynamics.

1.13 Heisenberg Quantum Mechanics

The first formulation of quantum mechanics, which is conceptually autonomous and logically consistent, is Heisenberg's matrix mechanics. The novel ideas published by Heisenberg in a research paper paved the way for a complete departure from the classical description of atomic physics, and advanced a new formulation of the laws of microphysics. According to the Heisenberg formulation, the spectral lines and their intensities are the observables, not the Bohr orbits. Heisenberg advanced two additional assumptions in regard to dynamics of motion: (1) he proposed that the equations of motion should have the same mathematical form as the classical *Newton's second law*, and (2) he modified the Bohr-Sommerfeld quantization condition. Heisenberg and Born also assumed that the Hamilton canonical equations of motion were to be preserved. The basis of Heisenberg quantum mechanics had been the fundamental commutation relation $[p,q] = -i\hbar$, where q and p are time-dependent position and momentum matrices.

1.14 Dirac Theory of Quantum Mechanics

In 1928, Dirac proposed a formulation of quantum mechanics for an electron. In his theory spin emerges as a natural consequence of the relativistic treatment of quantum mechanics. The Dirac equation consists of both the principles of quantum mechanics and the theory of special relativity. It was the first quantum theory that accounts fully for special relativity and was validated to account for the fine details of the hydrogen spectrum in a completely rigorous manner.

Also, the Dirac quantum theory implied the existence of antimatter, which was at that time unsuspected and unobserved. The existence of antimatter was experimentally confirmed several years later. In contrast to the Schrödinger equation that is described in terms of one complex value dependent wave function, the wave function in Dirac theory involves four complex values. A theoretical justification to introduce several component wave functions by Pauli in his spin theory is provided by the Dirac theory. The Dirac equation for freely moving particle is as follows:

$$\left[c\alpha.p + \beta m_0 c^2\right]\psi(\mathbf{r},t) = i\hbar\frac{\partial}{\partial t}\psi(\mathbf{r},t) \tag{1.23}$$

The new elements in the Dirac equation are the 4 × 4 α_k and β matrices, and the four-component wave function, $\psi(r,t)$. m_0 is the rest mass of the electron.

1.15 Important Quantum Mechanical Parameters in SI Units

a. Bohr radius: $a_0 = \dfrac{4\pi\varepsilon_0\hbar^2}{m_0 e^2} = 5.29177\times10^{-11}$ meter.

b. Binding energy of hydrogen atom: $E_1 = \dfrac{m_0 e^4}{2(4\pi\varepsilon_0)^2\hbar^2} = 13.6057$ eV.

c. Bohr magneton: $\mu_B = \dfrac{e\hbar}{2m_0} = 9.274\times10^{-21}$ Joule/tesla.

d. Compton wavelength of an electron: $\lambda = \dfrac{2\pi\hbar}{m_0 c} = 2.43\times10^{-12}$ meter.

e. Classical electron radius: $r_e = \dfrac{e^2}{4\pi\varepsilon_0 m_0 c^2} = 2.82 \times 10^{-15}$ meter.

f. Electron rest mass energy: $E_0 = m_0 c^2 = 0.511$ MeV.

g. Proton rest mass Energy: $E_p = M_p c^2 = 938$ MeV.

h. Fine structure constant: $\dfrac{e^2}{4\pi\varepsilon_0 \hbar c} \approx \dfrac{1}{137}$.

i. Planck's constant: $\hbar = 1.05457 \times 10^{-34}$ Joule second.

j. Speed of light: $c = 2.99792 \times 10^8$ meter/second.

k. Rest mass of electron: $m_0 = 9.10939 \times 10^{-31}$ kg.

l. Charge on an electron: $-e = -1.60218 \times 10^{-19}$ Coulomb.

m. Mass of proton: $m_p = 1.67262 \times 10^{-27}$ kg.

n. Boltzmann constant: $k_B = 1.38066 \times 10^{-23}$ Joule/Kalvin.

1.16 Solved Examples

1. For the free electron described by $\psi(x,t) = Ae^{i(kx-\omega t)}$, where $k = 100\ \text{nm}^{-1}$, determine the de Broglie wavelength, momentum, kinetic energy, speed, and angular frequency.

SOLUTION

$$\lambda = \frac{2\pi}{k} = \frac{2 \times 3.1416}{100} = 6.2832 \times 10^{-2}\ \text{nm} = 6.2832 \times 10^{-11}\ \text{m}$$

$$\text{Momentum}, p = \frac{h}{\lambda} = \frac{6.626 \times 10^{-34}}{6.2832 \times 10^{-11}} = 1.054 \times 10^{-23}\ \text{kg m s}^{-1}.$$

$$\text{Speed}, v = \frac{p}{m} = 1.157 \times 10^7\ \text{m s}^{-1}$$

$$\text{Kinetic energy}, K = \frac{1}{2} mv^2 = \frac{p^2}{2m} = \hbar\omega$$

$$K = 6.0975 \times 10^{-17}\,\text{J} = 380.6\ eV = 3.806\ \text{KeV}$$

$$\text{Angular frequency},\ \omega = \frac{K}{\hbar} = vk = 1.157 \times 10^{18}\,s^{-1}$$

2. A particle of mass m is described by $\psi(x) = Ae^{-\alpha x^2}$ in a region where the energy of the particle is zero. Find the potential energy of the particle.

SOLUTION

The time-independent Schrödinger equation for a particle is: $\frac{-\hbar^2}{2m}\frac{d^2}{dx^2}\psi(x)+V(x)\psi(x)=E\psi(x)$.

In this case:

$$\frac{\hbar^2}{2m}\frac{d^2}{dx^2}\psi(x)=V(x)\psi(x)$$

$$\Rightarrow V(x)Ae^{-\alpha x^2}=\frac{\hbar^2}{2m}\frac{d^2}{dx^2}Ae^{-\alpha x^2}.$$

$$\Rightarrow V(x)=\frac{\hbar^2\alpha}{m}(2\alpha x-1)$$

3. Wave function $\psi(x)=Ae^{-\alpha x^2}$ describes the ground state of a quantum oscillator that satisfies the equation:

$$-\frac{\hbar^2}{2m}\frac{d^2\psi(x)}{dx^2}+\frac{1}{2}m\omega^2x^2\psi(x)=E\psi(x) \tag{1.24}$$

Calculate A,α and E and then show that it satisfies the Heisenberg uncertainty relation $\Delta x\Delta p\geq\frac{\hbar}{2}$, where $\Delta x=\sqrt{\left(\langle x^2\rangle-\langle x\rangle^2\right)}$ and $\Delta p=\sqrt{\left(\langle p^2\rangle-\langle p\rangle^2\right)}$.

SOLUTION

The oscillating particle can be found anywhere in the range of $-\infty\leq x\leq\infty$. Therefore, from Eqn. (1.19), we have

$$\int_{-\infty}^{\infty}\psi^*(x)\psi(x)dx=1\Rightarrow\int_{-\infty}^{\infty}A^2e^{-2\alpha x^2}dx=1\Rightarrow A^2\times\left(\frac{\pi}{2\alpha}\right)^{1/2}=1 \tag{1.25}$$

Therefore, $A=\left(\frac{2\alpha}{\pi}\right)^{1/4}$. Substituting the value of $\psi(x)$ in Eqn. (1.24), we get:

$$\left[\frac{\hbar^2\alpha}{m}-E+\left(\frac{1}{2}m\omega^2-\frac{2\alpha^2\hbar^2}{m}\right)x^2\right]\psi(x)=0 \tag{1.26}$$

For the non-vanishing wave function, this equation can be satisfied for $\alpha=\frac{m\omega}{2\hbar}$ and $E=\frac{\hbar^2\alpha}{m}=\frac{1}{2}\hbar\omega$, which gives $A=\left(\frac{m\omega}{\pi\hbar}\right)^{1/4}$.

We next calculate:

$\langle x\rangle=\int_{-\infty}^{\infty}\psi(x)^*x\psi(x)dx=\int_{-\infty}^{\infty}A^2x\,e^{-2\alpha x^2}dx=0$, as the integrand is the odd function of x,

and:

$$\langle x^2 \rangle = \int_{-\infty}^{\infty} \psi(x)^* x^2 \psi(x) dx$$
$$= \int_{-\infty}^{\infty} A^2 x^2 e^{-2\alpha x^2} dx = \frac{\hbar}{2m\omega} \tag{1.27}$$

Hence, $\Delta x = \sqrt{\frac{\hbar}{2m\omega}}$.

$$\langle p \rangle = -i\hbar \int_{-\infty}^{\infty} \psi(x)^* \frac{d\psi(x)}{dx} dx$$
$$= -i\hbar \int_{-\infty}^{\infty} A^2 e^{-\alpha x^2} \frac{de^{-}\alpha x^2}{dx} dx \tag{1.28}$$
$$= 2i\hbar\alpha \int_{-\infty}^{\infty} A^2 x e^{-2\alpha x^2} dx = 0$$

and:

$$\langle p^2 \rangle = -\hbar^2 \int_{-\infty}^{\infty} \psi(x)^* \frac{d^2\psi(x)}{dx^2} dx$$
$$= 2\hbar^2\alpha \int_{-\infty}^{\infty} A^2 \left(1 - 2\alpha x^2\right) e^{-2\alpha x^2} dx \tag{1.29}$$
$$= \frac{\hbar m\omega}{2}$$

Hence $\Delta p = \sqrt{\frac{\hbar m\omega}{2}}$. Therefore, $\Delta x \Delta p = \hbar / 2$, which is the optimum value in the uncertainty relation.

4. Consider two cases: (a) Zero point oscillation for a pendulum of mass, $m = 1$ kg and length, $l = 1$ m. (b) Motion of a tennis ball of mass $m = 0.1$ kg moving with speed $v = 0.5$ m/sec. Show that a quantum description is not required for these motions.

SOLUTION

a. For pendulum, having zero-point oscillation, average potential energy is $\check{V} = \frac{1}{4}E = \frac{1}{2}m\omega^2 A^2 = \frac{1}{4}\hbar\omega$, where A is the root mean square amplitude of zero-point oscillation, and $\omega = \sqrt{g/l}$. Therefore, $A = \sqrt{\hbar / 2m\omega}$,

$$\omega = \sqrt{9.8} = 3.13 \text{ sec}^{-1}.$$

$$A = \sqrt{\frac{1.05 \times 10^{-34}}{2 \times 3.13}} = 0.41 \times 10^{-17} \text{ m}.$$

b. For a tennis ball $\lambda = \frac{h}{p} = \frac{h}{mv} = \frac{6.626 \times 10^{-34}}{0.1 \times 0.5} = 1.32 \times 10^{-32}$ m.

Thus, it is concluded that the values of A for a pendulum and λ for a tennis ball are negligibly small to describe them quantum mechanically.

5. A metal surface when illuminated by the light sources of wavelengths λ and $0.5\,\lambda$ ejects the photoelectrons with a maximum kinetic energy of 2.00 eV and 6.00 eV, respectively. What is the work function of the metal?

SOLUTION

Let K_1 and K_2 be the maximum kinetic energies of ejected photoelectrons and ϕ be the work function of metal. Then:

$$K_1 = \frac{hc}{\lambda} - \phi \text{ and } K_2 = \frac{hc}{0.5\lambda} - \phi. \text{ Hence, } \phi = K_2 - 2K_1 = 2.0 \text{ eV.}$$

6. The X-rays undergo Compton scattering with a target. The energy of X-rays is 300 keV and the scattered X-rays are detected at 30° relative to the incident X-rays. Calculate (a) the energy of the scattered X-ray, (b) the energy of the recoiling electron, and (c) Compton shift at this angle.

SOLUTION

a. The energy of the scattered X-ray is given by Eqn. (1.9):

$$E' = \frac{E}{1 + \left(\frac{E}{m_0 c^2}\right)(1 - \cos\theta)},$$

$$m_0 c^2 = 9.11 \times (2.9979)^2 \times 10^{-15} \text{J} = 511.8 \text{ keV},$$

$$E' = \frac{300}{1 + \left(\frac{300}{511.08}\right)(1.0 - 0.866)} = 278.13 \text{ keV.}$$

b. The energy of the recoil electron: $E - E' = 21.87$ keV.

c. Equation (1.8) gives the wavelength shift:

$$\Delta\lambda = \lambda' - \lambda = \frac{\hbar}{m_0 c}(1 - \cos\theta)$$

where λ' and λ are the wavelengths after and before scattering, m_0 is rest mass of the electron, and θ is the scattering angle. The $\frac{\hbar}{m_0 c}$ is known as the Compton wavelength of the electron, which is equal to 2.43×10^{-12} meter. Therefore:

$$\Delta\lambda = 2.43 \times 10^{-12}(1.0 - 0.866) = 3.255 \times 10^{-13} \text{meter.}$$

7. Calculate the wavelength of maximum power emitted by a blackbody at 100 °C temperature.

SOLUTION

The wavelength of maximum power emitted by a blackbody at temperature T is given by Wien's displacement law, $\lambda_{max}T = 2898\ \mu m$ K. Therefore:

$$\lambda_{max} = \frac{2898}{373} = 7.77\ \mu\text{m}.$$

8. Use the Bohr model for the hydrogen atom and calculate the energy change when the atom moves from energy level n = 5 to energy level n = 2. Is it radiation emission or absorption? Find the wavelength for this energy change.

SOLUTION

The energy of nth level is given by Eqn. (1.7):

$$E_n = -\frac{me^4}{32\pi^2\varepsilon_0^2\hbar^2 n^2} = -\frac{e^2}{8\pi\varepsilon_0 a_0 n^2} = -\frac{R_H}{n^2}$$

where $R_H = \dfrac{e^2}{8\pi\varepsilon_0 a_0} = 2.179\times10^{-18}\text{Joules} = 13.6\text{ eV}$ is a constant.

$$\begin{aligned}\Delta E = E_2 - E_5 &= -13.6\left(\frac{1}{4} - \frac{1}{25}\right)\\ &= -2.856\text{ eV}\end{aligned}$$

since this photon is moving down in energy levels 5 to 2 and the value of ΔE is negative. Radiation is emitted. To calculate the wavelength of emitted radiation, we use $\Delta E = hv = hc/\lambda$, which implies $\lambda = hc/\Delta E$.

$$\lambda = \frac{6.626\times2.9979\times10^{-26}}{2.856\times1.602\times10^{-19}} = 4.34\times10^{-7}\text{m} = 434\text{ nm}$$

9. Show that $\psi(x) = Axe^{-x^2}$ is a wave function for the operator $\left(-\dfrac{d^2}{dx^2} + 4x^2\right)$. Normalize the wave function.

SOLUTION

$$\begin{aligned}\left[-\frac{d^2}{dx^2} + 4x^2\right]\psi(x) &= -\frac{d^2}{dx^2}\left(Axe^{-x^2}\right) + 4Ax^3e^{-x^2}\\ &= 4Ax^3\ e^{-x^2} - A\frac{d}{dx}\left[e^{-x^2} - 2x^2e^{-x^2}\right]\\ &= 4Ax^3\ e^{-x^2} - A\left[-2xe^{-x^2} - 4xe^{-x^2} + 4x^3\ e^{-x^2}\right]\\ &= 6Axe^{-x^2}\\ &= 6\psi(x)\end{aligned} \tag{1.30}$$

which shows that $\psi(x) = Axe^{-x^2}$ is the wave function of the operator $\left(-\frac{d^2}{dx^2} + 4x^2\right)$ with eigenvalue equal to 6.

The normalization condition $\psi(x)$ is:

$$\int_{-\infty}^{\infty} |\psi(x)|^2 \, dx = 1$$

$$\Rightarrow \int_{-\infty}^{\infty} A^2 x^2 e^{-2x^2} dx = \frac{A^2\sqrt{\pi/2}}{4}.$$

$$\Rightarrow A = 2\left(\frac{2}{\pi}\right)^{1/4}$$

$$\text{Therefore } \psi(x) = 2\left(\frac{2}{\pi}\right)^{1/4} xe^{-x^2} \tag{1.31}$$

1.17 Exercises

1. Show that $f(x)p - pf(x) = i\hbar\frac{\partial f(x)}{\partial x}$ and then prove that $xp - px = i\hbar$, where $p = -i\hbar\frac{d}{dx}$ and $f(x)$ is an arbitrary function x.
2. The helium ion can be treated as a hydrogen-like atom with ionization energy, $E_{ion} = 54.4\text{eV}$ in the ground state. Calculate the frequency ν and wavelength λ of electromagnetic radiation that will just ionize the helium ion.
3. On application of DC voltage V_0 across the two superconducting layers separated by a thin insulating layer, an oscillating current $I = I_0 \sin\left(\frac{2eV_0}{\hbar}t\right)$ is produced by the tunneling of paired electrons through the insulating layer. The phenomenon is known as the AC Josephson effect. Calculate the frequency of current for $V_0 = 2\times10^{-6}\,\text{Volts}$.
4. Use the Bohr-Sommerfeld quantization condition and calculate the velocity of the electron in the first orbit of the hydrogen atom. Treat the first orbit as a circle of radius equal to the Bohr radius.
5. The uncertainty in the position of a moving particle is one-fourth of its de Broglie wavelength. Show that the maximum possible uncertainty in its velocity is $\Delta v = \frac{v}{\pi}$.
6. A free particle moving along the x-axis is represented by wave function, $\psi(x,t) = Ae^{i(kx-\omega t)}$. Use the time-dependent Schrödinger equation to find the relation between ω and k and then calculate the group velocity v_g and phase velocity v_{ph}.

7. The work function of aluminum is 4.26 eV. Its surface is irradiated by an ultraviolet beam of wavelength 220 nm. Calculate the maximum kinetic energy of electrons ejected from its surface.

8. Find for what value of E, $\psi(r) = Ne^{-r/a_0}$ satisfies $-\left(\frac{\hbar^2}{2m}\nabla^2 + \frac{e^2}{4\pi\varepsilon_0 r}\right)\psi(r) = E\psi(r)$, where N is constant.

9. Calculate N with the use of the condition of normalization for the wave function $\psi(r) = Ne^{-r/a_0}$.

2

Wave Mechanics and Its Simple Applications

As stated in Chapter 1, Schrödinger developed quantum wave mechanics as a continuation of de Broglie's hypothesis. He formulated a second-order differential equation to explain the wave nature of matter and the particle associated to the wave. The Schrödinger equation assumes that a particle behaves as a wave and yields a solution in terms of wave function and the energy of the particle under consideration. Once the wave function is known, then everything about the particle can be deduced from the wave function.

2.1 Schrödinger Equation

This equation is used in two forms. In one of the forms, time explicitly appears to describe how the wave function of a particle evolves in time. This equation is referred to as the time-dependent Schrödinger equation. The other is the equation in which the time dependence has been dropped and the equation describes, among other things, what are allowed values of energies and hence it is known as the time-independent Schrödinger equation. However, these are not two independent equations. The time-independent equation can be derived readily from the time-dependent equation (except if the potentials are time dependent). A simple derivation of the time-dependent Schrödinger equation is presented in Section 1.7 and the time-dependent Schrödinger equation, for a particle moving under the influence of a field defined by potential energy, $V(r)$, is given by Eqn. (1.16):

$$\left[-\frac{\hbar^2}{2m}\nabla^2 + V(\mathbf{r})\right]\psi(\mathbf{r},t) = i\hbar\frac{\partial\psi(\mathbf{r},t)}{\partial t} \tag{2.1}$$

In the Schrödinger representation of quantum mechanics, the Hamiltonian, $H = -\frac{\hbar^2}{2m}\nabla^2 + V(\mathbf{r})$ and other operators are time independent. The time independence of the Hamiltonian allows one to factorize the wave function into space- and time-dependent parts. Time dependence enters to wave function via a complex exponential factor, $e^{-iEt/\hbar}$, and hence the time-dependent wave function is written as:

$$\psi(\mathbf{r},t) = \psi(\mathbf{r})e^{-iEt/\hbar} \tag{2.2}$$

Substitution of Eqn. (2.2) into Eqn. (2.1) yields:

$$\left[-\frac{\hbar^2}{2m}\nabla^2+V(\mathbf{r})\right]\psi(\mathbf{r})=E\psi(\mathbf{r}) \tag{2.3}$$

which is the time-independent Schrödinger equation, also known as the eigenvalue equation.

Solutions of Eqn. (2.1) describe the dynamical behavior of the particle, which is in some sense similar to the classical physics description obtained from Newton's equation $\boldsymbol{F}=m\boldsymbol{a}$. But, there is an important difference between the two descriptions. By solving Newton's equation, one can determine the position of a particle as a function of time, whereas solution of the Schrödinger equation gives a wave function $\psi(\mathbf{r},t)$ which is related to the probability of finding the particle in some region where space varies as a function of time. E, termed as the energy of the particle in Eqn. (2.3), is a free parameter and in principle it can have all possible values without any restriction at any stage, suggesting that to determine the wave function for a particle moving under the influence of a potential with some specific value of E, we have to solve Eqn. (2.3) for the corresponding wave function. In doing so, we find different solutions $\psi(\mathbf{r})$ for different choices of E, for a given potential. We can emphasize this fact by writing $\psi_E(\mathbf{r})$ as the solution associated with a value of E. But all $\psi_E(\mathbf{r})$ are not acceptable. To be physically acceptable, $\psi_E(\mathbf{r})$ must satisfy two conditions; (i) it must be normalizable to fulfill the criteria of probability interpretation of the wave function, (ii) $\psi_E(\mathbf{r})$ and its derivative must be continuous. The first condition leads to a rather remarkable property of physical systems described by Eqn. (2.3), which is the quantization of energy:

$$\int|\psi(\mathbf{r},t)|^2 d^3r=\int|\psi(\mathbf{r})|^2 d^3r=1 \tag{2.4}$$

To have the finite (unity) value of the integral, we must have $\psi_E(r)=0$, for $r\rightarrow\pm\infty$. Thus, physically acceptable solutions are those which satisfy the normalization condition. Wave functions that do not satisfy the normalization condition and the corresponding value of the energy are not physically acceptable. A particle can never be observed to have any energy other than these values, referred to as *allowed energies* of the particle.

We thus find that the probability interpretation of the wave function forces us to conclude that the allowed energies of a particle moving in a potential are restricted to certain discrete values, which are determined by the nature of the potential. The quantization of energy, a result of quantum mechanics, has enormous significance for determining the structure of atoms, and the properties of matter overall.

2.2 Bound States and Scattering States

Equation (2.3) can be solved for two types of particles: (i) particles trapped by an attractive potential into what is known as a bound state, and (ii) the particles that are free to travel to infinity, also known as scattering states. A particle trapped in an infinitely deep potential well, an electron in an atom trapped by the attractive potential due to a positively charged atomic nucleus, and a nucleon trapped within a nucleus by attractive nuclear forces are examples of bound states. In all these cases, the probability of finding the particle at infinity is zero, because the wave function vanishes at infinity. Thus, when a particle is trapped or confined to a limited region of space by an attractive potential, we obtain wave functions that satisfy the above boundary condition Eqn. (2.4), and their energies are quantized.

When a particle is not bound by any attractive potential or it is repelled by a repulsive potential, it is free to move as far as it likes in space. For such cases, we find that the wave function does not vanish at infinity, and its energy is not quantized. The problem then arises of how to reconcile this situation with the normalization condition, and the probability interpretation of the wave function. When wave function does not diverge or remains finite at infinity, a physical meaning can be assigned to such states as being idealized mathematical limiting cases which can still be dealt with in the same way as the bound state wave functions, provided some care is taken with the physical interpretation.

2.3 Probability Density, Probability Current, and Expectation Value

Current density related to a wave function $\psi(\mathbf{r},t)$ can be calculated, as $\psi(\mathbf{r},t)$ evolves with time in accordance with the Schrödinger equation:

$$i\hbar\frac{\partial\psi(\mathbf{r},t)}{\partial t}=H\psi(\mathbf{r},t) \tag{2.5}$$

The complex conjugation of Eqn. (2.5) is:

$$-i\hbar\frac{\partial\psi^*(\mathbf{r},t)}{\partial t}=H^*\psi^*(\mathbf{r},t) \tag{2.6}$$

Probability density is defined as $P=|\psi(\mathbf{r},t)|^2$ and then charge density is given by $\rho=qP$, where q is charge on the particle. We have:

$$\frac{\partial P}{\partial t}=\frac{\partial|\psi|^2}{\partial t}=\frac{\partial}{\partial t}\left(\psi^*\psi\right)=\frac{\partial\psi^*}{\partial t}\psi+\psi^*\frac{\partial\psi}{\partial t} \tag{2.7}$$

With the use of Eqns. (2.5) and (2.6), Eqn. (2.7) can be rewritten as:

$$\frac{\partial P}{\partial t}=\frac{1}{i\hbar}\left[\psi^*(H\psi)-(H\psi)^*\psi\right] \tag{2.8}$$

where $H=-\frac{\hbar^2}{2m}\nabla^2+V(\mathbf{r})=H^*$, if $V(\mathbf{r})$ is real. We then get:

$$\begin{aligned}\frac{\partial P}{\partial t}&=\frac{1}{i\hbar}\left[-\psi^*\frac{\hbar^2}{2m}\left(\nabla^2\psi\right)+\psi^*V(\mathbf{r})\psi+\frac{\hbar^2}{2m}\left(\nabla^2\psi^*\right)\psi-\psi V(\mathbf{r})\psi^*\right]\\&=\frac{1}{i\hbar}\left[-\psi^*\frac{\hbar^2}{2m}\left(\nabla^2\psi\right)+\frac{\hbar^2}{2m}\left(\nabla^2\psi^*\right)\psi\right]\\&=-\frac{i\hbar}{2m}\nabla.\left[\psi\nabla\psi^*-\psi^*\nabla\psi\right]\end{aligned} \tag{2.9}$$

which yields:

$$\frac{\partial P}{\partial t}+\frac{i\hbar}{2m}\nabla.\left[\psi\nabla\psi^{*}-\psi^{*}\nabla\psi\right]=0$$
$$\Rightarrow\frac{\partial P}{\partial t}+\nabla.\mathbf{S}=0$$

with:

$$\mathbf{S}=\frac{i\hbar}{2m}\left[\psi\nabla\psi^{*}-\psi^{*}\nabla\psi\right] \tag{2.10}$$

S is called probability current, describing the flow of probability. This also implies:

$$\partial\rho/\partial t+\nabla.\mathbf{J}=0 \tag{2.11}$$

Here, we have defined current density $\mathbf{J}=q\mathbf{S}$. Eqn. (2.11) is a well-known continuity equation.

To understand more about Eqn. (2.10), let us consider a system completely confined to a volume V so that nothing is going in and out at the surface. On integrating both terms of the equation over the entire volume, we get:

$$\oint\left(\frac{\partial P}{\partial t}+\nabla.\mathbf{S}\right)d^{3}\mathrm{r}=0$$
$$\Rightarrow\oint\frac{\partial P}{\partial t}d^{3}r=-\oint\nabla.\mathbf{S}d^{3}r=-\oint\mathbf{S}.d\mathbf{s} \tag{2.12a}$$

where, we used the Gauss theorem to convert volume integration to surface integration. $d\mathbf{s}$ is surface element defined outwards and integrated over the surface of the volume V. In this case, because we confined to system completely in the volume, there is nothing at the surface and therefore $\oint\mathbf{S}.d\mathbf{s}=0$. On interchanging the order of the volume integral and the time derivative over P, we write;

$$\frac{\partial}{\partial t}\oint Pd^{3}r=0 \tag{2.12b}$$

which states that $\oint Pd^{3}r=$ constant or a conserved quantity. If the volume does not confine the full system, the amount in $\oint Pd^{3}r$ is given by amount of flow out of volume, as is seen from Eqn. (2.12a). This means that **S**(**r**) and P(**r**) are the current and density, respectively, of a conserved quantity.

The definition of probability density allows us to calculate expectation value of an observable (operator), A. When a large number of measurements on A of a particle are made in a particular state, the average of different measured values is the *expectation value*. The $|\psi(\mathbf{r},t)|^{2}$ represents the probability of measurement on A of a particle at position vector **r** and time t. The expectation value (average of measurements) of A is defined by

$\langle A\rangle = \int \psi(\mathbf{r},t)^* A\psi(\mathbf{r},t)d^3r$ for normalized wave function. If $\psi(\mathbf{r},t)$ is not a normalized wavefunction, then $\langle A\rangle$ is defined as follows:

$$\langle A\rangle = \frac{\int \psi(\mathbf{r},t)^* A\psi(\mathbf{r},t)d^3r}{\int \psi(\mathbf{r},t)^* \psi(\mathbf{r},t)d^3r} \tag{2.13}$$

2.4 Simple Applications of Time-Independent Schrödinger Equation

To illustrate how the time-independent Schrödinger equation can be solved in practice, and what are some of the characteristics of its solutions, we discuss here the some of its well-known solutions.

2.4.1 Free Particle Motion

A particle that has no external forces acting on it (a constant potential energy V_0) is called a free particle. The one-dimensional (1D) motion of a particle of mass m is described by:

$$-\frac{\hbar^2}{2m}\frac{d^2\psi(x)}{dx^2} + V_0\psi(x) = E\psi(x) \tag{2.14}$$

$$\frac{d^2\psi(x)}{dx^2} = -\frac{2m(E-V_0)}{\hbar^2}\psi(x) = -k^2\psi(x) \tag{2.15}$$

with $k \pm \frac{\sqrt{2m(E-V_0)}}{\hbar}$. k has real values when $E > V_0$ and it has imaginary values for $E < V_0$. For real values of k, a generalized solution of Eqn. (2.15) is given by:

$$\psi(x) = Ae^{ikx} + Be^{-ikx} \tag{2.16}$$

which represents a traveling wave of kinetic energy, $\frac{\hbar^2k^2}{2m} = (E - V_0)$, classically allowed values of kinetic energy.

When $E < V_0$, values of k are imaginary, say $k = iK$, and the solution of Eqn. (2.15) is given as:

$$\psi(x) = Ce^{Kx} + De^{-Kx} \tag{2.17}$$

where $\frac{\hbar^2k^2}{2m} = -\frac{\hbar^2K^2}{2m} = -(V_0 - E)$, which are classically forbidden values of kinetic energy. We thus find that in the case of $E > V_0$, kinetic energies of the particle are classically allowed, whereas in the situation in which $E < V_0$, kinetic energies are classically forbidden. This means that a particle, which is rolling on a potential surface described by $V(x)$, can exist in the region where $V(x) < E$ and it must turn around as and when $V(x) \geq E$. The points where $V(x) = E$ (kinetic energies or velocities of the particle are zero) are known as the *classical turning points*.

For the traveling wave solutions given by Eqn. (2.16), probability density $P(x)$ is given by:

$$\begin{aligned} P(x) &= \psi(x,t)^* \psi(x,t) = \psi(x)^* e^{\frac{iEt}{\hbar}} \psi(x) e^{-\frac{iEt}{\hbar}} \\ &= A^2 + B^2 + 2AB \cos(2kx) \end{aligned} \tag{2.18}$$

for real values of A and B. Equation (2.18) represents a standing wave. If we consider a wave is traveling along in only one direction, say the $+x$-direction, $P(x) = A^2$, which is independent of position, implying that the particle is equally likely to be anywhere in space; it is completely delocalized.

Solution of Eqn. (2.14) for $E < V_0$ is given by Eqn. (2.17) and

$$\begin{aligned} P(x) &= \psi(x,t)^* \psi(x,t) = \psi(x)^* e^{\frac{iEt}{\hbar}} \psi(x) e^{-\frac{iEt}{\hbar}} \\ &= A^2 e^{2Kx} + B^2 e^{-2Kx} + 2AB \cosh(2Kx) \end{aligned} \tag{2.19}$$

which suggests that the probability density of finding the particle in regions where $V(x) > E$ is nonzero, implying that particle penetrates to classically forbidden region (barrier). Thus, quantum mechanical solutions yield a non-zero probability of finding the particle in the classically forbidden region. Quantum mechanics allows a particle to penetrate (tunnel through) a barrier (classically forbidden region) and emerge on the other side of the barrier.

2.4.2 Infinite Potential Well (Particle in a Box)

Consider a single particle of mass m confined to within a region, $-a/2 < x < a/2$, of potential energy $V(x) = V_0 < E$, bounded by infinitely high potential barriers, i.e. $V(x) = \infty$ at the walls at $x = -\frac{a}{2}$ and $x = \frac{a}{2}$. Potential is written as:

$$V(x) = \begin{cases} V_0 < E & \text{for} \frac{-a}{2} < x < \frac{a}{2} \\ \infty & \text{when}\, x > \left|\frac{a}{2}\right| \end{cases} \tag{2.20}$$

The probability of finding the particle outside the well is zero; hence wave function must vanish at boundaries, i.e. $\psi\left(\pm\frac{a}{2}\right) = 0$, otherwise $V(x)\psi x)$ will be infinite outside the well, which has no meaning. Solutions of the 1D Schrödinger equation inside the well $(-a/2 < x < a/2)$ are given by Eqn. (2.16) and $\frac{\hbar^2 k^2}{2m} = E - V_0$. At the boundaries we have:

$$\psi\left(-\frac{a}{2}\right) = Ae^{-ika/2} + Be^{ika/2} = 0 \tag{2.21}$$

$$\psi\left(\frac{a}{2}\right) = Ae^{ika/2} + Be^{-ika/2} = 0 \tag{2.22}$$

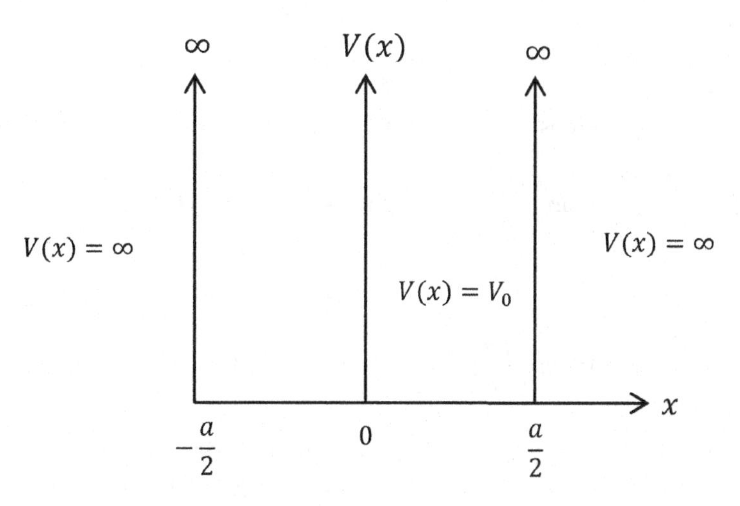

FIGURE 2.1
Infinite potential well.

The sum of Eqns. (2.21) and (2.22) yields:

$$(A+B)\cos\left(\frac{ka}{2}\right)=0 \tag{2.23}$$

while subtraction of Eqn. (2.21) from (2.22) gives:

$$i(A-B)\sin\left(\frac{ka}{2}\right)=0 \tag{2.24}$$

Two simple cases for which conditions Eqns. (2.23) and (2.24) can be satisfied are (i) both A and B are zero, and (ii) $A=B$ or $A=-B$. The first case gives a trivial solution which is physically unacceptable. For the second case, we have:

$$\cos\left(\frac{ka}{2}\right)=0, \text{ when } A=B \tag{2.25}$$

$$\sin\left(\frac{ka}{2}\right)=0, \text{ for } A=-B \tag{2.26}$$

Both the conditions (Eqns 2.25 and 2.26) must be satisfied. $k=\frac{n\pi}{a}$ satisfies Eqn. (2.25) for odd values of n $(=1,3,5,7,\ldots.)$ and it satisfies Eqn. (2.26) for even values of n $(=2,4,6\ldots\ldots\ldots)$. Note that energy E and momentum $\hbar k$ of a particle take discrete values. Each of values of n corresponds to a quantum state of the system. For the nth quantum state, we have:

$$E_n=V_0+\frac{\hbar^2(n\pi)^2}{2m\,a^2} \tag{2.27}$$

and,

$$\psi_n(x) = \begin{cases} A\cos\left(\dfrac{n\pi x}{a}\right) & \text{for odd } n-\text{values, } n = 1,3,5,7........ \\ A\sin\left(\dfrac{n\pi x}{a}\right) & \text{for even } n-\text{values, } n = 2,4,6,8........ \end{cases} \tag{2.28}$$

or,

$$\psi_n(x) = \sin\left(\frac{n\pi}{a}\left(x + \frac{a}{2}\right)\right) \text{ for } n = 1,2,3,4,5......... \tag{2.29}$$

E_n and $\psi_n(x)$ are called energy eigenvalue and eigenfunction, respectively. E_1, E_2 and E_3 are energy values and ψ_1, ψ_2 and ψ_3 are corresponding wave functions for the ground state, first exited state, and second excited state. The constant A can be determined by using the normalization condition on the wave function:

$$\begin{aligned} &\int_{-a/2}^{a/2} \psi_n(x)^* \psi_n(x)dx = 1 \\ &\Rightarrow \int_{-a/2}^{a/2} A^2 \sin^2\left(\frac{n\pi}{a}\left(x + \frac{a}{2}\right)\right)dx = 1 \\ &\Rightarrow A = \sqrt{2/a} \end{aligned} \tag{2.30}$$

It is to be noted that the normalization condition can be met for a range of complex amplitudes, $A = e^{i\theta}\sqrt{2/a}$, in which the phase is arbitrary, which implies that the outcome of a measurement, which is proportional to $|\psi(x)|^2$ about the particle position, is invariant under a global phase factor. Also, note that

$$\int_{-a/2}^{a/2} \psi_m(x)^* \psi_n(x)dx = \delta_{mn} \tag{2.31}$$

where, δ_{mn} is known as the Kronecker delta and it is defined as:

$$\delta_{mn} = \begin{cases} 1 & \text{when } m = n \\ 0 & \text{otherwise} \end{cases} \tag{2.32}$$

Wave functions that satisfy Eqn. (2.31) are called orthonormal wave functions.

The infinite potential well is a valuable model because it shows in a simple way how energy quantization takes place in nature. This model describes very accurately the quantum character of such systems of electrons trapped in a block of metal, quantum wire, and nanoparticle or gas molecules contained in a bottle. The potential experienced by an electron as it approaches the edges of a block of metal, quantum wire, and nanoparticle or as experienced by a gas molecule as it approaches the walls of its container are effectively infinite, provided particles have sufficiently low kinetic energy as compared to the height of these potential barriers.

2.4.3 The Finite Potential Well

When a particle inside potential well has an energy comparable to the height of the potential barriers, use of the infinite well potential model is not justified, because chances of the particle escaping the well are there. The motion of the particle can be discussed both classically and quantum mechanically, though, as we expect, the quantum mechanical treatment is not necessarily consistent with our classical physics-based expectations. We describe here the quantum mechanical properties of a particle in a finite potential well depicted in Fig. 2.2. We choose a well in which the potential is symmetric about its origin, and we also choose zero potential energy at the bottom of the well. To find bound state solutions, $E < V(x)$, let us describe the potential:

$$V(x) = \begin{cases} 0 \text{ for } -\frac{a}{2} < x < \frac{a}{2} \\ V_0 > E \text{ for } x > \left|\frac{a}{2}\right| \end{cases} \tag{2.33}$$

For cases of $-\frac{a}{2} > x > \frac{a}{2}$, classical motion is forbidden because kinetic energy is negative. However, quantum mechanically motion is allowed and it is described by the Schrödinger equation:

$$\frac{d^2\psi(x)}{dx^2} = \frac{2m(V_0 - E)}{\hbar^2}\psi(x) = \alpha^2\psi(x) \tag{2.34}$$

Where α^2 is a positive quantity. A general solution of Eqn. (2.34) is:

$$\psi(x) = Ae^{\alpha x} + Be^{-\alpha x} \tag{2.35}$$

In the region of $x < -a/2$, inclusion of a term having $e^{-\alpha x}$ leads to $\psi(x) \to \infty$, for $x \to -\infty$, which is an unacceptable solution. Hence,

$$\psi(x) = Ae^{\alpha x} \quad \text{for } x < -a/2 \tag{2.36}$$

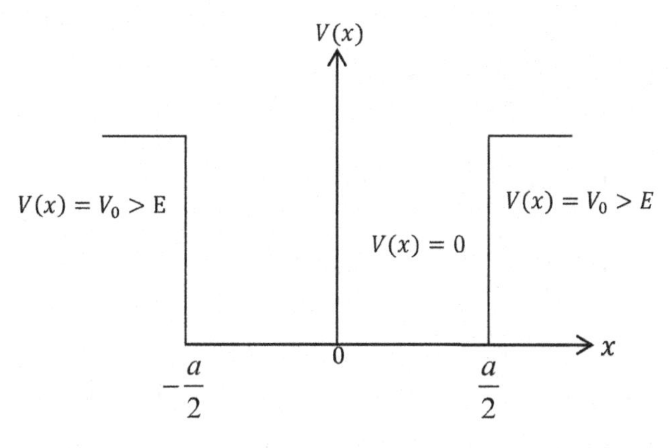

FIGURE 2.2
Schematic diagram for finite potential well.

Similarly, for $x > a/2$, we cannot include $e^{\alpha x}$ in our solution, and therefore,

$$\psi(x) = Be^{-\alpha x} \quad \text{for } x > a/2 \tag{2.37}$$

Inside the well, $-a/2 < x < a/2$, the Schrödinger equation is:

$$\frac{d^2\psi(x)}{dx^2} = -k^2\psi(x), \text{ with } k = \sqrt{\frac{2mE}{\hbar^2}} \tag{2.38}$$

A general solution can be given by:

$$\psi(x) = C\ cos(kx) + D\ sin(kx) \tag{2.39}$$

It is to be noted that Eqns. (2.39) and (2.16) convey the same meaning. The allowed solutions of the Schrödinger equation are those which satisfy necessary boundary conditions: wave function and its first order derivatives must be continuous at all values of x. Thus, in order to satisfy Eqns. (2.37), (2.38) and (2.39) everywhere, we require that $\psi(x)$ and $\psi'(x) = \frac{d\psi}{dx}$ be continuous at the walls of the well.

$$\psi\left(-\frac{a}{2}\right): \quad Ae^{-\alpha a/2} = C\cos\left(\frac{ka}{2}\right) - D\ \sin\left(\frac{ka}{2}\right) \tag{2.40}$$

$$\psi'\left(-\frac{a}{2}\right): \quad \alpha\ Ae^{-\alpha a/2} = kC\sin\left(\frac{ka}{2}\right) + kD\ \cos\left(\frac{ka}{2}\right) \tag{2.41}$$

$$\psi\left(\frac{a}{2}\right): \quad Be^{-\alpha a/2} = C\cos\left(\frac{ka}{2}\right) + D\ \sin\left(\frac{ka}{2}\right) \tag{2.42}$$

$$\psi'\left(\frac{a}{2}\right): \quad -\alpha\ Be^{-\frac{\alpha a}{2}} = -kC\sin\left(\frac{ka}{2}\right) + kD\ \cos\left(\frac{ka}{2}\right) \tag{2.43}$$

Equations (2.40) to (2.43) form a matrix equation:

$$\begin{bmatrix} 1 & 0 & -\cos\left(\frac{ka}{2}\right) & \sin\left(\frac{ka}{2}\right) \\ \alpha & 0 & -\sin\left(\frac{ka}{2}\right) & -k\cos\left(\frac{ka}{2}\right) \\ 0 & 1 & -\cos\left(\frac{ka}{2}\right) & -\sin\left(\frac{ka}{2}\right) \\ 0 & -\alpha & \sin\left(\frac{ka}{2}\right) & -k\cos\left(\frac{ka}{2}\right) \end{bmatrix} \begin{bmatrix} Ae^{-\frac{\alpha a}{2}} \\ Be^{-\frac{\alpha a}{2}} \\ C \\ D \end{bmatrix} = \begin{bmatrix} 0 \\ 0 \\ 0 \\ 0 \end{bmatrix} \tag{2.44}$$

The four equations can be solved to find the allowed discrete energy eigenvalues E_n of the states confined within the quantum well. A non-trivial solution of matrix Eqn. (2.44) is obtained when the determinant of the 4×4 matrix is zero:

$$\begin{vmatrix} 1 & 0 & -\cos\left(\frac{ka}{2}\right) & \sin\left(\frac{ka}{2}\right) \\ \alpha & 0 & -\sin\left(\frac{ka}{2}\right) & -k\cos\left(\frac{ka}{2}\right) \\ 0 & 1 & -\cos\left(\frac{ka}{2}\right) & -\sin\left(\frac{ka}{2}\right) \\ 0 & -\alpha & \sin\left(\frac{ka}{2}\right) & -k\cos\left(\frac{ka}{2}\right) \end{vmatrix} = 0 \tag{2.45a}$$

Simplification gives that the determinant goes to zero when:

$$\left[k\sin\left(\frac{ka}{2}\right) - \alpha\cos\left(\frac{ka}{2}\right)\right]\left[k\cos\left(\frac{ka}{2}\right) + \alpha\sin\left(\frac{ka}{2}\right)\right] = 0 \tag{2.45b}$$

which means, either

$$\alpha = k\ \tan\left(\frac{ka}{2}\right) \tag{2.46a}$$

or

$$\alpha = -k\ \cot\left(\frac{ka}{2}\right) \tag{2.46b}$$

Solutions presented by Eqns. (2.40) to (2.46) are found less frequently in quantum mechanics textbooks, but these are quite useful in applications like the double quantum well, the modeling of bonding in molecules, and in an infinite array of quantum wells that represents a periodic solid.

As the potential is symmetric, an easier way to get Eqns. (2.46a) and (2.46b), which exists in textbooks on quantum mechanics, is as follows: Addition and subtraction of Eqns. (2.40) and (2.42) give:

$$(A+B)e^{-\frac{\alpha a}{2}} = 2C\ \cos\left(\frac{ka}{2}\right) \tag{2.47a}$$

$$(A-B)e^{-\frac{\alpha a}{2}} = -2D\ \sin\left(\frac{ka}{2}\right) \tag{2.47b}$$

Addition and subtraction of Eqns. (2.41) and (2.43) yield:

$$\alpha(A-B)e^{-\frac{\alpha a}{2}} = 2kD\ cos\left(\frac{ka}{2}\right) \tag{2.48a}$$

$$\alpha(A+B)e^{-\frac{\alpha a}{2}} = 2kC\ sin\left(\frac{ka}{2}\right) \tag{2.48b}$$

Division of Eqn. (2.48b) by (2.47a) gives Eqn. (2.46a) provided $A \neq -B$ and $C \neq 0$. Similarly, we get Eqn. (2.46b) on dividing Eqn. (2.48a) by Eqn. (2.47b), when $A \neq B$ and $D \neq 0$. It is to be noted that Eqns. (2.46a) and (2.46b) cannot be satisfied simultaneously, because in that case $\tan^2\left(\frac{ka}{2}\right) = -1$, which cannot be satisfied for real values of k and a.

For the case of $A = B$ and $D = 0$, allowed energy values within the well are given by Eqn. (2.46a) and the wave functions are:

$$\psi(x) = \begin{cases} Ae^{\alpha x} & \text{when} \quad x < -a/2 \\ C\cos(kx) & \text{for} \quad -\frac{a}{2} < x < a/2 \\ Be^{-\alpha x} & \text{when} \quad x > \frac{a}{2} \end{cases} \tag{2.49a}$$

Wave function in this case is symmetric with respect to x; $\psi(x) = \psi(-x)$. In other words, it is an even function.

When $A = -B$ and $C = 0$, allowed energies within the well are given by Eqn. (2.46b) and the wave functions are:

$$\psi(x) = \begin{cases} Ae^{\alpha x} & \text{when} \quad x < -a/2 \\ D\sin(kx) & \text{for} \quad -\frac{a}{2} < x < a/2 \\ Be^{-\alpha x} & \text{when} \quad x > \frac{a}{2} \end{cases} \tag{2.49b}$$

Wave function is antisymmetric, $\psi(-x) = -\psi(x)$ in this case. It is an odd function.

To find the energy levels that correspond to symmetric solutions, we solve numerically or graphically Eqn. (2.46a), while for energy levels that correspond to antisymmetric solutions, Eqn. (2.46b) is solved. A simple graphical solution method is as follows: define, $\lambda = ka/2$ and $\mu = \alpha a/2$. Then, Eqns. (2.64a) and (2.46b) are:

$$\mu = \lambda\ \tan(\lambda) \tag{2.50a}$$

and,

$$\mu = -\lambda\ \cot(\lambda) \tag{2.50b}$$

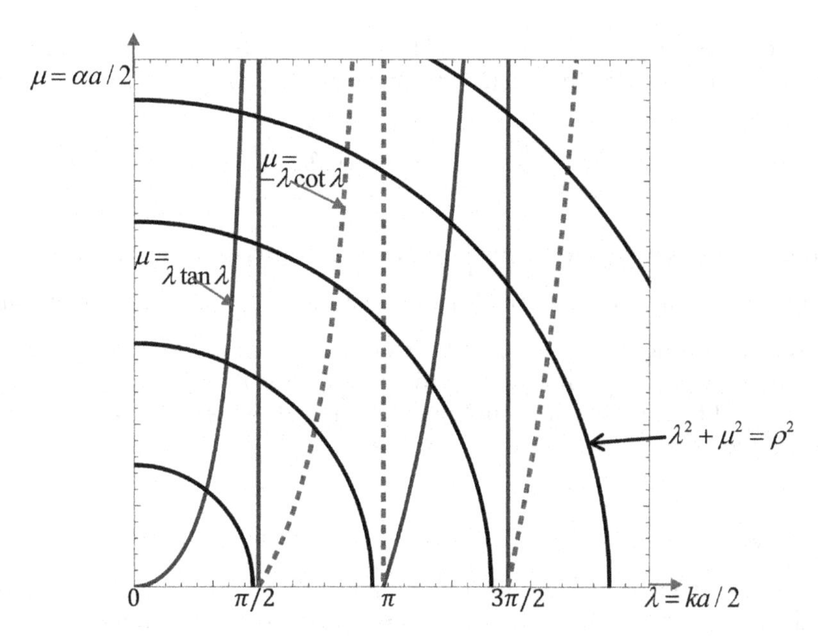

FIGURE 2.3
Graphical solution of Eqns. (2.50a) and (2.50b). Solid curves are $\mu = \lambda \tan\lambda$ and dashed curves are $\mu = -\lambda \cot\lambda$.

with,

$$\lambda^2 + \mu^2 = \frac{a^2}{4}\left(k^2 + \alpha^2\right) = \frac{mV_0a^2}{2\hbar^2} \tag{2.51}$$

Equation (2.51) represents a circle of radius $\rho = \frac{a}{\hbar}\sqrt{mV_0/2}$ in (λ, μ) -space. Equations (2.50a), (2.50b) and (2.51) are plotted as a function of λ in Fig. 2.3.

Intersections of the plots of $\mu = \lambda \tan(\lambda)$ and $\lambda^2 + \mu^2 = \rho^2$ curves in (μ, λ) -plane for different values of ρ; $\rho = 1\left(\frac{2\hbar^2}{m} = V_0a^2\right)$, $\rho = 2\left(\frac{8\hbar^2}{m} = V_0a^2\right)$, $\rho = 2\sqrt{3}\left(\frac{24\hbar^2}{m} = V_0a^2\right)$ etc. provide even (symmetric) solutions, while, intersections of $\mu = -\lambda\cot(\lambda)$ and $\lambda^2 + \mu^2 = \rho^2$ for different values of ρ yield odd (antisymmetric) solutions. The number of bound states depends on the height and width of the well through V_0a^2 factor. As is seen from Fig. 2.3, there is one bound state corresponding to the even (symmetric) wave function for $0 < \rho < \frac{\pi}{2}$, and there are two bound states, one corresponding to the even (symmetric) wave function and one corresponding to the odd (antisymmetric) solution, for $\frac{\pi}{2} < \rho < \pi$. Similarly, there exist three bound states for $\pi < \rho < \frac{3\pi}{2}$; two of these correspond to the symmetric solution and one corresponds to the antisymmetric solution. On increasing the values of ρ, the number of bound states energy levels increase successively. It is to be noted that there exists at least one bound state for the given value of V_0, and there is no degeneracy in the solutions. When the intersection of the circle with Eqn. (2.46a) is represented by $\lambda_1, \lambda_2, \lambda_3 \ldots.$, we have:

$$\lambda_n^2 = \frac{ma^2}{2\hbar^2} E_n \Rightarrow E_n = \frac{\hbar^2}{2m} (2\lambda_n / a)^2 \quad (2.52).$$

The case of $E > V_0$ does not provide any bound state. In this case, particle energy exceeds the height of the well and $k^2 = \frac{2m}{\hbar^2}(E - V_0)$ is positive everywhere giving the sinusoidal solution of the Schrödinger equation in all regions, and probability density is distributed all over the space. A particle is incident on the potential well, and it would keep going as there is no probability of reflection from the potential well. This is the case of scattering of a particle from the potential well.

2.4.4 Step Potential

We consider a case of a particle, initially traveling in a region of space of a constant potential, that suddenly moves into a region of different but again constant potential, as is shown in Fig. 2.4. The potential can be expressed as a piecewise function:

$$V(x) = \begin{cases} V_1 & \text{for} \quad x < 0 \\ V_2 & \text{when} \quad x \geq 0 \end{cases} \quad (2.53).$$

A particle moving from $-x$ to $+x$ faces a step potential by going from potential energy V_1 to V_2 at the origin. If particle energy $E > V_2$, it can continue to propagate arbitrarily far to the right, with an increased wavelength (decreased wave vector and momentum) across the step. But, if the particle has energy below the step $E < V_2$, it is classically forbidden to be found in the region of $x > 0$. However, the quantum mechanical description allows the particle to tunnel into this region and thus have a non-zero probability to be found in the region of $x > 0$. In both cases, there exists a finite reflectivity.

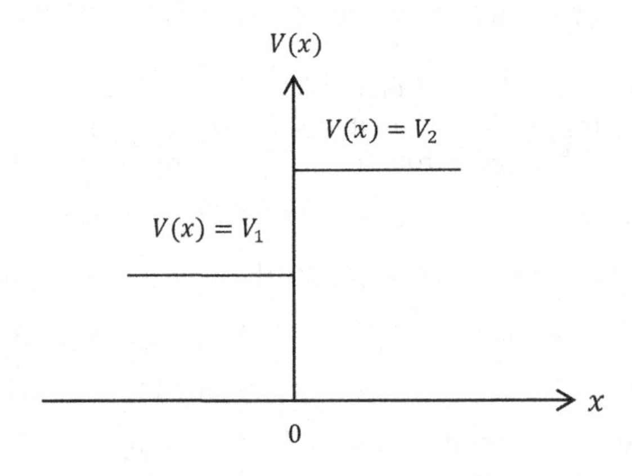

FIGURE 2.4
Potential step.

For the given energy of a particle E there are two cases of interest; (i) E is greater than both V_1 and V_2, and (ii) E is greater than V_1 but less than V_2.

Case-1: E is greater than both V_1 and V_2

The solution of the 1D Schrödinger equation in region-I ($x<0$), and in region-II ($x>0$), can be written as:

$$\psi_1(x) = Ae^{ik_1x} + Be^{-ik_1x} \tag{2.54a}$$

with $k_1 = \left(\frac{2m}{\hbar^2}(E-V_1)\right)^{1/2}$ for $x<0$. And,

$$\psi_2(x) = Ce^{ik_2x} + De^{-ik_2x} \tag{2.54b}$$

where, $k_2 = \left(\frac{2m}{\hbar^2}(E-V_2)\right)^{1/2}$, when $x>0$. The coefficients (A, B, C and D) are to be determined. Both terms in $\psi_1(x)$ are to be retained, as region-I can have both incident and reflected waves. However, we would not have a reflected wave in region-II, and hence, we take $D=0$ for writing $\psi_2(x)$. Equations (2.54a) and (2.54b) must satisfy the boundary conditions (the wave function and its first derivative must be continuous) at $x=0$. Therefore:

$$\psi_1(0) = \psi_2(0) \Rightarrow A + B = C \tag{2.55a}$$

$$\psi_1'(0) = \psi_2'(0) \Rightarrow ik_1A - ik_1B = ik_2C \tag{2.55b}$$

Elimination of C from Eqns. (2.55a) and (2.55b) gives:

$$\begin{aligned} k_1(A-B) &= k_2(A+B) \\ \Rightarrow \frac{B}{A} &= \frac{k_1}{k_2}\left(1-\frac{B}{A}\right)-1 \\ \Rightarrow \frac{B}{A} &= \frac{1-k_2/k_1}{1+k_2/k_1} \end{aligned} \tag{2.56}$$

with

$$\frac{k_2}{k_1} = \sqrt{\frac{1-V_2/E}{1-V_1/E}} \tag{2.57}$$

B/A is the reflection coefficient of the barrier. The reflectivity of the barrier, which corresponds to the ratio of probabilities, is given by:

$$R = \left|\frac{B}{A}\right|^2 = \left|\frac{1-k_2/k_1}{1+k_2/k_1}\right|^2 \tag{2.58}$$

Due to the conservation of the particle number, the transmissivity is given by:

$$T = 1 - R = 1 - \left| \frac{1 - k_2 / k_1}{1 + k_2 / k_1} \right|^2 \tag{2.59}$$

It is to be noted that on going from region-I to region-II, the de Broglie wavelength of the particle with energy E changes and becomes longer for an increased potential step. We see that in the quantum analogue of scattering, the energy of the particle is not quantized, and the wave function that describes the scattering of a particle of a given energy does not decrease as $x \to \pm\infty$; therefore everything that leads to the quantization of energy for a bound particle does not apply here. Wave functions given by Eqns. (2.54a) and (2.54b) do not go to zero as $x \to \pm\infty$, implying that they do not satisfy the normalization condition – the integral over $|\psi|^2$ will always diverge. Then, how does one maintain the claim that the wave function must have a probability interpretation if one of the principal requirements, the normalization condition, does not hold true? A wave function that cannot be normalized to unity is inconsistent with the probability interpretation of the wave function, and hence it is not physically permitted. Nevertheless, it is possible to work with such wave functions by reinterpreting the wave function and calling $|\psi(x,t)|^2$ *particle flux*. We thus find that there can be quantum mechanical descriptions which agree with our classical intuition, but there also occur new kinds of phenomena that have no explanation within classical physics.

Case-2: *E* is greater than V_1 but less than V_2

In this case, the 1D Schrödinger equation in region-I ($x < 0$) has a sinusoidal solution that can be given by Eqn. (2.54a).

However, in region-II ($x > 0$), the solution of the 1D Schrödinger equation cannot be given by Eqn. (2.54b). A general solution in this is:

$$\psi_2(x) = Ce^{\alpha x} + De^{-\alpha x} \tag{2.60}$$

where, $\alpha = \left(\frac{2m}{\hbar^2}(V_2 - E) \right)^{1/2}$. Inclusion of term $Ce^{\alpha x}$ in the solution in region-II would lead to an unacceptable solution because the probability of finding the particle in this region will go infinity as $x \to \infty$. Therefore, we should set $C = 0$. Applying again the boundary conditions at $x = 0$, we obtain:

$$A + B = D \tag{2.61a}$$

and,

$$ik(A - B) = \alpha D \tag{2.61b}$$

Elimination of D from Eqns. (2.61a) and (2.61b) gives:

$$\frac{B}{A} = \frac{ik - \alpha}{ik + \alpha} \tag{2.62}$$

which is the reflection coefficient of the barrier. The reflectivity of the barrier is thus given by:

$$R = \left|\frac{B}{A}\right|^2 = \left(\frac{ik-\alpha}{ik+\alpha}\right)\left(\frac{-ik-\alpha}{-ik+\alpha}\right) = 1 \tag{2.63}$$

We thus find that though the particle has nonzero probability to penetrate the classically forbidden region, it is still being totally reflected away, and there is no transmission to the step potential.

Note that the depth at which the particle penetrates into the classically forbidden region is given by the distance, Δx from $x = 0$, at which the probability drops by $1/e$.

$$P(\Delta x) = |\psi_2|^2 = D^2 e^{-2\alpha\Delta x} = D^2 e^{-1} \tag{2.64a}$$

which gives:

$$\Delta x = \frac{1}{2\alpha} = \frac{\hbar}{2\sqrt{2m(V_2 - E)}} \tag{2.64b}$$

2.4.5 Finite Potential Barrier and Tunneling

Another useful example of solving the 1D Schrödinger equation is the case of the finite potential barrier of height V_0 and width a, shown in Fig. 2.5. The potential energy is defined by:

$$V(x) = \begin{cases} V_0 & \text{for } -a/2 < x < a/2 \\ 0 & \text{when } x > |a/2| \end{cases} \tag{2.65}$$

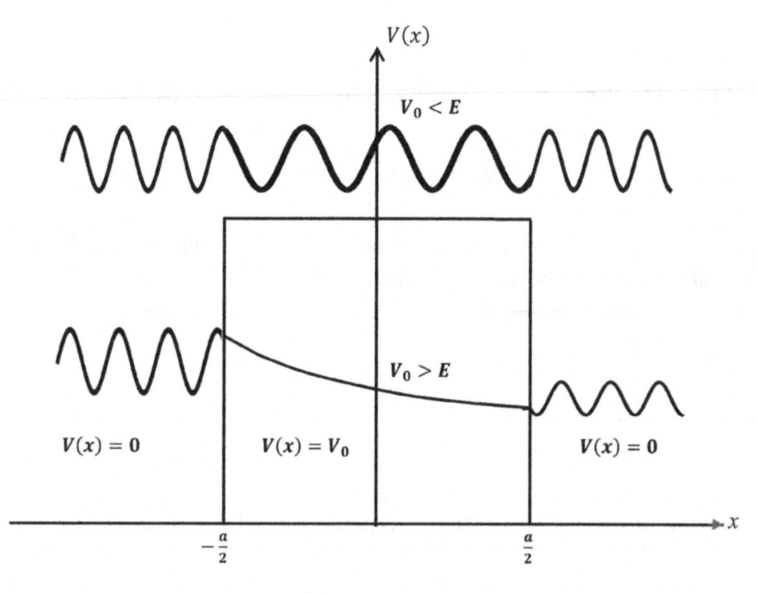

FIGURE 2.5
Finite potential barrier.

There can again be two types of behavior occurring. Case-1, where $E > V_0$, and case-2 where $E < V_0$. In case-1, the particle will have some reflection and some transmission as expected classically. Case-2 is of more interest, because classical physics predictions would be that the particle should be totally reflected and there should not be any transmission, which means that there should be zero probability to find the particle on the right-hand side of the barrier. However, quantum mechanics provide a finite (nonzero) probability of finding the particle on the right-hand side of the barrier, and it can continue to propagate to $+\infty$. This simple model is the precursor for discussing the decay of atomic nuclei as well as other quantum tunneling effects such as those associated with a scanning electron microscope.

The particle is moving from $-x$ to $+x$ across the barrier centred on the origin. In region-I $(x < -a/2)$ and in region-III $(x > a/2)$, $V(x)$ is zero and hence the general solution of the 1D Schrödinger equation will be a sinusoidal solution like that given by Eqn. (2.54a). However, if the particle initially starts from the left-hand side of the barrier $(x < -a/2)$, then it is on the right side of the barrier $(x > a/2)$, and there is no way to have e^{-ikx} in the solution, because this region cannot have a wave traveling to the $-x$ direction. Therefore:

$$\psi_1(x) = Ae^{ikx} + Be^{-ikx}, \text{ for } (x < -a/2) \tag{2.66}$$

$$\psi_3(x) = Fe^{ikx}, \text{ when } (x > a/2) \tag{2.67}$$

with $k = \left(\frac{2mE}{\hbar^2}\right)^{1/2}$. In region-II $(-a/2 < x < a/2)$, the solution of the 1D Schrödinger equation is to be worked out for two possible cases: (i) $E > V_0$, and (ii) $E < V_0$.

Case-1: $E > V_0$

The solution of the 1D Schrödinger equation will be similar to that given by Eqn. (2.66):

$$\psi_2(x) = Ce^{iKx} + De^{-iKx} \tag{2.68}$$

with $K = \sqrt{2m(E - V_0)}/\hbar$. The wave associated with the particle will have reflection as well as transmission at both boundaries of the barrier.

Case-2: $E < V_0$

The general solution of the 1D Schrödinger equation can be written as:

$$\psi_2(x) = Ce^{\alpha x} + De^{-\alpha x} \tag{2.69}$$

where $\alpha = \sqrt{2m(V_0 - E)}/\hbar$. Application of boundary conditions on wave functions given by Eqns. (2.66), (2.67) and (2.88) provides:

$$\psi_1\left(\frac{-a}{2}\right) = \psi_2\left(\frac{-a}{2}\right): \quad Ae^{-ika/2} + Be^{ika/2} = Ce^{-\alpha a/2} + De^{\alpha a/2} \tag{2.70a}$$

$$\psi_1'\left(\frac{-a}{2}\right) = \psi_2'\left(\frac{-a}{2}\right): \quad k\left(Ae^{-\frac{ika}{2}} - Be^{\frac{ika}{2}}\right) = \alpha\left(Ce^{-\frac{\alpha a}{2}} - De^{\frac{\alpha a}{2}}\right) \tag{2.70b}$$

$$\psi_2\left(\frac{a}{2}\right) = \psi_3\left(\frac{a}{2}\right): \quad Ce^{\alpha a/2} + De^{-\alpha a/2} = Fe^{ika/2} \tag{2.70c}$$

$$\psi_2'\left(\frac{a}{2}\right) = \psi_3'\left(\frac{a}{2}\right): \quad \alpha\left(Ce^{\frac{\alpha a}{2}} - De^{-\frac{\alpha a}{2}}\right) = ikFe^{ika/2} \tag{2.70d}$$

Equations (2.70a) to (2.70d) are to be solved to find out reflection (B/A) and transmission (F/A) coefficients from the barrier. We first divide Eqns. (2.70c) and (2.70d) by A and then solve these for C/A and D/A in terms of F/A to get:

$$\frac{C}{A} = \frac{(\alpha + ik)F}{2\alpha A} e^{ika/2} e^{-\alpha a/2} \tag{2.71a}$$

$$\frac{D}{A} = \frac{(\alpha - ik)F}{2\alpha A} e^{ika/2} e^{\alpha a/2} \tag{2.71b}$$

Next, first dividing by A and then subtracting Eqn. (2.71b) from (2.71a), we obtain B/A in terms of C/A and D/A:

$$\frac{B}{A} = -\frac{(\alpha - ik)C}{2ikA} e^{-\frac{\alpha a}{2}} e^{-\frac{ika}{2}} + \frac{(\alpha + ik)D}{2ikA} e^{\alpha a/2} e^{-ika/2} \tag{2.72}$$

On substituting values of C/A and D/A from Eqns. (2.71a) and (2.71b) into Eqn. (2.72), we obtain:

$$\frac{B}{A} = \frac{F}{A} \frac{\left(\alpha^2 + k^2\right)}{2ik\alpha} sinh(\alpha a) \tag{2.73}$$

Addition of Eqns. (2.70a) and (2.70b), after dividing by A, gives:

$$e^{-ika/2} = \frac{(\alpha + ik)C}{2ikA} e^{-\alpha a/2} - \frac{(\alpha - ik)D}{2ikA} e^{\alpha a/2} \tag{2.74}$$

Substitution of C/A and D/A from Eqns. (2.71a) and (2.71b) into Eqn. (2.74) yields:

$$e^{-ika/2} = \frac{F}{A}\left[\frac{(\alpha+ik)^2}{4ik\alpha}e^{-\alpha a} - \frac{(\alpha-ik)^2}{4ik\alpha}e^{\alpha a}\right]e^{ika/2}$$
$$\Rightarrow \frac{F}{A} = \frac{4ik\alpha e^{-ika}}{\left[(\alpha+ik)^2 e^{-\alpha a} - (\alpha-ik)^2 e^{\alpha a}\right]} \tag{2.75}$$

which gives:

$$\frac{B}{A} = \frac{2(\alpha^2+k^2)e^{-ika}\sinh(\alpha a)}{\left[(\alpha+ik)^2 e^{-\alpha a} - (\alpha-ik)^2 e^{\alpha a}\right]} \tag{2.76}$$

The reflectance R and transmittance T are the given by:

$$R = \left|\frac{B}{A}\right|^2 = \frac{2(\alpha^2+k^2)^2\sinh^2(\alpha a)}{\left[(\alpha^2+k^2)^2\cosh(2\alpha a) - (\alpha^4+k^4-6k^2a^2)\right]} \tag{2.77a}$$

and

$$T = \left|\frac{F}{A}\right|^2 = \frac{8k^2\alpha^2}{\left[(\alpha^2+k^2)^2\cosh(2\alpha a) - (\alpha^4+k^4-6k^2a^2)\right]} \tag{2.77b}$$

In case of weak transmission $(\alpha a \gg 1)$, the transmittance can be approximated by taking $\cosh(2\alpha a) \approx \frac{e^{2\alpha a}}{2}$ and then on dropping the smaller terms in the denominator, we get:

$$T \approx \frac{(4\alpha k)^2}{(k^2+\alpha^2)^2}e^{-2\alpha a} \tag{2.78}$$

2.4.6 Relevance of Free Particle, Potential Wells, and Potential Barriers

Examples illustrated above may seem somewhat abstract and not directly related to any realistic problems that one finds in the real world. However, these are not useless simplified cases of the 1D Schrödinger equation, as one might think. Rather, these are first-order approximations to scenarios that are encountered in the real world. These simplified solutions can give us insight into the behavior (both qualitative and quantitative) of actual physical systems. For example, (i) several properties of some metals can be understood with the use of free particle motion, (ii) the quantum tunneling model through the step barrier gives reasonable approximations for quantum tunneling effects in a variety of physical phenomena, such as the scanning electron microscope, (iii) the bound states for the infinite potential well demonstrate the concept of bound states, quantized energy levels, and temporal evolution of energy eigenstate superposition, which are linked to numerous real world scenarios such as the quantum wire, nanoparticle and hydrogen atom. Also, the propagation of a free particle, propagation of a particle across a potential step, and tunneling are all analogous to the propagation of light through various media with varying refractive indices. Particle-in-a-box solutions can be used to study the optical cavity.

2.5 Periodic Solids and their Band Structures

For many solids, each electron effectively experiences a similar average potential as experienced by all other electrons. This allows one to use an independent electron approximation and study the electronic properties of a solid by employing the one electron Schrödinger equation instead of solving something like the 10^{22} body problem. The exact form of the average potential may not be known, but it is expected to relate closely with the isolated atomic potential of an atom, which is the part of the solid. Further, many of the solids are crystalline, with a periodic lattice. Because the ground state electronic structure must also be periodic, with the same charge distribution in each unit cell, the average potential is periodic too; $V(\boldsymbol{r}) = V(\boldsymbol{r}+\boldsymbol{R})$, where $\boldsymbol{R}$ is a vector joining the same points in two different unit cells. Individual electron wave functions must also reflect this periodicity by satisfying a condition referred to as Bloch's theorem, which states that the wave functions of the one electron Hamiltonian $H = \frac{p^2}{2m} + V(r)$ can be chosen to have the form of a plane wave times a function with periodicity of the lattice:

$$\psi_{n\mathbf{k}}(\mathbf{r}) = e^{i\mathbf{k}.\mathbf{R}} u_{n\mathbf{k}}(\mathbf{r}) \tag{2.79a}$$

with

$$u_{nk}(\boldsymbol{r}+\boldsymbol{R}) = u_{nk}(\boldsymbol{r}) \tag{2.79b}$$

where the subscript n refers to a state associated with wave vector $\boldsymbol{k}$. Combining Eqns. (2.79a) and (2.79b), Bloch's theorem is restated as:

$$\psi_{nk}(\mathbf{r}+\mathbf{R}) = e^{ik.R}\psi_{nk}(\mathbf{r}) \tag{2.80}$$

The equation $\mathrm{H}\psi_{\mathrm{nk}}(\mathbf{r}) = \mathrm{E}_{\mathrm{nk}}\psi_{\mathrm{nk}}(\mathbf{r})$ is solved to find energy eigenvalues E_{nk}. A plot of E_{nk} versus k is referred as the band structure of a solid. For a 1D periodic structure with length L joining the same points in two different unit cells, Eqn. (2.79a) can be rewritten as:

$$\psi_{nk}(x) = e^{iqx} u_{nk}(x) \tag{2.81a}$$

with

$$u_{nk}(x+L) = u_{nk}(x) \tag{2.81b}$$

2.5.1 The Kronig-Penney Model

The Kronig-Penney model describes the electron in a 1D periodic potential. The possible eigenvalues that can be occupied by the electron and corresponding eigenstates are determined by solving a 1D Schrödinger equation. Many electronic properties of a periodic solid can be illustrated by using a periodic array of quantum wells, separated by barriers. A periodic structure having wells, each of width a and barriers each having width b and height V_0 is shown in Fig. 2.6.

Equations (2.81a) and (2.81b) describe the wave function in the 1D periodic potential, with $L = nd$, where $d = a + b$ is period of potential. Within a well, $V(x) = 0$. For the first well on the right-hand side to origin, $0 < x < a$, and hence the solution of the 1D Schrödinger equation is:

$$\psi_1(x) = Ae^{ikx} + Be^{-ikx} \tag{2.82}$$

with $k = \left(\frac{2mE}{\hbar^2}\right)^{1/2}$. The general solution of the 1D Schrödinger equation to the first barrier on the left-hand side of the origin $(-b < x < 0)$ is:

$$\psi_2(x) = Ce^{\alpha x} + De^{-\alpha x} \tag{2.83}$$

Where $\alpha = \sqrt{2m(V_0 - E)}/\hbar$ because $E < V_0$ inside the barrier. Two of the boundary conditions are applied at $x = 0$ and the other two at $x = a$. At $x = 0$:

$$\psi_1(0) = \psi_2(0): \qquad A + B = C + D \tag{2.84a}$$

$$\psi_1'(0) = \psi_2'(0): \qquad ik(A - B) = \alpha(C - D) \tag{2.84b}$$

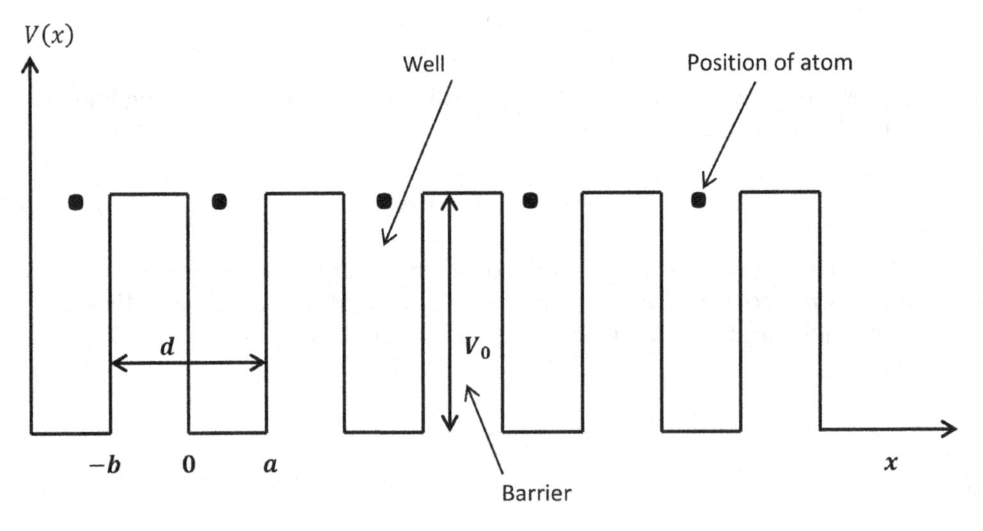

FIGURE 2.6
The Kronig-Penney periodic potential.

To apply boundary conditions at $x = a$, we have to make use of the Bloch theorem, Eqn. (2.80), for the 1D case, which becomes:

$$\psi(x+L) = e^{iqL}\psi(x) \tag{2.85}$$

where q is the wave vector. $x = a$ is related to $x = -b$ via the Bloch Theorem, Eqn. (2.85). Therefore by choosing $L = a + b$, we have:

$$\psi_1(a) = e^{iq(a+b)}\psi_2(-b): \quad Ae^{ika} + Be^{-ika} = \left(Ce^{-\alpha b} + De^{\alpha b}\right)e^{e^{iq(a+b)}} \tag{2.86a}$$

$$\psi_1'(a) = e^{iq(a+b)}\psi_2'(-b): \quad ik\left(Ae^{ika} - Be^{-ika}\right) = \alpha\left(Ce^{-\alpha b} - e^{\alpha b}\right)e^{e^{iq(a+b)}} \tag{2.86b}$$

Equations (2.84a), (2.84b), (2.86a), and (2.86b) are to be solved simultaneously to determine the energy eigenvalue E and the coefficients A, B, C, and D. These 4-equations are represented by the matrix equation:

$$\begin{bmatrix} 1 & 1 & -1 & -1 \\ ik & -ik & -\alpha & \alpha \\ e^{ika} & e^{-ika} & -e^{-\alpha b+iq(a+b)} & -e^{\alpha b+iq(a+b)} \\ ike^{ika} & -ike^{-ika} & -\alpha e^{-\alpha b+iq(a+b)} & \alpha e^{\alpha b+iq(a+b)} \end{bmatrix} \begin{bmatrix} A \\ B \\ C \\ D \end{bmatrix} = \begin{bmatrix} 0 \\ 0 \\ 0 \\ 0 \end{bmatrix} \tag{2.87}$$

A nontrivial solution of Eqn. (2.87) demands that the determinant of the 4 × 4 matrix must vanish. Hence:

$$\begin{vmatrix} 1 & 1 & -1 & -1 \\ ik & -ik & -\alpha & \alpha \\ e^{ika} & e^{-ika} & -e^{-\alpha b+iq(a+b)} & -e^{\alpha b+iq(a+b)} \\ ike^{ika} & -ike^{-ika} & -\alpha e^{-\alpha b+iq(a+b)} & \alpha e^{\alpha b+iq(a+b)} \end{vmatrix} = 0 \tag{2.88}$$

Simplification of Eqn. (2.88) yields:

$$\cos\{q(a+b)\} = \cos(ka)\cosh(\alpha b) + \frac{1}{2}\left(\frac{\alpha}{k} - \frac{k}{\alpha}\right)\sin(ka)\sinh(\alpha b) \tag{2.89}$$

Equation (2.89) is to be solved numerically or graphically to find energy eigenvalues for a 1D periodic solid. The left-hand side of the equation varies between –1 to +1. This suggests that the numerical values of the terms on right-hand side, which varies between –1 to +1, are acceptable (allowed) values. The values which are beyond the range of –1 to +1 are forbidden values. As is seen from Fig. 2.7, the plot of terms on the right-hand side: $f(y) = \cos(ka)\cosh(\alpha b) + \frac{1}{2}\left(\frac{\alpha}{k} - \frac{k}{\alpha}\right)\sin(ka)\sinh(\alpha b)$ as a function of $y = E/V_0$, is a continuous curve having some portions of the curve (near the dip and peak) outside the range of –1 to +1.

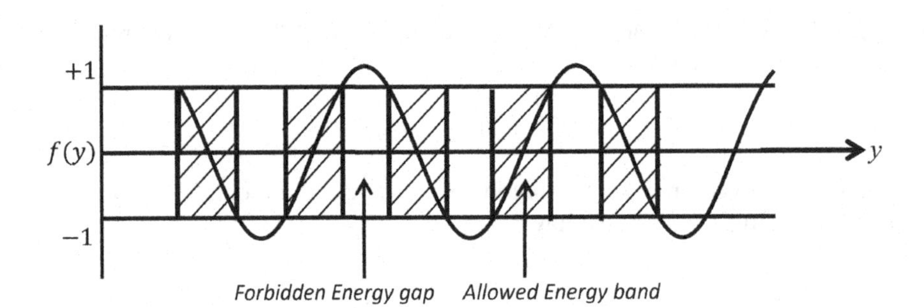

FIGURE 2.7
Graphical evaluation of energy gaps in the Kronig-Penney model.

Energies belonging to the portion of the curve outside the range of –1 to +1 are forbidden energy values. No solutions of the Schrödinger equation can be found for values of energy that lie beyond the range of –1 to +1. We thus find that the periodicity of potential along with the condition defined by Eqn. (2.81b) leads to the concept of energy gaps (forbidden energies). We deduce some important results from Eqn. (2.89):

a. On choosing $ka = n\pi$, we find that the right-hand side of the equation goes to $\pm\cosh(\alpha b)$, since $\sin(n\pi) = 0$ and $\cos(n\pi) = \pm 1$. The magnitude of $\cosh(\alpha b)$ is always greater than 1 for states that are bound within the well, suggesting that there can be no solutions of Eqn. (2.89) for $ka = n\pi$. This implies that no solutions are possible for $E = \frac{\hbar^2}{2m}\frac{(n\pi)^2}{a^2}$, likewise for neighboring values of energy. The q – values that correspond to $ka = n\pi$ are forbidden values of the wave vector.
b. For $\alpha b \to \infty$, allowed energy levels from Eqn. (2.89) should reduce to those of the finite potential well. Equation (2.89) can be rewritten as:

$$\cos(ka) + \frac{1}{2}\left(\frac{\alpha}{k} - \frac{k}{\alpha}\right)\sin(ka)\tanh(\alpha b) = \cos\{q(a+b)\}/\cosh(\alpha b) \tag{2.90}$$

which for $\alpha b \to \infty$ $(\tanh(\alpha b) \to 1 \text{ and } \cosh(\alpha b) \to \infty)$ reduces to:

$$\cos(ka) + \frac{1}{2}\left(\frac{\alpha}{k} - \frac{k}{\alpha}\right)\sin(ka) = 0 \tag{2.91a}$$

With the use of standard trigonometric identities, we find that Eqn. (2.45) reduces to Eqn. (2.91a). Equation (2.45) is expanded as:

$$k^2 \sin\left(\frac{ka}{2}\right)\cos\left(\frac{ka}{2}\right) - k\alpha\cos^2\left(\frac{ka}{2}\right) + k\alpha\sin^2\left(\frac{ka}{2}\right) - \alpha^2\sin\left(\frac{ka}{2}\right)\cos\left(\frac{ka}{2}\right) = 0$$

$$\Rightarrow \frac{1}{2}\left(\alpha^2 - k^2\right)\sin(ka) + k\alpha\cos(ka) = 0 \tag{2.91b}$$

$$\Rightarrow \cos(ka) + \frac{1}{2}\left(\frac{\alpha}{k} - \frac{k}{\alpha}\right)\sin(ka) = 0$$

2.5.2 Confined States in Quantum Wells, Wires, and Dots

Quantum wells, wires, and dots offer the possibility of band structure engineering and wave function engineering, whereby composition and hence electronic and opto-electronic properties can be tailored to optimize for specific applications. Such band structure features are not possible in bulk semiconductors. The development of experimental techniques such as molecular beam epitaxy and metal-organic vapor phase epitaxy has made it possible to produce semiconductor quantum wells, wires, and dots structures.

A layered structure of *GaAs* and *AlGaAs*, where few angstrom thick layers of *GaAs* and of *AlGaAs* are grown alternatively along one direction, say z-direction, exhibits a well–barrier structure in the conduction and valence band, as is seen in Fig. 2.8.

Such a structure is periodic in the $x-y$ plane and therefore Bloch's theorem holds in the $x-y$ plane. If we consider the states at the center of the 2D Brillouin zone $(k_x = k_y = 0)$, the band edge mismatch implies that electrons in the conduction band and holes in the valence band see a quantum well potential. The allowed energy levels for electrons can then be found by solving the 1D Schrödinger equation:

$$\left[-\frac{\hbar^2}{2m_e^*}\frac{d^2}{dz^2}+V(z)\right]\psi(z)=E\psi(z) \tag{2.92}$$

where m_e^* is the electron effective mass $\psi(z)$ is the electron envelope function, and $V(z)$ is the band edge distribution potential. To calculate electron (hole) states in such a quantum well, we require that at the interface of the well and barrier layers, the wave functions of the well $\psi_w(z)$ and of the barrier $\psi_b(z)$ satisfy the conditions:

$$\psi_w = \psi_b \tag{2.93a}$$

and

$$\frac{1}{m_e^*}\frac{d\psi_w}{dz}=\frac{1}{m_h^*}\frac{d\psi_b}{dz} \tag{2.93b}$$

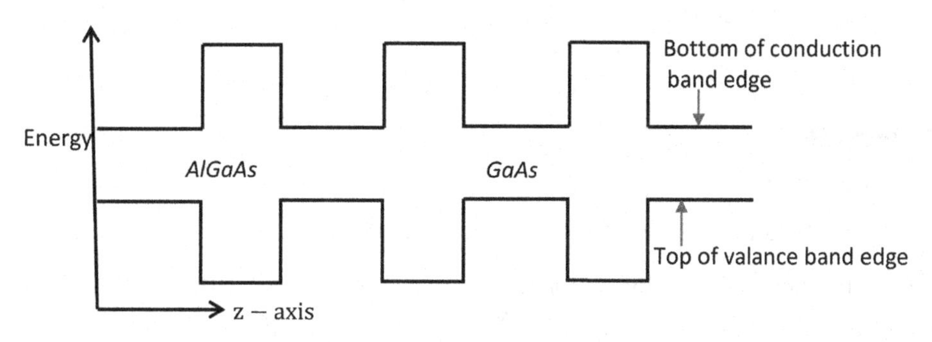

FIGURE 2.8
Simplified bottom of conduction band and top of valance band diagram for GaAs/AlGaAs heterostructure.

where, m_h^* is the effective mass of a hole. Equations (2.93) are the generalization of boundary conditions on wave function, to include the effect of mass change on crossing the boundary. Equation (2.93b) ensures the conservation of probability current density between different layers. Following the procedure laid down in Section 2.4.5, the confined energies are given by:

$$k \tan\left(\frac{ka}{2}\right) = \alpha\, m_e^* / m_h^* \tag{2.94a}$$

for states corresponding to symmetric solutions, while for antisymmetric solutions, energies are given by:

$$-k \cot\left(\frac{ka}{2}\right) = \alpha\, m_e^* / m_h^* \tag{2.94b}$$

where a is well width, $k^2 = \frac{2\, m_e^* E}{\hbar^2}$, and $\alpha^2 = 2\, m_h^* (2\Delta E_c - E) / \hbar^2$. The ΔE_c is the conduction band energy offset (difference between the energies of the conduction band edges of $AlGaAs$ and $GaAS$).

The above analysis can easily be extended to structures of quantum wires and quantum dots. If we assume infinite confining potential ($V = \infty$) in the barrier for rectangular wells, wires, and dots, then the allowed energy states of confined electrons are given by:

$$E_r\left(k_x, k_y\right) = \frac{\hbar^2 r^2}{8\, m_e^* L_z^2} + \frac{\hbar^2}{2\, m_e^*}\left(k_x^2 + k_y^2\right), \text{ quantum well} \tag{2.95a}$$

$$E_{r,s}\left(k_x\right) = \frac{\hbar^2}{8\, m_e^*}\left(\frac{r^2}{L_z^2} + \frac{s^2}{L_y^2}\right) + \frac{\hbar^2 k_x^2}{2\, m_e^*}, \text{ quantum wire} \tag{2.95b}$$

$$E_{r,s,t} = \frac{\hbar^2}{8\, m_e^*}\left(\frac{r^2}{L_z^2} + \frac{s^2}{L_y^2} + \frac{t^2}{L_x^2}\right), \text{ quantum dot(box)} \tag{2.95c}$$

where, $L_x, L_y,$ and L_z are confining lengths along $x-$, $y-$, and $z-$ direction. r,s,t = 1, 2, 3...... are quantum numbers.

2.6 Solved Examples

1. For a particle in an infinite potential well constrained along the $x-y$ axes, the wave function is given by $\psi(x,y) = \frac{2}{\sqrt{L_x L_y}} \sin(k_x x)\sin(k_y y)$ with $k_x = n_x\pi / L_x$ and $k_y = n_y\pi / L_y$, where n_x and n_y take values 1,2,3...... L_x and L_y are confining lengths. Show that wave function is normalized.

SOLUTION

The condition of normalization demands:

$$\int_0^{L_y}\int_0^{L_x} \psi^*(x.y)\psi(x,y)dxdy = 1 \tag{2.96}$$

We therefore have:

$$\begin{aligned}
&\int_0^{L_y}\int_0^{L_x} \psi^*(x.y)\psi(x,y)dxdy \\
&= \frac{4}{(L_xL_y)}\int_0^{L_y}\int_0^{L_x} dxdy\left(\sin(k_xx)\sin(k_yy)\right)^2 \\
&= \left(\frac{2}{L_x}\int_0^{L_x}\sin^2(k_xx)dx\right)\left(\frac{2}{L_y}\int_0^{L_x}\sin^2(k_yy)dy\right) \\
&= \left(\frac{1}{L_x}\int_0^{L_x}\{1-\cos(2k_xx)\}\,dx\right)\left(\frac{1}{L_y}\int_0^{L_y}\{1-\cos(2k_yy)\}\,dy\right) \\
&= 1
\end{aligned}$$

2. A particle of mass m moves in a 1D infinite potential well of length L, having boundaries at $x = 0$ and $x = L$. $V(x) = 0$ for $0 < x < L$, and $V(x) = \infty$, elsewhere. Show that the normalized wave functions of the Hamiltonian are given by $\psi(x) = \sqrt{\frac{2}{L}}\sin(kx)$ with $k = n\pi / L$, and energy eigenvalues are given by $E_n = \hbar^2n^2\pi^2 / 2mL^2$; quantum number n takes the values $n = 1,2,3\ldots\ldots.$
   a. Calculate the probability to find the particle in the range of $0 < x \le L/4$, when the particle is in the nth state. Show that probability depends on n.
   b. For what value of n, does the probability have the maximum value?
   c. Assume that ψ is a superposition of two eigenstates $\psi = a\psi_n + b\psi_m$ at time $t = 0$. What is ψ at time t and what energy expectation value does ψ have at time t, and how does this relate to its value at $t = 0$?
   d. What are the probabilities of measuring E_n and E_m?

SOLUTION

In this case:

$$\frac{d^2\psi(x)}{dx^2} + k^2\psi(x) = 0 \tag{2.97a}$$

which has the solution:

$$\psi(x) = Ae^{ikx} + Be^{-ikx} \tag{2.97b}$$

where $k^2 = 2mE/\hbar^2$. Conditions given by the Eqns. (2.21) and (2.22) are now modified to

$$\psi(0)=0: \qquad A+B=0 \tag{2.98a}$$

$$\psi(L)=0: \qquad Ae^{ikL}+Be^{-ikL}=0 \tag{2.98b}$$

Condition in Eqn. (2.98a) suggests that $B=-A$ and therefore Eqn. (2.98b) yields:

$$A\sin(kL)=0 \tag{2.98c}$$

For $A \neq 0$, condition in Eqn. (2.98c) can only be satisfied for $kL = n\pi$, where $n = 1,2,3,4.......$ Therefore, $E_n = \hbar^2 n^2 \pi^2 / 2mL^2$. Equation (2.97b) reduces to $\psi(x) = 2iA\ sin(kx)$. Normalization of wave function requires that:

$$\int_0^L |\psi(x)|^2\,dx = 1$$

$$\Rightarrow -4|A|^2 \int_0^L sin^2(kx)dx = 1 \tag{2.99}$$

$$\Rightarrow A = -i\frac{1}{\sqrt{2L}}$$

Therefore:

$$\psi_n(x) = \sqrt{\frac{2}{L}}\sin\left(\frac{n\pi x}{L}\right) \tag{2.100}$$

a. Probability density is defined by $P(x) = |\psi|^2$. Therefore, the probability to find a particle in the range of $0 < x \leq L/4$, when it is in the nth state is:

$$\int_0^{\frac{L}{4}} |\psi_n(x)|^2\,dx = \frac{2}{L}\int_0^{\frac{L}{4}} sin^2(kx)dx$$

$$= \frac{1}{L}\int_0^{\frac{L}{4}} [1-cos(2kx)]dx$$

$$= \frac{1}{4} - \frac{1}{2kL}\sin\left(\frac{kL}{2}\right)$$

$$= \frac{1}{4} - \frac{1}{2n\pi}\sin\left(\frac{n\pi}{2}\right),$$

we thus find that probability depends on n. Its value will be ¼ for even values of n. For odd values of n, probability will be given by $\left(\frac{1}{4} - \frac{1}{2n\pi}\right)$ for $n = 1,5,9,13...$, while it will be given by $\left(\frac{1}{4} + \frac{1}{2n\pi}\right)$, for $n = 3,7,11,15.......$

b. The probability to find a particle in the range of $0 < x \le L/4$ will have maximum value for $n = 3$. The maximum value is $\frac{1}{4} + \frac{1}{6\pi} \approx 0.303$.

c. We are given:

$\psi(x,0) = a\psi_n(x) + b\psi_m(x)$. The time-dependent wave function is:

$$\begin{aligned} \psi(x,t) &= \psi(x,0)e^{-\frac{iHt}{\hbar}} \\ &= a\psi_n(x)e^{-\frac{iE_nt}{\hbar}} + b\psi_m(x)e^{-\frac{iE_mt}{\hbar}} \end{aligned} \tag{2.101}$$

The expectation value of the Hamiltonian H at time t, can be given by:

$$H = \int_0^L \psi^*(x,t)H\psi(x,t)dx \tag{2.102}$$

It is to be noted here that ψ_n and ψ_m are eigenstates of the Schrödinger equation. Hence, $H\psi_n = E_n\psi_n$ and $H\psi_m = E_m\psi_m$. Also, these are orthogonal wave functions:

$$\int_0^L \psi_m^*(x)\psi_n(x)dx = \delta_{mn} \tag{2.103}$$

Therefore:

$$\begin{aligned} H &= \int_0^L \left(a\psi_n e^{-\frac{iE_nt}{\hbar}} + b\psi_m e^{-\frac{iE_mt}{\hbar}}\right)^* H\left(a\psi_n e^{-\frac{iE_nt}{\hbar}} + b\psi_m e^{-\frac{iE_mt}{\hbar}}\right)dx \\ &= \int_0^L \left(a\psi_n e^{-\frac{iE_nt}{\hbar}} + b\psi_m e^{-\frac{iE_mt}{\hbar}}\right)^* \left(aH\psi_n e^{-\frac{iE_nt}{\hbar}} + bH\psi_m e^{-\frac{iE_mt}{\hbar}}\right)dx \\ &= \int_0^L \left(a\psi_n e^{-\frac{iE_nt}{\hbar}} + b\psi_m e^{-\frac{iE_mt}{\hbar}}\right)^* \left(aE_n\psi_n e^{-\frac{iE_nt}{\hbar}} + bE_m\psi_m e^{-\frac{iE_mt}{\hbar}}\right)dx \\ &= |a|^2 E_n \int_0^L \psi_n^*(x)\psi_n(x)dx + |b|^2 E_m \int_0^L \psi_m^*(x)\psi_m(x)dx \\ &\quad + a^*bE_m \int_0^L \psi_n^*(x)\psi_m(x)e^{-\frac{i(E_m-E_n)t}{\hbar}}dx + ab^*E_n \int_0^L \psi_m^*(x)\psi_n(x)e^{-\frac{i(E_n-E_m)t}{\hbar}}dx \end{aligned} \tag{2.104}$$

which simplifies to:

$$H = |a|^2 E_n + |b|^2 E_m \tag{2.105}$$

The value of H at $t = 0$ is going to be same as that given by Eqn. (2.105), as the last two terms in Eqn. (2.104), which have time-dependent terms, do not contribute. Hence, the expectation value of the Hamiltonian is independent of time or $\frac{\partial}{\partial t}H = 0$.

d. The probabilities of measuring E_n and E_m are; $|a|^2$ and $|b|^2$.

3. For infinite well potential, define the above in problem-2, calculate $\langle x \rangle$, $\langle x^2 \rangle$, $\langle p \rangle$, and $\langle p^2 \rangle$, for the lowest quantum state. Use the definition $\Delta A = \left(\langle A^2 \rangle - \langle A \rangle^2\right)^{1/2}$ to define the uncertainty ΔA. Calculate Δx and Δp and verify the Heisenberg uncertainty relation that is $\Delta x \Delta p \geq \hbar / 2$.

SOLUTION

The wave function for the lowest quantum state is given by $\psi(x) = \sqrt{\frac{2}{L}} \sin(kx)$, with $k = \pi / L$. Notice that $\psi(x)^* = \psi(x)$; hence:

$$\begin{aligned}\langle x \rangle &= \int_0^L \psi(x)^* x \psi(x) dx = \frac{2}{L} \int_0^L x \sin^2(kx) dx \\ &= \frac{1}{L} \int_0^L x(1 - \cos(2kx)) dx = \frac{L}{2}\end{aligned} \tag{2.106a}$$

$$\begin{aligned}\langle x^2 \rangle &= \int_0^L \psi(x)^* x^2 \psi(x) dx = \frac{2}{L} \int_0^L x^2 sin^2(kx) dx \\ &= \frac{1}{L} \int_0^L x^2 (1 - \cos(2kx)) dx \\ &= \frac{L^2}{3} - \frac{L^2}{2\pi^2}\end{aligned} \tag{2.106b}$$

Therefore:

$$\begin{aligned}\Delta x &= \left(\langle x^2 \rangle - \langle x \rangle^2\right)^{1/2} = \left(\frac{L^2}{3} - \frac{L^2}{2\pi^2} - \frac{L^2}{4}\right)^{1/2} \\ &= L\left(\frac{1}{12} - \frac{1}{2\pi^2}\right)^{1/2}\end{aligned} \tag{2.107}$$

$$\begin{aligned}\langle p \rangle &= \int_0^L \psi(x)^* p \psi(x) dx \\ &= -i\hbar \frac{2}{L} \int_0^L \sin(kx) \frac{d}{dx}(\sin(kx)) dx \\ &= 0\end{aligned} \tag{2.108}$$

$$\begin{aligned}\langle p^2 \rangle &= \int_0^L \psi(x)^* p^2 \psi(x) dx \\ &= -\hbar^2 \frac{2}{L} \int_0^L \sin(kx) \frac{d^2}{dx^2}\{\sin(kx)\} dx \\ &= \hbar^2 k^2 \frac{2}{L} \int_0^L sin^2(kx) dx = \frac{\hbar^2 k^2}{L} \int_0^L (1 - \cos(2kx)) dx \\ &= \hbar^2 k^2 = \hbar^2 \pi^2 / L^2\end{aligned} \tag{2.109}$$

Hence:

$$\Delta p = \left(\langle p^2 \rangle - \langle p \rangle^2\right)^{1/2} = \left(\frac{\hbar^2\pi^2}{L^2} - 0\right)^{1/2}$$
$$= \frac{\hbar\pi}{L}$$

Therefore:

$$\Delta x \Delta p = L\sqrt{\left(\frac{1}{12} - \frac{1}{2\pi^2}\right)}\frac{\hbar\pi}{L}$$
$$= 0.58\hbar > \frac{\hbar}{2}$$

4. An electron is confined in the ground state of an infinite potential well of width 10^{-10} cm. Its energy in ground state is 40 eV. Calculate (a) the energies of the electron in its first, second, and third excited states; (b) the average force on the walls of the box when the electron is in the ground state.

SOLUTION

a. From problem-2 above, we have $E_n = \hbar^2 n^2 \pi^2 / 2mL^2$. $L = 10^{-10}$ m. For ground state ($n = 1$) and $E_1 = \hbar^2\pi^2 / 2mL^2 = 40$ eV. Therefore, the energy of the nth state in terms of the energy of the ground state is $E_n = 40n^2$ eV. Then the energies of the first, second, and third excited states are $E_2 = 160$ eV, $E_3 = 360$ eV, and $E_4 = 640$ eV.

b. The average force on the walls of the well is given by $F = -\left\langle \frac{\partial H}{\partial L} \right\rangle$. Also, $H\psi_n = E_n\psi_n$, which can be rewritten as:

$$(H - E_n)\,\psi_n = 0 \tag{2.110}$$

On differentiating Eqn. (2.110) with L, we obtain:

$$\left(\frac{\partial H}{\partial L} - \frac{\partial E_n}{\partial L}\right)\psi_n + (H - E_n)\frac{\partial \psi_n}{\partial L} = 0 \tag{2.111}$$

Hence:

$$\psi_n^*(E_n - H)\frac{\partial \psi_n}{\partial L} = \psi_n^*\left(\frac{\partial H}{\partial L} - \frac{\partial E_n}{\partial L}\right)\psi_n \tag{2.112}$$

Since the Hamiltonian H is real and Hermitian, we can write:

$$\int \psi_n^*(E_n - H)\frac{\partial \psi_n}{\partial L}dx = \int \frac{\partial \psi_n}{\partial L}(E_n - H)^*\,\psi_n^*\,dx,$$

which is zero, since $\left((H-E_n)\psi_n\right)^* = 0$. Therefore, integration of the right-hand side of Eqn. (2.112) gives:

$$\int \psi_n^* \left(\frac{\partial H}{\partial L} - \frac{\partial E_n}{\partial L} \right) \psi_n \, dx = \left\langle \frac{\partial H}{\partial L} \right\rangle - \left\langle \frac{\partial E_n}{\partial L} \right\rangle = 0 \tag{2.113}$$

Hence,

$$-\left\langle \frac{\partial E_n}{\partial L} \right\rangle = \hbar^2 n^2 \pi^2 / mL^3.$$

For the ground state (n = 1), force on the walls of the well is: $F = \frac{\hbar^2\pi^2}{mL^3} = \frac{2E_1}{L}$, which gives:

$$F = \frac{2 \times 40 \times 1.602 \times 10^{19}}{10^{10}} = 1.28 \times 10^{-7} \text{ Newton}.$$

5. Given an electron in a 1D potential well of width 0.8 nm and depth 6 eV, find the number of bound states present in the well.

SOLUTION

As discussed in Section 2.4.3, the value of the parameter $\rho = \frac{a}{\hbar}\sqrt{mV_0/2}$ determines the number of the bound state in a finite potential well. In this case, we have:

$$\rho = \frac{a}{\hbar}\sqrt{mV_0/2} = \frac{8 \times 10^{-10}}{1.0546 \times 10^{-34}} \sqrt{\frac{\left(9.11 \times 10^{-31} \times 6 \times 1.602 \times 10^{-19}\right)}{2}}$$
$$= 5.02$$

Since, $\frac{3\pi}{2} < \rho < 2\pi$, the number of bound states will be 5, 3 corresponding to symmetric solutions and 2 corresponding to antisymmetric solutions.

6. A beam of 14 eV electrons is incident on a potential barrier of height 35 eV and width 0.06 nm. Calculate the transmittance with the use of both exact and approximate formulae and estimate the error in approximation.

SOLUTION

We are given $E = 14 \times 1.602 \times 10^{-19}$ Joules, $V_0 = 35 \times 1.602 \times 10^{-19}$ Joules and $a = 6 \times 10^{-11}$ m. Therefore:

$$ka = a\sqrt{2mE/\hbar^2} = 1.21,$$

$$\alpha a = a\sqrt{2m(V_0 - E)/\hbar^2} = 1.48.$$

Transmittance with the use of Eqn. (2.77b) is:

$$T = \frac{2(2k\alpha)^2}{\left(k^2+\alpha^2\right)^2 \cosh(2\alpha a) - \left(k^4+\alpha^4-6k^2\alpha^2\right)}$$

$$T = \frac{25.656}{13.355 \times 9.675 - 12.3} = 0.219.$$

With the use of Eqn. (2.78):

$$T \approx \frac{\left(4k^2\alpha^2\right)^2}{\left(k^2+\alpha^2\right)^2} e^{-2\alpha a} = \frac{51.13}{13.555} \times 0.0518 = 0.196.$$

Error in approximation is around 10.5%.

7. Consider a stream of particles of mass m, each moving in the positive $x-$ direction with kinetic energy E toward a potential jump located at $x = 0$. The potential is zero for $x < 0$ and $3E/4$ for $x > 0$. What fraction of the particles is reflected at $x = 0$?

SOLUTION

The Schrödinger equation for $x \le 0$ is:

$$\frac{d^2\psi_1}{dx^2} + k_1^2\psi_1 = 0 \tag{2.114a}$$

where $k_1^2 = 2mE/\hbar^2$. For $x > 0$, we have:

$$\frac{d^2\psi_2}{dx^2} + k_2^2\psi_2 = 0 \tag{2.114b}$$

with $k_2^2 = \dfrac{2m\left(E - \dfrac{3E}{4}\right)}{\hbar^2} = \dfrac{mE}{2\hbar^2} = \dfrac{k_1^2}{4}$. Hence, $k_2 = k_1/2$. The solution of Eqn. (2.114) is given by:

$$\psi_1(x) = Ae^{ik_1x} + Be^{-ik_1x} \tag{2.115a}$$

Since, there is no reflected wave in the region of $x > 0$, we have:

$$\psi_2(x) = Ce^{ik_2x} \tag{2.115b}$$

Application of boundary conditions at $x = 0$ gives:

$$A + B = C \tag{2.116a}$$

$$ik_1(A - B) = ik_2C \tag{2.116b}$$

which give $r = \dfrac{B}{A} = 1/3$. Hence $R = |B/A|^2 = 1/9$. We thus find that one-ninth of the particles are reflected at $x = 0$.

8. **Nearly Free electron model**: With reference to the Kronig-Penney model described above, consider a periodic potential $V(x)$ defined by:

$$V(x) = \begin{bmatrix} V_0 & \text{for } -\frac{b}{2} < x < \frac{b}{2} \\ 0 & \text{when } \frac{b}{2} < x < a + \frac{b}{2} \end{bmatrix} \tag{2.117}$$

Wave functions for this potential can be written as a linear combination of the free electron wave functions $\psi_{nq}(x) = \frac{1}{\sqrt{L}} e^{iqx} \sum_{m=-\infty}^{\infty} \left(A_{mnq} e^{2\pi i m x / L} \right)$, with $L = a + b$. For the potential defined by Eqn. (2.117) most of the coefficients A_{mnq} are zero. Assume that wave functions are nonzero only for two states that correspond to $m = 0$ and $m = 1$. We then have:

$$\psi_q(x) = \frac{A}{\sqrt{L}} e^{iqx} + \frac{B}{\sqrt{L}} e^{i(q-g)x} \tag{2.118}$$

with $g = 2\pi / L$. Calculate the energy eigenvalues of the 1D Schrödinger equation and show that there exists an energy gap (forbidden energy) at $q = \pm \frac{\pi}{L}$.

SOLUTION

The 1D Schrödinger equation is:

$$\left[-\frac{\hbar^2}{2m} \frac{d^2}{dx^2} + V(x) \right] \psi_q(x) = E \psi_q(x) \tag{2.119}$$

After substituting $\psi_q(x)$ from Eqn. (2.118) into Eqn. (2.119), we multiply it by $\frac{1}{\sqrt{L}} e^{-iqx}$ from the left and then integrate between $-\frac{b}{2}$ and $a + b/2$ (over the unit cell) to get:

$$\begin{aligned} &\frac{1}{L} \int_{-b/2}^{b/2} e^{-iqx} \left[-\frac{\hbar^2}{2m} \frac{d^2}{dx^2} + V_0 \right] \left(A e^{iqx} + B e^{i(q-g)x} \right) dx \\ &+ \frac{1}{L} \int_{b/2}^{a+b/2} e^{-iqx} \left[-\frac{\hbar^2}{2m} \frac{d^2}{dx^2} \right] \left(A e^{iqx} + B e^{i(q-g)x} \right) dx \\ &= \frac{E}{L} \int_{-b/2}^{a+b/2} e^{-iqx} \left(A e^{iqx} + B e^{i(q-g)x} \right) dx \end{aligned} \tag{2.120}$$

Using the orthonormal property of two wave functions $\frac{1}{L} \int_{-b/2}^{a+b/2} e^{-iqx} e^{i(q-g)x} dx = \delta_{g,0} = 0$ since $g \neq 0$, we get:

$$A \left(\frac{\hbar^2 q^2}{2m} + \frac{b V_0}{L} \right) + \frac{B V_0}{\pi} \sin\left(\frac{\pi b}{L} \right) = AE \tag{2.121}$$

while, multiplying from the left with $\frac{1}{\sqrt{L}}e^{-i(q-g)x}$ Eqn. (2.119) and then integrating between $-\frac{b}{2}$ and $a+b/2$, yields:

$$B\left(\frac{\hbar^2(q-g)^2}{2m}+\frac{bV_0}{L}\right)+\frac{AV_0}{\pi}\sin\left(\frac{\pi b}{L}\right)=BE \tag{2.122}$$

Equations (2.121) and (2.122) can be combined to give:

$$\begin{bmatrix} \left(\frac{\hbar^2 q^2}{2m}+\frac{bV_0}{L}-E\right) & \frac{V_0}{\pi}\sin\left(\frac{\pi b}{L}\right) \\ \frac{V_0}{\pi}\sin\left(\frac{\pi b}{L}\right) & \left(\frac{\hbar^2(q-g)^2}{2m}+\frac{bV_0}{L}-E\right) \end{bmatrix}\begin{bmatrix} A \\ B \end{bmatrix}=\begin{bmatrix} 0 \\ 0 \end{bmatrix} \tag{2.123}$$

The nontrivial solution of Eqn. (2.123) requires that:

$$\begin{vmatrix} \left(\frac{\hbar^2 q^2}{2m}+\frac{bV_0}{L}-E\right) & \frac{V_0}{\pi}\sin\left(\frac{\pi b}{L}\right) \\ \frac{V_0}{\pi}\sin\left(\frac{\pi b}{L}\right) & \left(\frac{\hbar^2(q-g)^2}{2m}+\frac{bV_0}{L}-E\right) \end{vmatrix}=0 \tag{2.124}$$

which simplifies to:

$$E_q^{\pm}=\frac{bV_0}{L}+\frac{\hbar^2}{2m}\left[(\pi/L)^2+(q-\pi/L)^2\right]\pm\sqrt{\left(\frac{\hbar^2\pi}{mL}\right)^2(q-\pi/L)^2+\left(\frac{V_0}{\pi}\right)^2 sin^2\left(\frac{\pi b}{L}\right)} \tag{2.125}$$

These are the energy of the two lowest states. The energy difference $(E_q^+-E_q^-)$ when plotted as a function q exhibits maximum value at $q=0$ and minimum value at $q=\frac{\pi}{L}$. Also, it is independent of q at $q=\frac{\pi}{L}$. The minimum value of energy difference $(E_q^+-E_q^-)$ that occurs at $=\frac{\pi}{L}$ is known as the energy band gap E_g between the two lowest lying states in a periodic solid. Substituting $q=\frac{\pi}{L}$ in Eqn. (2.125), we obtain:

$$E_g=E_{\pi/L}^+-E_{\frac{\pi}{L}}^-=\frac{2V_0}{\pi}\sin\left(\frac{\pi b}{L}\right) \tag{2.126}$$

It is to be noted that for $V_0=0$, Eqns. (2.121) and (2.122) describe the energy as a function of q for a free electron. Let us rewrite these equations as follows:

$$E_1=\frac{\hbar^2 q^2}{2m} \tag{2.127a}$$

$$E_2=\frac{\hbar^2(q-g)^2}{2m} \tag{2.127b}$$

On plotting of E_2 and E_1 as functions of q, we find (i) $E_1 \to 0$ and E_2 tends to a maximum value, at $q = 0$, and (ii) at $q = 2\pi / L$, $E_2 \to 0$ and E_1 takes the value of E_2 at $q = 0$. Also, at $q = \pi / L$, $E_2 = E_1$ meaning that two curves intersect each other. It can therefore be inferred that including the weak potential V_0 keeps separated E_2 and E_1 from each other even at $q = \pi / L$. Thus, the energy gap appears at the point of intersection, in the absence of the periodic potential.

9. Consider the motion of a particle of mass m in the 1D potential defined by:

$$V(x) = \begin{bmatrix} \infty & \text{for } x < 0 \\ 0 & \text{when } 0 \le x \le a \\ V_0 & \text{for } x > a \end{bmatrix}$$

Show that bound state energies $(E < V_0)$ are given by the equation:

$$\tan\left(\frac{\sqrt{2mE}}{\hbar} a\right) = -\sqrt{\left(\frac{E}{V_0 - E}\right)}.$$

SOLUTION

We are given:

$$\frac{d^2\psi_1}{dx^2} + \frac{2mE}{\hbar^2}\psi_1 = 0 \text{ for } 0 \le x \le a \tag{2.128a}$$

$$\frac{d^2\psi_2}{dx^2} - \frac{2m(V_0 - E)}{\hbar^2}\psi_2 = 0 \text{ for } x > a \tag{2.128b}$$

The requirements that $\psi_1 = 0$ at $x = 0$ and $\psi_2 \to 0$ for $x \to \infty$ suggest that we can take: $\psi_1 = A sin(kx)$ and $\psi_2 = Be^{-\alpha x}$, with $k^2 = \frac{2mE}{\hbar^2}$ and $\alpha^2 = \frac{2m(V_0 - E)}{\hbar^2}$. Applying boundary conditions at $x = a$, we get:

$$\psi_1(a) = \psi_2(a) \Rightarrow A\sin(ka) = Be^{-\alpha a} \tag{2.129a}$$

$$\psi_1'(a) = \psi_2'(a) \Rightarrow Ak\ cos(ka) = -B\alpha e^{-\alpha a} \tag{2.129b}$$

On dividing Eqn. (2.129a) by Eqn. (2.129b) we obtain:

$$\tan(ka) = -\frac{k}{\alpha} \Rightarrow \tan\left(\frac{\sqrt{2mE}}{\hbar} a\right) = -\sqrt{\left(\frac{E}{V_0 - E}\right)}.$$

2.7 Exercises

1. A particle of mass 4.5×10^{-31} kg is confined to a one dimensional infinite square potential well of width 5 nm. It makes the transition from first excited state to the ground state. Calculate the wavelength of the emitted photon.
2. An electron confined to a one dimensional infinite square potential well makes the transition from $n=3$ to $n=1$ energy level. The frequency of the emitted photon is 2.5×10^{14} Hz. Find the width of the well.
3. A particle is in a state described by the wave function $\psi(r,0)=Ne^{-\alpha r}$, where α is a real constant. (a) Calculate the normalization factor N. (b) Calculate $\psi(\mathbf{r},t)$ and show that the probability density $|\psi(r,t)|^2$ is isotropic, independent of θ and ϕ.
4. Different states of a particle confined to a 1D infinite square potential well of width L are given by: $\psi_n(x)=\sqrt{\frac{2}{L}}\sin\left(\frac{n\pi x}{L}\right)$. Take $L=2$ nm and then calculate the probability of finding the particle in between 0.2 nm to 0.4 nm, in the ground state ($n=1$).
5. Show that $\psi_n(x)$ belong to different levels (different values of n) in the above exercise-4 are orthonormal $\int_0^L \psi_m^*(x)\psi_n(x)dx=\delta_{mn}$.
6. A free particle is represented by $\psi(\mathbf{r},t)=e^{i(\mathbf{k}.\mathbf{r}-\omega t)}$. Show its probability current $\mathbf{S}$ is equal to its group velocity $\mathbf{v}_g=\frac{d\omega}{d\mathbf{k}}$.
7. Calculate the probability current $\mathbf{S}$ when $\psi(\mathbf{r})=\frac{e^{ikr}}{r}$ and compare your answer with that obtained in exercise 6.
8. Show that for the 1D bound particle case $\frac{d}{dt}\int_{-\infty}^{\infty}|\psi(x,t)|^2\,dx=0$.
9. If $\psi(x)=Ae^{-x^2/a^2}$ is the wave function for the 1D Schrödinger equation, $\left[-\frac{\hbar^2}{2m}\frac{d^2}{dx^2}+V(x)\right]\psi(x)=E\psi(x)$, what should be the values of $V(x)$ and E?
10. Consider a particle whose normalized wave function is:

$$\psi(x)=\begin{cases}2\beta^{3/2}xe^{-\beta x}\text{for } x\geq 0\\ 0\text{ for } x<0\end{cases}$$

(a) Find for what value of x probability density is maximum. (b) Calculate expectation values $\langle x\rangle$, $\langle x^2\rangle$, $\langle p\rangle$ and $\langle p^2\rangle$ and show that they satisfy the Heisenberg uncertainty relation.

11. The Hamiltonian and wave function of a particle are given by $H=-\frac{\hbar^2}{2m}\left[\frac{d^2}{dx^2}-\frac{4x^2}{a^4}\right]$ and $\psi(x)=\frac{x}{a}e^{-x^2/a^2}$. Find the energy of the particle.

3

Matrix Formulation of Quantum Mechanics

In 1925, matrix mechanics was developed by Heisenberg while wave mechanics and the non-relativistic Schrödinger equation were invented by Schrödinger. While Schrödinger's approach requires integral and differential calculus, matrices and vectors are the basis of Heisenberg's representation of quantum mechanics. It was subsequently shown that the two approaches were equivalent. One of the most fundamental concepts in quantum theory lies in the idea $xp \neq px$, which is not followed by pure numbers but by operators and matrices. Here, x and p are the 1D position and the momentum. Our understanding of physical properties being represented by numbers is altered on being represented by operators and matrices. We know that matrices have the same non-commutativity as operators and they too can demonstrate the mathematical formalism of quantum theory. In fact, Heisenberg's original approach to quantum theory, which is called matrix mechanics, was to find matrices representing x and p, such that $xp - px = i\hbar$.

In the Schrödinger representation, wave function is time dependent and it evolves in accordance to the Schrödinger equation:

$$\left[-\frac{\hbar^2}{2m}\nabla^2 + V(\mathbf{r})\right]\psi(\mathbf{r},t) = i\hbar\frac{\partial\psi(\mathbf{r},t)}{\partial t} \tag{3.1a}$$

and an observable (operator) is treated independent of time. However, in the Heisenberg representation of quantum mechanics the vector representing a quantum state does not change with time, while an observable A satisfies:

$$\frac{dA}{dt} = \frac{1}{i\hbar}[A,H] + \left(\frac{\partial A}{\partial t}\right)_{classical} \tag{3.1b}$$

where the second term on the right-hand side will be zero if A does not have explicit time dependence. The Heisenberg Eqn. (3.1b) becomes an equation in Hamiltonian mechanics (similar to classical physics) on replacing the commutator by the Poisson bracket. Also, the Lorentz invariance is manifested in the Heisenberg picture. Therefore, it can be inferred that the Heisenberg picture is more natural and fundamental than the Schrödinger picture.

3.1 Matrices and their Basic Algebra

A two dimensional array of numbers (real or complex) or of functions is called the matrix:

$$A=\begin{bmatrix} a_{11} & a_{12} & \cdots & \cdots & a_{1n} \\ a_{21} & a_{22} & \cdots & \cdots & a_{2n} \\ a_{31} & a_{32} & \cdots & \cdots & a_{3n} \\ \cdots & \cdots & \cdots & \cdots & \cdots \\ \cdots & \cdots & \cdots & \cdots & \cdots \\ a_{m1} & a_{m2} & \cdots & \cdots & a_{mn} \end{bmatrix} \text{ or } F(x)=\begin{bmatrix} f_{11}(x) & f_{12}(x) & \cdots & \cdots & f_{1n}(x) \\ f_{21}(x) & f_{22}(x) & \cdots & \cdots & f_{2n}(x) \\ \cdots & \cdots & \cdots & \cdots & \cdots \\ \cdots & \cdots & \cdots & \cdots & \cdots \\ \cdots & \cdots & \cdots & \cdots & \cdots \\ f_{m1}(x) & f_{m2}(x) & \cdots & \cdots & f_{mn}(x) \end{bmatrix} \quad (3.2)$$

where, the horizontal and vertical lines are called rows and columns, respectively, of the matrix. The dimensions of a matrix are defined with the number of rows first and then the number of columns. An entry of a matrix is represented by a_{ij} or $f_{ij}(x)$ with $1 \le i \le m$ and $1 \le j \le n$. There is no universal convention to start indices i and j. Some of the programming languages start at zero. The elements, a_{ii} or f_{ii} are called diagonal elements, and the line on which they appear is termed the principal diagonal. A matrix with one row and n columns is called a row matrix, and the matrix of one column and m rows is called a column matrix.

A matrix that consists of an equal number of rows and columns is termed a *square matrix*. A matrix whose each and every elements set is equal to zero is called a null matrix or zero matrix that satisfies the property $0+A=A+0=A; A-A=0$.

A matrix that has all elements on the main diagonal set to 1 and all other elements set to 0 is known as a unit matrix. It satisfies $MI_n = M$ for any $m \times n$ matrix M; then I_n is a unit matrix of order n. A matrix where elements on the main diagonal set equal to a constant value and all other elements are set to 0 is called a constant matrix.

The algebra of matrices and operators differs from that of numbers. Matrices of the same order can only be added or subtracted. For given $m \times n$ matrices A and B, the sum $A+B$ is a $m \times n$ matrix computed by adding corresponding elements $(A \pm B)_{ij} = a_{ij} \pm b_{ij}$.

A *scalar multiplication* cA is computed by multiplying each matrix element by scalar c: $c(cA)_{ij} = ca_{ij}$. Two matrices can be multiplied if and only if the number of columns of the left matrix is the same as the number of rows of the right matrix. If A is an $m \times n$ matrix and then B is has to be an $n \times p$ matrix, to have the *matrix product* AB of order $m \times p$:

$$(AB)_{ij} = \sum_k a_{ik} b_{kj} = a_{i1}b_{1j} + a_{i2}b_{2j} + \dots\dots + a_{in}b_{nj} \quad (3.3a)$$

for each pair (ij). Two matrices A and B are said to commute if $AB = BA$, for well-defined products AB and BA. However, commutativity does *not* generally hold, which means $AB \ne BA$.

The transpose of a matrix A of order $m \times n$ is a matrix of order $n \times m$ denoted as A^{tr} or A^{T} or $\tilde{A}$, and it is formed by turning rows into columns and columns into rows: $\tilde{A}_{ij} = A_{ji}$ for all indices i and j. Transposition follows: $(A+B)^T = A^T + B^T$ and $(AB)^T = B^T A^T$.

A matrix formed by replacing each element by its complex conjugate is called a Complex Conjugation Matrix, represented by A^*. The complex conjugation of the product: $(AB)^* = B^* A^*$. A matrix formed by performing transposition and complex conjugation operations, one after other, is called a Hermitian Conjugation Matrix or conjugate adjoint matrix, and it is represented by $A^{\dagger}$. It holds that $(A+B)^{\dagger} = A^{\dagger} + B^{\dagger}$ and $(AB)^{\dagger} = B^{\dagger}A^{\dagger}$.

The sum of diagonal elements is termed as the trace of the matrix:

$$tr(A) = \sum_{i} a_{ii} \tag{3.3b}$$

The trace holds the rules: (a) $\mathrm{tr}(kA) = k\ \mathrm{tr}(A)$, where k is a real number, (b) $\mathrm{tr}(A+B) = \mathrm{tr}(A) + \mathrm{tr}(B)$, and $\mathrm{tr}(AB) = \mathrm{tr}(BA)$.

Every square matrix reduces to single number or expression of scalars, which is called the **determinant of the matrix**. It is defined as follows:

$$|A| = \begin{vmatrix} a_{11} & a_{12} & \cdots & \cdots & a_{1n} \\ a_{21} & a_{22} & \cdots & \cdots & a_{2n} \\ a_{31} & a_{32} & \cdots & \cdots & a_{3n} \\ \cdots & \cdots & \cdots & \cdots & \cdots \\ \cdots & \cdots & \cdots & \cdots & \cdots \\ a_{n1} & a_{n2} & \cdots & \cdots & a_{nn} \end{vmatrix} = \sum (-1)^{\sigma} a_{1i_1} a_{2i_2} a_{3i_3} \ldots a_{ni_n} \tag{3.4}$$

where the summation goes over all the permutations $i_1, i_2, i_3 \ldots i_n$ of indices $1, 2, 3, \ldots, n$ and σ is the number of inversions in the permutation $i_1, i_2, i_3 \ldots i_n$ of the row indices. Some notable properties of the determinant are: (i) The determinants of a matrix and of its transpose are equal, i.e. $|A| = |A^T|$. (ii) Multiplying all the elements of a row (column) of the determinant by a number is equivalent to the determinant multiplied by the same number. (iii) On interchanging two rows (columns) of the determinant, the determinant changes its sign. (iv) If two rows (columns) of the determinant are identical, then the determinant is equal to zero. (v) If each element in the row (column) of the determinant is a sum of two numbers, then the determinant expands into the sum of two determinants. (vi) The determinant will not change if a scalar multiple to a row (column) is added to another row (column). (vii) For arbitrary matrices A and B of order n it holds that $|AB| = |A||B|$.

The determinant *theorem of expansion by cofactors* is given as:

$$a_{i1}A_{k1} + a_{i2}A_{k2} + \ldots + a_{in}A_{kn} = |A|\delta_{ik} \tag{3.5}$$

$$a_{1i}A_{1k} + a_{2i}A_{2k} + \ldots + a_{ni}A_{nk} = |A|\delta_{ik} \tag{3.6}$$

where the Kronecker delta δ_{ik} is defined as:

$$\delta_{ik} = \begin{cases} 1 & i = k \\ 0 & i \neq k \end{cases} \tag{3.7}$$

The symbol A_{ik} represent the product of the number $(-1)^{i+k}$ and the determinant of the $(n-1) \times (n-1)$ matrix obtained by deleting the *i*-th row and *k*-th column of the given matrix.

3.2 Bra and Ket Notations

P. A. M. Dirac created a powerful and concise formalism of quantum mechanics, which is referred to as the Dirac notation or bra-ket (bracket) notation. Two distinctly different computational approaches, wave mechanics and matrix mechanics, to quantum theory are equivalent, each with its strengths in certain applications. The quantum mechanical calculation can be set in Dirac's notation in the first step. Then, depending on which method is most expedient computationally, one chooses one of the matrix mechanics and wave mechanics. Quantum states are preferred to describe the bra-ket notations. They are also used to denote abstract vectors and linear functionals in mathematics. The symbol $|\ \rangle$ is called ket and the conjugate adjoint of it is called bra, and it is denoted by $\langle\ |$. Bra and ket are derived from the word bracket.

3.3 Vectors and Vector Space

A *vector* is characterized by a length and a direction, and a *vector space* is a set of vectors closed under addition and multiplication. Vectors are added and multiplied together, and are multiplied by constants, including complex numbers. A vector is represented by any of the symbols $\mathbf{A}, \vec{A}, |A\rangle$ and $\langle A|$. A linear vector space is defined as an abstract set of vectors, which satisfy the following algebraic axioms:

1. For every pair of vectors $|A\rangle$ and $|B\rangle$, there exists a unique vector $|C\rangle = |A\rangle + |B\rangle$, and the vector addition is commutative as well as associative, that is $|A\rangle + |B\rangle = |B\rangle + |A\rangle$ and $(|A\rangle+|B\rangle)+|C\rangle=|A\rangle+(|B\rangle+|C\rangle)$.
2. The origin of vector space is zero or the null vector $|0\rangle$, which is defined by $|A\rangle+|0\rangle=|A\rangle$ and associated with $|A\rangle$; there exists -$|A\rangle$ so that $|A\rangle-|A\rangle=|0\rangle$.
3. There exists a vector called unit vector defined by $|1\rangle$, which is of unit length.
4. Elements of vector space can be multiplied by real or complex numbers, and they follow the rule: $a(|A\rangle+|B\rangle)=a|A\rangle+a|B\rangle$; $(a+b)|A\rangle=a|A\rangle+b|A\rangle$, where a and b are real or complex numbers.
5. Multiplication by numbers 0 and 1 is defined as: $0|A\rangle=0$ and $1|A\rangle=|A\rangle$.

3.3.1 Linearly Independent Vectors

N-vectors forming a set $\{|x_i\rangle\}$ are said to be linearly independent of each other if they satisfy:

$$c_1|x_1\rangle+c_2|x_2\rangle+c_3|x_3\rangle+.........+c_N|x_N\rangle=0 \tag{3.8}$$

if and only if $c_1=c_2=c_3.........=c_N=0$.

And, if

$$c_1|x_1\rangle+c_2|x_2\rangle+c_3|x_3\rangle+.........+c_N|x_N\rangle=0 \tag{3.9}$$

for nonzero values of $c_1, c_2, c_3, ...c_N$, vectors of set $\{|x_i\rangle\}$ are called linearly dependent. Every linearly dependent set contains at least one vector that is a linear combination of others:

$$|x_i\rangle = (c_1|x_1\rangle + c_2|x_2\rangle + c_3|x_3\rangle + ... + c_N|x_N\rangle)/c_i \tag{3.10}$$

for $c_i \neq 0$.

3.3.2 Orthogonal and Orthonormal Vectors

For any two vectors there exists an associated complex number called the inner or scalar product $\langle X|Y\rangle$. The positive square root of the inner product of a vector with itself is called the norm of the vector: $|X| = \sqrt{\langle X|X\rangle}$.

Two vectors are called orthogonal vectors if their inner product is zero.

$$\langle X|Y\rangle = 0 \tag{3.11}$$

There exists a set $\{|y_i\rangle\}$ of orthogonal vectors if $\langle y_i|y_j\rangle = 0$ for $i \neq j$, for $1 \leq (i,j) \leq N$. On replacing $|x_i\rangle$ by $|y_i\rangle$ for $1 \leq i \leq N$ in Eqn. (3.8) and then taking inner products with $|y_1\rangle$, $|y_2\rangle$, $|y_3\rangle ......$ $|y_N\rangle$; with imposing the condition of orthogonality we find that Eqn. (3.8) can be satisfied if and only if $c_1 = c_2 = c_3 .... = c_N = 0$. Thus, orthogonal vectors are linearly independent vectors, too. However, the converse is not true; linearly independent vectors need not necessarily be orthogonal.

The N-vectors belonging to the set $\{|e_i\rangle\}$, $1 \leq i \leq N$, are said to be orthonormal to each other if they satisfy: $\langle e_i|e_j\rangle = \delta_{ij}$. The norm of each is unity. As is obvious orthonormal vectors are orthogonal, too. On replacing $|x_i\rangle$ by $|e_i\rangle$ for $1 \leq i \leq N$ in Eqn. (3.8) and then taking inner products with $|e_1\rangle$, $|e_2\rangle$, $|e_3\rangle ......$ $|e_N\rangle$ and then applying $\langle e_i|e_j\rangle = \delta_{ij}$, we find that Eqn. (3.8) can be satisfied if and only if $c_1 = c_2 = c_3 .... = c_N = 0$, suggesting that orthonormal vectors are also linearly independent.

Linearly independent, orthogonal, and orthonormal vectors have the property that every vector that belongs to a linear vector space can be expressed as (i) a linear combination of linearly independent vectors, (ii) a linear combination of orthogonal vectors, and (iii) a linear combination of orthonormal vectors. The set of linearly independent vectors $\{|x_i\rangle\}$ or orthogonal vectors $\{|y_i\rangle\}$ or orthonormal vectors $\{|e_i\rangle\}$ can be used as a coordinate system or basis of vector space. The representation of any vector in terms of vectors of set $\{|x_i\rangle\}$ or $\{|y_i\rangle\}$, or $\{|e_i\rangle\}$ is unique. Suppose $|A\rangle = \sum_{i=1}^{N} a_i|e_i\rangle$ and also $|A\rangle = \sum_{i=1}^{N} a_i'|e_i\rangle$.

Then $0 = \sum_{i=1}^{N}(a_i - a_i')|e_i\rangle$, which can be true if and only if $a_i = a_i'$. The numbers a_i, which are uniquely defined, can be real or complex. They are called components or expansion coefficients of $|A\rangle$ relative to the basis vectors $\{|e_i\rangle\}$. The number of vectors N in a set $\{|e_i\rangle\}$ can be finite or infinite. Accordingly, vector space is called finite dimensional or infinite-dimensional space. For an infinite-dimensional space: $|A\rangle = \sum_{i=1}^{\infty} a_i|e_i\rangle$. The $a_i = \langle e_i|A\rangle$, whether $|A\rangle$ belongs to finite N-dimensional or infinite-dimensional vector space. Every vector, including a basis vector, has components equal to the dimensionality of the vector space. For example:

$$|e_i\rangle = 0|e_1\rangle + 0|e_2\rangle + 0|e_3\rangle + ... + 1|e_i\rangle + ... + 0|e_N\rangle \tag{3.12}$$

3.3.3 Abstract Representation of a Vector

In an N-dimensional space, there is one-to-one correspondence between the vectors in the set $\{|e_i\rangle\}$ and the expansion coefficients $(a_1, a_2, a_3......a_N)$. Also, the linear combination and inner product of two vectors $|A\rangle$ and $|B\rangle$ involve only the expansion coefficients with respect to basis $\{|e_i\rangle\}$:

$$|A\rangle + |B\rangle = (a_1 + b_1)|e_1\rangle + (a_2 + b_2)|e_2\rangle + ... + (a_N + b_N)|e_N\rangle \tag{3.13}$$

$$\langle A|B\rangle = (a_1^* b_1 + a_2^* b_2 + a_3^* b_3 + ... + a_N^* b_N) \tag{3.14}$$

Every relation among vectors has its exact copy in relation among representative expansion coefficients of vectors with respect to basis $\{|e_i\rangle\}$. Because of this it is possible to develop an entire theory of vectors in terms of representative expansion coefficients, without making further reference to basis vectors. Thus, intrinsic properties of a vector space, which are independent of basis, are physically or mathematically important. Physically significant properties of a vector space in quantum mechanics are independent of the choice of coordinate system or basis vectors. Equation (3.14) can equivalently be written in terms of row and column matrices as follows:

$$\begin{aligned}\langle A|B\rangle &= (a_1^* b_1 + a_2^* b_2 + a_3^* b_3 + ... + a_N^* b_N) \\ &= [a_1^* \quad a_2^* \quad a_3^* .... \quad a_N^*] \begin{bmatrix} b_1 \\ b_2 \\ b_3 \\ \vdots \\ b_N \end{bmatrix} \\ &= A^\dagger B\end{aligned} \tag{3.15}$$

3.3.4 Outer Product of Vectors

The product of bra and ket vectors is called the inner product, while the product of ket and bra vectors $|A\rangle\langle B|$ is termed the outer product. The inner product is a scaler quantity, while the outer product is a matrix as is shown below:

$$\begin{aligned}|A\rangle\langle B| &= \begin{bmatrix} a_1 \\ a_2 \\ a_3 \\ \vdots \\ a_N \end{bmatrix} \begin{bmatrix} b_1^* & b_2^* & b_3^* & ..... & b_n^* \end{bmatrix} \\ &= \begin{bmatrix} a_1 b_1^* & a_1 b_2^* & a_1 b_3^* & \cdots & a_1 b_n^* \\ a_2 b_1^* & a_2 b_2^* & a_2 b_3^* & \cdots & a_2 b_n^* \\ a_3 b_1^* & a_3 b_2^* & a_3 b_3^* & \cdots & a_3 b_n^* \\ \vdots & \vdots & \vdots & \ddots & \vdots \\ a_n b_1^* & a_n b_2^* & a_n b_3^* & \cdots & a_n b_n^* \end{bmatrix}\end{aligned} \tag{3.16}$$

Since, each column of $|A\rangle\langle B|$ represents a vector, the outer product is an analogue of the vector product of two vectors.

The orthonormal vectors, which can be chosen as basis vectors for an N-dimensional vector space, are:

$$|e_1\rangle = \begin{bmatrix} 1 \\ 0 \\ 0 \\ \cdot \\ \cdot \\ 0 \end{bmatrix}, \; |e_2\rangle = \begin{bmatrix} 0 \\ 1 \\ 0 \\ \cdot \\ \cdot \\ 0 \end{bmatrix}, \; |e_3\rangle = \begin{bmatrix} 0 \\ 0 \\ 1 \\ \cdot \\ \cdot \\ 0 \end{bmatrix} \ldots\ldots\ldots \; |e_N\rangle = \begin{bmatrix} 0 \\ 0 \\ 0 \\ \cdot \\ \cdot \\ 1 \end{bmatrix} \tag{3.17}$$

Depending upon the requirements for the solution of a problem, we can choose any of the sets of linearly independent, orthogonal, or orthonormal vectors as the basis vectors. A set of linearly independent vectors can be converted to a set of orthogonal vectors, which then can be converted to a set of orthonormal vectors.

3.4 Gram-Schmidt Method for Orthogonalization of Vectors

A set of orthogonal vectors is obtainable from a set of linearly independent vectors. The Gram-Schmidt method is a procedure that takes a set $\{|U_i\rangle\}$ of linearly independent vectors and constructs a set $\{|V_i\rangle\}$ of orthogonal vectors.

We take $|V_1\rangle = |U_1\rangle$ and then express the next orthogonal vector $|V_2\rangle$ as a leaner combination of $|U_2\rangle$ and $|V_1\rangle$:

$$|V_2\rangle = |U_2\rangle + a_{21}|V_1\rangle \tag{3.18}$$

On taking the inner product of $|V_2\rangle$ with $|V_1\rangle$, we get:

$$\langle V_1|V_2\rangle = \langle V_1|U_2\rangle + a_{21}\langle V_1|V_1\rangle \tag{3.19}$$

which gives:

$$a_{21} = -\langle V_1|U_2\rangle / |V_1|^2 \tag{3.20}$$

Similarly define:

$$|V_3\rangle = |U_3\rangle + a_{32}|V_2\rangle + a_{31}|V_1\rangle \tag{3.21}$$

Then take inner products of $|V_3\rangle$ with $|V_2\rangle$ and $|V_1\rangle$ to obtain:

$$a_{32} = -\langle V_2|U_3\rangle / |V_2|^2 \tag{3.22}$$

$$a_{31} = -\langle V_1|U_3\rangle / |V_1|^2 \tag{3.23}$$

By looking at Eqns. (3.20), (3.22) and (3.23), we can generalize to:

$$|V_i\rangle = |U_i\rangle + a_{i,i-1}|V_{i-1}\rangle + a_{i,i-2}|V_{i-2}\rangle + \ldots\ldots + a_{i1}|V_1\rangle \tag{3.24}$$

with:

$$a_{ij} = -\langle V_j | U_i \rangle / |V_j|^2 \tag{3.25}$$

where $2 \le i \le N$ and $1 \le j \le N-1$. A set of orthonormal vectors $\{|W_i\rangle\}$ can then be formed from orthogonal vectors by dividing each orthogonal vector by its norm: $|W_i\rangle = |V_i\rangle / |V_i|$.

3.5 Schwarz Inequality

The Schwarz inequality also known as the Cauchy-Schwarz inequality is a useful inequality. The norm of the scalar product of two vectors is $|X||Y||\cos\theta| \le |X||Y|$, because $|\cos\theta| \le 1$, the generalization of which leads to the Schwarz inequality:

$$|\langle X|Y\rangle|^2 \le \langle X|X\rangle\langle Y|Y\rangle \quad \text{or} \quad |\langle X|Y\rangle| \le |X||Y| \tag{3.26}$$

The two sides are equal if and only if $|X\rangle$ and $|Y\rangle$ are linearly independent or in the geometrical sense they are parallel. To prove the Schwarz inequality, let us define $|Z\rangle = |X - \lambda Y\rangle$, where λ can be taken as a complex number. We then have:

$$\begin{aligned} 0 &\le |Z|^2 = \langle X - \lambda Y | X - \lambda Y\rangle \\ &\Rightarrow 0 \le \langle X|X\rangle - \lambda\langle X|Y\rangle - \lambda^*\langle Y|X\rangle + |\lambda|^2\langle Y|Y\rangle \end{aligned} \tag{3.27}$$

Taking partial differentiation with λ^*, we get:

$$\lambda = \langle Y|X\rangle / |Y|^2, \quad \text{and} \quad \lambda^* = \langle X|Y\rangle / |Y|^2 \tag{3.28}$$

Substitution of values of λ and λ^* yields:

$$0 \le |X|^2 - |\langle X|Y\rangle|^2 / |Y|^2 \tag{3.29}$$

which is true if:

$$|\langle X|Y\rangle|^2 \le |X|^2 |Y|^2 \tag{3.30}$$

or equivalently:

$$|\langle X|Y\rangle| \le |X||Y| \tag{3.31}$$

For the case of the Euclidean space, Eqn. (3.30) is:

$$\left(\sum_{i=1}^{n}(x_i y_i)\right)^2 \leq \left(\sum_{i=1}^{n} x_i^2\right)\left(\sum_{i=1}^{n} y_i^2\right) \quad (3.32)$$

In terms of square integrable complex-valued functions, the Schwarz inequality is expressed as:

$$\left|\int f(x)g(x)\right|^2 dx \leq \left(\int |f(x)|^2 dx\right)\left(\int |g(x)|^2 dx\right) \quad (3.33)$$

3.6 Linear Transformation of Vectors

For an $m \times n$ matrix A and a vector $|X\rangle$ having n-components, there exists a uniquely defined vector $|Z\rangle = A|X\rangle$ that has m-components. The transforming matrix has the number of columns equal to the number of components in $|X\rangle$ and the number of rows equal to the number of components in vector $|Z\rangle$. The transformation is linear if: $\mathrm{A}(a|X\rangle + b|Y\rangle) = a\mathrm{A}|X\rangle + b\mathrm{A}|Y\rangle$ for the complex numbers a and b. Transformation with use of the unit matrix leaves a vector unchanged, while transformation by the null matrix yields a zero vector: $I|X\rangle = |X\rangle$ and $0|X\rangle = 0$.

3.6.1 Eigenvalues and Eigenvectors of a Matrix

If for a square matrix A of order-n $A|X\rangle = \lambda|X\rangle$, where λ is a real or complex scalar quantity, then matrix A is said to satisfy an eigenvalue equation. The λ and $|X\rangle$ are called the eigenvalue and eigenvector of matrix A, respectively. The equation $A|X\rangle = \lambda|X\rangle$ can also be written as:

$$(A - \lambda I)|X\rangle = 0 \quad (3.34)$$

I being the unit matrix of order n. Note that the null vector is an eigenvector for every square matrix. In the following, we want to find the non-trivial eigenvectors. The Eqn. (3.34) presents a system of homogenous linear algebraic equations, which has a non-trivial solution if the matrix $(A - \lambda I)$ of the system is singular.

$$|(A - \lambda I)| = 0 \quad (3.35)$$

The Eqn. (3.35) is known as the *characteristic equation* of the matrix, and the polynomial:

$$\rho(\lambda) = |(A - \lambda I)| \quad (3.36)$$

is called the *characteristic polynomial.* The Eqn. (3.35) is an algebraic equation of order n in terms of λ.

$$\begin{vmatrix} a_{11}-\lambda & a_{12} & a_{13} & \cdots & a_{1n} \\ a_{21} & a_{22}-\lambda & a_{23} & \cdots & a_{2n} \\ a_{31} & a_{32} & a_{33}-\lambda & \cdots & a_{3n} \\ \cdots & \cdots & \cdots & \cdots & \cdots \\ a_{n1} & a_{n2} & a_{n3} & \cdots & a_{nn}-\lambda \end{vmatrix} = 0 \tag{3.37}$$

which can also be expressed as:

$$\rho(\lambda) = \lambda^n + C_1\lambda^{n-1} + C_2\lambda^{n-2} + C_3\lambda^{n-3} + \ldots\ldots\ldots + C_n \tag{3.38}$$

The solution of Eqn. (3.38) yields n-values of λ, and hence matrix A has n-eigenvalues. The set of all eigenvalues, $\lambda_1, \lambda_2, \lambda_3 \ldots\ldots\ldots \lambda_n$ is called the *spectrum* of eigenvalues of the matrix. Thus, there are always n, not necessarily distinct, eigenvalues. Equal eigenvalues that correspond to multiple roots are called *degenerate*. Each of n-eigenvalues corresponds to an eigenvector: The different eigenvectors that belong to the same eigenvalue are called degenerate eigenvectors. Some important properties of eigenvalues and eigenvectors are as follows:

1. If $\lambda_1, \lambda_2, \lambda_3, \lambda_4 \ldots\ldots \lambda_n$ are the eigenvalues of the matrix A of order n, then $\lambda_1^k, \lambda_2^k, \ldots\ldots \lambda_n^k$ too are the eigenvalues of the matrix A^k.
   Proof: Multiply both sides of $A|X_i\rangle = \lambda_i|X_i\rangle$ by A^{k-1} to get:

$$\begin{aligned} A^k|X_i\rangle &= \lambda_i A^{k-1}|X_i\rangle \\ &= \lambda_i A^{k-2}(A|X_i\rangle) \\ &= \lambda_i^2 A^{k-2}|X_i\rangle \\ &= \lambda_i^3 A^{k-3}|X_i\rangle \\ &\vdots \\ &= \lambda_i^k|X_i\rangle \end{aligned} \tag{3.39}$$

2. The eigenvectors of the matrix A, which correspond to different eigenvalues, are linearly independent.
   Proof: Let $|X_1\rangle, |X_2\rangle, |X_3\rangle, \ldots\ldots |X_n\rangle$ be the eigenvectors of the matrix corresponding to the different eigenvalues $\lambda_1, \lambda_2, \lambda_3, \lambda_4 \ldots\ldots \lambda_n$. It is to be shown that the eigenvectors are linearly independent. Let us take:

$$|X\rangle = c_1|X_1\rangle + c_2|X_2\rangle + c_3|X_3\rangle + \ldots\ldots\ldots + c_N|X_N\rangle \tag{3.40}$$

   On operating each term of Eqn. (3.40) by matrix A, we obtain:

$$\begin{aligned} A|X\rangle &= A\big(c_1|X_1\rangle + c_2|X_2\rangle + c_3|X_3\rangle + \ldots\ldots\ldots + c_N|X_N\rangle\big) \\ &= c_1\lambda_1|X_1\rangle + c_2\lambda_2|X_2\rangle + c_3\lambda_3|X_3\rangle + \ldots\ldots\ldots + c_N\lambda_N|X_N\rangle \end{aligned} \tag{3.41}$$

Multiply each term of (3.40) by λ_1 and subtract the resulting equation from Eqn. (3.41) to get:

$$A|X\rangle - \lambda_1|X\rangle = c_1(\lambda_1 - \lambda_1)|X_1\rangle + c_2(\lambda_2 - \lambda_1)|X_2\rangle + c_3(\lambda_3 - \lambda_1)|X_3\rangle + \\ \ldots\ldots + c_j(\lambda_j - \lambda_1)|X_j\rangle + \ldots + c_N(\lambda_N - \lambda_1)|X_N\rangle \tag{3.42}$$

Since λ_1 is one of the eigenvalues of matrix A, we should have:

$$A|X\rangle - \lambda_1|X\rangle = 0 \tag{3.43}$$

$\lambda_1, \lambda_2, \lambda_3, \lambda_4 \ldots\ldots \lambda_n$ are distinct eigenvalues. Hence to satisfy Eqn. (3.43), we must have $c_2 = c_3 \ldots\ldots\ldots = c_N = 0$. Similarly, on multiplying each term of Eqn. (3.40) by λ_2 and then subtracting from Eqn. (3.42), it can be proved that $c_1 = c_3 \ldots\ldots\ldots = c_N = 0$, and so on. It therefore proves that eigenvectors $|X_1\rangle, |X_2\rangle, |X_3\rangle, \ldots\ldots. |X_n\rangle$ are linearly independent.

3. If the eigenvalues of the matrices A and B are nondegenerate and the matrices A and B are commutative, then they have common eigenvectors.

   Proof:

   Take $A|X\rangle = \lambda|X\rangle$ and multiply both sides from left with matrix B to obtain:

$$BA|X\rangle = B(\lambda|X\rangle) = \lambda(|BX\rangle) \tag{3.44}$$

Since matrices A and B commute, $BA = AB$, therefore:

$$BA|X\rangle = AB|X\rangle = \lambda(|BX\rangle) \tag{3.45}$$

Hence, if $|X\rangle$ is an eigenvector of the matrix A corresponding to eigenvalue λ, then $B|X\rangle$ is also an eigenvector of the matrix A corresponding to the same eigenvalue, and then the vectors $|X\rangle$ and $B|X\rangle$ are collinear: $B|X\rangle = \mu|X\rangle$. Therefore, the eigenvector $|X\rangle$ of A corresponding to the eigenvalue λ is also the eigenvector of matrix B corresponding to eigenvalue μ. In a similar manner, it can be shown that each eigenvector of the matrix B is an eigenvector of the matrix A.

3.6.2 Numerical Method to Find Eigenvalue and Eigenvector

The methods of diagonalization of a matrix are used to find the eigenvalues of a matrix. A simple iterative method can be used to find the largest eigenvalue and its corresponding eigenvector of a matrix. Consider a matrix A of order-n and vector

$$|X\rangle = \begin{bmatrix} x_1 \\ x_2 \\ x_3 \\ \vdots \\ x_n \end{bmatrix} \tag{3.46a}$$

Start with $x_1 = 1, x_2 = 0, x_3 = 0, \ldots\ldots, x_n = 0$ and calculate:

$$A|X\rangle = |Y\rangle = \begin{bmatrix} y_1 \\ y_2 \\ y_3 \\ \vdots \\ y_n \end{bmatrix} = y_1 \begin{bmatrix} 1 \\ z_2 \\ z_3 \\ \vdots \\ z_n \end{bmatrix} \tag{3.46b}$$

where, $z_i = \frac{y_i}{y_1}$ for $2 \leq i \leq n$. Redefine $|X\rangle$ as:

$$|X\rangle = \begin{bmatrix} 1 \\ z_2 \\ z_3 \\ \vdots \\ z_n \end{bmatrix} \tag{3.46c}$$

and then calculate $A|X\rangle$. Repeat the process until the value of $|X\rangle$ is the same, within the allowed error, in two successive iterations. At that stage y_1 is the eigenvalue and $|X\rangle$ is the eigenvector of matrix A.

3.7 Inverse Matrix

A $n \times n$ matrix A is invertible if and only if there exists a matrix B such that:

$$AB = I_n \ (= BA). \tag{3.47a}$$

The matrix B is known as the *inverse matrix* of A, commonly represented by A^{-1}. Invertible matrices are precisely those matrices whose determinant is nonzero. The elements of A^{-1} are given by $(A^{-1})_{ij} = \frac{\tilde{\alpha}_{ij}}{|A|}$, where α_{ij} is termed the cofactor of element a_{ij} of matrix A, and it is defined by: $\alpha_{ij} = (-1)^{i+j} \times$ (determinant formed after removing ith row and jth column of $|A|$).

For two matrices C and D of the same order:

$$(CD)^{-1} = D^{-1}C^{-1} \tag{3.47b}$$

If $|X\rangle$ is an eigenvector of a matrix A corresponding to the eigenvalue λ, then the same eigenvector $|X\rangle$ is also the eigenvector of the inverse matrix A^{-1} corresponding to the eigenvalue $1/\lambda$. This can be seen on multiplying both the sides of the $A|X\rangle = \lambda|X\rangle$ from the left by A^{-1}; $A^{-1}A|X\rangle = \lambda A^{-1}|X\rangle \Rightarrow A^{-1}|X\rangle = (1/\lambda)|X\rangle$.

An inverse matrix can also be obtained numerically. One of the numerical methods for matrix inversion is the Gauss-Jordan elimination method. To find the inverse matrix with the use of the Gauss-Jordan elimination method, one operates on the matrix:

$$\begin{bmatrix} A & I \end{bmatrix} \equiv \begin{bmatrix} a_{11} & \cdots & a_{1n} & 1 & 0 & \cdots & 0 \\ a_{21} & \cdots & a_{2n} & 0 & 1 & \cdots & 0 \\ \vdots & \cdots & \vdots & \vdots & \vdots & \cdots & \vdots \\ a_{n1} & \cdots & a_{nn} & 0 & 0 & \cdots & 1 \end{bmatrix} \tag{3.48}$$

where I is the unit matrix. By applying the Gaussian elimination method, Eqn. (3.48) is transformed to:

$$\begin{bmatrix} I & B \end{bmatrix} \equiv \begin{bmatrix} 1 & 0 & \cdots & 0 & b_{11} & \cdots & b_{1n} \\ 0 & 1 & \cdots & 0 & b_{21} & \cdots & b_{2n} \\ \vdots & \vdots & \cdots & \vdots & \vdots & \cdots & \vdots \\ 0 & 0 & \cdots & 1 & b_{n1} & \cdots & b_{nn} \end{bmatrix} \tag{3.49}$$

The matrix:

$$\begin{bmatrix} b_{11} & \cdots & b_{1n} \\ b_{21} & \cdots & b_{2n} \\ \vdots & \cdots & \vdots \\ b_{n1} & \cdots & b_{nn} \end{bmatrix} \tag{3.50}$$

is then A^{-1}. Reduction of matrix A to a diagonal matrix can be obtained by row-operation: $a_{ij} \Rightarrow a_{ij} - \left(\frac{a_{ik}}{a_{kk}}\right) a_{kj}$ where i, j and k vary between 1 and n. The corresponding operation applied to matrix I is: $b_{ij} \Rightarrow \delta_{ij} - \left(\frac{a_{ik}}{a_{kk}}\right)\delta_{kj}$ and then divide b_{ij} by a_{ii} to get the resultant matrix B. One finds that the procedure is numerically unstable unless pivoting (exchanging rows and columns as appropriate) is used. A good choice is picking the largest available element as the pivot.

3.8 Orthogonal Matrix

A square matrix Q whose transposition and inverse are equal is known as an orthogonal matrix:

$$\tilde{Q} = Q^{-1} \quad \text{or} \quad \tilde{Q}Q = I \tag{3.51}$$

Linear transformation of a vector by an orthogonal matrix preserves the vector length: $\langle V\tilde{Q}|QV\rangle = \langle V|V\rangle$, where Q can represent a linear translation, rotations, reflections, and the combinations of rotation and reflection in a finite dimensional space. Orthogonal matrices are important for a number of theoretical and practical reasons, and these are widely used in physics. For example, $\begin{bmatrix} 1 & 0 \\ 0 & 1 \end{bmatrix}$ is used for identity transformation, while $\begin{bmatrix} 1 & 0 \\ 0 & -1 \end{bmatrix}$ gives reflection across the x-axis.

The determinant of an orthogonal matrix is either +1 or –1: $|\tilde{Q}Q| = |\tilde{Q}||Q| = 1 \Rightarrow |Q| = \pm 1$ because $|Q| = |\tilde{Q}|$. The eigenvectors of an orthogonal matrix are orthogonal and an eigenvalue is either +1 or –1.

3.9 Hermitian Matrix

A matrix that satisfies; $\tilde{A} = A$ or $a_{ij} = a_{ji}$ is known as a real symmetric matrix. While a matrix that obeys $A^\dagger = A$ or $a_{ij} = a^*_{ji}$ is termed as a Hermitian or self-adjoint matrix. Real symmetric and Hermitian matrices are the commonest matrices in quantum mechanics as most of the Hamiltonians can be represented this way. The elements of a Hermitian matrix are symmetric about the principal diagonal. Hermitian matrices have an important application in quantum physics in the form of representations of measurement in Heisenberg's quantum mechanics. There is a Hermitian matrix that corresponds to each observable parameter of a physical system. The eigenvalues of a matrix are the possible values that can result from a measurement of that parameter. The eigenvectors of a matrix are the corresponding states of the system for a measurement. Each measurement yields precisely one real value and leaves the system in only one of a set of mutually orthogonal states. Therefore, the measured values must correspond to the eigenvalues of the Hermitian matrix, and the resulting states should be the eigenstate. A matrix in which elements about the diagonal are such that $a_{ij} = -a^*_{ji}$ or $A^\dagger = -A$ is known as a skew Hermitian or anti-Hermitian matrix.

For a Hermitian matrix, eigenvalues are real and the eigenvectors are orthogonal. If λ and $|X\rangle$ are the eigenvalue and eigenvector of a Hermitian matrix H, then we write:

$$H|X\rangle = \lambda|X\rangle \tag{3.52}$$

Taking the Hermitian conjugate of both sides, then noting that multiplication by a scalar is commutative and $H^\dagger = H$, we get:

$$\langle X|H = \lambda^* \langle X| \tag{3.53}$$

Multiply Eqn. (3.52) from the right by $\langle X|$ and Eqn. (3.53) from the left with $|X\rangle$ to get:

$$\langle X|A|X\rangle = \lambda\langle X|X\rangle \tag{3.54a}$$

$$\langle X|A|X\rangle = \lambda^* \langle X|X\rangle \tag{3.54b}$$

The left-hand sides of Eqns. (3.54a) and (3.54b) are equal; therefore $(\lambda - \lambda^*)\langle X|X\rangle = 0$. For a nonzero vector, $\langle X|X\rangle \neq 0$, and hence $\lambda = \lambda^*$ which is possible if λ is real. However, its converse may not be true; a matrix with real eigenvalues is not necessarily Hermitian. For example, matrix $\begin{bmatrix} a+i\alpha & b+i\beta \\ c+i\gamma & d+i\delta \end{bmatrix}$ has real eigenvalues but it is not a Hermitian matrix. Next consider the case where λ_1 and λ_2 are two distinct eigenvalues, which correspond to eigenvectors $|X_1\rangle$ and $|X_2\rangle$, respectively, of Hermitian matrix H:

$$A|X_1\rangle = \lambda_1|X_1\rangle \tag{3.55a}$$

$$A|X_2\rangle = \lambda_2|X_2\rangle \tag{3.55b}$$

Take the Hermitian conjugate of Eqn. (3.55a) and then multiply both sides by $|X_2\rangle$:

$$\langle X_1|A|X_2\rangle = \lambda_1\langle X_1|X_2\rangle \tag{3.56a}$$

where we used $\lambda_1^* = \lambda_1$. Multiply both sides of Eqn. (3.54b) by $\langle X_1|$ to get:

$$\langle X_1|A|X_2\rangle = \lambda_2\langle X_1|X_2\rangle \tag{3.56b}$$

On subtracting Eqn. (3.56b) from Eqn. (3.56a), we obtain:

$$\langle X_1|X_2\rangle(\lambda_1 - \lambda_2) = 0 \tag{3.57}$$

Since eigenvalues are distinct $(\lambda_1 \neq \lambda_2)$, $\langle X_1|X_2\rangle$ must vanish implying that $|X_1\rangle$ and $|X_2\rangle$ are orthogonal.

3.10 Unitary Matrix

A square matrix of order n whose product with its conjugate transpose is a unit matrix $U^\dagger U = UU^\dagger = I_n$ is known as a unitary matrix. Another definition of a unitary matrix is given as follows: a matrix whose columns are the eigenvectors of a Hermitian matrix is a unitary matrix. Also, a matrix whose conjugate transpose is equal to its inverse is a unitary matrix. There is no difference between a unitary matrix and an orthogonal matrix when elements are real. For a unitary matrix, the inner product $\langle X|Y\rangle = \langle X|U^\dagger U|Y\rangle = \langle UX|UY\rangle$ for all *complex* vectors $|X\rangle$ and $|Y\rangle$. We thus find that the transformation of a vector by a unitary matrix does not change its norm. For a unitary matrix U of order n, the following holds:

1. U is unitary then $U^\dagger$ is unitary.
2. If the columns of U form an orthogonal basis of a vector space, then the rows of U too form an orthonormal basis with respect to this inner product.
3. The vectors represented by the columns of U are orthonormal vectors.
4. Eigenvalues of U are complex numbers of absolute value 1. Let us say that U satisfies the eigenvalue equation:

$$U|X\rangle = \lambda|X\rangle \tag{3.58}$$

The conjugate transposition of it is:

$$\langle X|U^{\dagger} = \lambda^{*}\langle X| \tag{3.59}$$

which gives:

$$\begin{aligned}&\langle X|U^{\dagger}U|X\rangle = \lambda^{*}\lambda\langle X|X\rangle \\ &\Rightarrow \langle X|X\rangle = |\lambda|^{2}\langle X|X\rangle\end{aligned} \tag{3.60}$$

Therefore, $|\lambda|^{2} = 1$, which suggests that the eigenvalues of a unitary matrix are complex numbers of absolute value 1.

5. The determinant of a unitary matrix is ±1. If $|U| = d$ then $|UU^{\dagger}| = |U||U^{\dagger}| = d^{2} = 1$, which gives $d = \pm 1$.

3.11 Diagonalization of a Matrix

Diagonalization of a matrix plays an important role in the determination of eigenvalues and eigenvectors, which are key aspects of quantum mechanics. A matrix of eigenvalues is a diagonal matrix, and hence, finding the eigenvalues is equivalent to diagonalization of a matrix.

A triangular matrix precedes to a diagonal matrix. When all the matrix elements below the main diagonal are zero, $a_{ij} = 0$ for $i > j$, the matrix is called an upper triangular matrix, while a lower triangular matrix is the other way around, $a_{ij} = 0$ for $i < j$. The eigenvalues of both upper triangular and lower triangular matrices are also the elements of the main diagonal. The *characteristic equation* of an upper triangular matrix A is:

$$\begin{vmatrix} a_{11}-\lambda & a_{12} & \cdots & a_{1n} \\ 0 & a_{22}-\lambda & \cdots & a_{2n} \\ \vdots & \vdots & \cdots & \vdots \\ 0 & 0 & \cdots & a_{nn}-\lambda \end{vmatrix} = 0 \tag{3.61}$$

Expansion of the determinant along the columns gives:

$$(a_{11}-\lambda)(a_{22}-\lambda)(a_{33}-\lambda).......(a_{nn}-\lambda) = 0. \tag{3.62}$$

A matrix with all off-diagonal elements equal to zero is called a diagonal matrix.

$$D = \begin{bmatrix} d_{11} & 0 & \cdots & 0 \\ 0 & d_{22} & \cdots & 0 \\ \vdots & \vdots & \cdots & 0 \\ 0 & 0 & \cdots & d_{nn} \end{bmatrix} \tag{3.63}$$

To have a *non-singular* diagonal matrix all diagonal elements must be different from zero. The inverse of a diagonal matrix can be found by replacing all diagonal elements by their respective reciprocals and the multiplication of two diagonal matrices results in a diagonal matrix: $D_1D_2 = D_2D_1$. Important propositions related to diagonalization of a matrix are as follows:

1. If a matrix A has *n linearly independent eigenvectors,* $|X_1\rangle, |X_2\rangle, |X_3\rangle, ...|X_n\rangle$ corresponding to the eigenvalues $\lambda_1, \lambda_2, \lambda_3, \lambda_4.....\lambda_n$ then the matrix can be expressed in the form $A = S\Lambda S^{-1}$, where Λ is diagonal matrix having $\lambda_1, \lambda_2, \lambda_3, \lambda_4.....\lambda_n$ along the main diagonal and the matrix S is the matrix whose columns 1,2,3....n represent vectors $|X_1\rangle, |X_2\rangle, |X_3\rangle, ...|X_n\rangle$, respectively:

$$\Lambda = \begin{bmatrix} \lambda_1 & 0 & .... & 0 \\ 0 & \lambda_2 & ... & 0 \\ ... & ... & ... & ... \\ 0 & 0 & ... & \lambda_n \end{bmatrix}, \text{ and } S = \begin{bmatrix} x_{11} & x_{12} & ... & x_{1n} \\ x_{21} & x_{22} & ... & x_{2n} \\ ... & ... & ... & ... \\ x_{n1} & x_{n2} & ... & x_{nn} \end{bmatrix}; |X_i\rangle = \begin{bmatrix} x_{1i} \\ x_{2i} \\ ... \\ x_{ni} \end{bmatrix} \quad (3.64)$$

To prove this, we multiply both sides of $A = S\Lambda S^{-1}$ by S from the right to obtain $AS = S\Lambda$. Then:

$$\begin{aligned} AS &= A\left[|X_1\rangle....|X_i\rangle....|X_n\rangle\right] \\ &= \left[A|X_1\rangle....A|X_i\rangle....A|X_n\rangle\right] \\ &= \left[\lambda_1|X_1\rangle....\lambda_2|X_i\rangle....\lambda_n|X_n\rangle\right] \end{aligned} \quad (3.65)$$

and

$$\begin{aligned} S\Lambda &= \left[|X_1\rangle....|X_i\rangle....|X_n\rangle\right] \begin{bmatrix} \lambda_1 & 0 & ... & 0 \\ 0 & \lambda_2 & ... & 0 \\ ... & ... & ... & ... \\ 0 & 0 & ... & \lambda_n \end{bmatrix} \\ &= \left[\lambda_1|X_1\rangle....\lambda_2|X_i\rangle....\lambda_n|X_n\rangle\right] \end{aligned} \quad (3.66).$$

Therefore, $A = S\Lambda S^{-1}$. This also proves that $\Lambda = S^{-1}AS$. Thus, a matrix A is diagonalizable if there exists a non-singular matrix P and a diagonal matrix Λ such that $\Lambda = P^{-1}AP$. Note that an $n \times n$ matrix may fail to be diagonalizable if (i) all roots of its characteristic equation are not real numbers, and (ii) it does not have n-linearly independent eigenvectors.

2. A Hermitian matrix H can be diagonalized using a unitary matrix U. Which means that in $\Lambda = P^{-1}AP$, if A is Hermitian then P is a unitary matrix. To prove this, let us take $\Lambda = P^{-1}HP$, where Λ is the diagonal matrix of the eigenvalues of

$H = H^\dagger$, which are real. Therefore, Λ too is a Hermitian matrix: $\Lambda = \Lambda^\dagger$. On taking conjugate transposition of both sides we get:

$$\begin{aligned} &\Lambda^\dagger = (P^{-1}HP)^\dagger \\ &\Rightarrow \Lambda = P^\dagger H (P^{-1})^\dagger \\ &\text{or} \\ &P^{-1}HP = P^\dagger H (P^{-1})^\dagger \end{aligned} \tag{3.67}$$

This implies $P^{-1} = P^\dagger$ and $(P^{-1})^\dagger = P$, which is the property of a unitary matrix: $U^\dagger = U^{-1}$. Therefore, $\Lambda = U^\dagger H U$, where Λ is a diagonal matrix having eigenvalues of H along its principle diagonal.

3. Two commuting Hermitian matrices can be diagonalized by the same unitary transformation. Let us say that a unitary transformation brings two commuting matrices A and B to diagonal matrices Λ and Λ':

$$\Lambda = U^\dagger A U \tag{3.68a}$$

$$\Lambda' = U^\dagger B U \tag{3.68b}$$

Then $\Lambda\Lambda' - \Lambda'\Lambda = U^\dagger (AB - BA) U = 0$, because of $AB - BA = 0$, a commuting property.

3.12 Cayley-Hamilton Theorem

If A is a matrix of order n and $\rho(\lambda) = |A - \lambda I|$ then $\rho(A) = 0$, meaning that every matrix satisfies its own characteristic equation.

Proof:

Putting $\lambda = A$ in Eqn. (3.38), we get:

$$\rho(A) = A^n + c_1 A^{n-1} + c_2 \lambda A^{n-2} + c_3 A^{n-3} + \ldots\ldots + c_n = 0 \tag{3.69}$$

Let $B(\lambda)$ be the matrix of the cofactors of determinant $|A - \lambda I|$:

$$B(\lambda) = \begin{bmatrix} b_{11}(\lambda) & b_{12}(\lambda) & \ldots & \ldots & b_{1n}(\lambda) \\ b_{21}(\lambda) & b_{22}(\lambda) & \ldots & \ldots & b_{2n}(\lambda) \\ \ldots & \ldots & \ldots & \ldots & \ldots \\ \ldots & \ldots & \ldots & \ldots & \ldots \\ \ldots & \ldots & \ldots & \ldots & \ldots \\ b_{m1}(\lambda) & b_{m2}(\lambda) & \ldots & \ldots & b_{nn}(\lambda) \end{bmatrix} \tag{3.70a}$$

where $b_{ij}(\lambda)$ is the cofactor of the (i, j) element of matrix $(A - \lambda I)$. The $b_{ij}(\lambda)$ is the polynomial of order $(n-1)$. Therefore:

$$B(\lambda) = B_0 + B_1\lambda + B_2\lambda^2 + B_3\lambda^3 + \ldots\ldots B_{n-1}\lambda^{n-1} \tag{3.70b}$$

Each of the matrices $B_0, B_1, B_2, \ldots B_{n-1,}$ is a matrix of order n. The inverse of matrix $(A - \lambda I)$ is defined as:

$$\begin{aligned}
&(A-\lambda I)^{-1} = \tilde{B}(\lambda) / |A-\lambda I| \\
&or \\
&(A-\lambda I)^{-1}|A-\lambda I| = \tilde{B}(\lambda) = \tilde{B}_0 + \tilde{B}_1\lambda + \tilde{B}_2\lambda^2 + \tilde{B}_3\lambda^3 + \ldots\ldots \tilde{B}_{n-1}\lambda^{n-1}
\end{aligned} \tag{3.71a}$$

where $\tilde{B}(\lambda)$ is a transposition of $B(\lambda)$. Multiply both sides of Eqn. (3.71a) from the left by: $(A - \lambda I)$:

$$|A-\lambda I| = (A-\lambda I)(\tilde{B}_0 + \tilde{B}_1\lambda + \tilde{B}_2\lambda^2 + \tilde{B}_3\lambda^3 + \ldots\ldots \tilde{B}_{n-1}\lambda^{n-1}) \tag{3.71b}$$

Hence,

$$\rho(\lambda) = A\tilde{B}_0 + \lambda(A\tilde{B}_1 - \tilde{B}_0) + \lambda^2(A\tilde{B}_2 - \tilde{B}_1) + \ldots\ldots + \lambda^{n-1}(A\tilde{B}_{n-1} - \tilde{B}_{n-2}) + \lambda^n \tilde{B}_{n-1} \tag{3.71c}$$

Hence, $\rho(A) = 0$.

3.13 Bilinear, Quadratic, and Hermitian Forms

Consider vectors $|X\rangle$ and $|Y\rangle$ in an N-dimensional vector space:

$$|Y\rangle = \begin{bmatrix} y_1 \\ y_2 \\ \vdots \\ y_n \end{bmatrix} \quad \text{and} \quad |X\rangle = \begin{bmatrix} x_1 \\ x_2 \\ \vdots \\ x_n \end{bmatrix} \tag{3.72}$$

The expression $\langle Y|A|X\rangle$ where A is a square matrix of order n, is known as a bilinear form.

$$\langle Y|A|X\rangle = \begin{bmatrix} y_1^* & y_2^* & y_3^* & \ldots & y_n^* \end{bmatrix} \begin{bmatrix} a_{11} & a_{12} & a_{13} & \ldots & a_{1n} \\ a_{21} & a_{22} & a_{23} & \ldots & a_{2n} \\ a_{31} & a_{32} & a_{33} & \ldots & a_{3n} \\ \vdots & \vdots & \vdots & \vdots & \vdots \\ a_{n1} & a_{n2} & a_{n3} & \ldots & a_{nn} \end{bmatrix} \begin{bmatrix} x_1 \\ x_2 \\ x_3 \\ \vdots \\ x_n \end{bmatrix} \tag{3.73}$$

$$= \sum_{(i,j)=1}^{n} y_i^* a_{ij} x_j$$

On replacing Y by X in bilinear form $\langle Y|A|X\rangle$, we get:

$$\langle X|A|X\rangle = \sum_{(i,j)=1}^{n} x_i^* a_{ij} x_j \tag{3.74}$$

$\langle X|A|X\rangle$ is termed a quadratic form. The quadratic expression is analogous to the expectation value of an operator A, and to the scalar quadratic expression αx^2.

If the matrix A in Eqn. (3.74) is a Hermitian matrix, then it is called a Hermitian form:

$$\langle X|H|X\rangle = \langle X|H|X\rangle^{\dagger} \tag{3.75}$$

3.14 Change of Basis

Choice of basis vectors for representing a quantum state is purely arbitrary, as physical measurements do not depend on our choice of basis vectors. If $\{|e_i\rangle\}$ is set of orthonormal basis vectors in a N-dimensional vector space, then another set $\{|e_i'\rangle\}$ of orthonormal vectors can also be the basis vectors for the same vector space. Orthonormal vectors of set $\{|e_i'\rangle\}$ can be constructed by linear combinations of vectors of set $\{|e_i\rangle\}$. Each vector of set $\{|e_i'\rangle\}$ can be expressed as a linear combination of vectors of set $\{|e_i\rangle\}$ as follows:

$$|e_i'\rangle = \sum_{j=1}^{n} u_{ij}^* |e_j\rangle \tag{3.76}$$

in which u_{ij} are suitably chosen complex numbers. Since vectors of new and old basis sets are orthonormal, we have $\langle e_i|e_j\rangle = \delta_{ij}$ and $\langle e_i'|e_j'\rangle = \delta_{ij}$.

$$\begin{aligned} \delta_{ij} &= \langle e_i'|e_j'\rangle = \sum_{k,l} u_{ik} u_{jl}^* \langle e_k|e_l\rangle \\ &= \sum_{k,l} u_{ik} u_{jl}^* \delta_{kl} = \sum_{k} u_{ik} u_{jk}^* \end{aligned} \tag{3.77a}$$

Also:

$$\begin{aligned} \delta_{ij} &= \langle e_i|e_j\rangle = \langle e_i|UU^{\dagger}|e_j\rangle \\ &= \sum_{k=1} \langle e_i|U|e_k\rangle\langle e_k|U^{\dagger}|e_j\rangle \\ &= \sum_{k=1} \langle e_i|U|e_k\rangle\langle e_j|U|e_k\rangle^{\dagger} \\ &= \sum_{k=1} u_{ik} u_{jk}^* \end{aligned} \tag{3.77b}$$

In other words, matrix u_{ik} and u_{kj}^* are (i, j) elements of matrices U and $U^\dagger$, respectively, and $UU^\dagger = I$. It is also apparent that a necessary and sufficient condition is that the vectors of set $\{|e_i'\rangle\}$ are orthonormal and U is a unitary matrix.

3.14.1 Unitary Transformations

Let us say that vector $|X\rangle$ is expressed with respect to basis $\{|e_i\rangle\}$ and $\{|e_i'\rangle\}$ as follows:

$$|X\rangle = \sum_i x_i |e_i\rangle \tag{3.78a}$$

and

$$|X\rangle = \sum_i x_i' |e_i'\rangle \tag{3.78b}$$

We then have:

$$\begin{aligned} \sum_i x_i |e_i\rangle &= \sum_i x_i' |e_i'\rangle \\ &= \sum_{i,j} x_i' u_{ij}^* |e_j\rangle \\ &= \sum_{i,j} x_j' u_{ji}^* |e_i\rangle \end{aligned} \tag{3.79}$$

Hence:

$$\begin{aligned} &\sum_i \left(x_i - \sum_j x_j' u_{ji}^* \right) |e_i\rangle = 0 \\ &\Rightarrow x_i = \sum_j u_{ji}^* x'_j \end{aligned} \tag{3.80}$$

which provides a connection between components x_i and x'_i. Assigning 1 to n to both i and j we obtain:

$$\begin{bmatrix} x_1 \\ x_2 \\ \vdots \\ x_n \end{bmatrix} = \begin{bmatrix} u_{11}^* & u_{21}^* & \cdots & u_{n1}^* \\ u_{12}^* & u_{22}^* & \cdots & u_{n2}^* \\ \vdots & \vdots & \vdots & \vdots \\ u_{1n}^* & u_{2n}^* & \cdots & u_{nn}^* \end{bmatrix} \begin{bmatrix} x_1' \\ x_2' \\ \vdots \\ x_n' \end{bmatrix} \tag{3.81}$$

or

$$\begin{aligned} &|X\rangle = U^\dagger |X'\rangle \\ &\Rightarrow U|X\rangle = UU^\dagger |X'\rangle = |X'\rangle \end{aligned} \tag{3.82}$$

Next consider the case of vector transformation in the vector space of basis $\{|e_i\rangle\}$:

$$|Y\rangle = A|X\rangle \tag{3.83a}$$

The corresponding transformation in the vector space of basis $\{|e_i'\rangle\}$ can be represented as:

$$|Y'\rangle = A'|X'\rangle \tag{3.83b}$$

where A' is the transforming matrix in the vector space of basis $\{|e_i'\rangle\}$. The objective is to find a relation between A and A'. From Eqn. (3.82):

$$|Y'\rangle = U|Y\rangle \text{and} |X'\rangle = U|X\rangle \tag{3.84}$$

Then from Eqn. (3.83b):

$$\begin{aligned} &U|Y\rangle = A'U|X\rangle \\ &\Rightarrow UA|X\rangle = A'U|X\rangle \\ &\Rightarrow A = U^{\dagger}A'U \end{aligned} \tag{3.85a}$$

which implies:

$$A' = UAU^{\dagger} \tag{3.85b}$$

Also:

$$\langle X'|Y'\rangle = \langle X|U^{\dagger}U|Y\rangle = \langle X|Y\rangle \tag{3.86}$$

which shows that the inner product of two vectors, representing two eigenstates, is independent of the choice of basis vectors.

For commuting matrices A and B, we have:

$$\begin{aligned} &UABU^{\dagger} = UBAU^{\dagger} \\ &\Rightarrow UAU^{\dagger}UBU^{\dagger} = UBU^{\dagger}UAU^{\dagger} \\ &\Rightarrow A'B' = B'A' \end{aligned} \tag{3.87}$$

3.15 Infinite-dimensional Space

The theory of N-dimensional vector space can be extended to infinite-dimensional space by taking $N \to \infty$ limit. Each vector of infinite-dimensional space then has an infinite number of components:

$$|X\rangle = \begin{bmatrix} x_1 \\ x_2 \\ x_3 \\ \vdots \\ \vdots \end{bmatrix} \tag{3.88}$$

The inner product of two vectors is then defined by:

$$\langle X|Y\rangle = x_1^* y_1 + x_2^* y_2 + x_3^* y_3 + ... + x_i^* y_i + ... \tag{3.89}$$

The properties of an infinite-dimensional space can be inferred by generalization of the definitions and theorems of N-dimensional space. However, arguments such as the diagonalization of a matrix, which depends upon the assumption that N is finite must be re-examined. The assumption or theorem that N linearly-independent vectors span a space of N dimensions loses its meaning for $N \to \infty$. A set of finitely many orthonormal vectors $\{|e_i\rangle\}$ is said to be complete if every vector $|X\rangle$ can be expressed as a convergent sum:

$$|X\rangle = \sum_{i=1}^{\infty} x_i |e_i\rangle \tag{3.90}$$

which implies:

$$\langle X|X\rangle = \sum_{i=1}^{\infty} |x_i|^2 \text{, with } x_i = \langle e_i|X\rangle \tag{3.91}$$

This is known as the completeness relation for a set $\{|e_i\rangle\}$. This relation is true for every vector in the vector space.

3.16 Hilbert Space

The functions $f(x)$ and $g(x)$ are square-integrable in the interval (a,b) if the norms:

$$(f,f) = \int_a^b |f(x)^2| dx \tag{3.92a}$$

and

$$(g,g) = \int_a^b |g(x)^2| dx \tag{3.92b}$$

exist. The Schwarz-inequality for functions $|(f,g)|^2 \le (f,f)(g,g)$ suggests that the inner product of two functions (f,g) exists for square-integrable functions, and hence a linear combination $h(x) = af(x) + bg(x)$ is too square-integrable. The functions behave like vectors, and vector space defined this way is called *Hilbert space*. The continuous variable $a \le x \le b$ in this space takes the place of discrete variable i of the linear vector space, defined earlier. The vector $|f\rangle$ has infinitely uncountable components, because $f(x)$ corresponds to uncountable values of x. A linear transformation that corresponds to $|Y\rangle = A|X\rangle$ in the Hilbert space is:

$$g(x) = \int_a^b A(x,x')f(x')dx' \tag{3.93}$$

The continuous matrix $A(x,x')$ is a function of two variables. An identity transformation like $|X\rangle = \mathbf{1}|X\rangle$ is defined by:

$$f(x) = \int_a^b \delta(x-x')f(x')dx' \tag{3.94}$$

The $\delta(x-x')$ is the Dirac delta function that is discussed in Annexure B. We thus find that every definition of N-dimensional space has an analogue in Hilbert space.

A function is called a normalized function if its norm is equal to unity. Two normalized functions whose inner product is equal to zero are known as orthonormal functions. A Hilbert space can consist of a set of N-orthonormal functions $\{\phi_i(x)\}$. They are said to fulfill the criteria of completeness and closure if:

$$(\phi_i,\phi_j) = \int_a^b \phi_i^*(x)\phi_j(x)dx = \delta_{ij} \tag{3.95}$$

$$\sum_i \phi_i^*(x)\phi_i(x') = \delta(x-x') \tag{3.96}$$

3.16.1 Basis Vectors in Hilbert Space

We can define a set of N-orthonormal vectors $\{|\phi_i(x)\rangle\}$, where each vector is a continuous function of x. The functions $\phi_1(x),\phi_2(x),\phi_3(x),........\phi_n(x)$ satisfy Eqn. (3.96). The ket vectors $|\phi_1\rangle,|\phi_2\rangle,|\phi_3\rangle,.......|\phi_n\rangle$ can then be used as basis vectors to define a vector space, known as a *Hilbert* vector space. A quantum mechanical state in this vector space is then defined by:

$$|\psi\rangle = \sum_i c_i|\phi_i\rangle \tag{3.97a}$$

$$\text{with } c_i = \langle\phi_i|\psi\rangle. \tag{3.97b}$$

Therefore,

$$|\psi\rangle = \sum_i \langle\phi_i|\psi\rangle|\phi_i\rangle \tag{3.98}$$

where, $c_1, c_2, ...c_n$ are complex or real components of $|\psi\rangle$ along the basis vectors.

The quantum state of a system is represented by the wave function $\psi(\mathbf{r})$ in coordinate space, and by $\psi(\mathbf{p})$ in momentum space. The above definitions, Eqns. (3.97) and (3.98), of vector

Hilbert space are generalized on replacing x either by $\mathbf{r}$ or by $\mathbf{p}$ for coordinate (momentum) space. An important property of vector Hilbert space is: $\langle\phi|\varphi\rangle = \int d^3r\phi^*(\mathbf{r})\varphi(\mathbf{r}) = \int d^3p\phi^*(\mathbf{p})\varphi(\mathbf{p})$.

3.16.2 Quantum States and Operators in Hilbert Space

The wave functions defining the quantum states are represented as vectors in linear vector Hilbert space. The wave function $\psi_a(r)$, where a is representative of all quantum numbers, is denoted by $|\psi_a\rangle$ or simply by $|a\rangle$ as the vector. The $|\psi_a\rangle$ can be expressed in terms of basis kets as follows:

$$|\psi_a\rangle = \sum_i c_i|\phi_i\rangle \quad \text{or} \quad |a\rangle = \sum_i c_i|i\rangle \tag{3.99}$$

which in abstract representation becomes:

$$|a\rangle = \begin{bmatrix} c_1 \\ c_2 \\ c_3 \\ \cdots \\ c_n \end{bmatrix} \tag{3.100}$$

An operator A takes the wave function $\psi(\mathbf{r})$ to a new function $\xi(r)$:

$$|\xi\rangle = A|\psi\rangle \tag{3.101}$$

Expressing $|\xi\rangle$ and $|\psi\rangle$ in terms of basis kets $|\phi_1\rangle, |\phi_2\rangle, |\phi_3\rangle, \ldots\ldots |\phi_n\rangle$, we get:

$$\begin{aligned} \sum_i b_i|\phi_i\rangle &= A\sum_i c_i|\phi_i\rangle \\ &= \sum_i c_i A|\phi_i\rangle \end{aligned} \tag{3.102}$$

Taking the inner product with $|\phi_j\rangle$:

$$\sum_i b_i\langle\phi_j|\phi_i\rangle = \sum_i c_i\langle\phi_j|A|\phi_i\rangle \tag{3.103}$$

Since, $\langle\phi_j|\phi_i\rangle = \delta_{ij}$, Eqn. (3.103) simplifies to:

$$b_j = \sum_i c_i A_{ji} \tag{3.104}$$

where, $A_{ji} = \langle\phi_j|A|\phi_i\rangle$ for $1 \le (i,j) \le n$. Thus:

$$\begin{bmatrix} b_1 \\ b_2 \\ b_3 \\ \vdots \\ b_n \end{bmatrix} = \begin{bmatrix} A_{11} & A_{12} & A_{13} & \cdots & A_{1n} \\ A_{21} & A_{22} & A_{23} & \cdots & A_{2n} \\ A_{31} & A_{32} & A_{33} & \cdots & A_{3n} \\ \vdots & \vdots & \vdots & \vdots & \vdots \\ A_{n1} & A_{n2} & A_{n3} & \cdots & A_{nn} \end{bmatrix} \begin{bmatrix} c_1 \\ c_2 \\ c_3 \\ \vdots \\ c_n \end{bmatrix} \tag{3.105}$$

Equations (3.101) and (3.105) are equivalent. In Eqn. (3.105), a quantum state is represented by a vector while the operator is represented by a matrix. If A is a Hermitian operator, then the corresponding matrix too is Hermitian and hence $A_{ij} = A^*_{ji}$. If $|\phi_i\rangle$, for $1 \le i \le n$, are eigenvectors of operator A, then:

$$A|\phi_i\rangle = \alpha_i|\phi_i\rangle \tag{3.106}$$

or $\langle\phi_j|A|\phi_i\rangle = \alpha_i\delta_{ji}$ and the resulting matrix is a diagonal matrix. Thus, the matrix representation of an operator with respect to its own eigenvectors is a diagonal matrix and the matrix elements are the eigenvalues of the operator.

3.16.3 Schrödinger Equation in Matrix Form

The time-dependent Schrödinger equation is:

$$i\hbar\frac{\partial}{\partial t}|\psi\rangle = H|\psi\rangle \tag{3.107}$$

As discussed in Section 2.1, the time independence of the Schrödinger Hamiltonian allows us to factorize $\psi(\mathbf{r},t)$ into time-dependent and space dependable parts. We therefore can write: $\psi(\mathbf{r},t) = \sum_j c_j(t)\phi_j(\mathbf{r})$. Substituting $|\psi\rangle = \sum_j c_j|\phi_j\rangle$ in Eqn. (3.107) and then taking the inner product with $|\phi_i\rangle$, we obtain:

$$\frac{dc_i(t)}{dt} = -\frac{-i}{\hbar}\sum_i c_j(t)H_{ij} \tag{3.108a}$$

or:

$$\begin{bmatrix} \dot{c}_1 \\ \dot{c}_2 \\ \dot{c}_3 \\ \vdots \\ \dot{c}_n \end{bmatrix} = \begin{bmatrix} H_{11} & H_{12} & H_{13} & \cdots & H_{1n} \\ H_{21} & H_{22} & H_{23} & \cdots & H_{2n} \\ H_{31} & H_{32} & H_{33} & \cdots & H_{3n} \\ \vdots & \vdots & \vdots & \vdots & \vdots \\ H_{n1} & H_{n2} & H_{n3} & \cdots & H_{nn} \end{bmatrix} \begin{bmatrix} c_1 \\ c_2 \\ c_3 \\ \vdots \\ c_n \end{bmatrix} \tag{3.108b}$$

where $H_{ij} = \langle\phi_i|H|\phi_j\rangle$.

The time-independent Schrödinger equation $H\psi(\mathbf{r}) = E\psi(\mathbf{r})$, with the use of $\psi(\mathbf{r}) = \sum_i a_i\phi_i(\mathbf{r})$, can be written in matrix form as follows:

$$a_jE = \sum_i a_iH_{ji} \tag{3.109a}$$

or:

$$\begin{bmatrix} (H_{11}-E) & H_{12} & H_{13} & \cdots & H_{1n} \\ H_{21} & (H_{22}-E) & H_{23} & \cdots & H_{2n} \\ H_{31} & H_{32} & (H_{33}-E) & \cdots & H_{3n} \\ \vdots & \vdots & \vdots & \vdots & \vdots \\ H_{n1} & H_{n2} & H_{n3} & \cdots & (H_{nn}-E) \end{bmatrix} \begin{bmatrix} a_1 \\ a_2 \\ a_3 \\ \vdots \\ a_n \end{bmatrix} = 0 \tag{3.109b}$$

3.17 Statement of Assumptions of Quantum Mechanics

1. Every physical system corresponds to a linear vector space, where each vector of the space represents a possible quantum state of the system. The principle of superposition is implied by this assumption.
2. Each physically measurable property of an observable of the system corresponds to a linear operator, and the eigenvalues of the operator are the possible results of measurement on the observable. Thus, the normalized eigenvectors of a Hermitian operator that corresponds to an observable span the entire state-space, and they can be used as basis vectors. Consequently, any $|\psi\rangle$ can be expressed as a linear combination of eigenvectors of the Hermitian operator. If $|u_1\rangle, |u_2\rangle ......|u_n\rangle$ are eigenvectors and $\alpha_1, \alpha_2, ...\alpha_n$ are corresponding eigenvalues of the operator, then:

$$A|u_i\rangle = \alpha_i |u_i\rangle \tag{3.110}$$

and:

$$|\psi\rangle = \sum_i c_i |u_i\rangle \tag{3.111}$$

If an observation of a quantity that corresponds to an operator A is made in the state $|\psi\rangle$ of the system then the relative probability of obtaining the result α_i is:

$$P(\alpha_i) = |c_i|^2 = |\langle u_i|\psi\rangle|^2 \tag{3.112}$$

The expectation value of operator A in the state of $|\psi\rangle$ is given by:

$$\langle A\rangle = \frac{\langle\psi|A|\psi\rangle}{\langle\psi|\psi\rangle} = \frac{\sum_i \alpha_i |c_i|^2}{\sum_i |c_i|^2} \tag{3.113}$$

if $|u_i\rangle$ are orthonormal vectors.
3. The $|\psi\rangle$ can be a simultaneous eigenvector of two operators A and B if they commute. This means that simultaneous precise measurements of two observables A and B are possible in the state of $|\psi\rangle$, if they commute. When they do not commute, the uncertainties in simultaneous measurements (ΔA and ΔB) are related as follows:

$$\Delta A\,\Delta B \geq \frac{1}{2}|\langle C\rangle| \tag{3.114a}$$

where:

$$[A, B] = iC \tag{3.114b}$$

Equation (3.114a) is known as the general principle of uncertainty.

3.18 General Uncertainty Principle

The quantities corresponding to two non-commuting operators cannot be well defined in the same quantum state. If for two Hermitian operators A and B:

$$[A,B] = iC \tag{3.115}$$

with C too as a nonzero Hermitian operator, then eigenvectors of A are not eigenvectors of B. A linear combination of several eigenvectors of A is the eigenvectors of B. The results of measurements performed on B and on A define the probability distributions of finite widths--ΔA for A and ΔB for B, as follows:

$$\Delta A = \left\{\left\langle (A - \langle A \rangle)^2 \right\rangle\right\}^{1/2} \text{ and } \Delta B = \left\{\left\langle (B - \langle B \rangle)^2 \right\rangle\right\}^{1/2} \tag{3.116}$$

For any state $|\psi\rangle$, we then have:

$$\Delta A \, \Delta B \geq \frac{1}{2} |\langle C \rangle| \tag{3.117}$$

where: $\langle C \rangle = \langle \psi | C | \psi \rangle$. Equation (3.117) is the generalized statement of the uncertainty principle.

Proof:

Let us define $\bar{A} = A - \langle A \rangle$ and $\bar{B} = B - \langle B \rangle$. Evaluate the commutator:

$$\begin{aligned}
[\bar{A}, \bar{B}] &= (A - \langle A \rangle)(B - \langle B \rangle) - (B - \langle B \rangle)(A - \langle A \rangle) \\
&= AB - A\langle B \rangle - \langle A \rangle B + \langle A \rangle \langle B \rangle - BA + B\langle A \rangle + \langle B \rangle A - \langle B \rangle \langle A \rangle \\
&= AB - BA \\
&= [A, B]
\end{aligned} \tag{3.118}$$

Note that $\langle A \rangle$ is scalar while A is an operator. Therefore, $\bar{A}$ and $\bar{B}$ too are Hermitian.

Construct an operator $R = \bar{A} + i\lambda\bar{B}$, where λ is a real scalar quantity. Let us say $|\xi\rangle = R|\psi\rangle$. Since the norm square of a vector can never be negative, we have:

$$\langle \xi | \xi \rangle = \langle \psi | R^{\dagger} R | \psi \rangle \geq 0 \tag{3.119}$$

Therefore:

$$\begin{aligned}
&\langle \psi | (\bar{A} + i\lambda\bar{B})^{\dagger} (\bar{A} + i\lambda\bar{B}) | \psi \rangle \geq 0 \\
&\Rightarrow \langle \psi | (\bar{A} - i\lambda\bar{B})(\bar{A} + i\lambda\bar{B}) | \psi \rangle \geq 0 \\
&\Rightarrow \langle \psi | (\bar{A}^2 + \lambda^2 \bar{B}^2 + i\lambda[\bar{A}, \bar{B}]) | \psi \rangle \geq 0 \\
&\Rightarrow \langle \bar{A}^2 \rangle + \lambda^2 \langle \bar{B}^2 \rangle - \lambda \langle C \rangle \geq 0
\end{aligned} \tag{3.120}$$

We have made use of Eqn. (3.118) in driving Eqn. (3.120). Find the minimum value of λ by taking partial differentiation of Eqn. (3.120):

$$\frac{\partial}{\partial\lambda}\left(\langle\bar{A}^2\rangle+\lambda^2\langle\bar{B}^2\rangle-\lambda\langle C\rangle\right)=0 \tag{3.121a}$$

which gives:

$$\lambda=\frac{\langle C\rangle}{2\langle\bar{B}^2\rangle} \tag{3.121b}$$

substitution of which in Eqn. (3.120), yields:

$$\left\langle(A-\langle A\rangle)^2\right\rangle\left\langle(B-\langle B\rangle)^2\right\rangle\geq\frac{1}{4}\langle C\rangle^2 \tag{3.122a}$$

or:

$$(\Delta A)(\Delta B)=\frac{1}{2}|\langle C\rangle| \tag{3.122b}$$

3.19 One Dimensional Harmonic Oscillator

The problem of the harmonic oscillator is a prominent problem of physics. Many of the physical phenomena are represented by the harmonic oscillator. For example, the Maxwell field can be represented as a linear combination of harmonic oscillators of different frequencies. The electromagnetic fields can then be represented by the Hamiltonian describing a set of harmonic oscillators. As discussed in Section 3.16, one can always construct a Hilbert space that corresponds to the given system, which consists of different oscillators. All possible eigenstates of the system then are represented by the different vectors of that Hilbert space.

We here consider the case of a single harmonic oscillator, whose Hamiltonian is given by:

$$H=\frac{p^2}{2m}+\frac{1}{2}m\omega^2x^2 \tag{3.123}$$

where ω is the natural frequency of the oscillator. The eigenvalue equation for a single oscillator is:

$$H|\psi\rangle=E|\psi\rangle \tag{3.124}$$

where E is the energy eigenvalue and $|\psi\rangle$ is the state vector.

Let us define new operators in terms of operators x and p as follows:

$$a=\frac{i}{\sqrt{2m\hbar\omega}}(p-im\omega x) \tag{3.125a}$$

and:

$$a^{\dagger} = \frac{-i}{\sqrt{2m\hbar\omega}}(p + im\omega x) \tag{3.125b}$$

which gives:

$$a - a^{\dagger} = ip\sqrt{\frac{2}{m\hbar\omega}}, \text{and}\, a + a^{\dagger} = \sqrt{\frac{2m\omega}{\hbar}}x \tag{3.125c}$$

With the use of $[x, p] = i\hbar$, we obtain:

$$\begin{aligned} aa^{\dagger} &= \frac{1}{2m\hbar\omega}\left[p^2 + m^2\omega^2x^2 - im\omega(xp - px)\right] \\ &= \frac{1}{2m\hbar\omega}\left[p^2 + m^2\omega^2x^2 + \hbar m\omega\right] \end{aligned} \tag{3.126a}$$

and:

$$\begin{aligned} a^{\dagger}a &= \frac{1}{2m\hbar\omega}\left[p^2 + m^2\omega^2x^2 + im\omega(xp - px)\right] \\ &= \frac{1}{2m\hbar\omega}\left[p^2 + m^2\omega^2x^2 - \hbar m\omega\right] \end{aligned} \tag{3.126b}$$

Therefore:

$$aa^{\dagger} - a^{\dagger}a = \left[a, a^{\dagger}\right] = 1 \tag{3.127}$$

and:

$$(aa^{\dagger} + a^{\dagger}a) = \frac{1}{m\hbar\omega}\left[p^2 + m^2\omega^2x^2\right] \tag{3.128a}$$

Thus:

$$\begin{aligned} H &= \frac{1}{2}(aa^{\dagger} + a^{\dagger}a)\hbar\omega \\ &= \left(a^{\dagger}a + \frac{1}{2}\right)\hbar\omega \end{aligned} \tag{3.128b}$$

Let us evaluate the commutators $[a, H]$ and $\left[a^{\dagger}, H\right]$:

$$[a, H] = \hbar\omega(aa^{\dagger}a - a^{\dagger}aa) = \hbar\omega(aa^{\dagger} - a^{\dagger}a)a\hbar\omega = a\hbar\omega \tag{3.129a}$$

$$\left[a^{\dagger}, H\right] = \hbar\omega(a^{\dagger}a^{\dagger}a - a^{\dagger}aa^{\dagger}) = \hbar\omega a^{\dagger}(a^{\dagger}a - aa^{\dagger}) = -a^{\dagger}\hbar\omega \tag{3.129b}$$

H is Hermitian; therefore $a^{\dagger}a = \left(H/\hbar\omega - \frac{1}{2}\right)$ too is Hermitian. We define $|\phi\rangle = a|\psi\rangle$ and calculate its norm square:

$$\langle\phi|\phi\rangle = \langle\psi|a^{\dagger}a|\psi\rangle = \langle\psi|(H/\hbar\omega - 1/2)|\psi\rangle = (E/\hbar\omega - 1/2)\langle\psi|\psi\rangle \tag{3.130}$$

The norm of any vector can never be negative, therefore:

$$\langle\phi|\phi\rangle \geq 0 \tag{3.131}$$

Thus, if $|\psi\rangle$ is not a null vector, $\langle\psi|\psi\rangle \neq 0$ then the minimum value that can be assigned to E is $E_0 = \frac{1}{2}\hbar\omega$, which corresponds to the minimum value of $\langle\phi|\phi\rangle = 0$ in Eqn. (3.131). It can then be inferred that the E_0 corresponds to an eigenvector $|\psi_0\rangle$ and no state exists below the state $|\psi_0\rangle$. Hence:

$$a|\psi_0\rangle = 0 \tag{3.132}$$

Further:

$$\begin{aligned} H|\phi\rangle &= Ha|\psi\rangle \\ &= (aH - a\hbar\omega)|\psi\rangle \\ &= (E - \hbar\omega)a|\psi\rangle \end{aligned} \tag{3.133}$$

which means that $a|\psi\rangle$ is an eigenvector of H, and corresponds to the eigenvalue $E - \hbar\omega$. In a similar manner:

$$\begin{aligned} Ha^2|\psi\rangle &= Ha(a|\psi\rangle) \\ &= (aH - a\hbar\omega)(a|\psi\rangle) \\ &= aH\left(a|\psi\rangle\right) - \hbar\omega a^2|\psi\rangle \\ &= a(E - \hbar\omega)a|\psi\rangle - \hbar\omega a^2|\psi\rangle \\ &= (E - 2\hbar\omega)a^2|\psi\rangle \end{aligned} \tag{3.134}$$

Equations (3.133) and (3.134) can be generalized to:

$$Ha^n|\psi\rangle = (E - n\hbar\omega)a^n|\psi\rangle \tag{3.135}$$

We thus find that eigenvalues $E, (E - \hbar\omega), (E - 2\hbar\omega)$ and ...... $(E - n\hbar\omega)$ correspond to eigenvectors $|\psi\rangle, a|\psi\rangle, a^2|\psi\rangle$ and .... $a^n|\psi\rangle$, respectively.

In a similar manner:

$$\begin{aligned} Ha^{\dagger}|\psi\rangle &= (a^{\dagger}H + a^{\dagger}\hbar\omega)|\psi\rangle \\ &= (E + \hbar\omega)|\psi\rangle \end{aligned} \tag{3.136a}$$

and by successive operation with $a^\dagger$ on $a^\dagger|\psi\rangle$, we can construct $(a^\dagger)^n|\psi\rangle$ which on operating by H gives:

$$H(a^\dagger)^n|\psi\rangle = (E+n\hbar\omega)(a^\dagger)^n|\psi\rangle \tag{3.136b}$$

We again notice that eigenvalues E, $(E+\hbar\omega)$, $(E+2\hbar\omega)$ and ...... $(E+n\hbar\omega)$ correspond to eigenvectors $|\psi\rangle, a^\dagger|\psi\rangle, (a^\dagger)^2|\psi\rangle$ and .... $(a^\dagger)^n|\psi\rangle$, respectively.

We thus infer that operation of a on $|\psi\rangle$, which corresponds to energy eigenvalue E, takes it to a lower energy eigenvalue $(E-\hbar\omega)$ state, while operation by $a^\dagger$ on $|\psi\rangle$ takes it to a higher energy eigenvalue $(E+\hbar\omega)$ state. The operators a and $a^\dagger$ are therefore termed as lowering (annihilation) and raising (creation) operators.

As stated before, $E-n\hbar\omega$ in Eqn. (3.135) cannot be smaller than $\frac{1}{2}\hbar\omega$. We therefore have $E-n\hbar\omega=\frac{1}{2}\hbar\omega$. Thus, the energy eigenvalue of the nth state of the harmonic oscillator is:

$$E_n = \left(n+\frac{1}{2}\right)\hbar\omega \tag{3.137}$$

where $n=0,1,2.........$ The state that corresponds to the lowest eigenvalue $\frac{1}{2}\hbar\omega$ is $|\psi_0\rangle$. Further higher eigenvalue states can then be constructed by successive operation of $a^\dagger$ on $|\psi_0\rangle$.

Let us say that the quantum state that corresponds to E_n is represented by eigenvector $|\psi_n\rangle$ and we have:

$$H|\psi_n\rangle = E_n|\psi_n\rangle \tag{3.138}$$

If we assume that eigenvectors are normalized then $\langle\psi_m|\psi_n\rangle = \langle m|n\rangle = \delta_{mn}$, here $\langle m|n\rangle$ is an abbreviated form of $\langle\psi_m|\psi_n\rangle$. From Eqns. (3.133) and (3.136a), we notice that $a|\psi_n\rangle$ and $a^\dagger|\psi_n\rangle$ correspond to eigenvalues $(E_n-\hbar\omega)$ and $(E_n+\hbar\omega)$, respectively. Also, as per Eqns. (3.137) and (3.138), $(E_n-\hbar\omega)$ and $(E_n+\hbar\omega)$ belong to $|\psi_{n-1}\rangle$ and $|\psi_{n+1}\rangle$, respectively. It can therefore be concluded that vectors $a|\psi_n\rangle$ and $|\psi_{n-1}\rangle$, which have the same eigenvalue, are collinear, and hence we can write:

$$a|\psi_n\rangle = \alpha_n|\psi_{n-1}\rangle \quad \text{or} \quad a|n\rangle = \alpha_n|n-1\rangle \tag{3.139a}$$

Similarly, we can write:

$$a^\dagger|\psi_n\rangle = \beta_n|\psi_{n+1}\rangle \quad \text{or} \quad a^\dagger|n\rangle = \beta_n|n+1\rangle \tag{3.139b}$$

where α_n and β_n are scalars and can be determined as follows:

$$\begin{aligned}|\alpha_n|^2 &= |\alpha_n|^2\langle n-1|n-1\rangle \\ &= \langle(n-1)|\alpha_n^*\alpha_n|(n-1)\rangle \\ &= \langle n|a^\dagger a|n\rangle \\ &= n\end{aligned} \tag{3.140a}$$

Here we have made use of $\langle n-1|n-1\rangle = \langle n|n\rangle = 1$ and the Hermitian conjugation of Eqn. (3.139a). Similarly, we write:

$$\begin{aligned}|\beta_n|^2 &= |\beta_n|^2 \langle n+1|n+1\rangle \\ &= \langle n+1|\beta_n^*\beta_n|n+1\rangle \\ &= \langle n|aa^\dagger|n\rangle \\ &= (n+1)\end{aligned} \tag{3.140b}$$

We therefore obtain:

$$\alpha_n = \sqrt{n}, \quad \text{and} \quad \beta_n = \sqrt{n+1} \tag{3.141}$$

which gives:

$$a|\psi_n\rangle = \sqrt{n}|\psi_{n-1}\rangle \tag{3.142a}$$

$$a^\dagger|\psi_n\rangle = \sqrt{n+1}|\psi_{n+1}\rangle \tag{3.142b}$$

As is obvious, Eqns. (3.142) are not the eigenvalue equations. Equation (3.132a) with the use of Eqn. (3.125a) gives:

$$\left(\frac{d}{dx} + \frac{m\omega}{\hbar}x\right)\psi_0(x) = 0 \tag{3.143a}$$

which is a first order differential equation, whose solution can be given by:

$$\psi_0(x) = Ne^{-m\omega x^2/2\hbar} \tag{3.143b}$$

where N is a constant, which is determined by applying the normalization condition:

$$\begin{aligned}&\int_{-\infty}^{\infty} |\psi_0(x)|^2\, dx = 1 \\ &\Rightarrow 2N^2 \int_0^{\infty} e^{-m\omega x^2/\hbar} dx = 1 \\ &\Rightarrow N = \left(\frac{m\omega}{\hbar\pi}\right)^{1/4}\end{aligned} \tag{3.143c}$$

Equation (3.142b) along with Eqn. (3.125b) is then used to determine:

$$\begin{aligned}|\psi_1\rangle = a^\dagger|\psi_0\rangle &= \frac{\hbar}{\sqrt{2m\hbar\omega}}\left(-\frac{d}{dx} + \frac{m\omega}{\hbar}x\right)Ne^{-m\omega x^2/2\hbar} \\ &= N\sqrt{\frac{2m\omega}{\hbar}}xe^{-m\omega x^2/2\hbar}\end{aligned} \tag{3.143d}$$

Also from Eqn. (3.125c), we can write:

$$a\left|\psi_n\right\rangle + a^\dagger\left|\psi_n\right\rangle - \sqrt{\frac{2m\omega}{\hbar}}x\left|\psi_n\right\rangle = 0 \tag{3.144}$$

which with the use of Eqn. (3.142) give:

$$\sqrt{n+1}\left|\psi_{n+1}\right\rangle = \sqrt{\frac{2m\omega}{\hbar}}x\left|\psi_n\right\rangle - \sqrt{n}\left|\psi_{n-1}\right\rangle \tag{3.145}$$

Equation (3.145) is the recurrence relation, which can be used to derive $\left|\psi_2\right\rangle$ after knowing $\left|\psi_0\right\rangle$ and $\left|\psi_1\right\rangle$. Thus, all eigenstates of the 1D harmonic oscillator can be generated with the use of Eqn. (3.145). The $\psi_n(\rho)$ is then given by:

$$\psi_n(\rho) = \frac{1}{\sqrt{(2^n n!)}}\left(\frac{m\omega}{\hbar\pi}\right)^{1/4} e^{-\rho^2/2} H_n(\rho) \tag{3.146}$$

where $\rho = \sqrt{\frac{m\omega}{\hbar}}x$. $H_n(\rho)$ is the Hermite polynomial of degree n. The first few Hermite polynomials are: $H_0(\rho) = 1$, $H_1(\rho) = 2\rho$, $H_2(\rho) = 4\rho^2 - 2$.

The matrix representation of various operators used to describe the 1D harmonic oscillator can be given by employing the following essential properties of operators a, $a^\dagger$, and $a^\dagger a$:

$$a_{mn} = \langle m|a|n\rangle = \sqrt{n}\delta_{m,n-1} \tag{3.147a}$$

$$a^\dagger_{mn} = \langle m|a^\dagger|n\rangle = \sqrt{n+1}\delta_{m,n+1} \tag{3.147b}$$

and:

$$\left(a^\dagger a\right)_{mn} = \langle m|a^\dagger a|n\rangle = n\delta_{m,n} \tag{3.147c}$$

The matrices representing a and $a^\dagger$ are then given by:

$$a = \begin{bmatrix} 0 & \sqrt{1} & 0 & 0 & \cdots \\ 0 & 0 & \sqrt{2} & 0 & \cdots \\ 0 & 0 & 0 & \sqrt{3} & \cdots \\ 0 & 0 & 0 & 0 & \sqrt{4} \\ \vdots & \vdots & \vdots & \vdots & \vdots \end{bmatrix} \tag{3.148a}$$

and:

$$a^\dagger = \begin{bmatrix} 0 & 0 & 0 & 0 & \cdots \\ \sqrt{1} & 0 & 0 & 0 & \cdots \\ 0 & \sqrt{2} & 0 & 0 & \cdots \\ 0 & 0 & \sqrt{3} & 0 & \cdots \\ \vdots & \vdots & \vdots & \vdots & \vdots \end{bmatrix} \tag{3.148b}$$

The operators x and p are represented by the matrices:

$$x=\sqrt{\frac{\hbar}{2m\omega}}\begin{bmatrix} 0 & \sqrt{1} & 0 & 0 & 0 & \cdots \\ \sqrt{1} & 0 & \sqrt{2} & 0 & 0 & \cdots \\ 0 & \sqrt{2} & 0 & \sqrt{3} & 0 & \cdots \\ 0 & 0 & \sqrt{3} & 0 & \sqrt{4} & \cdots \\ \vdots & \vdots & \vdots & \vdots & \vdots & \vdots \end{bmatrix} \tag{3.149a}$$

$$p=\sqrt{\frac{m\hbar\omega}{2}}\begin{bmatrix} 0 & -i\sqrt{1} & 0 & 0 & 0 & \cdots \\ i\sqrt{1} & 0 & -i\sqrt{2} & 0 & 0 & \cdots \\ 0 & i\sqrt{2} & 0 & -i\sqrt{3} & 0 & \cdots \\ 0 & 0 & i\sqrt{3} & 0 & -i\sqrt{4} & \cdots \\ \vdots & \vdots & \vdots & \vdots & \vdots & \vdots \end{bmatrix} \tag{3.149b}$$

The Hamiltonian is represented by the diagonal matrix:

$$H=\hbar\omega\begin{bmatrix} \frac{1}{2} & 0 & 0 & 0 & \cdots \\ 0 & \frac{3}{2} & 0 & 0 & \cdots \\ 0 & 0 & \frac{5}{2} & 0 & \cdots \\ 0 & 0 & 0 & \frac{7}{2} & \cdots \\ \vdots & \vdots & \vdots & \vdots & \vdots \end{bmatrix} \tag{3.150}$$

Note that $|\psi_0\rangle, |\psi_1\rangle, |\psi_2\rangle \ldots |\psi_n\rangle \ldots$ are orthonormal vectors, which can be used as basis vectors for a Hilbert space that corresponds to a system of large number of harmonic oscillators.

3.20 Solved Examples

1. Show that the vectors:

$$|A_1\rangle=\begin{bmatrix} 1 \\ 0 \\ -1 \end{bmatrix}, |A_2\rangle=\begin{bmatrix} 0 \\ 1 \\ 0 \end{bmatrix} \text{ and } |A_3\rangle=\begin{bmatrix} 1 \\ 0 \\ 1 \end{bmatrix},$$

are (a) linearly independent, (b) orthogonal, (c) but not orthonormal.

SOLUTION

a. The condition for linear independence is that $c_1|A_1\rangle + c_2|A_2\rangle + c_3|A_3\rangle = 0$ can only be satisfied for $c_1 = c_2 = c_3 = 0$. We have:

$$\begin{bmatrix} c_1 \\ 0 \\ -c_1 \end{bmatrix} + \begin{bmatrix} 0 \\ c_2 \\ 0 \end{bmatrix} + \begin{bmatrix} c_3 \\ 0 \\ c_3 \end{bmatrix} = \begin{bmatrix} 0 \\ 0 \\ 0 \end{bmatrix}, \text{which gives:}$$

$$c_1 + c_3 = 0, \tag{3.151a}$$

$$c_2 = 0, \tag{3.151b}$$

$$c_3 - c_1 = 0. \tag{3.151c}$$

Addition and subtraction of Eqns. (3.151a) and (3.151c) give $c_1 = c_2 = c_3 = 0$.

b. The orthogonality condition demands that $\langle A_i|A_j\rangle = 0$ for $1 \leq (i, j) \leq 3$:

$$\langle A_1|A_2\rangle = \begin{bmatrix} 1 & 0 & -1 \end{bmatrix} \begin{bmatrix} 0 \\ 1 \\ 0 \end{bmatrix} = 0,$$

similarly:

$$\begin{bmatrix} 0 & 1 & 0 \end{bmatrix} \begin{bmatrix} 1 \\ 0 \\ 1 \end{bmatrix} = 0$$

and $\begin{bmatrix} 1 & 0 & 1 \end{bmatrix} \begin{bmatrix} 1 \\ 0 \\ -1 \end{bmatrix} = 0.$

Hence, it is proved that vectors $|A_1\rangle, |A_2\rangle$ and $|A_3\rangle$ are orthogonal.

c. To satisfy the orthonormal condition, the requirement in addition to the orthogonality condition is that the norm of each vector should be unity. $|A_1| = \sqrt{1^2 + 0^2 + (-1)^2} = \sqrt{2}$, $|A_2| = \sqrt{1}$, and $|A_3| = \sqrt{2}$; therefore $|A_1\rangle$, $|A_2\rangle$ and $|A_3\rangle$ are not orthonormal. Division of $|A_1\rangle$ and $|A_3\rangle$ by $\sqrt{2}$ will provide a set of orthonormal vectors.

2. Show that the vectors

$$|A\rangle = \begin{bmatrix} 1 \\ 0 \\ i \end{bmatrix}, \ |B\rangle = \begin{bmatrix} i \\ 0 \\ -i \end{bmatrix}, \ \text{and} \ |C\rangle = \begin{bmatrix} i \\ 1 \\ 0 \end{bmatrix}$$

are linearly independent but not orthogonal. Orthogonalize them using the Gram-Schmidt method.

SOLUTION

We have:

$$a\begin{bmatrix} 1 \\ 0 \\ i \end{bmatrix} + b\begin{bmatrix} i \\ 0 \\ -i \end{bmatrix} + c\begin{bmatrix} i \\ 1 \\ 0 \end{bmatrix} = \begin{bmatrix} 0 \\ 0 \\ 0 \end{bmatrix} \tag{3.152a}$$

which gives:

$$a + ib + ic = 0 \tag{3.152b}$$

$$c = 0 \tag{3.152c}$$

$$ia - ib = 0 \tag{3.152d}$$

Equation (3.152d) gives $a = b$. Substitution of $c = 0$ and $a = b$ in Eqn. (3.152b) gives $(1+i)b = 0$, which means $a = 0 = b$. Hence, vectors are linearly independent. The inner products of $|A\rangle$, $|B\rangle$ and $|C\rangle$ are:

$$\begin{aligned} \langle A|B\rangle &= \begin{bmatrix} 1 & 0 & -i \end{bmatrix}\begin{bmatrix} i \\ 0 \\ -i \end{bmatrix} = i - 1 \neq 0, \\ \langle B|C\rangle &= \begin{bmatrix} -i & 0 & i \end{bmatrix}\begin{bmatrix} i \\ 1 \\ 0 \end{bmatrix} = 1 \neq 0, \text{ and} \\ \langle C|A\rangle &= \begin{bmatrix} -i & 1 & 0 \end{bmatrix}\begin{bmatrix} 1 \\ 0 \\ i \end{bmatrix} = -i \neq 0 \end{aligned} \tag{3.153}$$

Hence $|A\rangle$, $|B\rangle$ and $|C\rangle$ are not orthogonal. To orthogonalize these, we choose a set of three vectors $|U\rangle$, $|V\rangle$ and $|W\rangle$, and then use the Gram-Schmidt method of orthogonalization. We take $|U\rangle = |A\rangle$, and $|V\rangle = |B\rangle + a|U\rangle$. On imposing the condition that $|V\rangle$ and $|U\rangle$ are orthogonal $\langle U|V\rangle = 0$, we obtain $a = -\langle U|B\rangle / |U|^2$, which gives:

$$a = (1-i)/2, \quad \text{and} \quad |V\rangle = \frac{1}{2}\begin{bmatrix} 1+i \\ 0 \\ 1-i \end{bmatrix}.$$

We next take $|W\rangle = |C\rangle + b|V\rangle + c|U\rangle$ and then we use $\langle U|W\rangle = 0$ and $\langle V|W\rangle = 0$ to determine b and c. This yields $b = -\langle V|C\rangle / |V|^2$ and $c = -\langle U|C\rangle / |U|^2$, which provides $b = -(1+i)/2$ and $c = -i/2$. Therefore:

$$|W\rangle = \begin{bmatrix} i \\ 1 \\ 0 \end{bmatrix} - \frac{(1+i)}{4}\begin{bmatrix} 1+i \\ 0 \\ 1-i \end{bmatrix} - \frac{i}{2}\begin{bmatrix} 1 \\ 0 \\ i \end{bmatrix} = \begin{bmatrix} 0 \\ 1 \\ 0 \end{bmatrix} \tag{3.154}$$

Three orthogonalized vectors are:

$$|U\rangle = \begin{bmatrix} 1 \\ 0 \\ i \end{bmatrix}, |V\rangle = \frac{1}{2}\begin{bmatrix} 1+i \\ 0 \\ 1-i \end{bmatrix} \quad \text{and} \quad |W\rangle = \begin{bmatrix} 0 \\ 1 \\ 0 \end{bmatrix} \tag{3.155}$$

As is seen, vectors $|V\rangle$ and $|W\rangle$ are orthonormal vectors, while $|U\rangle$ is not. To orthonormalize $|U\rangle$, divide it by its norm $\sqrt{2}$.

3. Construct a 3×3 matrix A that transforms the vectors:

$$|X_1\rangle = \begin{bmatrix} 1 \\ 0 \\ -1 \end{bmatrix}, \ |X_2\rangle = \begin{bmatrix} 0 \\ 1 \\ 0 \end{bmatrix} \quad \text{and} \quad |X_3\rangle = \begin{bmatrix} 1 \\ 0 \\ 1 \end{bmatrix}$$

to the vectors:

$$|Y_1\rangle = \begin{bmatrix} 2 \\ 0 \\ 0 \end{bmatrix}, \ |Y_2\rangle = \begin{bmatrix} 1 \\ 0 \\ 1 \end{bmatrix} \quad \text{and} \quad |Y_3\rangle = \begin{bmatrix} 1 \\ 0 \\ -1 \end{bmatrix}$$

SOLUTION

The transforming equation is $|Y_i\rangle = A|X_i\rangle$ for $1 \le i \le 3$. Therefore:

$$\begin{bmatrix} 2 \\ 0 \\ 0 \end{bmatrix} = \begin{bmatrix} a_{11} & a_{12} & a_{13} \\ a_{21} & a_{22} & a_{23} \\ a_{31} & a_{32} & a_{33} \end{bmatrix}\begin{bmatrix} 1 \\ 0 \\ -1 \end{bmatrix} \Rightarrow \left\{ \begin{array}{c} 2 = a_{11} - a_{13} \\ 0 = a_{21} - a_{23} \\ 0 = a_{31} - a_{33} \end{array} \right\} \quad \text{for } i = 1 \tag{3.156a}$$

$$\begin{bmatrix} 1 \\ 0 \\ 1 \end{bmatrix} = \begin{bmatrix} a_{11} & a_{12} & a_{13} \\ a_{21} & a_{22} & a_{23} \\ a_{31} & a_{32} & a_{33} \end{bmatrix}\begin{bmatrix} 0 \\ 1 \\ 0 \end{bmatrix} \Rightarrow \left\{ \begin{array}{c} 1 = a_{12} \\ 0 = a_{22} \\ 1 = a_{32} \end{array} \right\} \quad \text{when } i = 2 \tag{3.156b}$$

$$\begin{bmatrix} 1 \\ 0 \\ -1 \end{bmatrix} = \begin{bmatrix} a_{11} & a_{12} & a_{13} \\ a_{21} & a_{22} & a_{23} \\ a_{31} & a_{32} & a_{33} \end{bmatrix}\begin{bmatrix} 1 \\ 0 \\ 1 \end{bmatrix} \Rightarrow \left\{ \begin{array}{c} 1 = a_{11} + a_{13} \\ 0 = a_{21} + a_{23} \\ -1 = a_{31} + a_{33} \end{array} \right\} \quad \text{for } i = 3 \tag{3.156c}$$

the solution of which yields:

$a_{11} = \frac{3}{2}, a_{13} = a_{31} = a_{33} = -\frac{1}{2}, \ a_{22} = a_{21} = a_{23} = 0, \ a_{12} = a_{32} = 1$. The matrix is:

$$A = \begin{bmatrix} \frac{3}{2} & 1 & \frac{-1}{2} \\ 0 & 0 & 0 \\ \frac{-1}{2} & 1 & \frac{-1}{2} \end{bmatrix} \tag{3.157}$$

4. A vector space is represented by basis vectors:

$$\begin{bmatrix} 1 \\ 0 \\ 0 \end{bmatrix}, \begin{bmatrix} 0 \\ 1 \\ 0 \end{bmatrix} \text{ and } \begin{bmatrix} 0 \\ 0 \\ 1 \end{bmatrix}.$$

Construct a matrix that transforms these to another set of basis vectors:

$$\begin{bmatrix} 1/\sqrt{2} \\ 0 \\ i/\sqrt{2} \end{bmatrix}, \begin{bmatrix} (1+i)/2 \\ 0 \\ (1-i)/2 \end{bmatrix}, \text{ and } \begin{bmatrix} 0 \\ 1 \\ 0 \end{bmatrix}.$$

Is the transforming matrix unitary?

SOLUTION

We have:

$$\frac{1}{\sqrt{2}}\begin{bmatrix} 1 \\ 0 \\ i \end{bmatrix} = \begin{bmatrix} a_{11} & a_{12} & a_{13} \\ a_{21} & a_{22} & a_{23} \\ a_{31} & a_{32} & a_{33} \end{bmatrix}\begin{bmatrix} 1 \\ 0 \\ 0 \end{bmatrix} \Rightarrow \left\{ \begin{array}{c} \frac{1}{\sqrt{2}} = a_{11} \\ 0 = a_{21} \\ \frac{i}{\sqrt{2}} = a_{31} \end{array} \right\} \tag{3.158a}$$

$$\frac{1}{2}\begin{bmatrix} 1+i \\ 0 \\ 1-i \end{bmatrix} = \begin{bmatrix} a_{11} & a_{12} & a_{13} \\ a_{21} & a_{22} & a_{23} \\ a_{31} & a_{32} & a_{33} \end{bmatrix}\begin{bmatrix} 0 \\ 1 \\ 0 \end{bmatrix} \Rightarrow \left\{ \begin{array}{c} \frac{1+i}{2} = a_{12} \\ 0 = a_{22} \\ \frac{1-i}{2} = a_{32} \end{array} \right\} \tag{3.158b}$$

$$\begin{bmatrix} 0 \\ 1 \\ 0 \end{bmatrix} = \begin{bmatrix} a_{11} & a_{12} & a_{13} \\ a_{21} & a_{22} & a_{23} \\ a_{31} & a_{32} & a_{33} \end{bmatrix}\begin{bmatrix} 0 \\ 0 \\ 1 \end{bmatrix} \Rightarrow \left\{ \begin{array}{c} 0 = a_{13} \\ 1 = a_{23} \\ 0 = a_{33} \end{array} \right\} \tag{3.158c}.$$

The transforming matrix is:

$$A = \begin{bmatrix} 1/\sqrt{2} & (1+i)/2 & 0 \\ 0 & 0 & 1 \\ i/\sqrt{2} & (1-i)/2 & 0 \end{bmatrix} \tag{3.159a}$$

and:

$$A^{\dagger} = \begin{bmatrix} 1/\sqrt{2} & 0 & -i/\sqrt{2} \\ (1-i)/2 & 0 & (1+i)/2 \\ 0 & 1 & 0 \end{bmatrix}, \tag{3.159b}$$

which give:

$$AA^{\dagger} = \begin{bmatrix} 1 & 0 & 0 \\ 0 & 1 & 0 \\ 0 & 0 & 1 \end{bmatrix} \tag{3.159c}$$

This shows that the transforming matrix is a unitary matrix.

5. Find the eigenvalues and eigenvectors of the matrix:

$$A = \begin{bmatrix} 1 & 0 & 0 \\ 1 & 2 & 3 \\ 1 & 0 & 3 \end{bmatrix}.$$

SOLUTION

The characteristic equation of the matrix is:

$$\begin{vmatrix} 1-\lambda & 0 & 0 \\ 1 & 2-\lambda & 3 \\ 1 & 0 & 3-\lambda \end{vmatrix} = 0 \tag{3.160}$$

Simplification of the determinant gives $(1-\lambda)(2-\lambda)(3-\lambda)=0$, whose solution yields three eigenvalues: $\lambda_1 = 1, \lambda_2 = 2$ and $\lambda_3 = 3$. The eigenvector corresponding to the eigenvalue $\lambda_1 = 1$ is obtained from:

$$\begin{bmatrix} 0 & 0 & 0 \\ 1 & 1 & 3 \\ 1 & 0 & 2 \end{bmatrix} \begin{bmatrix} x_{11} \\ x_{21} \\ x_{31} \end{bmatrix} = \begin{bmatrix} 0 \\ 0 \\ 0 \end{bmatrix} \tag{3.161a}$$

which provides two independent equations:

$$\begin{aligned} x_{11} + x_{21} + 3x_{31} &= 0 \\ x_{11} + 2x_{31} &= 0 \end{aligned} \tag{3.161b}$$

Here the number of degrees of freedom is 1, and therefore by choosing $x_{31} = 1$, we get the eigenvector corresponding to the eigenvalue $\lambda_1 = 1$:

$$|X_1\rangle = \begin{bmatrix} -2 \\ -1 \\ 1 \end{bmatrix} \tag{3.161c}$$

To obtain the eigenvector that belongs to $\lambda_2 = 2$ we have:

$$\begin{bmatrix} -1 & 0 & 0 \\ 1 & 0 & 3 \\ 1 & 0 & 1 \end{bmatrix} \begin{bmatrix} x_{12} \\ x_{22} \\ x_{32} \end{bmatrix} = \begin{bmatrix} 0 \\ 0 \\ 0 \end{bmatrix} \Rightarrow -x_{12} = 0\ ,\ \ x_{12} + 3x_{32} = 0, \text{ and } x_{12} + x_{32} = 0.$$

This suggests that x_{12} and x_{32} are zero and x_{22} can be assigned an arbitrary value. We therefore can take:

$$|X_2\rangle = \begin{bmatrix} 0 \\ 1 \\ 0 \end{bmatrix} \tag{3.161d}$$

The eigenvector corresponding to $\lambda_3 = 3$ is obtained from:

$$\begin{bmatrix} -2 & 0 & 0 \\ 1 & -1 & 3 \\ 1 & 0 & 0 \end{bmatrix} \begin{bmatrix} x_{13} \\ x_{23} \\ x_{33} \end{bmatrix} = \begin{bmatrix} 0 \\ 0 \\ 0 \end{bmatrix} \Rightarrow x_{13} = 0,\ x_{13} - x_{23} + 3x_{33} = 0 \tag{3.161e}$$

which yields $x_{23} = 3x_{33}$. By choosing $x_{33} = 1$, we get:

$$|X_3\rangle = \begin{bmatrix} 0 \\ 3 \\ 1 \end{bmatrix} \tag{3.161f}$$

6. For the matrix:

$$A = \begin{bmatrix} 4 & 2 & 0 & 4 \\ 0 & 2 & -1 & 0 \\ 0 & 0 & 3 & 3 \\ 0 & 4 & 0 & 7 \end{bmatrix}$$

if $\lambda_1 = 4$, $\lambda_2 = 1$, and $\lambda_3 = 6$ are three eigenvalues of A. Find the fourth eigenvalue of the matrix and its determinant.

SOLUTION

Since the trace of any matrix is equal to the sum of all eigenvalues, we have: $4+2+3+7 = 4+1+6+\lambda_4 \Rightarrow \lambda_4 = 5$. The determinant of a matrix is equal to the multiple of all eigenvalues and $(-1)^n$. Therefore $|A| = \lambda_1\lambda_2\lambda_3\lambda_4 = 4\times1\times6\times5 = 120$.

7. The eigenvalues and corresponding eigenvectors of a 3×3 matrix are given by:

$$\lambda_1 = 3,\ |X_1\rangle = \begin{bmatrix} -1 \\ -1 \\ 1 \end{bmatrix},\ \lambda_2 = -3,\ |X_2\rangle = \begin{bmatrix} 2 \\ 1 \\ 0 \end{bmatrix} \text{ and } \lambda_3 = 5,\ |X_3\rangle = \begin{bmatrix} 0 \\ 0 \\ 1 \end{bmatrix},$$

so, find the matrix.

SOLUTION

We know that a matrix A can be written as $A = S\Lambda S^{-1}$, where:

$$\Lambda = \begin{bmatrix} 3 & 0 & 0 \\ 0 & -3 & 0 \\ 0 & 0 & 5 \end{bmatrix} \text{ and } S = \begin{bmatrix} -1 & 2 & 0 \\ -1 & 1 & 0 \\ 1 & 0 & 1 \end{bmatrix}.$$

Computation of the inverse of S gives:

$$S^{-1} = \begin{bmatrix} 1 & -2 & 0 \\ 1 & -1 & 0 \\ -1 & 2 & 1 \end{bmatrix}$$

Therefore:

$$A = \begin{bmatrix} -1 & 2 & 0 \\ -1 & 1 & 0 \\ 1 & 0 & 1 \end{bmatrix} \begin{bmatrix} 3 & 0 & 0 \\ 0 & -3 & 0 \\ 0 & 0 & 5 \end{bmatrix} \begin{bmatrix} 1 & -2 & 0 \\ 1 & -1 & 0 \\ -1 & 2 & 1 \end{bmatrix}$$

$$A = \begin{bmatrix} -1 & 2 & 0 \\ -1 & 1 & 0 \\ 1 & 0 & 1 \end{bmatrix} \begin{bmatrix} 3 & -6 & 0 \\ -3 & 3 & 0 \\ -5 & 10 & 5 \end{bmatrix} = \begin{bmatrix} -9 & 12 & 0 \\ -6 & 9 & 0 \\ -2 & 4 & 5 \end{bmatrix}$$

8. For the matrix,

$$A = \begin{bmatrix} -4 & -6 \\ 3 & 5 \end{bmatrix}.$$

find the non-singular matrices S and S^{-1} to compute $\Lambda = S^{-1}AS$. Find Λ^4 and then show that $A^4 = S\Lambda^4 S^{-1}$. Compute A^4.

SOLUTION

We first need to find the eigenvalues and eigenvectors of A. The characteristic equation of A is:

$$|\lambda I - A| = \begin{vmatrix} \lambda + 4 & 6 \\ -3 & \lambda - 5 \end{vmatrix} = (\lambda - 2)(\lambda + 1) = 0 \tag{3.162a}$$

giving $\lambda_1 = 2$ and $\lambda_2 = -1$. The eigenvector that corresponds to $\lambda_1 = 2$ is:

$$A|X_1\rangle = 2|X_1\rangle \Rightarrow \begin{bmatrix} -6 & -6 \\ 3 & 3 \end{bmatrix} \begin{bmatrix} x_{11} \\ x_{21} \end{bmatrix} = 0 \tag{3.162b}$$

$$\Rightarrow (x_{11} + x_{21}) = 0$$

which can be satisfied for $|X_1\rangle = r\begin{bmatrix} -1 \\ 1 \end{bmatrix}$, where r is an arbitrary constant. When $\lambda_2 = -1$ we have:

$$A|X_2\rangle = -|X_2\rangle \Rightarrow \begin{bmatrix} -3 & -6 \\ 3 & 6 \end{bmatrix} \begin{bmatrix} x_{12} \\ x_{22} \end{bmatrix} = 0 \tag{3.162c}$$

$$\Rightarrow (x_{12} + 2x_{22}) = 0$$

whose solution can be $|X_2\rangle = t\begin{bmatrix} -2 \\ 1 \end{bmatrix}$, where t, too, is an arbitrary constant. Two linearly independent eigenvectors of matrix are $\begin{bmatrix} -1 \\ 1 \end{bmatrix}$ and $\begin{bmatrix} -2 \\ 1 \end{bmatrix}$. We thus have:

$$S = \begin{bmatrix} -1 & -2 \\ 1 & 1 \end{bmatrix}, \; \Lambda = \begin{bmatrix} 2 & 0 \\ 0 & -1 \end{bmatrix} \text{ and } \; S^{-1} = \begin{bmatrix} 1 & 2 \\ -1 & -1 \end{bmatrix} \tag{3.163a}$$

and:

$$S^{-1}AS = \begin{bmatrix} -1 & -2 \\ 1 & 1 \end{bmatrix}\begin{bmatrix} -4 & -6 \\ 3 & 5 \end{bmatrix}\begin{bmatrix} 1 & 2 \\ -1 & -1 \end{bmatrix} = \begin{bmatrix} 2 & 0 \\ 0 & -1 \end{bmatrix} \tag{3.163b}$$

Direct multiplication yields:

$$\Lambda^4 = \begin{bmatrix} 2^4 & 0 \\ 0 & (-1)^4 \end{bmatrix}.$$

Also, we find that:

$$\Lambda^4 = \left(S^{-1}AS\right)\left(S^{-1}AS\right)\left(S^{-1}AS\right)\left(S^{-1}AS\right) = S^{-1}A^4S \tag{3.164a}$$

which implies $A^4 = S\Lambda^4 S^{-1}$:

$$\begin{aligned} A^4 &= \begin{bmatrix} -1 & -2 \\ 1 & 1 \end{bmatrix}\begin{bmatrix} 2^4 & 0 \\ 0 & (-1)^4 \end{bmatrix}\begin{bmatrix} 1 & 2 \\ -1 & -1 \end{bmatrix} \\ &= \begin{bmatrix} -1 & -2 \\ 1 & 1 \end{bmatrix}\begin{bmatrix} 16 & 32 \\ -1 & -1 \end{bmatrix} \\ &= \begin{bmatrix} -14 & -30 \\ 15 & 31 \end{bmatrix} \end{aligned} \tag{3.164b}$$

It is to be noted here that for any $n \times n$ diagonalizable matrix A, $\Lambda = S^{-1}AS$, then

$$A^k = S\Lambda^k S^{-1}, \; k = 1, 2, \ldots$$

where:

$$\Lambda^k = \begin{bmatrix} \lambda_1^k & 0 & \cdots & 0 \\ 0 & \lambda_2^k & \cdots & 0 \\ \vdots & \vdots & \ddots & \vdots \\ 0 & 0 & \cdots & \lambda_n^k \end{bmatrix}. \tag{3.164c}$$

9. Show that scalar product of two vectors is invariant under unitary transformation.

SOLUTION

Let us say that after unitary transformation two vectors $|A\rangle$ and $|B\rangle$ are:

$$|A'\rangle = U|A\rangle \tag{3.165a}$$

$$|B'\rangle = U|B\rangle \tag{3.165b}$$

Notice that $\langle B'| = \langle B|U^{\dagger}$, and then calculate the inner product of $|A'\rangle$ and $|B'\rangle$:

$$\langle B'|A'\rangle = \langle B|U^{\dagger}U|A\rangle = \langle B|A\rangle \tag{3.165c}$$

Hence, we find that the scalar (inner) product of two vectors is invariant under unitary transformation.

10. Show that every Hermitian matrix can give rise to a unitary matrix.

SOLUTION

For Hermitian matrix $H^{\dagger} = H$, take the matrix:

$$U = (H - iI)(H + iI)^{-1} \tag{3.166b}$$

where I is a unit matrix. We then have:

$$U^{\dagger} = \left((H + iI)^{-1}\right)^{\dagger}(H - iI)^{\dagger} = (H + iI)(H - iI)^{-1} \tag{3.166b}$$

which yields $UU^{\dagger} = I$. This proves that every Hermitian matrix can give rise to a unitary matrix.

11. Consider the case of a 1D harmonic oscillator, where for ground state $\psi_0(x) = \frac{\sqrt{\alpha}}{(\pi)^{1/4}}e^{-\alpha^2x^2/2}$, with $\alpha = \left(\frac{m\omega}{\hbar}\right)^{1/2}$. Calculate $\psi_1(x)$, $\psi_2(x)$ and then generalize it for $\psi_n(x)$.

SOLUTION:

We know that operation by $a^{\dagger}$ on a state vector takes it to the next eigenvalue state: $\psi_1(x) = a^{\dagger}\psi_0(x)$, where:

$$a^{\dagger} = \frac{1}{i\sqrt{2m\hbar\omega}}(p + im\omega x) = \frac{1}{i\sqrt{2m\hbar\omega}}\left(-i\hbar\frac{d}{dx} + im\omega x\right) = \frac{-1}{\alpha\sqrt{2}}\left(\frac{d}{dx} - \alpha^2 x\right) \tag{3.167a}$$

Thus:

$$\begin{aligned}\psi_1(x) &= -\frac{1}{\alpha\sqrt{2}}\left(\frac{d}{dx} - \alpha^2 x\right)\psi_0(x) \\ &= -\frac{1}{\sqrt{2\alpha}(\pi)^{1/4}}\left(\frac{d}{dx} - \alpha^2 x\right)e^{-\alpha^2x^2/2} \\ &= \left(\frac{1}{\sqrt{2\alpha\sqrt{\pi}}}\right)2\alpha^2 x e^{-\alpha^2x^2/2} = \left(\frac{\alpha}{2\sqrt{\pi}}\right)^{1/2} H_1(\alpha x)e^{-\alpha^2x^2/2}\end{aligned} \tag{3.167b}$$

With the use of Eqn. (3.145), we get:

$$
\begin{aligned}
|\psi_2\rangle &= \alpha x|\psi_1\rangle - \frac{1}{\sqrt{2}}|\psi_0\rangle \\
&= \left(\frac{1}{\sqrt{2\alpha}(\pi)^{1/4}}\right)2\alpha^3 x^2 e^{-\alpha^2x^2/2} - \frac{1}{(\pi)^{1/4}}\sqrt{\frac{\alpha}{2}}e^{-\alpha^2x^2/2} \\
&= \frac{1}{(\pi)^{1/4}}\sqrt{\frac{\alpha}{2}}\left(2\alpha^2x^2 - 1\right)e^{-\alpha^2x^2/2} \\
&= \frac{1}{2(\pi)^{1/4}}\sqrt{\frac{\alpha}{2}}H_2(\alpha x)e^{-\alpha^2x^2/2} \\
&= \left(\frac{\alpha}{\sqrt{\pi}}\frac{1}{2^2(2!)}\right)^{1/2} H_2(\alpha x)e^{-\alpha^2x^2/2},
\end{aligned}
\tag{3.168}
$$

Generalization of $\psi_2(x)$ and $\psi_1(x)$ yields:

$$
\psi_n(x) = \left(\frac{\alpha}{\sqrt{\pi}}\frac{1}{2^n(n!)}\right)^{1/2} H_n(\alpha x)e^{-\alpha^2x^2/2} \tag{3.169}
$$

12. Show that operators a and $a^\dagger$ are not Hermitian while operator $a^\dagger a$ is Hermitian.

SOLUTION

For a Hermitian matrix A, $A = A^\dagger \Rightarrow A_{ij} = A^*_{ji}$. The matrix elements of a and $a^\dagger$ are:

$$
a_{mn} = \langle m|a|n\rangle = \sqrt{n}\langle m|n-1\rangle = \sqrt{n}\delta_{m,n-1} \tag{3.170a}
$$

and:

$$
(a_{nm})^* = \langle n|a|m\rangle^* = \left(\sqrt{m}\langle n|m-1\rangle\right)^* = \sqrt{m}\delta_{n,m-1} \tag{3.170b}
$$

which yield $a_{mn} \neq (a_{nm})^*$, and hence a is not a Hermitian operator. Further:

$$
\left(a^\dagger\right)_{mn} = \langle m|a^\dagger|n\rangle = \sqrt{n+1}\langle m|n+1\rangle = \sqrt{n+1}\delta_{m,n+1}, \tag{3.171}
$$

and:

$$
\begin{aligned}
\left(a^\dagger_{nm}\right)^* &= \langle n|a^\dagger|m\rangle^* = \sqrt{m+1}\langle n|m+1\rangle \\
&= \sqrt{m+1}\delta_{n,m+1} \neq a^\dagger_{nm}
\end{aligned}
\tag{3.172}
$$

We thus find that $a^\dagger$ is not a Hermitian operator.
The matrix elements of operator $a^\dagger a$ are:

$$
\left(a^\dagger a\right)_{mn} = \langle m|a^\dagger a|n\rangle = n\delta_{mn} \tag{3.173}
$$

and:

$$\left\{\left(a^{\dagger}a\right)_{nm}\right\}^{*} = \langle n|a^{\dagger}a|m\rangle^{*} = m\delta_{nm} \tag{3.174}$$

Since, m and n are integer number, $\left(a^{\dagger}a\right)_{nm} = \left(a^{\dagger}a\right)_{mn}$. Hence $a^{\dagger}a$ is a Hermitian operator.

13. Use the Heisenberg equation and show that the time dependence of annihilation and creation operators for a 1D harmonic oscillator can be expressed as: $a(t) = a(0)e^{-i\omega t}$ and $a^{\dagger}(t) = a^{\dagger}(0)e^{i\omega t}$.

SOLUTION

From Eqn. (3.125), we have:

$$a = \frac{i}{\sqrt{2m\hbar\omega}}(p - im\omega x) \quad \text{and} \quad a^{\dagger} = \frac{-i}{\sqrt{2m\hbar\omega}}(p + im\omega x) \tag{3.175}$$

The Heisenberg equation for time-independent operators gives:

$$\begin{aligned}\frac{dp}{dt} &= \frac{1}{i\hbar}\left[p, H\right] = \frac{1}{i\hbar}\left[p, \frac{p^2}{2m} + \frac{1}{2}m\omega^2 x^2\right] \\ &= \frac{1}{i\hbar}\left(\frac{p^3}{2m} + \frac{1}{2}m\omega^2 px^2 - \frac{p^3}{2m} - \frac{1}{2}m\omega^2 x^2 p\right) \\ &= \frac{m\omega^2}{2i\hbar}\left(px^2 - x^2 p\right) \\ &= \frac{m\omega^2}{2i\hbar}\left(px^2 - xpx + xpx - x^2 p\right) \\ &= \frac{m\omega^2}{2i\hbar}(-2i\hbar x)\end{aligned} \tag{3.176a}$$

which implies $\frac{dp}{dt} = -m\omega^2 x$. And:

$$\begin{aligned}\frac{dx}{dt} &= \frac{1}{i\hbar}[x, H] = \frac{1}{i\hbar}\left[x, \frac{p^2}{2m} + \frac{1}{2}m\omega^2 x^2\right] \\ &= \frac{1}{i\hbar}\left(\frac{xp^2}{2m} + \frac{1}{2}m\omega^2 x^3 - \frac{p^2 x}{2m} - \frac{1}{2}m\omega^2 x^3\right) \\ &= \frac{1}{2i\hbar m}\left(xp^2 - p^2 x\right) \\ &= \frac{1}{2i\hbar m}\left(xp^2 - pxp + pxp - p^2 x\right) \\ &= \frac{1}{2i\hbar m}\left(2i\hbar p\right)\end{aligned} \tag{3.176b}$$

which means $\frac{dx}{dt} = \frac{p}{m}$. Therefore:

$$\begin{aligned}\frac{da}{dt} &= \frac{i}{\sqrt{2m\hbar\omega}}\left(\frac{dp}{dt} - im\omega\frac{dx}{dt}\right) \\ &= \frac{i}{\sqrt{2m\hbar\omega}}\left(-m\omega^2 x - i\omega p\right) \\ &= -i\omega a\end{aligned} \tag{3.177a}$$

and:

$$\begin{aligned}\frac{da^{\dagger}}{dt} &= \frac{-i}{\sqrt{2m\hbar\omega}}\left(\frac{dp}{dt} + im\omega\frac{dx}{dt}\right) \\ &= \frac{-i}{\sqrt{2m\hbar\omega}}\left(-m\omega^2 x + i\omega p\right) \\ &= i\omega a^{\dagger}\end{aligned} \tag{3.177b}$$

The solution of Eqns. (3.177a) and (3.177b) can always be written as $a(t) = a(0)e^{-i\omega t}$ and $a^{\dagger}(t) = a^{\dagger}(0)e^{i\omega t}$. Note that though a and $a^{\dagger}$ are time dependent, the Hamiltonian $H = \left(a^{\dagger}a + \frac{1}{2}\right)\hbar\omega$ is independent of time.

14. Consider the matrix:

$$A = \begin{bmatrix} 2 & \sqrt{\frac{2}{3}} & 0 \\ \sqrt{\frac{2}{3}} & 2 & \sqrt{\frac{1}{3}} \\ 0 & \sqrt{\frac{1}{3}} & 2 \end{bmatrix}.$$

(a) Is A a Hermitian matrix? If yes, find the Hermitian form $\langle X|A|X\rangle$. (b) Take the unitary matrix:

$$U = \begin{bmatrix} \frac{1}{\sqrt{3}} & \frac{1}{\sqrt{2}} & \frac{1}{\sqrt{6}} \\ \frac{1}{\sqrt{3}} & 0 & -\frac{2}{\sqrt{6}} \\ \frac{1}{\sqrt{3}} & -\frac{1}{\sqrt{2}} & \frac{1}{\sqrt{6}} \end{bmatrix}$$

and then compute $\Lambda = UAU^{\dagger}$ to show that a Hermitian matrix is diagonalized with the use of a unitary matrix. Verify that elements along the diagonal are eigenvalues of A.

SOLUTION

a. As can be seen $A^\dagger = A$; hence it is a Hermitian matrix satisfying: $\langle X|A|X\rangle = \langle X|A|X\rangle^\dagger$. We have $\langle X|A|X\rangle = \sum_{i=1}^{3}\sum_{j=1}^{3} x_i^* a_{ij} x_j = \sum_{i=1}^{3}\sum_{j=1}^{3} x_i a_{ij} x_j^*$, which implies that $x_i^* = x_i$. Therefore:

$$\begin{aligned}\langle X|A|X\rangle &= x_1\left(a_{11}x_1 + a_{12}x_2 + a_{13}x_3\right) + x_2\left(a_{21}x_1 + a_{22}x_2 + a_{23}x_3\right)\\ &\quad + x_3\left(a_{31}x_1 + a_{32}x_2 + a_{33}x_3\right)\\ &= 2\left(x_1^2 + x_2^2 + x_3^2\right) + 2\sqrt{\frac{2}{3}}x_1x_2 + 2\sqrt{\frac{1}{3}}x_2x_3\end{aligned} \tag{3.178a}$$

b. To compute $\Lambda = UAU^\dagger$, let us first calculate:

$$AU^\dagger = \begin{bmatrix} \frac{1}{\sqrt{3}} & \frac{1}{\sqrt{3}} & \frac{1}{\sqrt{3}} \\ \frac{1}{\sqrt{2}} & 0 & -\frac{1}{\sqrt{2}} \\ \frac{1}{\sqrt{6}} & -\frac{2}{\sqrt{6}} & \frac{1}{\sqrt{6}} \end{bmatrix} \tag{3.178b}$$

Next calculate:

$$\Lambda = UAU^\dagger = \begin{bmatrix} \frac{1}{\sqrt{3}} & \frac{1}{\sqrt{2}} & \frac{1}{\sqrt{6}} \\ \frac{1}{\sqrt{3}} & 0 & -\frac{2}{\sqrt{6}} \\ \frac{1}{\sqrt{3}} & -\frac{1}{\sqrt{2}} & \frac{1}{\sqrt{6}} \end{bmatrix} \begin{bmatrix} \sqrt{3} & \frac{2}{\sqrt{3}} & \frac{1}{\sqrt{3}} \\ \frac{3}{\sqrt{2}} & 0 & -\frac{1}{\sqrt{2}} \\ \sqrt{\frac{3}{2}} & -\frac{4}{\sqrt{6}} & \frac{1}{\sqrt{6}} \end{bmatrix} \tag{3.179a}$$

$$= \begin{bmatrix} 3 & 0 & 0 \\ 0 & 2 & 0 \\ 0 & 0 & 1 \end{bmatrix}$$

The eigenvalues of A are given by:

$$|A - \lambda I| = 0 \begin{vmatrix} 2-\lambda & \sqrt{\frac{2}{3}} & 0 \\ \sqrt{\frac{2}{3}} & 2-\lambda & \sqrt{\frac{1}{3}} \\ 0 & \sqrt{\frac{1}{3}} & 2-\lambda \end{vmatrix} = 0 \tag{3.179b}$$

$$\Rightarrow (2-\lambda)\left\{(2-\lambda)^2 - 1\right\} = 0$$

Solution of this gives three values of λ: $\lambda_1 = 3$, $\lambda_2 = 2$ and $\lambda_3 = 1$, which are the diagonal elements of matrix Λ.

3.21 Exercises

1. If $\lambda_1, \lambda_2, \lambda_3, \lambda_4 \ldots\ldots \lambda_n$ are the eigenvalues of matrix A, then show that the determinant $|A| = (-1)^n \lambda_1 \lambda_2 \lambda_3 \ldots\ldots \lambda_n$.
2. Show that if $\lambda_1, \lambda_2, \lambda_3, \lambda_4 \ldots\ldots \lambda_n$ are eigenvalues of the matrix A, then $\lambda_1 \pm \alpha, \lambda_2 \pm \alpha \ldots\ldots\ldots \lambda_n \pm \alpha$ are eigenvalues of the matrix $A \pm \alpha I$.
3. Show that if $|X\rangle$ is an eigenvector of a matrix A corresponding to the eigenvalue λ, then the same vector $|X\rangle$ is also an eigenvector of the matrix A^2 corresponding to the eigenvalue λ^2.
4. Consider the following three vectors:

$$|A_1\rangle = \begin{bmatrix} 25 \\ 64 \\ 144 \end{bmatrix}, |A_2\rangle = \begin{bmatrix} 5 \\ 8 \\ 12 \end{bmatrix} \text{ and } |A_3\rangle = \begin{bmatrix} 1 \\ 1 \\ 1 \end{bmatrix}.$$

   Are these linearly independent?
5. Consider the matrix and vector:

$$A = \begin{bmatrix} 0 & 1 & 1 & 0 \\ 1 & 0 & 0 & 1 \\ 1 & 0 & 0 & 1 \\ 0 & 1 & 1 & 0 \end{bmatrix} \text{ and } |X\rangle = \begin{bmatrix} -2 \\ 1 \\ -1 \\ 0 \end{bmatrix}.$$

   (a) Find the vector $|Y\rangle = A|X\rangle$. (b) Find a vector $|X\rangle$ such that $A|X\rangle = 2|X\rangle$.
6. Consider the matrices:

$$\sigma_1 = \begin{bmatrix} 0 & 1 \\ 1 & 0 \end{bmatrix}, \text{ and } \sigma_3 = \begin{bmatrix} 1 & 0 \\ 0 & -1 \end{bmatrix}.$$

$$\sigma_2 = \begin{bmatrix} 0 & -i \\ i & 0 \end{bmatrix}$$

   (a) Show that they satisfy the relations:

$$\sigma_1^2 = \sigma_2^2 = \sigma_3^2 = 1, \sigma_1\sigma_2 = i\sigma_3, \sigma_2\sigma_3 = i\sigma_1, \sigma_3\sigma_1 = i\sigma_2, \text{ and } \sigma_1\sigma_2 + \sigma_2\sigma_1 = 0.$$

   (b) Find eigenvalues of σ_1, σ_2, and σ_3.
7. Show that matrix

$$A = \begin{bmatrix} \frac{1}{\sqrt{3}} & \frac{1}{\sqrt{2}} & \frac{1}{\sqrt{6}} \\ \frac{1}{\sqrt{3}} & 0 & -\frac{2}{\sqrt{6}} \\ \frac{1}{\sqrt{3}} & -\frac{1}{\sqrt{2}} & \frac{1}{\sqrt{6}} \end{bmatrix}$$

   is an orthogonal matrix. Calculate its determinant.

8. Show that the matrix

$$A = \begin{bmatrix} 2 & \sqrt{\frac{2}{3}} & 0 \\ \sqrt{\frac{2}{3}} & 2 & \sqrt{\frac{1}{3}} \\ 0 & \sqrt{\frac{1}{3}} & 2 \end{bmatrix}$$

is a Hermitian matrix. Calculate its eigenvalues and eigenvectors and show that eigenvectors are orthogonal vectors.

9. $|X\rangle$ and $|Y\rangle$ are eigenkets of Hermitian matrix H. Under what condition will $|X\rangle + |Y\rangle$ also be an eigenket of H.

10. An observable is represented by the matrix:

$$A = \begin{bmatrix} 0 & \frac{1}{\sqrt{2}} & 0 \\ \frac{1}{\sqrt{2}} & 0 & \frac{1}{\sqrt{2}} \\ 0 & \frac{1}{\sqrt{2}} & 0 \end{bmatrix}.$$

Calculate the eigenvalues and the normalized eigenvectors of this observable. Is there degeneracy?

11. Show that the matrices

$$A = \begin{bmatrix} 1 & 0 & 0 \\ 0 & -1 & 0 \\ 0 & 0 & -1 \end{bmatrix} \text{ and } B = \begin{bmatrix} 2 & 0 & 0 \\ 0 & 0 & -2i \\ 0 & 2i & 0 \end{bmatrix}.$$

commute. Find eigenvalues of the matrices. Is there degeneracy?

12. Show that the vectors

$$|X\rangle = \begin{bmatrix} 1 \\ 0 \\ i \end{bmatrix} \text{ and } |Y\rangle = \frac{1}{2}\begin{bmatrix} 1+i \\ 0 \\ 1-i \end{bmatrix}$$

satisfy the Schwarz inequality.

13. Show that matrix $U = \frac{1}{\sqrt{2}}\begin{bmatrix} 1 & 1 \\ i & -i \end{bmatrix}$ is a unitary matrix. Calculate $|Y\rangle = U|X\rangle$ where $|X\rangle = \begin{bmatrix} 2 \\ 1 \end{bmatrix}$ and then show that $\langle Y|Y\rangle = \langle X|X\rangle$.

14. Show that the Hamiltonian of the 1D harmonic oscillator can be given by:

$$H = \left(aa^{\dagger} - \frac{1}{2}\right)\hbar\omega, \text{ and } H = \frac{1}{2}\left(a^{\dagger}a + aa^{\dagger}\right)\hbar\omega.$$

4

Transformations, Conservation Laws, and Symmetries

There would be no laws of physics without invariance principles. We rely on experimental results remaining the same from day to day and place to place. An invariance principle is intimately related to a conservation law and it reflects a basic symmetry. Analysis of various transformations makes the form of the laws involved invariant. For example, (i) the invariance of physical systems with respect to spatial translation (in other word that the laws of physics do not change with locations in space) gives the law of conservation of linear momentum. (ii) Invariance with respect to rotation gives the law of conservation of angular momentum. (iii) Invariance with respect to time translation gives the well-known law of conservation of energy. In quantum field theory, the gauge invariance of the electric potential and vector potential yields conservation of electric charge.

4.1 Translation in Space

In the case of translation in space, it is important to notice whether the *system* is being translated by vector $\boldsymbol{a}$ or the coordinate axes are being translated by $\boldsymbol{a}$, which would result in the opposite change in the wave function. Results for translating the wave function by $+\boldsymbol{a}$ would be the same as for translating the coordinate axes along with the apparatus and any external fields by $-\boldsymbol{a}$ relative to the wave function. These two equivalent operations are analogous to the time development in the Schrödinger picture and in the Heisenberg picture. In the Schrödinger picture, bras and kets change in time while operators do not. In the Heisenberg picture, operators develop with time while the bras and kets do not change. Therefore, there are two possible ways to deal with space translations: transform the bras and kets *or* transform the operators. We here consider the situation where bras and kets are transformed while operators are left unchanged. Let us consider a 1D case, where the translational is made by an operator $T(a)$ such that:

$$T(a)\psi(x) = \psi(x-a) \tag{4.1}$$

Thus, if $\psi(x)$ is a wave function centered at the origin, $\psi(x-a)$ is a wave function centered at the point a. To get the form of the operator $T(a)$, we make the Taylor series expansion of $\psi(x-a)$:

$$\begin{aligned}\psi(x-a) &= \psi(x) - a\frac{d\psi(x)}{dx} + \frac{a^2}{2!}\frac{d^2\psi(x)}{dx^2} - \dots\dots\dots \\ &= \left[1 - a\frac{d}{dx} + \frac{a^2}{2!}\frac{d^2}{dx^2} - \dots\dots\right]\psi(x) \\ &= e^{-a\frac{d}{dx}}\psi(x)\end{aligned} \tag{4.2}$$

Since $p = -i\hbar \frac{d}{dx}$, the translation operator is:

$$T(a) = e^{-iap/\hbar} \tag{4.3}$$

For two successive translations in space: $T(a')T(a)\psi(x) = \psi(x - a - a')$, which would be the same if we take $T(a)T(a')\psi(x)$. Therefore, $T(a')T(a) = T(a)T(a')$, meaning that the space translation operators are commutative. In the case of an infinitesimal translation by ε:

$$T(\varepsilon) = e^{-i\varepsilon p/\hbar} \approx 1 - \frac{i\varepsilon p}{\hbar} \tag{4.4}$$

The linear momentum p is said to be the generator of the translation. Once we have established that the momentum operator is the generator of spatial translations, its generalization to three dimensions is trivial:

$$T(\boldsymbol{a}) = e^{-i\boldsymbol{a}.\boldsymbol{p}/\hbar} \tag{4.5}$$

4.2 Translation in Time

The time translation operator $T(\tau)$ is an operator that it when acts on $\psi(t)$ gives the new wave function $\psi(t + \tau)$:

$$T(\tau)\psi(t) = \psi(t + \tau) \tag{4.6}$$

The Taylor series expansion of $\psi(t + \tau)$ gives:

$$\begin{aligned} T(\tau)\psi(t) = \psi(t+\tau) &= \psi(t) + \tau\frac{\partial}{\partial t}\psi(t) + \frac{\tau^2}{2!}\frac{\partial^2}{\partial t^2}\psi(t) + \ldots\ldots. \\ &= e^{\tau\frac{\partial}{\partial t}}\psi(t) \\ &= e^{\frac{-iH\tau}{\hbar}}\psi(t) \end{aligned} \tag{4.7}$$

Thus $T(\tau) = e^{-iH\tau/\hbar}$, where we have made use of:

$$i\hbar\frac{\partial}{\partial t}\psi(\mathbf{r},t) = H\psi(\mathbf{r},t) \tag{4.8}$$

For infinitesimal time translation, in a manner like that for space translation, we have:

$$T(\xi) = e^{\frac{-iH\xi}{\hbar}} \simeq 1 - \frac{-iH\xi}{\hbar} \tag{4.9}$$

4.3 Rotation in Space

In contrast to translations in space and time, rotations do not commute even for a classical system. Rotating a book through $\pi/2$, first about the z-axis and then about the x-axis, leaves it in a different orientation from that obtained by rotating from the same starting position, first $\pi/2$ about the x-axis and then $\pi/2$ about the z-axis. Though the commutator is of second order, still operations for smallest rotations do not commute.

The R-operators that represent rotations reflect this noncommutativity structure. This is seen by considering ordinary classical rotations of a real vector in three dimensional space. Consider that a vector $|V\rangle$ with components V_1, V_2 and V_3 is rotated by an angle θ about an arbitrary direction so that components become V_1', V_2' and V_3'; $|V'\rangle = R(\theta)|V\rangle$. The old and new components are related via a 3×3 orthogonal matrix: $R(\theta)$:

$$\begin{bmatrix} V_1' \\ V_2' \\ V_3' \end{bmatrix} = \begin{bmatrix} R_{11} & R_{12} & R_{13} \\ R_{21} & R_{22} & R_{23} \\ R_{31} & R_{32} & R_{33} \end{bmatrix} \begin{bmatrix} V_1 \\ V_2 \\ V_3 \end{bmatrix}, \text{ with } R\tilde{R} = I \tag{4.10}$$

The matrix for rotation of a vector by an angle θ about the x-axis is:

$$R_x(\theta) = \begin{pmatrix} 1 & 0 & 0 \\ 0 & \cos\theta & -\sin\theta \\ 0 & \sin\theta & \cos\theta \end{pmatrix} \tag{4.11a}$$

Similarly, the matrices representing the rotation of a vector by an angle θ about y and z axes, respectively are:

$$R_y(\theta) = \begin{pmatrix} \cos\theta & 0 & \sin\theta \\ 0 & 1 & 0 \\ -\sin\theta & 0 & \cos\theta \end{pmatrix} \tag{4.11b}$$

$$R_z(\theta) = \begin{pmatrix} \cos\theta & -\sin\theta & 0 \\ \sin\theta & \cos\theta & 0 \\ 0 & 0 & 1 \end{pmatrix} \tag{4.11c}$$

Let us consider the case of rotation by an infinitesimal angle ε. On ignoring the terms having ε^3 and higher powers, we obtain:

$$R_x(\varepsilon) = I + \varepsilon \begin{pmatrix} 0 & 0 & 0 \\ 0 & -\frac{\varepsilon}{2} & -1 \\ 0 & 1 & -\frac{\varepsilon}{2} \end{pmatrix} \tag{4.12a}$$

$$R_y(\varepsilon) = I + \varepsilon \begin{pmatrix} -\frac{\varepsilon}{2} & 0 & 1 \\ 0 & 0 & 0 \\ -1 & 0 & -\frac{\varepsilon}{2} \end{pmatrix} \tag{4.12b}$$

and:

$$R_z(\varepsilon) = I + \varepsilon \begin{pmatrix} -\frac{\varepsilon}{2} & -1 & 0 \\ 1 & -\frac{\varepsilon}{2} & 0 \\ 0 & 0 & 0 \end{pmatrix} \tag{4.12c}$$

The elementary matrix multiplication yields:

$$R_x(\varepsilon)R_y(\varepsilon) = I + \varepsilon \begin{bmatrix} -\frac{\varepsilon}{2} & 0 & 1 \\ \varepsilon & -\frac{\varepsilon}{2} & -1 \\ -1 & 1 & -\varepsilon \end{bmatrix}, \text{ and } R_y(\varepsilon)R_x(\varepsilon) = I + \varepsilon \begin{bmatrix} -\frac{\varepsilon}{2} & \varepsilon & 1 \\ 0 & -\frac{\varepsilon}{2} & -1 \\ -1 & 1 & -\varepsilon \end{bmatrix} \tag{4.13}$$

which exhibits that $R_x(\varepsilon)R_y(\varepsilon) \neq R_y(\varepsilon)R_x(\varepsilon)$, meaning that the infinitesimal rotations about different axes do not commute and we find that:

$$\left[R_x(\varepsilon), R_y(\varepsilon)\right] = \varepsilon^2 \begin{pmatrix} 0 & -1 & 0 \\ 1 & 0 & 0 \\ 0 & 0 & 0 \end{pmatrix} \tag{4.14a}$$

Ignoring the terms having ε^3 and ε^4, we can write:

$$R_z(\varepsilon^2) = I + \varepsilon^2 \begin{pmatrix} 0 & -1 & 0 \\ 1 & 0 & 0 \\ 0 & 0 & 0 \end{pmatrix} \tag{4.14b}$$

Multiplying $R_y(\varepsilon)R_x(\varepsilon)$ with $R_z(\varepsilon^2)$ from the left and then leaving the terms having ε^3 and ε^4, we get:

$$R_x(\varepsilon)R_y(\varepsilon) = R_z(\varepsilon^2)R_y(\varepsilon)R_x(\varepsilon) \tag{4.15a}$$

which suggest that the application of rotation operators on quantum mechanical kets must follow the pattern:

$$\left[R_x(\varepsilon)R_y(\varepsilon) - R_z(\varepsilon^2)R_y(\varepsilon)R_x(\varepsilon)\right]|\psi\rangle = 0 \tag{4.15b}$$

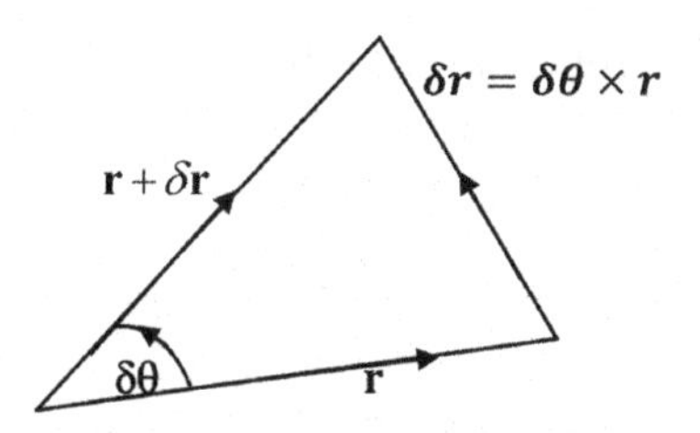

FIGURE 4.1
Schematic diagram for rotation of vector.

This in fact leads to very important results in quantum mechanics, as would be shown in forthcoming discussions.

Let us consider the rotation by an infinitesimal angle $\delta\boldsymbol{\theta}$ about an axis through the origin. After rotation, the system that is initially at $\boldsymbol{r}$ shifts to $\boldsymbol{r}+\delta\boldsymbol{r}$, where $\delta\boldsymbol{r} = \delta\boldsymbol{\theta}\times\boldsymbol{r}$.

We transform a wave function under this small rotation. Like the linear translation case $R(\delta\theta)\psi(\mathbf{r}) = \psi(\mathbf{r}-\delta\mathbf{r})$, where $R(\delta\theta)$ is an operator causing the rotation. The minus sign can again be understood from the discussions on linear translations in Section 4.1. The procedure is similar to that has been used to obtain Eqn. (4.5) gives us:

$$R(\delta\theta)\psi(\mathbf{r}) = \psi(\mathbf{r}-\delta\mathbf{r}) = e^{\frac{-i\delta\mathbf{r}.\mathbf{p}}{\hbar}}\psi(\mathbf{r}) \tag{4.16a}$$

Using $\delta\boldsymbol{r} = \delta\boldsymbol{\theta}\times\boldsymbol{r}$ and $\mathbf{L} = \mathbf{r}\times\mathbf{p}$, we get:

$$\begin{aligned} R(\delta\theta)\psi(\mathbf{r}) &= e^{\frac{-i\delta\mathbf{r}.\mathbf{p}}{\hbar}}\psi(\mathbf{r}) \\ &= e^{-i\delta\theta\times\mathbf{r}.\mathbf{p}/\hbar}\psi(\mathbf{r}) \\ &= \left(e^{-i\delta\theta.\mathbf{L}/\hbar}\right)\psi(\mathbf{r}) \end{aligned} \tag{4.16b}$$

which yields:

$$R(\delta\theta) = e^{\frac{-i\delta\theta.L}{\hbar}} \tag{4.17}$$

Note that a finite rotation by an angle $\boldsymbol{\theta}$ can be represented as multiplication together of a large number of $\delta\boldsymbol{\theta}$ operators, which is equivalent to replacing $\delta\boldsymbol{\theta}$ by $\boldsymbol{\theta}$ in the exponential.

For the case of infinitesimal rotation, Eqn. (4.16b) reduces to:

$$R(\delta\theta)\psi(\mathbf{r}) \simeq \left(1-\frac{i}{\hbar}\delta\theta.\mathbf{L}\right)\psi(\mathbf{r}) \tag{4.18}$$

Equation (4.17) establishes that the orbital angular momentum operator **L** is the generator of spatial rotations. The appropriately transformed wave function is generated by the action of $R(\delta\theta)$ on the original wave function on rotating our apparatus and the wave function with it. To clarify it further, we consider the case of rotating the system, and therefore the wave function, through an infinitesimal angle $\delta\theta_z$ about the z-axis. Let us denote the rotated wave function by $\psi_{rot}(x,y)$. We then have:

$$\begin{aligned}\psi_{rot}(x,y) &= \left[1-\frac{i}{\hbar}(\delta\theta_z)L_z\right]\psi(x,y)\\ &= \left[1-(\delta\theta_z)\left(x\frac{d}{dy}-y\frac{d}{dx}\right)\right]\psi(x,y)\\ &\simeq \psi\left(x+(\delta\theta_z)y, y-(\delta\theta_z)x\right)\end{aligned} \tag{4.19}$$

Thus, the value of the new wave function at (x,y) is the value of the old wave function at the point $\{x+(\delta\theta_z)y, y-(\delta\theta_z)x\}$.

4.4 Quantum Generalization of the Rotation Operator

It is well known that in quantum mechanics orbital angular momentum is not the total angular momentum. It is found experimentally that particles like the electron exhibit an internal angular momentum, called spin angular momentum. The total angular momentum, called $\mathbf{J}$, is the sum of the orbital angular momentum $\mathbf{L}$ and spin angular momentum $\mathbf{S}$; $\mathbf{J}=\mathbf{L}+\mathbf{S}$. The operator $\mathbf{J}$ is defined as the generator of rotations on any wave function, including possible spin components. Therefore, Eqn. (4.16b) is generalized to:

$$R(\delta\theta)\psi(\mathbf{r}) = \left(e^{-i\delta\theta.\mathbf{J}/\hbar}\right)\psi(\mathbf{r}) \tag{4.20}$$

which is similar to the equation for $\mathbf{L}$, whose components are written as differentials. Up to this point, we considered $\psi(\mathbf{r})$ a complex valued function of position. But the wave function at a point can have several components in some vector space. The rotation operator will operate in that space, and it is a differential operator too with respect to position. Then, the state vector $|\psi\rangle$ is a vector at each point, and the rotation of the system rotates this vector as well as moving it to a different value of $\mathbf{r}$. The $|\psi\rangle$, in general, has n-components at each point in space; $R(\boldsymbol{\delta\theta})$ is then a $n\times n$ matrix in the component space, and Eqn. (4.20) is the definition of $\mathbf{J}$ in that component space. This definition is used to study the properties of $\mathbf{J}$. For an infinitesimal angle ε, we can write: $R_x(\varepsilon)\simeq 1-\frac{i}{\hbar}\varepsilon J_x$; $R_y(\varepsilon)\simeq 1-\frac{i}{\hbar}\varepsilon J_y$ and $R_z(\varepsilon^2)\simeq 1-\frac{i}{\hbar}\varepsilon^2 J_x$, which on substituting into Eqn. (4.15b) yield:

$$\left[\left(1-\frac{i}{\hbar}\varepsilon J_x\right)\left(1-\frac{i}{\hbar}\varepsilon J_y\right)-\left(1-\frac{i}{\hbar}\varepsilon^2 J_z\right)\left(1-\frac{i}{\hbar}\varepsilon J_y\right)\left(1-\frac{i}{\hbar}\varepsilon J_x\right)\right]|\psi\rangle = 0 \tag{4.21}$$

All the zeroth and first-order terms in ε cancel, and the second-order term gives:

$$\left[J_x, J_y\right] = i\hbar J_z \tag{4.22a}$$

which is generalized to:

$$\left[J_i, J_j\right] = i\hbar\varepsilon_{ijk} J_k \tag{4.22b}$$

where the symbol ε_{ijk} is equal to +1, if ijk take values in cyclic order (123, 231, 312) and it is equal to –1 when ijk take values that are not in cyclic order. It equals to zero if any two of ijk are assigned the same value, for example, 112. As is seen, the use of Eqns. (4.15b) and (4.18) would give us:

$$\left[L_i, L_j\right] = i\hbar\varepsilon_{ijk} L_k \tag{4.22c}$$

for orbital angular momentum. Equations (4.22b) and (4.22c) constitute important formulas in quantum mechanics.

4.5 Invariance and Conservation Laws

Any of the three transformations (space translation, time translation, and rotation in space) changes a state vector $|\phi\rangle$, which is imagined to be tied to a system, to a new state vector $|\phi\rangle'$. Thus, the general form of a transformation is:

$$|\phi\rangle' = S|\phi\rangle \tag{4.23}$$

where S is a transforming operator that is equal to $e^{\frac{-iap}{\hbar}}$ for space translation, $e^{\frac{-i\tau H}{\hbar}}$ for time translation, and $e^{\frac{-i\delta\theta.\mathbf{J}}{\hbar}}$ for rotation. We thus are considering two systems of coordinates, one before transformation (without prime) and the other after transformation (with prime). In the first system of coordinates, suppose the application of operator F on $|\phi\rangle$ changes it to $|\chi\rangle$:

$$|\chi\rangle = F|\phi\rangle \tag{4.24a}$$

The corresponding procedure in the second system of coordinates, let us say, is represented by primed symbols to give:

$$|\chi\rangle' = F'|\phi\rangle' \tag{4.24b}$$

which, with the use of Eqn. (4.23), implies:

$$S|\chi\rangle = F'S|\phi\rangle \tag{4.25a}$$

The use of Eqn. (4.24a) then yields:

$$\begin{aligned} SF|\phi\rangle &= F'S|\phi\rangle \\ \Rightarrow SF &= F'S \end{aligned} \tag{4.25b}$$

or:

$$F' = SFS^{-1} \text{ and } F = S^{-1}F'S \tag{4.26}$$

Transformations described by Eqn. (4.26) are applicable to all operators including the Hamiltonian operator H. We therefore write:

$$H' = SHS^{-1} \tag{4.27}$$

where H is Hamiltonian in the system of coordinates before transformation, while H' is the Hamiltonian in the system of coordinates after transformation. The invariance of the Hamiltonian under transformations requires that $H' = H$, meaning that a Hamiltonian has the same form whether expressed in terms of original or transformed coordinates of the system. In the following we consider translation and rotation operations.

4.5.1 Infinitesimal Space Translation

For space translation $S = e^{-i\varepsilon p/\hbar}$, which in the case of an infinitesimal translation is given by: $S = 1 - \frac{-i\varepsilon \mathrm{p}}{\hbar}$. We apply Eqn. (4.26) to cases of 3-operators: (a) $F = x$, (b) $F = p$, and (c) $F = H$. Here p is linear momentum component along x-axis.

a. $x' = T^{-1}(\varepsilon)xT(\varepsilon) = e^{-i\varepsilon p/\hbar}xe^{i\varepsilon p/\hbar}$

$$\simeq \left(1 + \frac{i\varepsilon p}{\hbar}\right)x\left(1 - \frac{i\varepsilon p}{\hbar}\right) \simeq x + \frac{i\varepsilon}{\hbar}(px - xp) \simeq x + \varepsilon \tag{4.28a}$$

The new position coordinate after translation is $x + \varepsilon$. Thus, the position vector is not conserved during linear motion.

b.

$$\begin{aligned} p' &= T^{-1}(\varepsilon)pT(\varepsilon) \\ &= \left(e^{-i\frac{\varepsilon p}{\hbar}}\right)p\left(e^{i\frac{\varepsilon p}{\hbar}}\right) \\ &\simeq \left(1 - i\frac{\varepsilon p}{\hbar}\right)p\left(1 + i\frac{\varepsilon p}{\hbar}\right) \\ &\simeq p \end{aligned} \tag{4.28b}$$

where we ignored terms having ε^2 or higher powers of ε. This means that linear momentum remains unchanged during translation in space. In other words, momentum is conserved during the linear motion of the system. Equation (4.28b) can easily be generalized for a system of n-particles by saying that $p = p_1 + p_2 + p_3 + p_4 + \cdots\cdots\cdots + p_n$.

c.

$$\begin{aligned} H' &= T^{-1}(\varepsilon)HT(\varepsilon) \\ &= \left(e^{-i\frac{\varepsilon.\mathbf{p}}{\hbar}}\right)H\left(e^{i\frac{\varepsilon.\mathbf{p}}{\hbar}}\right) \\ &\simeq \left(1 - i\frac{\varepsilon.\mathbf{p}}{\hbar}\right)H\left(1 + i\frac{\varepsilon.\mathbf{p}}{\hbar}\right) \\ &\simeq H - \frac{i}{\hbar}\varepsilon.[\mathbf{p}, H] \end{aligned} \tag{4.28c}$$

where the term having ε^2 has been ignored. The requirement $H' = H$ demands that $[\mathbf{p}, H] = 0$. Use of the Heisenberg equation $[\mathbf{p}, H] = i\hbar \frac{d\mathbf{p}}{dt}$ implies $\frac{d\mathbf{p}}{dt} = 0$ or $\mathbf{p}$ is the constant of motion. Thus, the invariance of the Hamiltonian –shifting of the system and equipment (as a whole) from one place to another in space should not affect the measurements – demands that linear momentum be conserved under the linear motion of system. Note that Eqns. (4.28b) and (4.28c) convey the same message.

4.5.2 Infinitesimal Time Translation

Invariance of the measurements with time requires that $H' = H$. From Eqn. (4.27), we get:

$$\begin{aligned} H' &= \left(1 - \frac{i\tau H}{\hbar}\right) H \left(1 + \frac{i\tau H}{\hbar}\right) \\ &\simeq H \end{aligned} \tag{4.29}$$

on ignoring terms having τ^2. This suggests that the total energy of the system is conserved under time translation or there is no change in total energy with time.

4.5.3 Infinitesimal Rotation

For rotational motion $S = e^{-i\delta\theta.\mathbf{J}/\hbar}$. We here again consider two cases:

a.

$$\begin{aligned} \mathbf{J}' &= \left(e^{-i\delta\theta.\mathbf{J}/\hbar}\right) \mathbf{J} \left(e^{i\delta\theta.\mathbf{J}/\hbar}\right) \\ &\simeq \left(1 - \frac{i}{\hbar}\delta\theta.\mathbf{J}\right) \mathbf{J} \left(1 + \frac{i}{\hbar}\delta\theta.\mathbf{J}\right) \\ &\simeq \mathbf{J} \end{aligned} \tag{4.30a}$$

where we ignored terms having $(\delta\theta)^2$. This suggests that the total angular momentum of a system does not change during the rotation motion of the system.

b. We next take the transformation:

$$H' = \left(e \frac{-i\delta\theta.\mathbf{J}}{\hbar}\right) H \left(e \frac{i\delta\theta.\mathbf{J}}{\hbar}\right), \tag{4.30b}$$

which for the infinitesimal value of $\delta\theta$ reduces to:

$$\begin{aligned} H' &\simeq \left(1 - \frac{i\delta\theta.\mathbf{J}}{\hbar}\right) H \left(1 + \frac{i\delta\theta.\mathbf{J}}{\hbar}\right) \\ &\simeq H - \frac{i}{\hbar}\delta\theta.[\mathbf{J}, H] \end{aligned} \tag{4.30c}$$

The invariance of the Hamiltonian under rotational motion (rotating system and measuring equipment as a whole) demands that $[\mathbf{J}, H]$ should be zero or $\mathbf{J}$ and H should commute. Therefore, angular momentum $\mathbf{J}$ is conserved or it is a constant of motion under rotational motion.

4.5.4 Conservation of Charge

For the invariance of physical systems under a translation in the electrostatic potential, we require that the total electric charge on a system be a conserved quantity. The physics of a system only depends on potential *differences*. Quantum mechanically, we can define a charge operator $\hat{Q}$ which on operating upon a wave function ket $|\psi_q\rangle$ returns to an eigenvalue of q. The $|\psi_q\rangle$ describes a system of total charge q.

$$\hat{Q}|\psi_q\rangle = q|\psi_q\rangle \tag{4.31a}$$

If q is conserved, then $\hat{Q}$ and H must commute: $\left[\hat{Q}, H\right] = 0$. It is assured by the invariance under a global phase (or gauge) transformation like:

$$|\psi_q\rangle' = e^{i\varepsilon\hat{Q}}|\psi_q\rangle \tag{4.31b}$$

where ε is an arbitrary real parameter. The above continuous transformations led to additive conservation laws: The sum of all charges is conserved.

4.6 Parity and Space Inversion

Parity in physics refers to the relationship between an object or process and the mirror image of it. For example, any right-handed object produces a mirror image that is identical to it in every way except that the mirror image is left handed. A moving particle that spins in a clockwise manner will have a mirror image particle that is identical to it in every manner except that it spins anticlockwise. The law of conservation of parity states that for every real object or process there exists a mirror image that obeys the same physical laws. This concept played a very important role in quantum physics. Because of this law, it was inferred that all elementary particles and their interactions possessed mirror image counterparts, and therefore they exist. However, it has also been said and argued that parity may not conserve in weak interactions. It is conserved in the strong nuclear interactions and in the electromagnetic interactions. A parity operator p is an operator that describes the behavior of the wave function of any system of particles when the spatial coordinates (x, y, z) of the wave function are reflected through the origin to $(-x, -y, -z)$. The simultaneous flip in the sign of all spatial coordinates is therefore referred as a parity transformation or parity inversion:

$$P: \begin{pmatrix} x \\ y \\ z \end{pmatrix} \Rightarrow \begin{pmatrix} -x \\ -y \\ -z \end{pmatrix} \tag{4.32}$$

A reflection, mirror image, can be obtained by an inversion followed by rotation. Conversely, an inversion is same as three successive reflections with respect to three axes. Unlike translation and rotation, which can proceed continuously in terms of infinitesimal displacements, it is not possible to transform a system into a mirror image without distorting it into a final configuration that is very different physically from the original system. It is therefore not at all clear how quantities like electric charge, which is unrelated to coordinate displacements, should be treated in reflection if symmetry is to be preserved.

The equation $\boldsymbol{F} = m\boldsymbol{a}$ that equates two vectors for constant mass is invariant under parity in classical mechanics. The law of gravity too involves only vectors and hence it is invariant under parity. Angular momentum $\boldsymbol{L} = \boldsymbol{r} \times \boldsymbol{P}$, which is an axial vector, is also invariant under parity because of $P(\boldsymbol{L}) = P(\boldsymbol{r} \times \boldsymbol{P}) = (-\boldsymbol{r}) \times (-\boldsymbol{P}) = \boldsymbol{L}$. Classical variables that do not change under parity are the time, the energy of the particle, power, total angular momentum of the particle, the electric potential, the magnetic induction, the magnetization, energy density of the electromagnetic field, the Maxwell stress tensor, all masses, charges, coupling constants, and other physical constants. Classical variables which are negated on spatial inversion include the position, velocity, acceleration and linear momentum of a particle, the force on the particles, the electric field, the electric displacement vector, electric polarization, electromagnetic vector potential, and Poynting vector.

4.6.1 Parity Operator

If we ignore the spin motion, then the parity operator P is an inversion operator. The operation of P on the wave function of the system gives:

$$P\psi(\mathbf{r}) = \psi(-\mathbf{r}) \tag{4.33a}$$

In the case of a system of many particles, wave function depends on coordinates of n particles, and we write:

$$P\psi(\mathbf{r}_1, \mathbf{r}_2, \mathbf{r}_3, \ldots\ldots, \mathbf{r}_n) = \psi(-\mathbf{r}_1, -\mathbf{r}_2, -\mathbf{r}_3, \ldots\ldots, -\mathbf{r}_n) \tag{4.33b}$$

Also notice that:

$$P^2\psi(\mathbf{r}) = P\psi(-\mathbf{r}) = \psi(\mathbf{r}) \tag{4.34}$$

Let us say that P satisfies the eigenvalue equation:

$$P\psi(\mathbf{r}) = \lambda\psi(\mathbf{r}) \tag{4.35a}$$

where λ and $\psi(\mathbf{r})$ are the eigenvalue and eigenfunction of the parity operator. The parity operator and the Hamiltonian can have simultaneous eigenfunctions, if they commute. On operating with P, Eqn. (4.35a) goes to:

$$\begin{aligned} &P^2\psi(\mathbf{r}) = \lambda^2\psi(\mathbf{r}) = \psi(\mathbf{r}) \\ &\Rightarrow (\lambda^2 - 1)\psi(\mathbf{r}) = 0 \end{aligned} \tag{4.35b}$$

The nontrivial solution demands that $\lambda = \pm 1$. It is to be noted here that one can also have $P^2\psi(\mathbf{r}) = e^{i\varphi}\psi(\mathbf{r})$, since an overall phase is unobservable because of $|e^{i\varphi}\psi(\mathbf{r})|^2 = |\psi(\mathbf{r})|^2$. For the eigenstate of P that corresponds to $\lambda = 1$:

$$P\psi_e(\mathbf{r}) = \lambda\psi_e(\mathbf{r}) \equiv \psi_e(-\mathbf{r}) = \psi_e(\mathbf{r}) \tag{4.36}$$

Similarly, for the eigenstate of P that corresponds to $\lambda = -1$:

$$P\psi_o(\mathbf{r}) = \lambda\psi_o(\mathbf{r}) \equiv \psi_o(-\mathbf{r}) = -\psi_o(\mathbf{r}) \tag{4.37}$$

We thus find that all wave functions, which remain unchanged on replacing $\mathbf{r}$ by $-\mathbf{r}$, correspond to $\lambda = 1$, while all wave functions that change sign on replacing $\mathbf{r}$ by $-\mathbf{r}$ correspond to $\lambda = -1$.

Some of the important properties of parity operators are: (i) It is a Hermitian operator. (ii) It is a unitary operator, and its eigenstates are orthogonal. To prove that it is a Hermitian operator, we require that elements of the matrix representing the parity operator should follow:

$$\begin{aligned} &P_{ij} = P_{ji}^* \\ &\Rightarrow \langle\psi_i|P|\psi_j\rangle = \langle\psi_j|P|\psi_i\rangle^* \end{aligned} \tag{4.38}$$

for all eigenvectors $|\psi_i\rangle$ and $|\psi_j\rangle$. We take:

$$P_{ji}^* = \langle\psi_j|P|\psi_i\rangle^* = \left(\int\psi_j^*(\mathbf{r})P\psi_i(\mathbf{r})d^3r\right)^* = \int\psi_i^*(-\mathbf{r})\psi_j(\mathbf{r})d^3r \tag{4.39}$$

Letting $\mathbf{r}' = -\mathbf{r}$ and then noting that $\int_{-\infty}^{\infty} f(-x)dx = -\int_{\infty}^{-\infty} f(x')dx' = \int_{-\infty}^{\infty} f(x')dx'$, we get:

$$\begin{aligned} P_{ji}^* &= \int\psi_i^*(-\mathbf{r})\psi_j(\mathbf{r})d^3r \\ &= -\int_{\infty}^{-\infty}\int_{\infty}^{\infty}\int_{\infty}^{\infty}\psi_i^*(x',y',z')\psi_j(-x',-y',-z')dx'dy'dz' \\ &= \int_{-\infty}^{\infty}\int_{-\infty}^{\infty}\int_{-\infty}^{\infty}\psi_i^*(x',y',z')P\psi_j(x',y',z')dx'dy'dz' = P_{ij} \end{aligned} \tag{4.40}$$

To prove that P is a unitary operator and its eigenstates $|\psi_e\rangle$ and $|\psi_o\rangle$ are orthogonal, take:

$$\begin{aligned} &P|\psi_e\rangle = |\psi_e\rangle \\ &\Rightarrow \langle\psi_e|P^\dagger = \langle\psi_e| \end{aligned} \tag{4.41a}$$

and,

$$P|\psi_o\rangle = -|\psi_o\rangle \tag{4.41b}$$

Therefore:

$$\begin{aligned} &\langle\psi_e|P^\dagger P|\psi_e\rangle = \langle\psi_e|\psi_e\rangle \\ &\Rightarrow P^\dagger P = I \end{aligned} \tag{4.42a}$$

and:

$$\langle \psi_e | P^{\dagger} P | \psi_o \rangle = -\langle \psi_e | \psi_o \rangle \tag{4.42b}$$

$$\Rightarrow \langle \psi_e | \psi_o \rangle = -\langle \psi_e | \psi_o \rangle \tag{4.42c}$$

which implies that $\langle \psi_e | \psi_o \rangle = 0$, proving that wave functions are orthogonal.

As is noticed from Eqns. (4.5), (4.7), and (4.20), *the operators which cause linear translation, time translation, rotation, and reflection (parity operator) are unitary operators, because each of these satisfies the condition $AA^{\dagger} = 1$. Also, these are linear operators.* We next discuss the operator that is not a unitary and a linear operator.

4.7 Time-Reversal Operator

Let us first consider time reversal in classical mechanics. For example, motion of a charged particle with charge q and mass m in an electric **E** is described by:

$$m\frac{d^2\mathbf{r}}{dt^2} = q\mathbf{E}(\mathbf{r}) \tag{4.43}$$

If **r**(t) is a solution of Eqn. (4.43), then so is **r**(−t), as can be seen from the fact that the Eqn. (4.43) is second order in time, so that the two changes of sign coming from $t \to -t$ cancel out. However, this does not hold for magnetic forces, where the equations of motion include first order time derivatives:

$$m\frac{d^2\mathbf{r}}{dt^2} = \frac{q}{c}\frac{d\mathbf{r}}{dt} \times \mathbf{B}(\mathbf{r}) \tag{4.44}$$

To restore the invariance under time reversal, we have to replace **B** by −**B** along with $t \to -t$ in Eqn. (4.44), which means that the time reversed motion is physically allowed in the reversed magnetic field. We find that when a system interacting with external fields does not possess time reversal invariance by itself, inclusion of the charges and currents producing external fields in our definition of the system may restore the time reversal invariance. If the motions of all the charges in a closed system are reversed, then charge densities and currents transform according to $\rho \to \rho$ and $\mathbf{J} \to -\mathbf{J}$, while the electric and magnetic fields produced by these charges and currents transform according to $\mathbf{E} \to \mathbf{E}$ and $\mathbf{B} \to -\mathbf{B}$. Thus, electromagnetic effects are invariant under time reversal when these basic transformations along with Eqns. (4.43) and (4.44) are considered. Though, we have shown this only at the classical level, it is true at the quantum level as well.

A quantum mechanical equation that is analogous to the equations of motion of the charged particle under an electromagnetic field is the Schrödinger equation:

$$i\hbar\frac{\partial \psi_{\alpha}(\mathbf{r},t)}{\partial t} = \left[\frac{-\hbar^2}{2m}\nabla^2 + V(\mathbf{r})\right]\psi_{\alpha}(\mathbf{r},t) \tag{4.45}$$

Equation (4.45) describes the motion of a system in the quantum state $|\alpha\rangle$, at position $\mathbf{r}$ and time t. Here, α compositely represents all quantum numbers. As is seen, $\psi_\alpha(\mathbf{r},t)$ and $\psi^*_\alpha(\mathbf{r},-t)$ are solutions but $\psi_\alpha(\mathbf{r},-t)$ is not the solution of Eqn. (4.45). It is apparent that the complex conjugate of $\psi_\alpha(\mathbf{r},t)$ is necessary, because without it, the left-hand side of Eqn. (4.45) would change sign under $t \to -t$. In the Schrödinger representation, one can write $\psi_n(\mathbf{r},t) = u_n(\mathbf{r})e^{-iE_nt/\hbar}$ and $\psi^*_n(\mathbf{r},-t) = u^*_n(\mathbf{r})e^{-iE_nt/\hbar}$, when system is in the state of energy E_n. *We thus conjecture that the time reversal has something to do with complex conjugation.* As was shown in Sections 4.5 and 4.6, under symmetry operations such as translation, rotation, and parity, the inner product of two vectors is preserved, which means that if a symmetry operation changes $|\alpha\rangle$ to $|\alpha\rangle'$ and $|\beta\rangle$ to $|\beta\rangle'$, then:

$$\langle\beta|\alpha\rangle' = \langle\beta|\alpha\rangle \tag{4.46}$$

where, we have abbreviated $\langle\psi_\beta|\psi_\alpha\rangle$ by $\langle\beta|\alpha\rangle$. Eqn. (4.46) states that if $|\alpha\rangle$ is rotated and $|\beta\rangle$ is also rotated in same manner, then $\langle\beta|\alpha\rangle$ remains unchanged during rotation. Equation (4.46) arises from the fact that the translation, rotation, and parity operators are unitary operators: $|\alpha\rangle' = U|\alpha\rangle$ and $\langle\beta|' = \langle\beta|U^+$:

$$\langle\beta|\alpha\rangle' = \langle\beta|U^+U|\alpha\rangle = \langle\beta|\alpha\rangle \tag{4.47}$$

For a time reversed state $\psi_\alpha(\mathbf{r},-t) = \theta\psi_\alpha(\mathbf{r},t)$ or in short $|\alpha\rangle' = \theta|\alpha\rangle$, θ being the time reversal operator, we cannot fulfill the requirement Eqn. (4.46). However, a more generalized requirement:

$$\left|\langle\beta|\alpha\rangle'\right| = |\langle\beta|\alpha\rangle| \tag{4.48}$$

fulfills the Eqn. (4.46), and at the same time it works equally well for:

$$\langle\beta|\alpha\rangle' = \langle\beta|\alpha\rangle^* = \langle\alpha|\beta\rangle \tag{4.49}$$

Let us pursue the possibility shown in Eqn. (4.49), because we have said that time reversal has something to do with complex conjugation, in the case of the Schrödinger equation.

Transformations $|\alpha\rangle' = \theta|\alpha\rangle$ and $|\beta\rangle' = \theta|\beta\rangle$ are said to be antiunitary if they fulfill criteria Eqn. (4.49). And the operator θ is called an antilinear operator if:

$$\theta(c_1|\alpha\rangle + c_2|\beta\rangle) = c_1^*\theta|\alpha\rangle + c_2^*\theta|\beta\rangle \tag{4.50}$$

because for a linear operator, A:

$$A(c_1|\alpha\rangle + c_2|\beta\rangle) = c_1A|\alpha\rangle + c_2A|\beta\rangle \tag{4.51}$$

Thus, an antilinear operator does not commute with a constant because $\theta c = c^*\theta$. We thus infer that the product of two antilinear operators is linear, and the product of a linear with an antilinear operator is antilinear. In a more generalized manner, we say that *a product of operators is linear if the number of antilinear factors is even and it is antilinear when the number of antilinear factors is odd.*

The reason that the only quantities which are physically measurable are the absolute squares of scalar products allows us the relaxing of requirement Eqn. (4.46) to Eqn. (4.48). These are the probabilities, which are experimentally measurable. Its relevance for the discussion of symmetries in quantum mechanics is that a symmetry operation presumably must preserve the probabilities of all experimental outcomes. This implies that transformations in quantum mechanics must be implemented by means of either unitary or antiunitary operators. For example, rotations are implemented by means of unitary operators, as are most other symmetries commonly encountered in quantum mechanics. But time reversal is an exception, and it is implemented by means of antiunitary operators.

If the time reversed operation transforms an operator A to A' then $A' = \theta A \theta^{-1}$. $(\mathbf{r}_0, \mathbf{p}_0) \rightarrow (\mathbf{r}_0, -\mathbf{p}_0)$ is the initial condition of a motion transform under time reversal in classical mechanics. This allows postulating that the time reversal operator in quantum mechanics satisfies the conjugation relations:

$$\mathbf{r}' = \theta \mathbf{r} \theta^{-1} = \mathbf{r} \tag{4.52a}$$

$$\mathbf{p}' = \theta \mathbf{p} \theta^{-1} = -\mathbf{p} \tag{4.52b}$$

which results in:

$$\mathbf{L}' = \theta \mathbf{L} \theta^{-1} = -\mathbf{L}. \tag{4.53a}$$

Further, Eqn. (4.53a) can be generalized to be applicable to all kinds of angular momentum, orbital as well as spin:

$$\mathbf{J}' = \theta \mathbf{J} \theta^{-1} = -\mathbf{J} \tag{4.53b}$$

which is reasonable, because if we think of a simple model of a charged spinning particle as a charged–rotating sphere, then we see that reversing the motion will reverse both the angular momentum as well as the magnetic field produced by the spin. We now take the canonical commutation relations $\left[x_i, p_j\right] = i\hbar\delta_{ij}$ and conjugate with θ:

$$\theta\left[x_i, p_j\right]\theta^{-1} = \left[x_i, -p_j\right] = -i\hbar\delta_{ij} \tag{4.54a}$$

$$\theta(i\hbar)\theta^{-1} = -i\hbar \Rightarrow \theta(i\hbar) = (-i\hbar)\theta \tag{4.54b}$$

Equation (4.54b), which is in conformity with Eqn. (4.50), forces us to conclude that the time reversal operator θ is antilinear, so that $i\hbar$ on the right-hand side changes into $-i\hbar$ when θ is pulled through it.

4.7.1 Properties of Antilinear Operator

In the case of a linear operator A, $\left(\langle\alpha|A\right)\left(|\beta\rangle\right)$ is the same as $\left(\langle\alpha|\right)\left(A|\beta\rangle\right)$, and we conclude that the positioning of the parentheses is irrelevant; we can write it simply as $\langle\alpha|A|\beta\rangle$. In other words, we can think of A as acting either to the right or to left. However, this is not true for an antilinear operator. A complex-valued linear operator on kets and the value of a bra acting on a ket are usual scalar products, and we write $\langle\alpha|\beta\rangle$ for bra $\langle\alpha|$. To understand the act of the antilinear operator on bra, let us assume that the action of a given antilinear operator θ on kets is known, and we want to define its action on bras. For example, if $\langle\alpha|$ is a bra, we wish to define $\langle\alpha|\theta$. The antilinear operators cannot be treated like linear operators. $\langle\alpha|\theta$, a complex-valued operator acting on kets, is an antilinear operator. However, bras are supposed to be linear operators. Let us introduce a complex conjugation to make $\langle\alpha|\theta$ a linear operator on kets. Set:

$$(\langle\alpha|\theta)(|\beta\rangle) = \left[(\langle\alpha|)(\theta|\beta\rangle)\right]^* \tag{4.55}$$

The rule says that for an antilinear operator, it does matter whether the operator acts to the right or to the left in a matrix element. On changing the direction in which the operator acts, one must complex conjugate the matrix element. Therefore, in the case of antilinear operators, parentheses are necessary to indicate in which direction the operator acts. We next, consider the Hermitian conjugation for linear and antilinear operators. In the case of linear operators, the Hermitian conjugate is $\left(A|\beta\rangle\right)^\dagger = \langle\beta|A^\dagger$ for all kets $|\beta\rangle$ or equivalently by:

$$\langle\phi_\alpha|A^\dagger|\psi_\beta\rangle = \langle\psi_\beta|A|\phi_\alpha\rangle^\dagger \tag{4.56}$$

A similar definition can also work for antilinear operators, and we can set $\left(\theta|\beta\rangle\right)^\dagger = \langle\beta|\theta^\dagger$.

Note that $\theta^\dagger$ is an antilinear operator if θ is antilinear. Now, however, when we try to write an equation analogous to Eqn. (4.56), we must be careful about the parentheses. Thus, we have:

$$\langle\alpha|(\theta^\dagger|\beta\rangle) = \left[(\langle\beta|\theta)|\alpha\rangle\right]^\dagger \tag{4.57a}$$

or:

$$\langle\alpha|(\theta'|\beta\rangle) = \langle\beta|(\theta|\alpha\rangle) \tag{4.57b}$$

Equations (4.55), (4.57a), and (4.57b) summarize the principal rules for antilinear operators, which differ from those of linear operators.

We next consider antiunitary operators. An operator is antiunitary if it is antilinear and it satisfies: $\theta\theta^\dagger = \theta^\dagger\theta = 1$ as has been said earlier. Similar to the case of unitary operators, antiunitary operators too preserve the absolute values of scalar products.

Let us say that an antiunitary operator can be written as $\theta = UK$, where U is a unitary operator and K is a complex-conjugation operator, which forms complex conjugation of any coefficient that multiplies a ket from right:

$$Kc|\alpha\rangle = c^* K|\alpha\rangle \tag{4.58}$$

We examine the action of K on $|\psi_\alpha\rangle$, when it is expended in terms of base ket:

$$\begin{aligned} |\phi_\alpha\rangle' &= K|\phi_\alpha\rangle \\ &= K\sum_n c_n|u_n\rangle = K\sum_n \langle u_n|\phi_\alpha\rangle|u_n\rangle \\ &= \sum_n \langle u_n|\phi_\alpha\rangle^* K|u_n\rangle = \sum_n \langle\phi_\alpha|u_n\rangle|u_n\rangle \end{aligned} \tag{4.59}$$

Here, we made use of $c_n = \langle u_n|\phi_\alpha\rangle$, $\langle u_n|\phi_\alpha\rangle^* = \langle\phi_\alpha|u_n\rangle$ and have taken base ket:

$$|u_n\rangle = \begin{bmatrix} 0 \\ \vdots \\ 1 \\ \vdots \\ 0 \end{bmatrix} \tag{4.60}$$

which is real and does not change under the action of the complex–conjugation operator. We then take:

$$\begin{aligned} |\varphi_\alpha\rangle' &= \theta|\varphi_\alpha\rangle = UK\sum_m \langle u_m|\varphi_\alpha\rangle|u_m\rangle \\ &= \sum_m \langle u_m|\varphi_\alpha\rangle^* UK|u_m\rangle \\ &= \sum_m \langle\varphi_\alpha|u_m\rangle U|u_m\rangle = \sum_m U|u_m\rangle\langle u_m|\varphi_\alpha\rangle^* \end{aligned} \tag{4.61}$$

Also, consider:

$$\begin{aligned} |\psi_\beta\rangle' &= UK|\psi_\beta\rangle = UK\sum_n \langle u_n|\psi_\beta\rangle|u_n\rangle \\ &= \sum_n \langle u_n|\psi_\beta\rangle^* UK|u_n\rangle = \sum_n \langle u_n|\psi_\beta\rangle^* U|u_n\rangle \end{aligned} \tag{4.62}$$

The Hermitian adjoint of Eqn. (4.62) gives:

$$(\langle\psi_\beta|)' = \sum_n \langle u_n|\psi_\beta\rangle\langle u_n|U^\dagger \tag{4.63}$$

Equations (4.61) and (4.63) give:

$$\begin{aligned} \langle\psi_\beta|\varphi_\alpha\rangle' &= \sum_n\sum_m \langle u_n|\psi_\beta\rangle\langle u_n|U^\dagger U|u_m\rangle\langle u_m|\phi_\alpha\rangle^* \\ &= \sum_n\sum_m \langle u_n|\psi_\beta\rangle\langle u_m|\varphi_\alpha\rangle^*\delta_{nm} = \sum_n \langle u_n|\psi_\beta\rangle\langle\varphi_\alpha|u_n\rangle \\ &= \sum_n \langle u_n|u_n\rangle\langle\varphi_\alpha|\psi_\beta\rangle = \langle\varphi_\alpha|\psi_\beta\rangle = \langle\psi_\beta|\varphi_\alpha\rangle^* \end{aligned} \tag{4.64}$$

which is exactly the same as stated in Eqn. (4.49), and hence, it is proved that *an antiunitary operator can be expressed as a product of a unitary operator and a complex-conjugation operator.*

Spinless particles are described in terms of position r, linear momentum p, and orbital angular momentum L. Therefore, requirements given by Eqns. (4.52a) to (4.53b) are satisfied by taking $U = 1$ and then $\theta = K$:

$$\begin{aligned}
\theta \mathbf{r} \theta^{-1} &= K\mathbf{r}K^{-1} = \mathbf{r} \\
\theta \mathbf{p} \theta^{-1} &= K(-i\hbar\nabla)K^{-1} = (i\hbar\nabla)KK^{-1} = -\mathbf{p} \\
\theta \mathbf{L} \theta^{-1} &= K(\mathbf{r}\times\mathbf{p})K^{-1} = -\mathbf{L} \\
\theta \mathbf{J} \theta^{-1} &= \theta(\mathbf{L}+\mathbf{S})\theta^{-1} = -\mathbf{L}-\mathbf{S}
\end{aligned} \tag{4.65}$$

The Hamiltonian is invariant under time reversal, and hence, the time reversal operator commutes with the Hamiltonian. However, it cannot be said that θ is the constant of motion, because the equation $i\hbar\dfrac{dA}{dt} = [AH - HA]$ holds only for linear operators.

4.7.2 Time Reversal Operator for Non-zero Spin Particles

To satisfy Eqns. (4.52a) to (4.53b), we require that spin operator $\mathbf{S}$ should change sign under time reversal operation $\theta\mathbf{S}\theta^{-1} = -\mathbf{S}$. The explicit form of θ depends on chosen coordinate representation. We take standard representation $\mathbf{S} = \dfrac{\hbar}{2}\boldsymbol{\sigma}$, where $\boldsymbol{\sigma}$ represents the Pauli matrices:

$$\sigma_x = \begin{pmatrix} 0 & 1 \\ 1 & 0 \end{pmatrix}, \sigma_y = \begin{pmatrix} 0 & -i \\ i & 0 \end{pmatrix}, \text{ and } \sigma_z = \begin{pmatrix} 1 & 0 \\ 0 & -1 \end{pmatrix} \tag{4.66}$$

As is seen, $\mathbf{r}, S_x$ and S_z are real and $\mathbf{p}$ and S_y are imaginary, which means that $KS_x = S_xK$, $KS_y = -S_yK$, and $KS_z = S_zK$. On using $\theta = UK$, we obtain:

$$\begin{aligned}
&\theta S_x \theta^{-1} = UKS_xK^{-1}U^{-1} = US_xKK^{-1}U^{\dagger} = US_xU^{-1} \\
&\Rightarrow -S_x = US_xU^{\dagger}
\end{aligned} \tag{4.67a}$$

and:

$$\begin{aligned}
&\theta S_y \theta^{-1} = UKS_yK^{-1}U^{-1} = -US_yKK^{-1}U^{-1} = -US_yU^{-1} \\
&\Rightarrow S_y = US_yU^{\dagger}
\end{aligned} \tag{4.67b}$$

similarly:

$$-S_z = US_zU^{\dagger} \tag{4.67c}$$

We thus find that transformation with the unitary operator U leaves S_y invariant and changes the sign of S_x and S_z. Therefore, U should represent the rotation of the coordinate system around the y-axis through an angle of π. Such a rotation can be represented by $e^{i\pi S_y/\hbar}$ and hence $\theta = e^{\frac{i\pi S_y}{\hbar}} K$. For a spin $\frac{1}{2}$ particle:

$$e^{\frac{i\pi S_y}{\hbar}} = i\sigma_y \tag{4.68}$$

because iS_y or $i\sigma_y$ is real, K commutes with $e^{\frac{i\pi S_y}{\hbar}}$ or $Ke^{\frac{i\pi S_y}{\hbar}} = e^{\frac{i\pi S_y}{\hbar}} K$. Two successive operations of K on a scalar c give $K^2 c = Kc^* = c$, which means $K^2 = 1$. Also, operation of U^2 yields a rotation of 2π through the y-axis. And, the 2π-rotation leaves the wave function of a system of integer spin (2s is even) unchanged, while the wave function of a system of particles of half-integer spin (2s is odd) changes. This implies that $\theta^2 = 1$ for particles of integer spin and $\theta^2 = -1$ for a system of a half-integer spin. For the half-integer spin particle $\theta^2 = (i\sigma_y K)(i\sigma_y K) = -\sigma_y^2 = -1$. We have $\theta_n = (i\sigma_{1y})(i\sigma_{2y})(i\sigma_{3y})(i\sigma_{4y}).....(i\sigma_{ny})K$ for a system of n particles, and therefore $\theta_n^2 = (-1)^n$, which is 1 for the even value of n and -1 for the odd value of n.

Again, θ commutes with H and therefore the eigenket $|\psi_\alpha(r,t)\rangle$ is also the eigenket of θ_n and we can write:

$$\theta_n|\psi_\alpha(r,t)\rangle = c|\psi_\alpha(r,t)\rangle \tag{4.69a}$$

where c is the eigenvalue of θ_n. Operating on both sides of Eqn. (4.69a) with θ_n, we get:

$$\begin{aligned} \theta_n^2|\psi_\alpha(r,t)\rangle &= \theta_n c|\psi_\alpha(r,t)\rangle \\ &= c^*\theta_n|\psi_\alpha(r,t)\rangle = c^*c|\psi_\alpha(r,t)\rangle = |c|^2|\psi_\alpha(r,t)\rangle \end{aligned} \tag{4.69b}$$

which yields $|c|^2 = 1$, when $\theta_n^2 = 1$ and $|c|^2 = -1$ for $\theta_n^2 = -1$. $|c|^2 = 1$ yields eigenvalue $|c| = \pm 1$ that could be a possible solution if the system is non-degenerate, for $|c|^2 = -1$ results to no possible eigenvalue that can satisfy an eigenvalue equation, which suggests there cannot be degeneracy in the system.

The two eigenkets $|\psi_\beta\rangle$ and $\theta|\psi_\beta\rangle$ are orthogonal in the case of a system of half-integer spin particles. To prove it, we take $\langle\beta|\alpha\rangle' = \langle\beta|\alpha\rangle^* = \langle\alpha|\beta\rangle$, where $|\beta\rangle' = \theta|\beta\rangle$ and $|\alpha\rangle' = \theta|\alpha\rangle$. Then:

$$\begin{aligned} &\langle\beta|\alpha\rangle' = \langle\alpha|\beta\rangle \\ &\Rightarrow \langle\theta\beta|\theta\alpha\rangle = \langle\alpha|\beta\rangle \end{aligned} \tag{4.70a}$$

Use $|\alpha\rangle = \theta|\beta\rangle$ on left of Eqn. (4.70a) to get:

$$\langle\theta\beta|\theta^2\beta\rangle = \langle\theta\beta|\beta\rangle \tag{4.70b}$$

Since $\theta^2 = -1$ for the half-integer spin particle, the left-hand side of Eqn. (4.70b) is equal to $-\langle\theta\beta|\beta\rangle$ and hence:

$$-\langle\theta\beta|\beta\rangle = \langle\theta\beta|\beta\rangle \Rightarrow \langle\theta\beta|\beta\rangle = 0 \tag{4.70c}$$

Equation (4.70c) proves that $|\psi_\beta\rangle$ and $\theta|\psi_\beta\rangle$ are orthogonal.

4.8 Solved Examples

1. Show that the 1D harmonic oscillator Hamiltonian $H = \frac{p^2}{2m} + \frac{1}{2}kx^2$ is invariant under space inversion and its wave function kets can be divided into two categories.

SOLUTION

Under space inversion, x is changed to $-x$. The $H = \frac{p^2}{2m} + \frac{1}{2}kx^2$ remains unchanged on replacing x by $-x$, as it has x^2 and hence $H(-x) = H(x)$. We take $H(x)\psi(x) = E\psi(x)$. On applying space inversion and using $H(-x) = H(x)$, we get:

$$\begin{aligned} H(-x)\,\psi(-x)\rangle &= E\psi(-x) \\ \Rightarrow H(x)\,\psi(-x)\rangle &= E\psi(-x) \end{aligned} \tag{4.71a}$$

We thus find that both $|\psi(x)\rangle$ and $|\psi(-x)\rangle$ are wave function kets of $H(x)$ and hence these should be either parallel vectors or a scalar multiple of each other. Therefore, we can write:

$$|\psi(-x)\rangle = \lambda|\psi(x)\rangle \tag{4.71b}$$

On replacing x by $-x$ in Eqn. (4.71b), we get:

$$|\psi(x)\rangle = \lambda|\psi(-x)\rangle = \lambda^2|\psi(x)\rangle \tag{4.71c}$$

which yields $\lambda^2 = 1$ or $\lambda = \pm 1$. We can thus say that the wave function kets of a 1D harmonic oscillator can be divided into two groups; one of those belongs to $\lambda = 1$ and the other corresponds to $\lambda = -1$.

2. Show that spherical harmonics $Y_{lm}(\theta,\phi)$ are eigenfunctions of the parity operator.

SOLUTION

In spherical polar coordinates, the parity operator P changes: $r \to r$, $\theta \to \pi - \theta$ and $\phi \to \pi + \phi$. The operation of P on $Y_{lm}(\theta,\phi) = CP_l^m(\cos\theta)e^{im\phi}$, where C is constant, gives:

$$\begin{aligned} PY_{lm}(\theta,\phi) &= Y_{lm}(\pi-\theta,\pi+\phi) \\ &= CP_l^m(\cos(\pi-\theta))e^{im(\pi+\phi)} \\ &= CP_l^m(-\cos\theta)e^{im\phi}e^{im\pi} \\ &= CP_l^m(\cos\theta)(-1)^{l+m}e^{im\phi}(-1)^m \\ &= (-1)^l CP_l^m(\cos\theta)e^{im\phi} \end{aligned} \tag{4.72a}$$

where we used $(-1)^{2m} = 1$ and $P_l^m(-x) = (-1)^{l+m}P_l^m(x)$. Therefore,

$$PY_{lm}(\theta,\phi) = \lambda Y_{lm}(\theta,\phi) = (-1)^l Y_{lm}(\theta,\phi) \tag{4.72b}$$

Hence $\lambda = (-1)^l$, which could be +1 or −1 depending on whether l is even or odd.

3. An operator is represented by matrices A and A' before and after rotation, respectively. If A is Hermitian then show that A' too is Hermitian.

SOLUTION

From Eqn. (4.26), we have:

$$A' = SAS^{-1} \text{ with } S = e^{-i\delta\theta.\mathbf{j}/\hbar} \tag{4.73}$$

and $A^{\dagger} = A$. The Hermitian conjugate of A' is:

$$A'^{\dagger} = \left(SAS^{-1}\right)^{\dagger} = \left(S^{-1}\right)^{\dagger} A^{\dagger} S^{\dagger} \tag{4.74}$$

Since $SS^{-1} = I = SS^{\dagger}$, S is a unitary matrix that satisfies $S^{-1} = S^{\dagger}$. Hence:

$$\begin{aligned} A'^{\dagger} &= \left(S^{-1}\right)^{\dagger} A^{\dagger} S^{\dagger} \\ &= SAS^{-1} \\ &= A' \end{aligned}$$

4. Show that the parity operator commutes with the orbital angular momentum operator.

SOLUTION

The orbital angular momentum operator is $\mathbf{L} = \mathbf{r} \times \mathbf{p}$. Let us consider a state represented by $\psi(\mathbf{r})$. Then:

$$\begin{aligned} P\mathbf{L}\psi(\mathbf{r}) &= P(\mathbf{r} \times \mathbf{p})\psi(\mathbf{r}) \\ &= \left(-\mathbf{r} \times -\mathbf{p}\right)\psi(-\mathbf{r}) \\ &= \mathbf{L}P\psi(\mathbf{r}) \end{aligned}$$

which implies $(P\mathbf{L} - \mathbf{L}P)\psi(\mathbf{r}) = 0$ or $[P, \mathbf{L}] = 0$.

4.9 Exercises

1. A particle is described by the wave function $\psi(\mathbf{r},t) = Ae^{i(\mathbf{k}.\mathbf{r}-\omega t)}$. Show that $\psi(\mathbf{r},t)$ and $\psi^{*}(\mathbf{r},-t)$ are acceptable solutions of the Schrödinger equation, while $\psi(\mathbf{r},-t)$ is not an acceptable solution.
2. Calculate the eigenvalues and eigenvectors of the matrix $s_y = \frac{\hbar}{2}\begin{bmatrix} 0 & -i \\ i & 0 \end{bmatrix}$ and show that its eigenvectors are orthogonal.
3. Show that the time reversal operator commutes with the Hamiltonian.
4. Show that probability density $P = |\psi(\mathbf{r},t)|^2$ remains invariant under space inversion, rotation, reflection, and time reversal operations.

5. Matrices representing the rotation by an infinitesimal angle ε around the y-axis and z-axis are as follows:

$$R_y(\varepsilon)=\begin{pmatrix} 1-\dfrac{\varepsilon^2}{2} & 0 & \varepsilon \\ 0 & 1 & 0 \\ -\varepsilon & 0 & 1-\dfrac{\varepsilon^2}{2} \end{pmatrix} \text{ and } R_z(\varepsilon)=\begin{pmatrix} 1-\dfrac{\varepsilon^2}{2} & -\varepsilon & 0 \\ \varepsilon & 1-\dfrac{\varepsilon^2}{2} & 0 \\ 0 & 0 & 1 \end{pmatrix}.$$

Show that $R_y(\varepsilon)$ and $R_z(\varepsilon)$ do not commute.

5

Angular Momentum

Angular momentum plays a central role, both classically as well as quantum mechanically, in understanding the structure of atoms and other systems that involve rotational symmetry centered around the angular momentum. Like other observable quantities, angular momentum, too, is described by an operator in quantum mechanics, and it is a vector operator, like the linear momentum operator. In this chapter, we discuss various properties of angular momentum as a vector operator.

5.1 Orbital Angular Momentum

The classical definition of the orbital angular momentum of a particle with position vector $\mathbf{r}$ and conjugate linear vector momentum $\mathbf{p}$ about the origin is $\mathbf{L}=\mathbf{r}\times\mathbf{p}$. The 3-components of $\mathbf{L}$ are $L_x=(yp_z-zp_y)$, $L_y=(zp_x-xp_z)$, $L_z=(xp_y-yp_x)$. Components of the position and linear momentum, as operators, satisfy the fundamental commutation relations: $[p_i,p_j]=0$, $[x_i,x_j]=0$, $[x_i,p_j]=i\hbar\delta_{ij}$, where i and j stand for x, y, and z. The commutation relation for L_x and L_y is:

$$\begin{aligned}
[L_x,L_y] &= L_xL_y-L_yL_x \\
&= \left(yp_z-zp_y\right)\left(zp_x-xp_z\right)-\left(zp_x-xp_z\right)\left(yp_z-zp_y\right) \\
&= yp_z(zp_x)-yxp_z^2-z^2p_yp_x+zxp_yp_z \\
&\quad -zyp_xp_z+z^2p_xp_y+xyp_z^2-xp_z(zp_y) \\
&= yp_z(zp_x)-xp_z(zp_y)+zxp_yp_z-zyp_xp_z \\
&= -iy\hbar p_x+yzp_zp_x+ix\hbar p_y-xzp_zp_y+zxp_yp_z-zyp_xp_z \\
&= i\hbar(xp_y-yp_x) \\
&= i\hbar L_z
\end{aligned} \tag{5.1}$$

The commutators $\left[L_y,L_z\right]$ and $\left[L_z,L_x\right]$ can be found by changing x,y and z in cyclic order to have:

$$[L_x,L_y]=i\hbar L_z,\ [L_y,L_z]=i\hbar L_x,\ \text{and}\ [L_z,L_x]=i\hbar L_y \tag{5.2}$$

In quantum mechanics, the foundation of the theory of angular momentum is based on three commutation relations given by Eqn. (5.2). Other dynamical variables which are represented by the above three commuting relations are found to have identical properties to those of the components of an angular momentum. Next take:

$$\begin{aligned} \mathbf{L}\times\mathbf{L} &= \hat{i}(L_yL_z - L_zL_y) + \hat{j}(L_zL_x - L_xL_z) + \hat{k}(L_xL_y - L_yL_x) \\ &= i\hbar(\hat{i}L_x + \hat{j}L_y + \hat{k}L_z) = i\hbar\mathbf{L} \end{aligned} \tag{5.3}$$

For a system of N-particles with orbital angular momentum vectors $\mathbf{L}_i$ (where, i runs from 1 to N), each of these vectors satisfies:

$$\boldsymbol{L}_i \times \boldsymbol{L}_i = i\hbar\boldsymbol{L}_i \tag{5.4}$$

The orbital angular momentum operators of different particles represent the different degrees of freedom of the system, and hence we expect them to commute with each other. We therefore write: $\mathbf{L}_i \times \mathbf{L}_j + \mathbf{L}_j \times \mathbf{L}_i = 0$ for $i \neq j$ where i and j refer to different particles.

For total orbital angular momentum of the system $\mathbf{L} = \sum_{i=1}^{N} \mathbf{L}_i$:

$$\begin{aligned} \mathbf{L}\times\mathbf{L} &= \sum_{i=1}^{N}\mathbf{L}_i \times \sum_{j=1}^{N}\mathbf{L}_j \\ &= \sum_{i=1}^{N}\mathbf{L}_i \times \mathbf{L}_i + \frac{1}{2}\sum_{(i\neq j)=1}^{N}(\mathbf{L}_i \times \mathbf{L}_j + \mathbf{L}_j \times \mathbf{L}_i) \\ &= i\hbar\sum_{i=1}^{N}\mathbf{L}_i = i\hbar\mathbf{L} \end{aligned} \tag{5.5}$$

We thus find that the sum of two or more angular momentum vectors and the total orbital angular momentum of the system satisfies the same commutation relation as a primitive orbital angular momentum vector. *An important conclusion drawn from the commutation relations is that the three components of an angular momentum operator cannot be specified (or measured) simultaneously.* Once we specify one component, the values of the other two components become uncertain. It has been conventional to specify the z−component L_Z.

The magnitude squared of the orbital angular momentum $\mathbf{L}^2$ and L_z commutes one with the other:

$$\left[\mathbf{L}^2, L_z\right] = \left[L_x^2, L_z\right] + \left[L_y^2, L_z\right] + \left[L_z^2, L_z\right] \tag{5.6}$$

To prove this, take:

$$\begin{aligned} \left[L_x^2, L_z\right] &= L_x^2L_z - L_zL_x^2 \\ &= L_x^2L_z - L_xL_zL_x + L_xL_zL_x - L_zL_x^2 \\ &= L_x\left(L_xL_z - L_zL_x\right) - \left(L_zL_x - L_xL_z\right)L_x \\ &= -i\hbar(L_xL_y + L_yL_x) \end{aligned} \tag{5.7a}$$

Similarly:

$$
\begin{aligned}
\left[L_y^2, L_z\right] &= L_y^2 L_z - L_z L_y^2 \\
&= L_y^2 L_z - L_y L_z L_y + L_y L_z L_y - L_z L_y^2 \\
&= L_y\left(L_y L_z - L_z L_y\right) + \left(L_y L_z - L_z L_y\right) L_y \\
&= i\hbar(L_y L_x + L_x L_y)
\end{aligned} \tag{5.7b}
$$

and:

$$
\begin{aligned}
\left[L_x^2, L_x\right] &= L_x^2 L_x - L_x L_x^2 \\
&= 0
\end{aligned} \tag{5.7c}
$$

Therefore:

$$
\begin{aligned}
\left[\mathbf{L}^2, L_z\right] &= \left[L_x^2, L_z\right] + \left[L_y^2, L_z\right] + \left[L_z^2, L_z\right] \\
&= 0
\end{aligned} \tag{5.8}
$$

Similarly we can prove that $\mathbf{L}^2$ commutes with L_x and L_y, because there is nothing special about the z-axis. This suggests that one can specify the magnitude of an angular momentum vector along with one of its components (by convention, the z-component). The operators $L^+ = L_x + iL_y$ and $L^- = L_x - iL_y$ are also used in quantum mechanics. They commute with $\mathbf{L}^2$ and they satisfy the relations:

$$
\begin{aligned}
\left[L^+, L_z\right] &= L^+ L_z - L_z L^+ \\
&= L_x L_z + iL_y L_z - L_z L_x - iL_z L_y \\
&= -i\hbar L_y - \hbar L_x \\
&= -\hbar L^+
\end{aligned} \tag{5.9a}
$$

$$
\begin{aligned}
\left[L^-, L_z\right] &= L^- L_z - L_z L^- \\
&= L_x L_z - iL_y L_z - L_z L_x + iL_z L_y \\
&= -i\hbar L_y + \hbar L_x \\
&= \hbar L^-
\end{aligned} \tag{5.9b}
$$

and:

$$
\begin{aligned}
\left[L^+, L^-\right] &= L^+ L^- - L^- L^+ \\
&= (L_x + iL_y)(L_x - iL_y) - (L_x - iL_y)(L_x + iL_y) \\
&= -iL_x L_y + iL_y L_x - iL_x L_y + iL_y L_x \\
&= -2i(L_x L_y - L_y L_x) \\
&= 2\hbar L_z
\end{aligned} \tag{5.10a}
$$

From Eqn. (5.8), we then have:

$$\left[L^2, L^+\right] = \left[L^2, L^-\right] = 0 \tag{5.10b}$$

5.2 Eigenvalues of Angular Momentum

Though the procedure presented in this section is used to find eigenvalues for $\mathbf{L}^2$ and L_z, it is equally applicable to compute the eigenvalues for $\mathbf{J}^2$ and J_z as well as eigenvalues for $\mathbf{S}^2$ and S_z. The $\mathbf{J}$ and $\mathbf{S}$ are total angular momentum and spin angular momentum, respectively. We start with the assumption that the simultaneous eigenkets of L^2 and L_z are specified by two quantum numbers, l and m, and the wavefunction kets are denoted by $|l, m\rangle$. Then the quantum number m is defined by:

$$L_z |l, m\rangle = m\hbar |l, m\rangle \tag{5.11}$$

Thus, $m\hbar$ is the eigenvalue of L_z. Since the dimensions of $\hbar$ and L_z are same, it is possible to write such an equation. Because L_z is a Hermitian operator, m has to be a real number. We take:

$$\mathbf{L}^2 |l, m\rangle = \lambda_{lm}\hbar^2 |l, m\rangle \tag{5.12}$$

without loss of generality, where λ_{lm} is some real dimensionless function of l and m. We next take:

$$\begin{aligned} \langle l, m|(\mathbf{L}^2 - L_z^2)|l, m\rangle &= (\lambda_{lm}\hbar^2 - m^2\hbar^2)\langle l, m|l, m\rangle \\ &= \lambda_{lm}\hbar^2 - m^2\hbar^2 \end{aligned} \tag{5.13}$$

where we have assumed that $|l, m\rangle$ is a normalized ket. Also

$$\begin{aligned} \langle l, m|(\mathbf{L}^2 - L_z^2)|l, m\rangle &= \langle l, m|(L_x^2 + L_y^2)|l, m\rangle \\ &= \langle l, m|L_x^2|l, m\rangle + \langle l, m|L_y^2|l, m\rangle \end{aligned} \tag{5.14}$$

We know that for any general ket, $|\phi\rangle$ and a Hermitian operator A, $|\xi\rangle = A|\phi\rangle$ and $\langle\xi|\xi\rangle = \langle\phi|A^\dagger A|\phi\rangle = \langle\phi|A^2|\phi\rangle \geq 0$, since the norm square of a vector can never be negative. Therefore, from Eqn. (5.13), we conclude that $m^2 \leq \lambda_{lm}$. We next consider the effect of the L^+ on $|l, m\rangle$. $\left[\mathbf{L}^2, L^+\right] = 0$ implies that $\mathbf{L}^2 L^+ = L^+ \mathbf{L}^2$. Therefore:

$$\begin{aligned} \mathbf{L}^2 L^+ |l, m\rangle &= L^+ \mathbf{L}^2 |l, m\rangle \\ &= \hbar^2 \lambda_{lm} L^+ |l, m\rangle \end{aligned} \tag{5.15}$$

From Eqns. (5.12) and (5.15) we note that both $|l, m\rangle$ and $L^+|l, m\rangle$ belong to the same eigenvalue of $\mathbf{L}^2$. Thus, the action of the shift operator L^+ on a wave function ket does not affect the magnitude of the angular momentum. Further:

$$L_z L^+ |l, m\rangle = (L_z L_x + i L_z L_y)|l, m\rangle \tag{5.16}$$

Equation (5.2) gives $L_zL_x = L_xL_z + i\hbar L_y$ and $L_zL_y = L_yL_z - i\hbar L_x$. Therefore, Eqn. (5.16) reduces to:

$$\begin{aligned} L_zL^+|l,m\rangle &= (L_xL_z + i\hbar L_y + iL_yL_z + \hbar L_x)|l,m\rangle \\ &= (L^+L_z + \hbar L^+)|l,m\rangle \\ &= L^+m\hbar|l,m\rangle + \hbar L^+|l,m\rangle \\ &= (m+1)\hbar L^+|l,m\rangle \end{aligned} \tag{5.17a}$$

We thus find that $L^+|l,m\rangle$ is an eigenket of operator L_z having the eigenvalue $(m+1)\hbar$. Further note that from Eqn. (5.11), we get:

$$L_z|l,m+1\rangle = (m+1)\hbar|l,m+1\rangle \tag{5.17b}$$

Equations (5.17a) and (5.17b) imply that $L^+|l,m\rangle$ and $|l,m+1\rangle$ are collinear vectors and hence we write:

$$L^+|l,m\rangle = c_{lm}^+\hbar|l,m+1\rangle \tag{5.18}$$

where c_{lm}^+ is a number. We thus infer from the above discussion *that when the operator L^+ acts on a simultaneous eigenstate of $\boldsymbol{L}^2$ and L_z, the eigenvalue of $\boldsymbol{L}^2$ is* not *changed, while the eigenvalue of L_z is increased by $\hbar$*. For this reason, L^+ is termed a raising operator. Adopting a procedure similar to that used above, we find:

$$L^-|l,m\rangle = c_{lm}^-\hbar|l,m-1\rangle \tag{5.19}$$

Hence, L^- is called a lowering operator. The shift operators step the value of m up and down by unity each time, operating on one of the simultaneous eigenstates of $\boldsymbol{L}^2$ and L_z. At first sight, it appears that any value of m can be obtained by applying the shift operators for enough numbers of times. However, as per $m^2 \le \lambda_{lm}$, there is a definite upper bound to the values that can be taken by m^2. This upper bound is determined by the eigenvalue of $\boldsymbol{L}^2$. It follows that there is a maximum and a minimum possible value which can be assigned to m. Let us say that $m_{\max}$ and $m_{\min}$ represent upper and lower cut-offs. Now, there is no state with $m > m_{\max}$; any attempt to raise the value of m above its maximum value $m_{\max}$ will result to zero and hence:

$$L^+|l,m_{\max}\rangle = 0 \tag{5.20a}$$

This implies that:

$$L^-L^+|l,m_{\max}\rangle = 0 \tag{5.20b}$$

But $L^-L^+ = (L_x - iL_y)(L_x + iL_y) = \left(L_x^2 + L_y^2 + i(L_xL_y - L_yL_x)\right) = \left(L_x^2 + L_y^2 - \hbar L_z\right)$, which means that:

$$\left(L_x^2 + L_y^2 - \hbar L_z\right)|l,m_{\max}\rangle = 0 \tag{5.20c}$$

We write Eqn. (5.20c) as follows:

$$\begin{aligned}&\left(L_x^2+L_y^2+L_z^2\right)\left|l,m_{\max}\right\rangle=(L_z^2+\hbar L_z)\left|l,m_{\max}\right\rangle\\&\Rightarrow L^2\left|l,m_{\max}\right\rangle=(L_z^2+\hbar L_z)\left|l,m_{\max}\right\rangle=m_{\max}(m_{\max}+1)\hbar^2\left|l,m_{\max}\right\rangle\\&\Rightarrow L^2\left|l,m_{\max}\right\rangle=m_{\max}(m_{\max}+1)\hbar^2\left|l,m_{\max}\right\rangle\end{aligned}\tag{5.21}$$

Comparison between Eqns. (5.12) and (5.21) gives $\lambda_{lm_{\max}}=m_{\max}(m_{\max}+1)$. We, however, note that the successive operation of L^- on a wave function ket $\left|q,m_{\max}\right\rangle$ generates $\left|q,m_{\max}-1\right\rangle$, $\left|q,m_{\max}-2\right\rangle$, etc. Application of a lowering operator leaves unchanged the eigenvalue of L^2; hence all these states must correspond to the same value of λ_{lm}, namely $m_{\max}(m_{\max}+1)$.

$$\text{Thus, } \mathbf{L}^2\left|l,m\right\rangle=m_{\max}(m_{\max}+1)\hbar^2\left|l,m\right\rangle\tag{5.22}$$

Therefore, an unknown quantum number l can be equated to $m_{\max}$, without any loss of generality and we write:

$$\mathbf{L}^2\left|l,m\right\rangle=l(l+1)\hbar^2\left|l,m\right\rangle\tag{5.23}$$

and then:

$$\begin{aligned}\left\langle l,m\right|L^-L^+\left|l,m\right\rangle&=\left\langle l,m\right|\left\{\mathbf{L}^2-(L_z^2+\hbar L_z)\right\}\left|l,m\right\rangle\\&=\hbar^2\left\{l(l+1)-m(m+1)\right\}\end{aligned}\tag{5.24}$$

However, we also have:

$$\begin{aligned}\left\langle l,m\right|L^-L^+\left|l,m\right\rangle&=\left\langle l,m\right|L^-\hbar c_{lm}^+\left|l,m+1\right\rangle\\&=\left\langle l,m\right|\hbar c_{lm}^+L^-\left|l,m+1\right\rangle\\&=\hbar^2c_{lm}^+c_{l,m+1}^-\left\langle l,m\right|\left|l,m\right\rangle\\&=\hbar^2c_{lm}^+c_{l,m+1}^-\end{aligned}\tag{5.25}$$

Equations (5.24) and (5.25) yield:

$$c_{lm}^+c_{l,m+1}^-=l(l+1)-m(m+1)\tag{5.26}$$

Now take:

$$\begin{aligned}\left\langle l,m\right|L^-\left|l,m+1\right\rangle&=\left\langle l,m\right|L_x\left|l,m+1\right\rangle-i\left\langle l,m\right|L_y\left|l,m+1\right\rangle\\&=\left[\left\langle l,m+1\right|L_x^\dagger\left|l,m\right\rangle+i\left\langle l,m+1\right|L_y^\dagger\left|l,m\right\rangle\right]^\dagger\\&=\left[\left\langle l,m+1\right|L_x\left|l,m\right\rangle+i\left\langle l,m+1\right|L_y\left|l,m\right\rangle\right]^\dagger\end{aligned}\tag{5.27}$$

Here, we used the fact that L_x and L_y are Hermitian. Eqn. (5.27) further simplifies to:

$$\begin{aligned} \langle l,m|L^-|l,m+1\rangle &= \left[\langle l,m+1|L^+|l,m\rangle\right]^\dagger \\ \Rightarrow c^-_{l,m+1} &= \left(c^+_{l,m}\right)^* \end{aligned} \tag{5.28}$$

Equations (5.26) and (5.28) can be combined to give:

$$\left|c^+_{lm}\right|^2 = l(l+1) - m(m+1) \tag{5.29a}$$

which yields:

$$c^+_{lm} = \sqrt{l(l+1) - m(m+1)} \tag{5.29b}$$

It is to be noticed that $c^+_{l,m}$ is undetermined to an arbitrary phase factor; we can replace $c^+_{l,m}$, given above, by $c^+_{l,m}e^{i\phi}$ where ϕ is real, and we still satisfy Eqn. (5.28). An arbitrary but convenient choice that $c^+_{l,m}$ is real and positive has been made here. This is equivalent to choosing relative phases of the wave function kets $|l,m\rangle$. Similar calculation yields:

$$c^-_{lm} = \left(c^+_{l,m-1}\right)^* = \sqrt{l(l+1) - m(m-1)} \tag{5.30}$$

The inequality $m^2 \le \lambda_{lm}$ implies that m has a maximum and a minimum possible value. The maximum value of m is denoted by l. To find minimum value, let us start to lower the value of m below its minimum value $m_{\min}$. Since, there is no state below the state having m_{min}, we must have:

$$L^-|l,m_{\min}\rangle = c^-_{l,m_{\min}}|l,m_{\min}\rangle = 0 \tag{5.31}$$

which implies that $c^-_{l,m_{\min}} = 0$. Therefore, $l(l+1) - m_{\min}(m_{\min}-1) = 0$ or $m_{\min} = -l$. We therefore conclude that m takes a ladder of discrete values, each rung differing from its immediate neighbors by 1, with the top rung at l and the bottom rung at $-l$. The two possible choices for l are as follows: (i) it is an integer allowing m to take values $-l, \ldots. -2, -1, 0, 1, 2, \ldots. l$, or (ii) it is a half-integer which allows m to take the values $-l, \ldots -3/2, -1/2, 0, 1/2, 3/2, \ldots l$. *We will prove in the next section that l can be assigned only take integer values.* With the use of fundamental commutation relations along with the fact that L_x, L_y and L_z are Hermitian operators, we obtained that eigenvalues of $\mathbf{L}^2$ are given by $l(l+1)\hbar^2$, where l is an integer, or a half-integer. Also, we have shown that the eigenvalues of L_z can only take the values $m\hbar$ where m lies in the range of $-l, -l+1, \ldots, -1, 0, 1, \ldots l-1, l$. A normalized simultaneous wave function ket of $\mathbf{L}^2$ and L_z, belonging to the eigenvalues $l(l+1)\hbar^2$ and $m\hbar$, respectively is $|l,m\rangle$. With the use of Eqns. (5.18) and (5.29b), we write:

$$L^+|l,m\rangle = \hbar\sqrt{l(l+1) - m(m+1)}\,|l,m+1\rangle \tag{5.32a}$$

Use of Eqns. (5.19) and (5.30) gives:

$$L^-|l,m\rangle = \hbar\sqrt{l(l+1) - m(m-1)}\,|l,m-1\rangle \tag{5.32b}$$

5.3 Eigenfunctions of Orbital Angular Momentum

Three components of orbital angular momentum, expressed in Cartesian coordinates (x,y,z), are:

$$L_x = -i\hbar\left(y\frac{\partial}{\partial z} - z\frac{\partial}{\partial y}\right) \tag{5.33a}$$

$$L_y = -i\hbar\left(z\frac{\partial}{\partial x} - x\frac{\partial}{\partial z}\right) \tag{5.33b}$$

and,

$$L_z = -i\hbar\left(x\frac{\partial}{\partial y} - y\frac{\partial}{\partial x}\right) \tag{5.33c}$$

Use of standard spherical polar coordinates:

$$\begin{aligned} x &= r\cos\phi\sin\theta, \\ y &= r\sin\phi\sin\theta, \\ z &= r\cos\theta \end{aligned} \tag{5.34}$$

gives:

$$L_x = i\hbar\left(\sin\phi\frac{\partial}{\partial\theta} + \cot\theta\cos\phi\frac{\partial}{\partial\phi}\right) \tag{5.35a}$$

$$L_y = -i\hbar\left(\cos\phi\frac{\partial}{\partial\theta} - \cot\theta\sin\phi\frac{\partial}{\partial\phi}\right) \tag{5.35b}$$

$$L_z = -i\hbar\frac{\partial}{\partial\phi}. \tag{5.35c}$$

The shift operators, $L^{\pm}$ and $\mathbf{L}^2$ are then expressed as follows:

$$L^{\pm} = \pm\hbar e^{\pm i\phi}\left(\frac{\partial}{\partial\theta} \pm i\cot\theta\frac{\partial}{\partial\phi}\right) \tag{5.36a}$$

and

$$\mathbf{L}^2 = -\hbar^2\left\{\frac{1}{\sin\theta}\frac{\partial}{\partial\theta}\left(\sin\theta\frac{\partial}{\partial\theta}\right) + \frac{1}{\sin^2\theta}\frac{\partial^2}{\partial\phi^2}\right\} \tag{5.36b}$$

The eigenvalue equation for $\mathbf{L}^2$ is given by Eqn. (5.23). On combining Eqns. (5.23) and (5.36b), we get:

$$\left[\left(\frac{1}{\sin\theta}\frac{\partial}{\partial\theta}\left\{\sin\theta\frac{\partial}{\partial\theta}\right\}+\frac{1}{\sin^2\theta}\frac{\partial^2}{\partial\phi^2}\right)+l(l+1)\right]|l,m\rangle=0 \tag{5.37}$$

Equation (5.37) suggests that $|l,m\rangle$ depends on l, m, θ and ϕ, which is represented by $|Y_{lm}(\theta,\phi)\rangle$. We then have:

$$\left[\left(\sin\theta\frac{\partial}{\partial\theta}\left\{\sin\theta\frac{\partial}{\partial\theta}\right\}+\frac{\partial^2}{\partial\phi^2}\right)+l(l+1)\sin^2\theta\right]Y_{lm}(\theta,\phi)=0 \tag{5.38}$$

It is to be noted that a square integrable solution of Eqn. (5.38), which has to be single valued in real space, can exist if and only if l is an integer. Also, by writing $Y_{lm}(\theta,\phi)=\Theta(\theta)\Phi(\phi)$, Eqn. (5.38) becomes:

$$\frac{\sin\theta}{\Theta}\frac{\partial}{\partial\theta}\left(\sin\theta\frac{\partial\Theta}{\partial\theta}\right)+l(l+1)\sin^2\theta=-\frac{1}{\Phi}\frac{\partial^2\Phi}{\partial\phi^2} \tag{5.39}$$

In Eqn. (5.39), the left-hand side is a function of θ while the right-hand side is a function of ϕ. This can be valid if and only if each side is equal to a constant value. Hence, we write:

$$\begin{aligned}&-\frac{1}{\Phi}\frac{\partial^2\Phi}{\partial\phi^2}=m^2\\ &\Rightarrow\frac{\partial^2\Phi}{\partial\phi^2}+m^2\Phi=0\end{aligned} \tag{5.40}$$

whose normalized solution is: $\Phi=\frac{1}{\sqrt{2\pi}}e^{im\phi}$. We thus have:

$$\sin\theta\frac{\partial}{\partial\theta}\left(\sin\theta\frac{\partial\Theta}{\partial\theta}\right)+\left\{l(l+1)\sin^2\theta-m^2\right\}\Theta=0 \tag{5.41}$$

By taking $\xi=\cos\theta$, Eqn. (5.41) converts to:

$$\frac{d}{d\xi}\left((1-\xi^2)\frac{dP}{d\xi}\right)+\left(l(l+1)-\frac{m^2}{1-\xi^2}\right)P=0 \tag{5.42}$$

whose solutions are the associated Legendre functions $P_{lm}(\cos\theta)$. The solution of Eqn. (5.38) therefore is given by $Y_{lm}(\theta,\phi)=C_{lm}P_{lm}(\cos\theta)e^{im\phi}$. Since L^2 and L_z commute, $Y_{lm}(\theta,\phi)$ should also be the eigenfunction of L_z. We thus have:

$$L_zY_{lm}=-i\hbar\frac{\partial}{\partial\phi}\left(C_{lm}P_{lm}e^{im\phi}\right)=m\hbar Y_{lm} \tag{5.43}$$

The normalization condition:

$$\int_0^{\pi}\int_0^{2\pi} Y_{lm}^*(\theta,\phi)Y_{l'm'}(\theta,\phi)\sin\theta d\theta d\phi = \delta_{ll'}\delta_{mm'} \tag{5.44}$$

along with the requirement $Y_{l,-m} = (-1)^m Y_{lm}^*$, yields:

$$Y_{lm}(\theta,\phi) = \sqrt{\frac{(2l+1)(l-m)!}{4\pi(l+m)!}}(-1)^m P_{lm}(\cos\theta)e^{im\phi} \tag{5.45}$$

m is an integer that lies in the range $-l \le m \le l$. Thus, the wave function $\psi(r,\theta,\phi) = R(r)Y_l^m(\theta,\phi)$, where $R(r)$ is a radial function, has all the expected features of the wave function of a simultaneous eigenstate of L^2 and L_z belonging to the quantum numbers l and m. $Y_{lm}(\theta,\phi)$ are known as spherical harmonics. Note that a spherical harmonic wave function $Y_{lm}(\theta,\phi)$ is symmetric about the z-axis (independent of ϕ) whenever $m = 0$, and is spherically symmetric whenever $l = 0$ (since $Y_{00} = 1/\sqrt{4\pi}$). Therefore, by solving directly for the eigenfunctions of L^2 and L_z, in Schrödinger's representation, one can reproduce all the results of Section 5.2. However, the results of Section 5.2 are more generalized in nature than those obtained in this section, because they still apply when the quantum number l takes on half-integer values. Some of the spherical harmonics are as follows:

$$\begin{aligned}
Y_{00} &= \frac{1}{\sqrt{4\pi}} \\
Y_{10} &= \left(\frac{3}{4\pi}\right)^{1/2}\cos\theta;\; Y_{11} = -\left(\frac{3}{8\pi}\right)^{1/2}\sin\theta e^{i\phi};\; Y_{1,-1} = \left(\frac{3}{8\pi}\right)^{1/2}\sin\theta e^{-i\phi} \\
Y_{20} &= \left(\frac{5}{16\pi}\right)^{1/2}\left(3\cos^2\theta - 1\right) \\
Y_{21} &= -\left(\frac{15}{8\pi}\right)^{1/2}(\sin\theta\cos\theta)e^{i\phi};\; Y_{2,-1} = \left(\frac{15}{8\pi}\right)^{1/2}(\sin\theta\cos\theta)e^{-i\phi} \\
Y_{22} &= \left(\frac{15}{32\pi}\right)^{1/2}\sin^2\theta e^{2i\phi};\; Y_{2,-2} = \left(\frac{15}{32\pi}\right)^{1/2}\sin^2\theta e^{-2i\phi}
\end{aligned} \tag{5.46}$$

5.4 General Angular Momentum

The theory of orbital angular momentum, as discussed in Section 5.2, suggests that quantum number m varies between $-l$ *to* l, each value of m differing from immediate neighbors by unity, and there can be two possibilities of values of l. If angular momentum is taken as $\mathbf{L} = \mathbf{r} \times \mathbf{P}$, then the single-valued real space solution of $L^2 Y_{lm}(\theta,\phi) = \hbar^2 l(l+1)Y_{lm}(\theta,\phi)$ demands that l should take integer values. But, the general treatment presented in

Section 5.2 suggests that l can be assigned the half-integer values as well, with m to take the values $-l, \ldots -3/2, -1/2, 0, 1/2, 3/2, \ldots l$. This would be true if particles have, in addition to orbital angular momentum $(\boldsymbol{r} \times \boldsymbol{P})$, some intrinsic angular momentum that does not depend on spatial coordinates. *There exists a wealth of experimental evidence which suggests that the quantum mechanics cannot be completely specified by giving the wave function ψ as a function of the spatial coordinates only.* There exists intrinsic angular momentum called spin angular momentum, and the vector sum of orbital and spin angular momentums is known as the total angular momentum of system. Therefore, the general angular momentum is defined by $\mathbf{J} = \mathbf{L} + \mathbf{S}$, where $\mathbf{S}$ represents the spin angular momentum. Such an angular momentum operator commutes with the Hamiltonian, and consequently is constant of motion. For a particle with spin, total angular momentum in the rest frame is non-vanishing. The commutation relations given by Eqns. (5.2) and (5.3) are redefined for $\mathbf{J}$ and its components as follows:

$$[J_x, J_y] = i\hbar J_z,\ [J_y, J_z] = i\hbar J_x,\ \text{and}\ [J_z, J_x] = i\hbar J_y \tag{5.47a}$$

and:

$$\mathbf{J} \times \mathbf{J} = i\hbar \mathbf{J} \tag{5.47b}$$

Equations (5.8) and (5.10) can also be generalized to:

$$[\mathbf{J}^2, J_x] = [\mathbf{J}^2, J_y] = [\mathbf{J}^2, J_z] = 0 \tag{5.47c}$$

$$[\mathbf{J}^2, J^+] = [\mathbf{J}^2, J^-] = 0 \tag{5.47d}$$

where $J^+ = J_x + iJ_y$ and $J^- = J_x - iJ_y$. The eigenvalue equations for $\mathbf{J}^2$ and J_z are then given by:

$$\mathbf{J}^2 |j, m\rangle = j(j+1)\hbar^2 |j, m\rangle \tag{5.48a}$$

and:

$$J_z |j, m\rangle = m\hbar |j, m\rangle \tag{5.48b}$$

where j can take both integer and half integer values. The half integer values are $0, \frac{1}{2}, \frac{3}{2}, \frac{5}{2}, \ldots\ldots$ and m varies between $-j, -j+1, -j+2, \ldots\ldots -\frac{1}{2}, 0, \frac{1}{2}, \frac{3}{2}, \ldots . j-1, j$.

The generalization of Eqns. (5.32) give:

$$J^+ |j, m\rangle = \hbar\sqrt{j(j+1) - m(m+1)}\, |j, m+1\rangle \tag{5.49a}$$

and:

$$J^- |j, m\rangle = \hbar\sqrt{j(j+1) - m(m-1)}\, |j, m-1\rangle \tag{5.49b}$$

The matrix elements for $\mathbf{J}^2, J_z, J^+, J^-$ are given by:

$$\langle j', m' | \mathbf{J}^2 | j, m \rangle = j(j+1)\hbar^2 \delta_{j'j} \delta_{m'm} \tag{5.50a}$$

$$\langle j', m' | J_z | j, m \rangle = m\hbar \delta_{j'j} \delta_{m'm} \tag{5.50b}$$

$$\langle j', m' | J^+ | j, m \rangle = \hbar\sqrt{j(j+1) - m(m+1)} \delta_{jj'} \delta_{m', m+1} \tag{5.50c}$$

$$\langle j', m' | J^- | j, m \rangle = \hbar\sqrt{j(j+1) - m(m-1)} \delta_{jj'} \delta_{m', m-1} \tag{5.50d}$$

The matrix element of $J^- J^+$ is obtained from Eqn. (5.24) as:

$$\langle j, m | J^- J^+ | j, m \rangle = \hbar^2 \{ j(j+1) - m(m+1) \} \tag{5.51}$$

It is to be noted that $J_x = \dfrac{(J^+ + J^-)}{2}$ and $J_y = \dfrac{(J^+ - J^-)}{2i}$. Hence, the matrix elements of J_x and J_y can be obtained from that of J^+ and J^-. The matrix form of $\mathbf{J}^2, J_z, J^+, J^-$ can be deduced with the use of Eqns. (5.50). As is obvious from these equations, matrices representing $\mathbf{J}^2$ and J_z are diagonal matrices, while matrices that represent, J^+, J^-, J_x and J_y are not diagonal. Example matrices for $\mathbf{J}^2, J_z, J^+, J^-$ for $j = 1$, are given by:

$$J^2 \equiv \begin{bmatrix} 2\hbar^2 & 0 & 0 \\ 0 & 2\hbar^2 & 0 \\ 0 & 0 & 2\hbar^2 \end{bmatrix}, J_z \equiv \begin{bmatrix} \hbar & 0 & 0 \\ 0 & 0 & 0 \\ 0 & 0 & -\hbar \end{bmatrix}, J^+ \equiv \hbar \begin{bmatrix} 0 & \sqrt{2} & 0 \\ 0 & 0 & \sqrt{2} \\ 0 & 0 & 0 \end{bmatrix} \text{ and}$$

$$J^- \equiv \hbar \begin{bmatrix} 0 & 0 & 0 \\ \sqrt{2} & 0 & 0 \\ 0 & \sqrt{2} & 0 \end{bmatrix}.$$

5.5 Spin Angular Momentum

The Stern-Gerlach experiment pointed to another source of magnetic moment, which is proportional to the spin angular momentum $\mathbf{M}_s \propto \mathbf{S}$. From this and other experiments, it was concluded that each elementary particle has intrinsic angular momentum, known as spin angular momentum, and it is denoted by $\mathbf{S}$, which is different from the orbital angular momentum. Spin is a new degree of freedom in addition to the spatial coordinates (x, y, z). Unlike the spatial coordinates, spin only takes a discrete set of values. The proton also has spin of equal magnitude, but the magnetic momentum due to the proton spin is much smaller and it is neglected in the experiments. Since $\mathbf{S}$ is an angular momentum, it too satisfies the commutation relations:

$$[S_x, S_y] = i\hbar S_z, [S_y, S_z] = i\hbar S_x, \text{and} [S_z, S_x] = i\hbar S_y \tag{5.52a}$$

and:

$$\mathbf{S} \times \mathbf{S} = i\hbar\mathbf{S} \tag{5.52b}$$

Spin angular momentum has many properties in common with orbital angular momentum. However, there is one important difference. Unlike orbital angular momentum, spin angular momentum operators cannot be expressed in terms of position ($\boldsymbol{r}$) and linear momentum ($\boldsymbol{p}$) operators, since this identification depends on an analogy with classical mechanics. *The concept of spin has no analogy in classical physics; it is a purely quantum mechanical concept.* Consequently, the restriction that the quantum number l must take *integer* values is lifted for spin angular momentum, and the quantum number s can take *half-integer* values. All relations for angular momentum discussed in Sections 5.1 and 5.2 are applicable to $\mathbf{S}$. Equations (5.47) for $\mathbf{S}$ are:

$$\mathbf{S}^2 |s, s_z\rangle = s(s+1)\hbar^2 |s, s_z\rangle \tag{5.53a}$$

and

$$S_z |s, s_z\rangle = s_z\hbar |s, s_z\rangle \tag{5.53b}$$

where quantum number s can, in principle, take integer or half-integer values, and the quantum number s_z can only take the values $-s, -s+1......, -1, 0, 1, ...s-1, s$. There is a fixed magnitude of spin vector for each elementary particle, which is given by the quantum number s. However, the projection of the spin onto one axis, generally chosen to be the $z-$axis, is needed in addition to the coordinates (or momenta) to fully specify the state of the particle. To have a complete description of spin, one requires relativistic quantum mechanics. The magnetic moment of a charged particle is contributed by spin angular momentum. It is predicted from relativistic quantum mechanics that a particle of charge q with spin must possess a magnetic moment $\mu_s = \mathrm{g}q\mathbf{S}/2m$, where $\mathrm{g} = -2.0023$ for the electron.

5.5.1 Pauli Theory of Spin One-half Systems

Particles such as electrons, protons, and neutrons have spin $s = \frac{1}{2}$. In such a case, a projection onto the z-axis can take only two values, $s_z = \frac{1}{2}$ and $s_z = -\frac{1}{2}$. One therefore introduces a new variable ε, which takes two values $+(\uparrow)$ and $-(\downarrow)$. Thus, a full set of coordinates to describe a spin $\frac{1}{2}$ particle is represented by (x, y, z, ε), where spin operators $\mathbf{S}^2, S_z, S^+, S^-$ act only on the spin variable and therefore they commute with an operator that acts on the (x, y, z)-space. A complete set of commuting operators in spin variable space is formed by S^2 and S_z. The eigenvectors of $\mathbf{S}^2$ and S_z are specified by:

$$|+\rangle = \left|\frac{1}{2}, \frac{1}{2}\right\rangle \left(s = \frac{1}{2}, s_z = \frac{1}{2}\right) \tag{5.54a}$$

$$|-\rangle = \left|\frac{1}{2}, -\frac{1}{2}\right\rangle \left(s = \frac{1}{2}, s_z = -\frac{1}{2}\right) \tag{5.54b}$$

The eigenvectors are orthonormal $\langle +|-\rangle = 0$ and $\langle +|+\rangle = \langle -|-\rangle = 1$. A generalized spin state is:

$$|\chi\rangle = c^+|+\rangle + c^-|-\rangle \tag{5.55}$$

Equations (5.53) are reduced to:

$$\mathbf{S}^2|\pm\rangle = \frac{3}{4}\hbar^2|\pm\rangle \tag{5.56a}$$

$$S_z|\pm\rangle = \pm\frac{\hbar}{2}|\pm\rangle \tag{5.56b}$$

By choosing $|+\rangle$ and $|-\rangle$ as basis vectors for spin space, we write:

$$|\chi\rangle = \begin{pmatrix} c^+ \\ c^- \end{pmatrix} \tag{5.57a}$$

particularly when:

$$|+\rangle = \begin{pmatrix} 1 \\ 0 \end{pmatrix} \text{ and } |-\rangle = \begin{pmatrix} 0 \\ 1 \end{pmatrix} \tag{5.57b}$$

Then, all spin operators $\mathbf{S}^2, S_z, S^+, S^-$ are represented by 2×2 matrices. To find the matrix for S_z, we write:

$$S_z|+\rangle = \frac{\hbar}{2}|+\rangle \Rightarrow \begin{pmatrix} a & b \\ c & d \end{pmatrix}\begin{pmatrix} 1 \\ 0 \end{pmatrix} = \frac{\hbar}{2}\begin{pmatrix} 1 \\ 0 \end{pmatrix} \tag{5.58a}$$

and:

$$S_z|-\rangle = -\frac{\hbar}{2}|-\rangle \Rightarrow \begin{pmatrix} a & b \\ c & d \end{pmatrix}\begin{pmatrix} 0 \\ 1 \end{pmatrix} = -\frac{\hbar}{2}\begin{pmatrix} 0 \\ 1 \end{pmatrix} \tag{5.58b}$$

which gives: $a = \hbar/2$, $d = -\hbar/2$ and $c = b = 0$. Hence:

$$S_z = \frac{\hbar}{2}\begin{pmatrix} 1 & 0 \\ 0 & -1 \end{pmatrix} = \frac{\hbar}{2}\sigma_z \tag{5.58c}$$

where $\sigma_z = \begin{pmatrix} 1 & 0 \\ 0 & -1 \end{pmatrix}$. Equations (5.48) for spin angular momentum become:

$$S^+|s, s_z\rangle = \hbar\sqrt{s(s+1) - s_z(s_z+1)}\,|s, s_z+1\rangle \tag{5.59a}$$

and

$$S^-|s,s_z\rangle = \hbar\sqrt{s(s+1)-s_z(s_z-1)}\,|s,s_z-1\rangle \tag{5.59b}$$

Therefore:

$$S^+|+\rangle = 0 \Rightarrow \begin{pmatrix} a & b \\ c & d \end{pmatrix}\begin{pmatrix} 1 \\ 0 \end{pmatrix} = \begin{pmatrix} 0 \\ 0 \end{pmatrix} \tag{5.60a}$$

and

$$S^+|-\rangle = \hbar|-\rangle \Rightarrow \begin{pmatrix} a & b \\ c & d \end{pmatrix}\begin{pmatrix} 0 \\ 1 \end{pmatrix} = \hbar\begin{pmatrix} 0 \\ 1 \end{pmatrix} \tag{5.60b}$$

Solution of Eqns. (5.60) gives:

$$S^+ = \hbar\begin{pmatrix} 0 & 1 \\ 0 & 0 \end{pmatrix} \tag{5.61a}$$

Similarly, we obtain:

$$S^- = \hbar\begin{pmatrix} 0 & 0 \\ 1 & 0 \end{pmatrix} \tag{5.61b}$$

Since $S^+ = S_x + iS_y$ and $S^- = S_x - iS_y$, we have:

$$S_x = \frac{\hbar}{2}\begin{pmatrix} 0 & 1 \\ 1 & 0 \end{pmatrix} = \frac{\hbar}{2}\sigma_x \tag{5.62a}$$

and

$$S_y = \frac{\hbar}{2}\begin{pmatrix} 0 & -i \\ i & 0 \end{pmatrix} = \frac{\hbar}{2}\sigma_y \tag{5.62b}$$

The matrices $\sigma_x = \begin{pmatrix} 0 & 1 \\ 1 & 0 \end{pmatrix}$, $\sigma_y = \begin{pmatrix} 0 & -i \\ i & 0 \end{pmatrix}$, and $\sigma_z = \begin{pmatrix} 1 & 0 \\ 0 & -1 \end{pmatrix}$ are known as Pauli matrices, because Pauli was the first person to recognize the need of two component state vectors to explain some of the features of experimentally observed atomic spectrum. It is easy to verify that:

$$\sigma_x^2 = \sigma_y^2 = \sigma_z^2 = 1, \tag{5.63a}$$

$$\sigma_x\sigma_y = -\sigma_y\sigma_x = i\sigma_z, \sigma_y\sigma_z = -\sigma_z\sigma_y = i\sigma_x, \text{and}\, \sigma_z\sigma_x = -\sigma_x\sigma_z = i\sigma_y \tag{5.63b}$$

We thus see that Pauli matrices anticommute and the square of each matrix is equal to the unit matrix.

5.6 Addition of Angular Momentum

Addition of angular momentum operators is an important problem in quantum mechanics. It could be the addition of orbital angular momentum and spin angular momentum or the addition of general angular momentum of two particles (systems). We consider the case of addition of two sets of angular momentum operators $\mathbf{J}_1$ and $\mathbf{J}_2$, which are Hermitian and obey the fundamental commutation relations:

$$\mathbf{J}_1 \times \mathbf{J}_1 = i\hbar \mathbf{J}_1 \text{ and } \mathbf{J}_2 \times \mathbf{J}_2 = i\hbar \mathbf{J}_2 \tag{5.64}$$

$$\left[J_{ni}, J_{nj} \right] = i\hbar \varepsilon_{ijk} J_{nk} \tag{5.65}$$

where n takes values 1 or 2. The (i, j, k) stand for (x, y, z) and the symbol ε_{ijk} is equal to +1 if ijk take values in cyclic order (123, 231, 312), and it is equal to -1 when ijk take values not in cyclic order. It is equal to zero if any two of ijk are assigned the same value, for example 112. Also, we have:

$$[\mathbf{J}_n^2, J_{nx}] = [J_n^2, J_{ny}] = [J_n^2, J_{nz}] = 0 \tag{5.66}$$

We further say that the two groups of operators correspond to different degrees of freedom of the system, and therefore:

$$\left[J_{1i}, J_{2j} \right] = 0 \tag{5.67}$$

The $\mathbf{J}_1$ and $\mathbf{J}_2$ can also be an orbital angular momentum operator and a spin angular momentum operator, or the orbital angular momentum operators of two different particles in a multi-particle system. From Eqns. (5.47), we write:

$$\mathbf{J}_1^2 \left| j_1, m_1 \right\rangle = j_1(j_1 + 1)\hbar^2 \left| j_1, m_1 \right\rangle \tag{5.68a}$$

$$J_{1z} \left| j_1, m_1 \right\rangle = m_1 \hbar \left| j_1, m_1 \right\rangle \tag{5.68b}$$

$$\mathbf{J}_2^2 \left| j_2, m_2 \right\rangle = j_2(j_2 + 1)\hbar^2 \left| j_2, m_2 \right\rangle \tag{5.69a}$$

and

$$J_{2z} \left| j_2, m_2 \right\rangle = m_2 \hbar \left| j_2, m_1 \right\rangle \tag{5.69b}$$

where j_i $(i = 1, 2)$ can take the values $0, \frac{1}{2}, \frac{3}{2}, \frac{5}{2}, \ldots\ldots$ and m_i $(i = 1, 2)$ varies between $-j_i, -j_i + 1, -j_i + 2, \ldots\ldots -\frac{1}{2}, 0, \frac{1}{2}, \frac{3}{2}, \ldots . j_i - 1, j_i$.

Let us define the total angular momentum operator as:

$$\mathbf{J} = \mathbf{J}_1 + \mathbf{J}_2 \tag{5.70a}$$

with:

$$\mathbf{J} \times \mathbf{J} = i\hbar \mathbf{J} \tag{5.70b}$$

$\mathbf{J}_1$ and $\mathbf{J}_2$ are Hermitian operators and therefore $\mathbf{J}$ too is a Hermitian operator. Thus, $\mathbf{J}$ possesses all the expected properties of an angular momentum operator, like:

$$\mathbf{J}^2|j,m\rangle = j(j+1)\hbar^2|j,m\rangle \tag{5.71a}$$

and

$$J_z|j,m\rangle = m\hbar|j,m\rangle \tag{5.71b}$$

At this stage, we do not know how j and m can be expressed in terms of j_1, j_2, m_1 and m_2. We write $\mathbf{J}^2 = \mathbf{J}_1^2 + \mathbf{J}_2^2 + 2\mathbf{J}_1.\mathbf{J}_2$. And then with the use of Eqns. (5.64) and (5.65), we have:

$$\left[\mathbf{J}^2, \mathbf{J}_1^2\right] = \left[\mathbf{J}^2, \mathbf{J}_2^2\right] = 0 \tag{5.72}$$

This implies that the quantum numbers j_1, j_2, and j can all be measured simultaneously, which means that one can know the magnitude of the total angular momentum together with the magnitudes of the component angular momenta. However:

$$\left[\mathbf{J}^2, J_{1z}\right] \neq 0, \text{ and } \left[\mathbf{J}^2, J_{2z}\right] \neq 0 \tag{5.73}$$

which states that it is not possible to measure the quantum numbers m_1 and m_2 simultaneously with the quantum number j. Thus, we cannot determine the projections of the individual angular momenta along the $z-$axis and the magnitude of the total angular momentum simultaneously. From the above discussions, it became clear that we form two separate groups of mutually commuting operators: (i) one group is $\left(\mathbf{J}_1^2, \mathbf{J}_2^2, J_{1z}, J_{2z}\right)$ and (ii) another group is $\left(\mathbf{J}_1^2, \mathbf{J}_2^2, \mathbf{J}^2, J_z\right)$, which are incompatible with one another. We can define simultaneous wave function kets for each operator group. Let us denote the simultaneous wave function kets for $\left(\mathbf{J}_1^2, \mathbf{J}_2^2, J_{1z}, J_{2z}\right)$ by $|j_1, j_2, m_1, m_2\rangle$. We then have:

$$\mathbf{J}_1^2|j_1, j_2, m_1, m_2\rangle = j_1(j_1+1)\hbar^2|j_1, j_2, m_1, m_2\rangle \tag{5.74a}$$

$$\mathbf{J}_2^2|j_1, j_2, m_1, m_2\rangle = j_2(j_2+1)\hbar^2|j_1, j_2, m_1, m_2\rangle \tag{5.74b}$$

$$J_{1z}|j_1, j_2, m_1, m_2\rangle = m_1\hbar|j_1, j_2, m_1, m_2\rangle \tag{5.74c}$$

$$J_{2z}|j_1, j_2, m_1, m_2\rangle = m_2\hbar|j_1, j_2, m_1, m_2\rangle \tag{5.74d}$$

Similarly, simultaneous wave function kets for $\left(\mathbf{J}_1^2, \mathbf{J}_2^2, \mathbf{J}^2, J_z\right)$ are taken as $|j_1, j_2, j, m\rangle$ to write:

$$\mathbf{J}_1^2|j_1, j_2, j, m\rangle = j_1(j_1+1)\hbar^2|j_1, j_2, j, m\rangle \tag{5.75a}$$

$$\mathbf{J}_2^2|j_1, j_2, j, m\rangle = j_2(j_2+1)\hbar^2|j_1, j_2, j, m\rangle \tag{5.75b}$$

$$\mathbf{J}^2|j_1, j_2, j, m\rangle = j(j+1)\hbar^2|j_1, j_2, j, m\rangle \tag{5.75c}$$

$$J_z|j_1, j_2, j, m\rangle = m\hbar|j_1, j_2, j, m\rangle \tag{5.75d}$$

Each wave function ket is complete and normalized. Also, $|j_1, j_2, m_1, m_2\rangle$ and $|j_1, j_2, j, m\rangle$ are mutually orthogonal. Since, the operators $\mathbf{J}_1^2$ and $\mathbf{J}_2^2$ are common to both operator groups, we assume that the quantum numbers $j_1$and j_2 are known and hence we can always determine the magnitudes of the individual angular momenta. Additionally, one can either know the quantum numbers m_1 and m_2, or the quantum numbers j and m, but one cannot know both pairs of quantum numbers at the same time. We write a conventional completeness relation for both sets of wave function kets:

$$\sum_{m_1, m_2} |j_1, j_2, m_1, m_2\rangle\langle j_1, j_2, m_1, m_2| = 1 \tag{5.76a}$$

$$\sum_{j,m} |j_1, j_2, j, m\rangle\langle j_1, j_2, j, m| = 1 \tag{5.76b}$$

The right-hand sides of Eqns. (5.76) denote the identity operator in the ket-space, corresponding to the states of given j_1 and j_2. The summation is taken over all allowed values of m_1, m_2, j, and m. The incompatibility between two groups of operators means that if the system is in a simultaneous eigenstate of the former group, then, in general, it is not in an eigenstate of the latter, which also means that if the quantum numbers j_1, j_2, j, and m are known with certainty, then a measurement of the quantum numbers m_1 and m_2 will give a range of possible values. The completeness relation (Eqn. 5.76a) allows us to write:

$$|j_1, j_2, j, m\rangle = \sum_{m_1, m_2} \langle j_1, j_2, m_1, m_2 | j_1, j_2, j, m\rangle |j_1, j_2, m_1, m_2\rangle \tag{5.77}$$

We thus find that wave function kets of the first group of operators are the weighted sum of wave function kets of the second group.

5.6.1 Clebsch-Gordon Coefficients and their Properties

The weights $\langle j_1, j_2, m_1, m_2 | j_1, j_2, j, m\rangle$ are known as Clebsch-Gordon coefficients, which means that if the measurements on $\mathbf{J}_1^2, \mathbf{J}_2^2, \mathbf{J}^2$ and J_z in a state of the system are bound to give results, $j_1(j_1+1)\hbar^2$, $j_2(j_2+1)\hbar^2$, $j(j+1)\hbar^2$, and $m\hbar$, respectively, then the measurements over J_{1z} and J_{2z} will yield results $m_1 h$ and $m_2 h$, with the probability of $|\langle j_1, j_2, m_1, m_2 | j_1, j_2, m_1, m_2\rangle|^2$. The Clebsch-Gordon coefficients have many important properties.

1. The coefficients are zero unless $m = m_1 + m_2$. To prove it note that:

$$\begin{aligned} &(J_z - J_{1z} - J_{2z})|j_1, j_2, j, m\rangle = 0 \\ &\Rightarrow (m - m_1 - m_2)|j_1, j_2, j, m\rangle = 0 \end{aligned} \tag{5.78}$$

On operating with $\langle j_1, j_2, m_1, m_2|$ we get:

$$(m - m_1 - m_2)\langle j_1, j_2, m_1, m_2 | j_1, j_2, j, m\rangle = 0 \tag{5.79}$$

which proves the assertion, suggesting that z-components of different angular momenta are added algebraically.

2. Coefficients vanish unless $|j_1 - j_2| \le j \le j_1 + j_2$. We assume that $j_1 \ge j_2$. Since the largest values that can be assigned to m_1 and m_2 are j_1 and j_2, the largest possible value m can take is $j_1 + j_2$. Further, m varies between j and $-j$, and therefore j too can take the largest value equal to $j_1 + j_2$. Also, the number of values m_1 and m_2 can take are $(2j_1+1)$ and $(2j_2+1)$, respectively. Therefore, allowable independent wave function kets values of $|j_1, j_2, m_1, m_2\rangle$ to span the entire ket-space are $(2j_1+1)(2j_2+1)$. Since $|j_1, j_2, j, m\rangle$ span the same ket-space, $(2j_1+1)(2j_2+1)$ must also be the number of independent $|j_1, j_2, j, m\rangle$ kets. As is seen:

$$\sum_{j_1-j_2}^{j_1+j_2} (2j+1) = (2j_1+1)(2j_2+1) \tag{5.80}$$

which suggests that if $j_1 + j_2$ is the maximum value of j then $j_1 - j_2$ is its minimum value. It should be noted from Eqn. (5.70a) that $j = \left(j_1^2 + j_2^2 + 2j_1 j_2 \cos\theta\right)^{1/2}$, which takes a maximum value $j_1 + j_2$ when vectors are parallel and a minimum value $j_1 - j_2$ when vectors are anti-parallel.

3. The sum of the modulus squared of all of the Clebsch-Gordon coefficients is unity, which means:

$$\sum_{m_1, m_2} \left|\langle j_1, j_2, m_1, m_2 | j_1, j_2, m_1, m_2\rangle\right|^2 = 1 \tag{5.81}$$

From Eqn. (5.77), we write:

$$\begin{aligned} \langle j_1, j_2, j, m | j_1, j_2, j, m\rangle &= \sum_{m_1, m_2} \langle j_1, j_2, m_1, m_2 | j_1, j_2, j, m\rangle \langle j_1, j_2, j, m | j_1, j_2, m_1, m_2\rangle \\ \Rightarrow 1 &= \left|\sum_{m_1, m_2} \langle j_1, j_2, m_1, m_2 | j_1, j_2, j, m\rangle\right|^2 \end{aligned} \tag{5.82}$$

where we have made use of the fact that $|j_1, j_2, j, m\rangle$ are normalized kets.

5.6.2 Recursion Relations for Clebsch-Gordon Coefficients

Two recursion relations are followed by Clebsch-Gordon coefficients. Consider the lowering and rising operators $J^- = J_1^- + J_2^- = \left(J_{1x} - iJ_{1y}\right) + \left(J_{2x} - iJ_{2y}\right)$ and $J^+ = J_1^+ + J_2^+ = \left(J_{1x} + iJ_{1y}\right) + \left(J_{2x} + iJ_{2y}\right)$. We write:

$$J^- |j_1, j_2, j, m\rangle = \sum_{m_1', m_2'} (J_1^- + J_2^-) |j_1, j_2, m_1', m_2'\rangle \langle j_1, j_2, m_1', m_2' | j_1, j_2, j, m\rangle \tag{5.83}$$

or:

$$\begin{aligned}
&\sqrt{(j(j+1)-m(m-1))}\hbar\left|j_1,j_2,j,m-1\right\rangle \\
&=\sum_{m_1',m_2'}\left(j_1(j_1+1)-m_1'(m_1'-1)\right)^{1/2}\hbar\left|j_1,j_2,m_1'-1,m_2'\right\rangle\left\langle j_1,j_2,m_1',m_2'\right|j_1,j_2,j,m\rangle \\
&+\sum_{m_1',m_2'}\left(j_2(j_2+1)-m_2'(m_2'-1)\right)^{1/2}\hbar\left|j_1,j_2,m_1',m_2'-1\right\rangle\left\langle j_1,j_2,m_1',m_2'\right|j_1,j_2,j,m\rangle
\end{aligned} \tag{5.84}$$

Operating both sides from the left with $\langle j_1,j_2,m_1,m_2|$, we get:

$$\begin{aligned}
&\sqrt{(j(j+1)-m(m-1))}\left\langle j_1,j_2,m_1,m_2\right|j_1,j_2,j,m-1\rangle \\
&=\left(j_1(j_1+1)-m_1(m_1+1)\right)^{1/2}\left\langle j_1,j_2,m_1+1,m_2\right|j_1,j_2,j,m\rangle \\
&+\left(j_2(j_2+1)-m_2(m_2+1)\right)^{1/2}\left\langle j_1,j_2,m_1,m_2+1\right|j_1,j_2,j,m\rangle
\end{aligned} \tag{5.85}$$

On replacing J^- by J^+ in Eqn. (5.83) and then following a similar procedure we obtain:

$$\begin{aligned}
&\sqrt{(j(j+1)-m(m+1))}\left\langle j_1,j_2,j,m_1,m_2\right|j_1,j_2,j,m+1\rangle \\
&=\left(j_1(j_1+1)-m_1(m_1-1)\right)^{1/2}\left\langle j_1,j_2,m_1-1,m_2\right|j_1,j_2,j,m\rangle \\
&+\left(j_2(j_2+1)-m_2(m_2-1)\right)^{1/2}\left\langle j_1,j_2,m_1,m_2-1\right|j_1,j_2,j,m\rangle
\end{aligned} \tag{5.86}$$

Equations (5.85) and (5.86) are two important recursion relations to compute the Clebsch-Gordon coefficients.

5.6.3 Computation of Clebsch-Gordon Coefficients

Let us write matrix elements $\langle j_1,j_2,m_1,m_2|j_1,j_2,j,m\rangle$ as $\langle m_1m_2|jm\rangle$ in short. The $\langle m_1m_2|jm\rangle$ can have $(2j_1+1)(2j_2+1)$ values because m_1 and m_2 have $(2j_1+1)$ and $(2j_2+1)$ values, respectively. Therefore, the number of elements in the matrix of Clebsch-Gordon coefficients will be $(2j_1+1)(2j_2+1)$. Depending upon the value of m, this matrix further breaks into submatrices of smaller size. For example, when $j=j_1+j_2$ and $m=j_1+j_2$ they are going to be a submatrix of 1×1. There will a submatrix of 2×2 if $m=j_1+j_2-1$ and $j=j_1+j_2$ or $j=j_1+j_2-1$. The rank of these submatrices increases to reach a maximum value and then it decreases to 1. Generally, the first 1×1 submatrix is chosen to be +1:

$$\langle m_1m_2|jm\rangle=\langle j_1,j_2|j_1+j_2,j_1+j_2\rangle=1 \tag{5.87}$$

To compute the next 2×2 submatrix, take $m_1=j_1$ and $m_2=j_2-1$, $m=j_1+j_2$, and $j=j_1+j_2$. Then from Eqn. (5.85) we get:

$$\begin{aligned}
&(j_1+j_2)^{1/2}\langle j_1,j_2-1|(j_1+j_2),(j_1+j_2-1)\rangle=(j_2)^{1/2}\langle j_1,j_2|(j_1+j_2),(j_1+j_2)\rangle \\
&\Rightarrow\langle j_1,j_2-1|(j_1+j_2),(j_1+j_2-1)\rangle=\left(\frac{j_2}{j_1+j_2}\right)^{1/2}
\end{aligned} \tag{5.88}$$

Similarly, when $m_1 = j_1 - 1$ and $m_2 = j_2$, $m = j_1 + j_2$, and $j = j_1 + j_2$, we get:

$$\langle j_1 - 1, j_2 | (j_1 + j_2), (j_1 + j_2 - 1)\rangle = \left(\frac{j_1}{j_1 + j_2}\right)^{1/2} \tag{5.89}$$

The Eqns. (5.88) and (5.89) describe two elements for the case of $j = j_1 + j_2$ and $m = j_1 + j_2 - 1$. The other two elements are $\langle j_1, j_2 - 1 | (j_1 + j_2 - 1), (j_1 + j_2 - 1)\rangle$ and $\langle j_1 - 1, j_2 | (j_1 + j_2 - 1), (j_1 + j_2 - 1)\rangle$, which are evaluated with the use of the unitary nature of the transformation matrix, Eqn. (5.81). We obtain:

$$\begin{aligned} &\langle j_1, j_2 - 1 | (j_1 + j_2 - 1), (j_1 + j_2 - 1)\rangle \langle (j_1 + j_2), (j_1 + j_2 - 1) | j_1, j_2 - 1\rangle + \\ &\langle (j_1 + j_2), (j_1 + j_2 - 1) | j_1 - 1, j_2\rangle \langle j_1 - 1, j_2 | (j_1 + j_2 - 1), (j_1 + j_2 - 1)\rangle = 0 \end{aligned} \tag{5.90}$$

Since $\langle m_1 m_2 | jm\rangle = \langle jm | m_1 m_2\rangle^*$ matrix elements are real and positive, Eqn. (5.90) reduces to:

$$\begin{aligned} &\langle j_1, j_2 - 1 | (j_1 + j_2 - 1), (j_1 + j_2 - 1)\rangle \left(\frac{j_2}{j_1 + j_2}\right)^{1/2} + \\ &\langle j - 1_1, j_2 | (j_1 + j_2 - 1), (j_1 + j_2 - 1)\rangle \left(\frac{j_1}{j_1 + j_2}\right)^{1/2} = 0 \end{aligned} \tag{5.91}$$

Conventionally, matrix elements are real and positive, and hence we can write:

$$\langle j_1 - 1, j_2 | (j_1 + j_2 - 1), (j_1 + j_2 - 1)\rangle = -\left(\frac{j_2}{j_1 + j_2}\right)^{1/2} \tag{5.92}$$

$$\langle j_1, j_2 - 1 | (j_1 + j_2 - 1), (j_1 + j_2 - 1)\rangle = \left(\frac{j_1}{j_1 + j_2}\right)^{1/2} \tag{5.93}$$

The results of Eqns. (5.88), (5.89), (5.92), and (5.93) can be summarized as follows:

		$\lvert jm\rangle$	
m_1	m_2	$\lvert j_1 + j_2, j_1 + j_2 - 1\rangle$	$\lvert j_1 + j_2 - 1, j_1 + j_2 - 1\rangle$
j_1	$j_2 - 1$	$\left(\frac{j_2}{j_1 + j_2}\right)^{1/2}$	$\left(\frac{j_1}{j_1 + j_2}\right)^{1/2}$
$j_1 - 1$	j_2	$\left(\frac{j_1}{j_1 + j_2}\right)^{1/2}$	$-\left(\frac{j_2}{j_1 + j_2}\right)^{1/2}$

In a similar manner one can get the submatrices of 3×3 and 4×4 for the values of j and m.

5.7 Solved Examples

1. Show that matrices:

$$L_x = \frac{\hbar}{\sqrt{2}}\begin{bmatrix} 0 & 1 & 0 \\ 1 & 0 & 1 \\ 0 & 1 & 0 \end{bmatrix},\ L_y = \frac{\hbar}{\sqrt{2}}\begin{bmatrix} 0 & -i & 0 \\ i & 0 & -i \\ 0 & i & 0 \end{bmatrix},\ \text{and } L_z = \hbar\begin{bmatrix} 1 & 0 & 0 \\ 0 & 0 & 0 \\ 0 & 0 & -1 \end{bmatrix}$$

satisfy the relation $\left[L_x, L_y\right] = i\hbar L_z$.

SOLUTION

$$L_xL_y - L_yL_x = \frac{\hbar^2}{2}\begin{bmatrix} 0 & 1 & 0 \\ 1 & 0 & 1 \\ 0 & 1 & 0 \end{bmatrix}\begin{bmatrix} 0 & -i & 0 \\ i & 0 & -i \\ 0 & i & 0 \end{bmatrix} - \frac{\hbar^2}{2}\begin{bmatrix} 0 & -i & 0 \\ i & 0 & -i \\ 0 & i & 0 \end{bmatrix}\begin{bmatrix} 0 & 1 & 0 \\ 1 & 0 & 1 \\ 0 & 1 & 0 \end{bmatrix}$$

$$= \frac{\hbar^2}{2}\begin{bmatrix} i & 0 & -i \\ 0 & 0 & 0 \\ i & 0 & -i \end{bmatrix} - \frac{\hbar^2}{2}\begin{bmatrix} -i & 0 & -i \\ 0 & 0 & 0 \\ i & 0 & i \end{bmatrix}$$

$$= \frac{\hbar^2}{2}\begin{bmatrix} 2i & 0 & 0 \\ 0 & 0 & 0 \\ 0 & 0 & -2i \end{bmatrix} = i\hbar \times \hbar\begin{bmatrix} 1 & 0 & 0 \\ 0 & 0 & 0 \\ 0 & 0 & -1 \end{bmatrix}$$

$$= i\hbar L_z$$

It can therefore be said that the above matrices give 3×3 matrix representation of components of orbital angular momentum.

2. Show that the vectors $|-1\rangle = \begin{bmatrix} 0 \\ 0 \\ 1 \end{bmatrix},\ |0\rangle = \begin{bmatrix} 0 \\ 1 \\ 0 \end{bmatrix},\ |1\rangle = \begin{bmatrix} 1 \\ 0 \\ 0 \end{bmatrix}$ are the wave function kets for the matrix representing L_z in the above example 1.

SOLUTION

The eigenvalue equation for L_z is $L_z|Y\rangle = m\hbar|Y\rangle$. We therefore have:

$$L_z|-1\rangle = \hbar\begin{bmatrix} 1 & 0 & 0 \\ 0 & 0 & 0 \\ 0 & 0 & -1 \end{bmatrix}\begin{bmatrix} 0 \\ 0 \\ 1 \end{bmatrix} = -\hbar\begin{bmatrix} 0 \\ 0 \\ 1 \end{bmatrix} = -\hbar|-1\rangle \tag{5.94a}$$

$$L_z|0\rangle = \hbar\begin{bmatrix} 1 & 0 & 0 \\ 0 & 0 & 0 \\ 0 & 0 & -1 \end{bmatrix}\begin{bmatrix} 0 \\ 1 \\ 0 \end{bmatrix} = 0 \times \hbar\begin{bmatrix} 0 \\ 1 \\ 0 \end{bmatrix} = 0 \times \hbar|0\rangle = 0 \tag{5.94b}$$

and

$$L_z|1\rangle = \hbar\begin{bmatrix} 1 & 0 & 0 \\ 0 & 0 & 0 \\ 0 & 0 & -1 \end{bmatrix}\begin{bmatrix} 1 \\ 0 \\ 0 \end{bmatrix} = \hbar\begin{bmatrix} 1 \\ 0 \\ 0 \end{bmatrix} = \hbar|1\rangle \tag{5.94c}$$

3. For a system, the wave function ket $|\phi\rangle$ satisfies $L^2|\phi\rangle = l(l+1)\hbar^2|\phi\rangle$ and $L_z|\phi\rangle = m\hbar|\phi\rangle$. Calculate $\Delta L_y = \left\{\langle L_y^2\rangle - \langle L_y\rangle^2\right\}^{1/2}$ for the state of $|\phi\rangle$.

SOLUTION

Since L_z is Hermitian, we have $L_z|\phi\rangle = m\hbar|\phi\rangle$ and $\langle\phi|L_z = m\hbar\langle\phi|$. Therefore:

$$\langle L_z\rangle = \langle\phi|L_z|\phi\rangle = m\hbar\langle\phi|\phi\rangle = m\hbar \tag{5.95a}$$

$[L_z, L_x] = i\hbar L_y$ implies:

$$\begin{aligned}\langle L_y\rangle &= \frac{1}{i\hbar}\langle\phi|L_zL_x - L_xL_z|\phi\rangle \\ &= \frac{1}{i\hbar}\left[\langle\phi|L_zL_x|\phi\rangle - \langle\phi|L_xL_z|\phi\rangle\right] \\ &= -im\left[\langle\phi|L_x|\phi\rangle - \langle\phi|L_x|\phi\rangle\right] \\ &= 0\end{aligned} \tag{5.95b}$$

Further, symmetry in $x-y$ plane permits us to write $\langle L_x^2\rangle = \langle L_y^2\rangle$ and then:

$$\begin{aligned}\langle L_y^2\rangle &= \frac{1}{2}\langle(L_x^2 + L_y^2)\rangle \\ &= \frac{1}{2}\langle(L^2 - L_z^2)\rangle \\ &= \frac{1}{2}\langle\phi|L^2|\phi\rangle - \frac{1}{2}\langle\phi|L_z^2|\phi\rangle\end{aligned} \tag{5.96a}$$

Hence:

$$\langle L_y^2\rangle = \frac{1}{2}\left(l(l+1) - m^2\right)\hbar^2 \tag{5.96b}$$

Therefore:

$$\begin{aligned}\Delta L_y &= \left\{\langle L_y^2\rangle - \langle L_y\rangle^2\right\}^{1/2} \\ &= \hbar\left[\frac{l(l+1) - m^2}{2}\right]^{1/2}\end{aligned} \tag{5.97}$$

4. The state of an electron is described by $\psi = \frac{1}{2\sqrt{2\pi}}\left(e^{i\phi}\sin\theta + 2\cos\theta\right)f(r)$, where $\int_0^\infty |f(r)|^2 r^2 dr = 1$. (a) What are the possible results of a measurement on L_z in this state? (b) Calculate the expectation value of L_z. (c) Is ψ normalized?

SOLUTION

a. From Eqn. (5.46), we know that $Y_{10}(\theta,\phi) = \sqrt{\frac{3}{4\pi}}\cos\theta$, and $Y_{11}(\theta,\phi) = -\sqrt{\frac{3}{8\pi}}\sin\theta e^{i\phi}$, which gives $\psi = \frac{1}{\sqrt{3}}\left(\sqrt{2}Y_{10} - Y_{11}\right)f(r)$. Therefore, the given state has $l = 1$ and $m = 1$ and 0. The possible results of measurements on L_z are $+\hbar$ and 0.

b. The expectation value of L_z is:

$$\begin{aligned}
\langle L_z \rangle &= \int \psi^* L_z \psi d^3r \\
&= \frac{1}{3}\int_0^\infty\int_0^\pi\int_0^{2\pi} |f(r)|^2 r^2 dr\left(\sqrt{2}Y_{10} - Y_{11}\right)^* L_z\left(\sqrt{2}Y_{10} - Y_{11}\right)\sin\theta d\theta d\phi \\
&= \frac{\hbar}{3}\int_0^\pi\int_0^{2\pi}\left(\sqrt{2}Y_{10} - Y_{11}\right)^*(-Y_{11})\sin\theta d\theta d\phi \\
&= \frac{\hbar}{3}\int_0^\pi\int_0^{2\pi}|Y_{11}|^2\sin\theta d\theta d\phi - \frac{\sqrt{2}\hbar}{3}\int_0^\pi\int_0^{2\pi}Y_{10}^* Y_{11}\sin\theta d\theta d\phi \\
&= \frac{\hbar}{3}\langle Y_{11}|Y_{11}\rangle - \frac{\sqrt{2}\hbar}{3}\langle Y_{10}|Y_{11}\rangle \\
&= \frac{\hbar}{3}
\end{aligned} \tag{5.98}$$

In deriving the above results, we have used the fact that spherical harmonics are normalized and orthonormal.

c. The condition of normalization is $\langle\psi|\psi\rangle = \int\psi^*\psi d^3r = 1$

$$\begin{aligned}
\langle\psi|\psi\rangle &= \int\psi^*\psi d^3r \\
&= \frac{1}{3}\int_0^\infty\int_0^\pi\int_0^{2\pi}|f(r)|^2 r^2 dr\left(\sqrt{2}Y_{10} - Y_{11}\right)^*\left(\sqrt{2}Y_{10} - Y_{11}\right)\sin\theta d\theta d\phi \\
&= \frac{1}{3}\int_0^\pi\int_0^{2\pi}\left(\sqrt{2}Y_{10} - Y_{11}\right)^*\left(\sqrt{2}Y_{10} - Y_{11}\right)\sin\theta d\theta d\phi \\
&= \frac{1}{3}\int_0^\pi\int_0^{2\pi}2|Y_{10}|^2\sin\theta d\theta d\phi + \frac{1}{3}\int_0^\pi\int_0^{2\pi}|Y_{11}|^2\sin\theta d\theta d\phi - \frac{\sqrt{2}}{3}\int_0^\pi\int_0^{2\pi}\left(Y_{10}^*Y_{11} + Y_{11}^*Y_{10}\right)\sin\theta d\theta d\phi \\
&= \frac{2}{3}\langle Y_{10}|Y_{10}\rangle + \frac{1}{3}\langle Y_{11}|Y_{11}\rangle - \frac{\sqrt{2}}{3}\left(\langle Y_{10}|Y_{11}\rangle + \langle Y_{11}|Y_{10}\rangle\right) \\
&= 1
\end{aligned} \tag{5.99}$$

We thus find that ψ is normalized.

5. For a particle of spin $\frac{1}{2}$, consider an operator $A = 3S_x + 4S_z$, where S_x and S_z are spin angular momentum operators. Express A in matrix form and then find the energy eigenvalues and eigenvectors of the particle.

SOLUTION

The eigenvalue equation is $A|\psi\rangle = \lambda|\psi\rangle$. The matrices representing S_z and S_x are:

$$S_z = \frac{\hbar}{2}\sigma_z = \frac{\hbar}{2}\begin{bmatrix} 1 & 0 \\ 0 & -1 \end{bmatrix} \text{ and } S_x = \frac{\hbar}{2}\sigma_x = \frac{\hbar}{2}\begin{bmatrix} 0 & 1 \\ 1 & 0 \end{bmatrix} \tag{5.100}$$

Therefore:

$$A = 3\frac{\hbar}{2}\begin{bmatrix} 0 & 1 \\ 1 & 0 \end{bmatrix} + 4\frac{\hbar}{2}\begin{bmatrix} 1 & 0 \\ 0 & -1 \end{bmatrix} = \begin{bmatrix} 2\hbar & \frac{3\hbar}{2} \\ \frac{3\hbar}{2} & -2\hbar \end{bmatrix} \tag{5.101}$$

The eigenvalues are given by:

$$\begin{vmatrix} 2\hbar - \lambda & \frac{3\hbar}{2} \\ \frac{3\hbar}{2} & -2\hbar - \lambda \end{vmatrix} = 0$$

$$\Rightarrow -(2\hbar - \lambda)(2\hbar + \lambda) - \left(\frac{9\hbar^2}{4}\right) = 0 \tag{5.102}$$

$$\Rightarrow \lambda^2 = \frac{25\hbar^2}{4}$$

$$\Rightarrow \lambda = \pm\frac{5\hbar}{2}$$

Two eigenvalues are:

$\lambda_1 = \frac{5\hbar}{2}$ and $\lambda_2 = -\frac{5\hbar}{2}$. The eigenvector that corresponds to λ_1 is $|X_1\rangle = \begin{pmatrix} x_{11} \\ x_{21} \end{pmatrix}$, components of which are given by:

$$\frac{\hbar}{2}\begin{bmatrix} -1 & 3 \\ 3 & -9 \end{bmatrix}\begin{bmatrix} x_{11} \\ x_{21} \end{bmatrix} = 0$$

$$\Rightarrow -x_{11} + 3x_{21} = 0 \tag{5.103}$$

$$3x_{11} - 9x_{21} = 0$$

which along with the condition of normalization $|x_{11}|^2 + |x_{21}|^2 = 1$ gives:

$$x_{11} = \pm\frac{3}{\sqrt{10}} \text{ and } x_{21} = \pm\frac{1}{\sqrt{10}}$$
$$\text{or } |X_1\rangle = \pm\frac{1}{\sqrt{10}}\begin{bmatrix} 3 \\ 1 \end{bmatrix} \tag{5.104}$$

Similar calculations for the eigenvector $|X_2\rangle = \begin{pmatrix} x_{12} \\ x_{22} \end{pmatrix}$ that belongs to $\lambda_2 = -\frac{5\hbar}{2}$ yield:

$$|X_2\rangle = \pm\frac{1}{\sqrt{10}}\begin{bmatrix} 1 \\ 3 \end{bmatrix} \tag{5.105}$$

6. Components of arbitrary vectors A and B commute with those of σ. Show that:

$$(\sigma.\mathbf{A})(\sigma.\mathbf{B}) = \mathbf{A}.\mathbf{B} + i\sigma.(\mathbf{A}\times\mathbf{B})$$

SOLUTION

$$(\sigma.\mathbf{A}) = \begin{pmatrix} 0 & A_x \\ A_x & 0 \end{pmatrix} + \begin{pmatrix} 0 & -iA_y \\ iA_y & 0 \end{pmatrix} + \begin{pmatrix} A_z & 0 \\ 0 & -A_z \end{pmatrix} = \begin{pmatrix} A_z & A^- \\ A^+ & -A_z \end{pmatrix},$$

and

$(\sigma.\mathbf{B}) = \begin{pmatrix} B_z & B^- \\ B^+ & -B_z \end{pmatrix}$ where $A^{\pm} = A_x \pm iA_y$; $B^{\pm} = B_x \pm iB_y$. We therefore have:

$$(\sigma.\mathbf{A})(\sigma.\mathbf{B}) = \begin{pmatrix} A_zB_z + A^-B^+ & A_zB^- - A^-B_z \\ A^+B_z - A_zB^+ & A^+B^- + A_zB_z \end{pmatrix} \tag{5.106}$$

$$\mathbf{A}.\mathbf{B} = A_xB_x + A_yB_y + A_zB_z$$
$$= \begin{bmatrix} A_xB_x + A_yB_y + A_zB_z & 0 \\ 0 & A_xB_x + A_yB_y + A_zBz \end{bmatrix}$$

and,

$$i\sigma.(\mathbf{A}\times\mathbf{B}) = i\left[\sigma_x(A_yB_z - A_zB_y) + \sigma_y(A_zB_x - A_xB_z) + \sigma_z(A_xB_y - A_yB_x)\right]$$
$$= \begin{bmatrix} 0 & i \\ i & 0 \end{bmatrix}(A_yB_z - A_zB_y) + \begin{bmatrix} 0 & 1 \\ -1 & 0 \end{bmatrix}(A_zB_x - A_xB_z) + \begin{bmatrix} i & 0 \\ 0 & -i \end{bmatrix}(A_xB_y - A_yB_x).$$
$$= \begin{bmatrix} i(A_xB_y - A_yB_x) & A_zB^- - B_zA^- \\ A_zB^+ - B_zA^+ & -i(A_xB_y - A_yB_x) \end{bmatrix}$$

Therefore:

$$\mathbf{A}.\mathbf{B} + i\sigma.(\mathbf{A}\times\mathbf{B}) = \begin{pmatrix} A_zB_z + A^-B^+ & -B_zA^- + B^-A_z \\ -B^+A_z + B_zA^+ & A^+B^- + A_zB_z \end{pmatrix} \tag{5.107}$$

Hence $(\sigma.\mathbf{A})(\sigma.\mathbf{B}) = \mathbf{A}.\mathbf{B} + i\sigma.(\mathbf{A}\times\mathbf{B})$.

7. A system of two particles each having spin $-\frac{1}{2}$ is described by the Hamiltonian $H = 2\mathbf{S}_1.\mathbf{S}_2$, where $\mathbf{S}_1$ and $\mathbf{S}_2$ are two spin vectors. Calculate the energy eigenvalues of the Hamiltonian.

SOLUTION

Let us say that $|s,m\rangle$ is the eigenvector which satisfies $\mathbf{S}^2|s,m\rangle = s(s+1)\hbar^2|s,m\rangle$, where s can take two values, 1 and 0. $s=1$ is a triplet and symmetric state, while $s=0$, is a singlet and anti-symmetric state. For triplet states, $\mathbf{S}^2|s,m\rangle = 2\hbar^2|s,m\rangle$ and for a singlet state, $\mathbf{S}^2|s,m\rangle = 0$. Now:

$$\begin{aligned} \mathbf{S}^2 &= (\mathbf{S}_1 + \mathbf{S}_2)^2 \\ &= \mathbf{S}_1^2 + \mathbf{S}_2^2 + 2\mathbf{S}_1.\mathbf{S}_2 \\ &= \left(S_{1x}^2 + S_{1y}^2 + S_{1z}^2 + S_{2x}^2 + S_{2y}^2 + S_{2z}^2\right) + 2\mathbf{S}_1.\mathbf{S}_2 \\ &= 6\left(\frac{\hbar}{2}\right)^2 + 2\mathbf{S}_1.\mathbf{S}_2 \end{aligned} \tag{5.108}$$

Therefore:

$$\begin{aligned} H|s,m\rangle &= 2\mathbf{S}_1.\mathbf{S}_2|s,m\rangle \\ &= \left(\mathbf{S}^2 - \frac{3}{2}\hbar^2\right)|s,m\rangle \\ &= \left(s(s+1)\hbar^2 - \frac{3}{2}\hbar^2\right)|s,m\rangle \end{aligned} \tag{5.109}$$

which gives the eigenvalue of H. It is equal to $\frac{\hbar^2}{2}$ for each of the triplet states and $-\frac{3\hbar^2}{2}$ for the singlet state.

There are four states, three of which belong to ($s=1$, triplet states) and one belongs to a singlet state ($s=0$). The energy of each of triplet state is $\frac{1}{2}\hbar^2$, which corresponds to the eigenvectors $|1,1\rangle;|1,0\rangle$ and $|1,-1\rangle$, respectively. For the singlet state, the energy is $-\frac{3\hbar^2}{2}$ and the corresponding eigenvector is $|0,0\rangle$.

8. Evaluate $\left[L_x^2, L_y\right]$ and $\left[L_x, L_y^2\right]$.

SOLUTION

$$\begin{aligned}\left[L_x^2, L_y\right] &= L_x^2 L_y - L_y L_x^2 \\ &= L_x^2 L_y - L_x L_y L_x + L_x L_y L_x - L_y L_x^2 \\ &= L_x\left[L_x, L_y\right] + \left[L_x, L_y\right] L_x \\ &= i\hbar\left(L_x L_z + L_z L_x\right) \\ &= i\hbar\left(2L_x L_z + L_z L_x - L_x L_z\right) \\ &= -\hbar^2 L_y + 2i\hbar L_x L_z\end{aligned} \tag{5.110}$$

and

$$\begin{aligned}\left[L_x, L_y^2\right] &= L_x L_y^2 - L_y^2 L_x \\ &= L_x L_y^2 - L_y L_x L_y + L_y L_x L_y - L_y^2 L_x \\ &= \left[L_x, L_y\right] L_y + L_y\left[L_x, L_y\right] \\ &= i\hbar\left(L_z L_y + L_y L_z\right) \\ &= -\hbar^2 L_x + 2i\hbar L_y L_z\end{aligned} \tag{5.111}$$

9. Calculate Clebsch-Gordon coefficients for a system having $J_1 = \frac{1}{2}$ and $J_2 = \frac{1}{2}$.

SOLUTION

We must add the angular momenta of two spin one-half systems, such as two electrons at rest. We know that $|m_1| = |m_2| = \frac{1}{2}$ and $0 \le j \le 1$. Hence, j has two values, 1 and 0. The combination of two spin one-half systems would give either a spin-zero system or a spin-one system. The m has values $1, 0, -1$, for $j = 1$, while $m = 0$ when $j = 0$. Therefore, the Clebsch-Gordon coefficients form a 4×4 matrix. Also, the Clebsch-Gordon coefficients exist only when $m = m_1 + m_2$. Non-zero Clebsch-Gordon coefficients $\langle m_1, m_2 | j, m\rangle$ for $j = 1$ are:

$\left\langle \frac{1}{2}, \frac{1}{2} \middle| 1, 1\right\rangle, \left\langle \frac{1}{2}, -\frac{1}{2} \middle| 1, 0\right\rangle, \left\langle -\frac{1}{2}, \frac{1}{2} \middle| 1, 0\right\rangle$ and $\left\langle -\frac{1}{2}, -\frac{1}{2} \middle| 1, -1\right\rangle$. When $j = 0$ and $m = 0$, two nonzero possible values are $\left\langle -\frac{1}{2}, \frac{1}{2} \middle| 0, 0\right\rangle$ and $\left\langle \frac{1}{2}, -\frac{1}{2} \middle| 0, 0\right\rangle$.

Equation (5.87) yields $\left\langle \frac{1}{2}, \frac{1}{2} \middle| 1, 1\right\rangle = \left\langle -\frac{1}{2}, -\frac{1}{2} \middle| 1, -1\right\rangle = 1$. From Eqns. (5.88) and (5.89), we have $\left\langle -\frac{1}{2}, \frac{1}{2} \middle| 1, 0\right\rangle = \frac{1}{\sqrt{2}}$ and $\left\langle -\frac{1}{2}, \frac{1}{2} \middle| 1, 0\right\rangle = \frac{1}{\sqrt{2}}$. The Eqn. (5.93) gives

$\left\langle -\frac{1}{2},\frac{1}{2}\middle|0,0\right\rangle = -\frac{1}{\sqrt{2}}$ and $\left\langle \frac{1}{2},-\frac{1}{2}\middle|0,0\right\rangle = \frac{1}{\sqrt{2}}$. Other elements of the 4×4 matrix are zero because for these we cannot fulfill the condition: $m = m_1 + m_2$.

Following Eqn. (5.77), the above results can also be represented by matrix equations as follow:

$$\begin{bmatrix} |1,1\rangle \\ |1,0\rangle \\ |1,-1\rangle \\ |0,0\rangle \end{bmatrix} = \begin{bmatrix} 1 & 0 & 0 & 0 \\ 0 & \frac{1}{\sqrt{2}} & \frac{1}{\sqrt{2}} & 0 \\ 0 & 0 & 0 & 1 \\ 0 & \frac{1}{\sqrt{2}} & -\frac{1}{\sqrt{2}} & 0 \end{bmatrix} \begin{bmatrix} \left|\frac{1}{2},\frac{1}{2}\right\rangle \\ \left|\frac{1}{2},-\frac{1}{2}\right\rangle \\ \left|-\frac{1}{2},\frac{1}{2}\right\rangle \\ \left|-\frac{1}{2},-\frac{1}{2}\right\rangle \end{bmatrix} \tag{5.112}$$

10. Show that $S_xS_y + S_yS_x = 0$.

SOLUTION

We know that $S_x = \frac{\hbar}{2}\sigma_x = \frac{\hbar}{2}\begin{bmatrix} 0 & 1 \\ 1 & 0 \end{bmatrix}$ and $S_y = \frac{\hbar}{2}\sigma_y = \frac{\hbar}{2}\begin{bmatrix} 0 & -i \\ i & 0 \end{bmatrix}$. Therefore:

$$S_xS_y + S_yS_x = \frac{\hbar^2}{4}\left(\sigma_x\sigma_y + \sigma_y\sigma_x\right) \tag{5.113}$$

Let us evaluate:

$$\begin{aligned} \sigma_x\sigma_y + \sigma_y\sigma_x &= \begin{bmatrix} 0 & 1 \\ 1 & 0 \end{bmatrix}\begin{bmatrix} 0 & -i \\ i & 0 \end{bmatrix} + \begin{bmatrix} 0 & -i \\ i & 0 \end{bmatrix}\begin{bmatrix} 0 & 1 \\ 1 & 0 \end{bmatrix} \\ &= \begin{bmatrix} i & 0 \\ 0 & -i \end{bmatrix} + \begin{bmatrix} -i & 0 \\ 0 & i \end{bmatrix} = 0 \end{aligned} \tag{5.114}$$

Therefore $S_xS_y + S_yS_x = 0$.

5.8 Exercises

1. For an axially symmetric rotator, the Hamiltonian is given by $H = \frac{L^+L^-}{2I_1} + \frac{L_z^2}{2I_2}$. What are its eigenvalues?
2. The wave function of the particle moving under the influence of a spherically symmetric potential is given by $\psi(r) = \beta\left(x + y + 3z\right)e^{-\alpha\left(x^2+y^2+z^2\right)^{1/2}}$ where α and β are constants. Calculate $\mathbf{L}^2\psi(r)$. Is $\psi(r)$ the eigenfunction of $\mathbf{L}^2$? If yes, find the value of l. [Hint: write $\psi(r)$ and $\mathbf{L}^2$ in terms of spherical polar coordinates.]

3. Use $J_z|jm\rangle = m\hbar$ and $J^2|jm\rangle = j(j+1)\hbar^2$ and then show that $\langle J_x\rangle = \langle J_y\rangle = 0$ and $\langle J_x^2\rangle = \langle J_y^2\rangle = \frac{1}{2}\left[j(j+1) - m^2\right]\hbar^2$.
4. Calculate eigenvalues of $S_x = \frac{\hbar}{2}\begin{pmatrix} 0 & 1 \\ 1 & 0 \end{pmatrix}$ and show that $tr(S_x) = 0$ and determinant $|S_x| = -\frac{\hbar^2}{4}$.
5. The wave function $\psi(x,y,z) = Aze^{-\beta\left(x^2+y^2+z^2\right)^{1/2}}$ describes one quantum state of a system. Is it a definite angular momentum state? If yes, what are the values of l and m?
6. Use $L_x = \left(yp_z - zp_y\right)$ and evaluate $[L_x, x], [L_x, y], [L_x, z], [L_x, p_x], [L_x, p_y]$ and $[L_x, p_z]$. Show that $\left[L_x, r^2\right] = 0$ and $\left[L_x, p^2\right] = 0$.
7. Use Eqns. (5.11), (5.23), (5.32a), and (5.32b) to find the matrix representations for L^2, L_z, L_x and L_y for $l = 1$.
8. The state of an electron is described by the wave function $\psi(r,\theta,\phi) = \frac{2B}{\sqrt{4\pi}}\sin\theta\cos\phi f(r)$, with $\int_0^\infty |f(r)|^2 r^2 dr = 1$. (a) Use the normalization condition to find the value of constant, B. (b) Express the wave function in terms of spherical harmonics. (c) What are the possible results of measurement on L_z?
9. Consider the electron described in Exercise 8. (a) Find the probability of obtaining each of the two possible results of measurement on L_z. (b) Calculate the expectation value of L_z.

6

Schrödinger Equation for Central Potentials and 3D System

In Chapter 2, we solved the Schrödinger equation for one-dimensional problems. All physics problems cannot be understood using one-dimensional solutions. In this chapter, we would solve the Schrödinger equation and discuss it for three-dimensional systems such the hydrogen atom, the 3D potential well, and the cubic box. The hydrogen atom and three-dimensional potential wells are approximated by spherically symmetric potentials, which do not depend on the direction of the position vector.

6.1 Motion in a Central Field

The Hamiltonian for a particle of mass m moving in a spherically symmetric potential is written as:

$$H = \frac{p^2}{2m} + V(r) \tag{6.1}$$

where $V(r)$ is the spherically symmetric potential. In Schrödinger's representation, $\mathbf{p} = -i\hbar\nabla$. For spherically symmetric systems, it is more convenient to work in spherical polar coordinates (r,θ,ϕ). After expressing ∇^2 in terms of spherical polar coordinates, we get:

$$H = -\frac{\hbar^2}{2m}\frac{1}{r^2}\left[\frac{\partial}{\partial r}\left(r^2\frac{\partial}{\partial r}\right) + \left\{\frac{1}{\sin\theta}\frac{\partial}{\partial\theta}\left(\sin\theta\frac{\partial}{\partial\theta}\right) + \frac{1}{\sin^2\theta}\frac{\partial^2}{\partial^2\phi}\right\}\right] + V(r) \tag{6.2}$$

With the use of Eqn. (5.36b), Eqn. (6.2) goes to:

$$H = \frac{\hbar^2}{2m}\left[-\frac{1}{r^2}\frac{\partial}{\partial r}\left(r^2\frac{\partial}{\partial r}\right) + \frac{L^2}{\hbar^2 r^2}\right] + V(r) \tag{6.3}$$

Three components of orbital angular momentum, L_X, L_Y, and L_Z take the form of partial derivative operators in the angular coordinates, when written in terms of spherical polar coordinates; see Eqns. (5.35). As has been shown by Eqn. (5.8), L_X, L_Y, and L_Z commute with L^2. Therefore, it follows from Eqn. (6.3) that L_X, L_Y, and L_Z commute with the Hamiltonian, H, and hence:

$$[\mathbf{L}, H] = 0 \tag{6.4}$$

Also, $\mathbf{L}^2$ is a function of only (θ,ϕ); it does not depend on r. We conclude that:

$$\left[\mathbf{L}^2, H\right] = 0 \tag{6.5}$$

This means that both orbital angular momentum $\mathbf{L}$ and its squared $\mathbf{L}^2$ are constants of the motion, which is expected for a spherically symmetric potential. Let us now take the energy eigenvalue problem:

$$H|\psi\rangle = E|\psi\rangle \tag{6.6}$$

Since $\mathbf{L}^2$ and L_Z commute with each other and with the Hamiltonian, it is always possible to represent the state of the system in terms of the simultaneous eigenvectors of $\mathbf{L}^2$, L_z, and H. The most general form for the wave function, which would be a simultaneous eigenvector of $\mathbf{L}^2$, L_z, and H is:

$$\psi(r,\theta,\phi) = R_l(r)Y_{lm}(\theta,\phi) \tag{6.7}$$

substitution of which in Eqn. (6.3) along with the use of $\mathbf{L}^2 Y_{lm}(\theta,\phi) = l(l+1)\hbar^2 Y_{lm}(\theta,\phi)$ gives:

$$\left[\frac{\hbar^2}{2m}\left(-\frac{1}{r^2}\frac{d}{dr}\left(r^2\frac{d}{dr}\right) + \frac{l(l+1)}{r^2}\right) + V(r) - E\right]R_l(r) = 0 \tag{6.8}$$

Since there remains only one variable, the partial derivative with respect to r has been changed to a full derivative. Let us rewrite Eqn. (6.8) as:

$$\frac{1}{r^2}\frac{d}{dr}\left(r^2\frac{dR_l}{dr}\right) + \frac{2m}{\hbar^2}\left(E - V_{eff}(r)\right)R_l(r) = 0 \tag{6.9a}$$

with:

$$V_{eff}(r) = \left(V(r) + \frac{l(l+1)\hbar^2}{2mr^2}\right) \tag{6.9b}$$

$V_{eff}(r)$ is an effective potential, which is the sum of two terms. The term $l(l+1)\hbar^2/2mr^2$ represents the effect of centrifugal force in the equivalent one-dimensional problem. It is often called the centrifugal barrier potential and makes a dominatingly larger contribution to $V_{eff}(r)$ for $r \to 0$. When $V(r)$ is the coulombic attractive potential, the centrifugal barrier potential is negligibly small at large values of r. Equation (6.9a), which is equivalent to the one-dimensional Schrödinger equation, is known as the *Sturm-Liouville equation* for the function $R_l(r)$. From the general properties of this type of equation, it is well known that if $R_l(r)$ is required to be well behaved at $r = 0$ and as $r \to \infty$ then solutions can only exist for a discrete set of values of E, which are the energy eigenvalues. In general, the energy eigenvalues can depend on the quantum number l but are independent of the quantum number m.

6.2 Energy Eigenvalues of the Hydrogen Atom

We here consider the case of the hydrogen atom depicted in Fig. 6.1:

$$V(r) = -\frac{e^2}{4\pi\varepsilon_0 r} \tag{6.10}$$

Thus for the hydrogen atom, we have:

$$\frac{\hbar^2}{2\mu}\left\{-\frac{1}{r^2}\frac{d}{dr}\left(r^2\frac{dR_l}{dr}\right)+\frac{l(l+1)R_l}{r^2}\right\}-\left(E+\frac{e^2}{4\pi\varepsilon_0 r}\right)R_l = 0 \tag{6.11}$$

Note that in writing radial Eqn. (6.9a) for the hydrogen atom, m is replaced by μ because of the following: In the case of the hydrogen atom, both the electron having mass m_e and the proton of mass m_p rotate about a common center. This is equivalent to a particle of mass μ rotating about a fixed point, where $\mu = m_e m_p/(m_e + m_p)$ is termed the reduced mass of the electron. Hydrogen also has two isotopes, known as deuterium and tritium, having reduced masses of the electron, $\mu = 2m_e m_p/(m_e + 2m_p)$ and $\mu = 3m_e m_p/(m_e + 3m_p)$, respectively. Since $m_e << m_p$, our forthcoming solutions of Eqn. (6.11) remain valid for all isotopes of hydrogen. On taking $rR_l(r) = P(r)$, Eqn. (6.11) goes to:

$$\frac{d^2P(r)}{dr^2}-\frac{2\mu}{\hbar^2}\left(\frac{l(l+1)}{r^2}-\frac{e^2}{4\pi\varepsilon_0 r}-E\right)P(r) = 0 \tag{6.12}$$

which has the form of a one-dimensional Schrödinger equation for a particle of mass μ moving in the effective potential:

$$V_{eff}(r) = -\frac{e^2}{4\pi\varepsilon_0 r}+\frac{l(l+1)\hbar^2}{2\mu r^2} \tag{6.13}$$

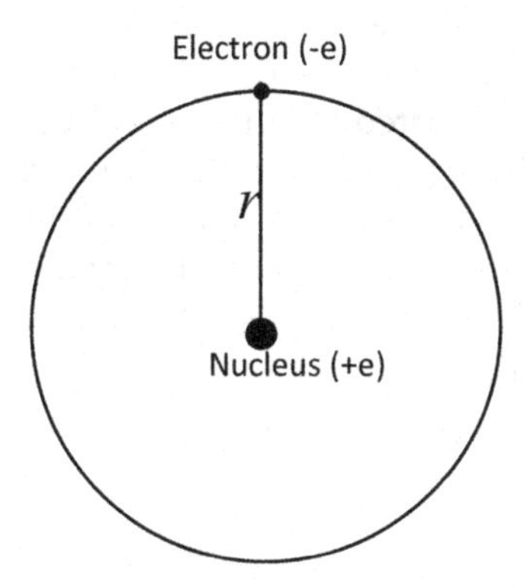

FIGURE 6.1
Schematic diagram for the hydrogen atom.

To find out the solution of Eqn. (6.12) for bound states (E is negative), let us define: $a = \sqrt{-\hbar^2/2\mu E}$ and $y = r/a$. Then for $r \to \infty$, Eqn. (6.12) reduces to:

$$\frac{d^2 P}{dy^2} - P = 0 \tag{6.14}$$

which has the solution $P(r) \propto e^{-y}$. We therefore take:

$$P(y) = f(y)e^{-y} \tag{6.15}$$

the use of which in Eqn. (6.12) yields:

$$\left(\frac{d^2}{dy^2} - 2\frac{d}{dy} - \frac{l(l+1)}{y^2} + \frac{2\mu e^2 a}{4\pi\varepsilon_0 y\hbar^2} \right) f(y) = 0 \tag{6.16}$$

The solution of Eqn. (6.16) can always be seen in a power series of the form:

$$f(y) = \sum_k c_k y^k \tag{6.17}$$

where c_k is constant in y. Substitution of Eqn. (6.17) into Eqn. (6.16) gives:

$$\sum_k c_k \left[k(k-1)y^{k-2} - 2ky^{k-1} - l(l+1)y^{k-2} + \frac{2\mu e^2 a}{4\pi\varepsilon_0 \hbar^2} y^{k-1} \right] = 0 \tag{6.18}$$

The coefficient of each power of y must separately go to zero to satisfy Eqn. (6.18). On equating to zero the coefficient of y^{k-2} we get:

$$c_k \{k(k-1) - l(l+1)\} = c_{k-1} \left\{ 2(k-1) - \frac{2\mu e^2 a}{4\pi\varepsilon_0 \hbar^2} \right\} \tag{6.19}$$

The wave function must be well behaved both at $y \to 0$ and $y \to \infty$. For $y \to 0$, this demands that the power law series (Eqn. 6.17) must terminate at some smallest positive value of k (> 0), to avoid the unphysical behavior of $f(y)$ for $y \to 0$. This is possible if and only if the expression under the bracket on the left-hand side of Eqn. (6.19) goes to zero at $k_{\min}$ so that no value of c_k having $k < k_{\min}$ exists, which means that $[k_{\min}(k_{\min} - 1) - l(l+1)] = 0$, and hence the very first term in the series is $c_{k_{\min}} y^{k_{\min}}$. There are two possibilities: $k_{\min} = -l$ or $k_{\min} = l+1$. The choice $k_{\min} = -l$ predicts unphysical behavior of the wave function at $y = 0$. We therefore conclude that $k_{\min} = l+1$ is the only physically acceptable solution. As is noted from Eqn. (6.17), there is a finite probability of finding the electron at the nucleus, for the state with $l = 0$. Whereas for $l > 0$ state, there is zero probability of finding the electron at the nucleus ($|\psi|^2 = 0$ at $r = 0$). Also, note that it is only possible to obtain sensible behavior of the wave function for $r \to 0$, if l is an integer.

The ratio of successive terms in the series (Eqn. 6.17) is $\frac{c_{k+1} y}{c_k}$. However, from Eqn. (6.19):

$$\frac{c_{k+1}}{c_k} = \frac{2k - \dfrac{2\mu e^2 a}{4\pi\varepsilon_0 \hbar^2}}{\{k(k+1) - l(l+1)\}} \tag{6.20}$$

which tends to $\frac{2}{k}$ and $\frac{c_{k+1}y}{c_k} \to \frac{2y}{k}$ for $k \to \infty$. The series $\sum_k (2y)^k / k!$ converges to e^{2y}, and the ratio of two successive terms in the series of e^{2y} is also equal to $\frac{2y}{k}$. We therefore conclude that $f(y) \to e^{2y}$ as $y \to \infty$. It then follows from Eqns. (6.15) and (6.16) that $R(r) \to \frac{e^{r/a}}{r}$ as $r \to \infty$. This does not permit us to have physically acceptable behavior of the wave function for $r \to \infty$, as $\int |\psi|^2 d^3r$ will not be finite. The only way to avoid this unphysical behavior of the wave function is to terminate series (Eqn. 6.17) at some maximum value of k, which means $c_{k_{max}+1} = 0$. To make c_{n+1} zero, we find from the recursion relation (Eqn. 6.20) that:

$$\frac{\mu e^2 a}{4\pi\varepsilon_0 \hbar^2} = n \tag{6.21}$$

And, the last term of the series is $c_n y^n$. Equation (6.21) with the use of $a = \sqrt{-\hbar^2 / 2\mu E}$ gives the quantized energy eigenvalues:

$$E_n = -\frac{\mu e^4}{2(4\pi\varepsilon_0)^2 \hbar^2 n^2} \tag{6.22}$$

and:

$$a = \frac{4\pi\varepsilon_0 \hbar^2}{\mu e^2} n = 5.29n \times 10^{-11} \text{ meters} \tag{6.23}$$

where n is a positive integer that must exceed the quantum number l $(n > l)$; otherwise there would be no terms in the series (Eqn. 6.17), as is seen from $n_{min} = l + 1$. Thus, the maximum value that can be assigned to l is $n - 1$. Since l is a positive integer, the minimum value that can be assigned to n is 1. n takes values 1,2, 3,…., in general. It is referred as a principal quantum number. A quantum state is completely specified by the set of the quantum numbers (n, l, m), where quantum numbers are restricted to follow $|m| \le l < n$. The $a_0 \left(= \frac{4\pi\varepsilon_0 \hbar^2}{\mu e^2} = 5.29 \times 10^{-11} \text{meters} \right)$ is known as the Bohr radius.

6.3 Wave Functions of the Hydrogen Atom

The $R_l(r)$ is a solution of radial Eqn. (6.11). Equation (6.16) with the use of Eqn. (6.21) becomes:

$$\frac{d^2 f}{dy^2} - 2\frac{df}{dy} - \frac{l(l+1)f}{y^2} + \frac{2n}{y} f = 0 \tag{6.24}$$

On substituting $f(y) = y^{l+1} w(y)$, we obtain:

$$\left[y\frac{d^2}{dy^2} + 2(l+1-y)\frac{d}{dy} + 2(n-l-1) \right] w(y) = 0. \tag{6.25}$$

Let us compare this equation with the equation that is satisfied by the associated Laguerre polynomial $L_p^k(\rho)$ of order k and degree $(p-k)$:

$$\left[\rho\frac{d^2}{d\rho^2}+(k+1-\rho)\frac{d}{d\rho}+(p-k)\right]L_p^k(\rho)=0. \tag{6.26}$$

where k and p are integers. Note that on taking $2y=\rho$, $2l+1=k$ and $n+l=p$, and then replacing $w(\rho)$ by $L_{n+l}^{2l+1}(\rho)$ Eqns. (6.25) become Eqn. (6.26). We therefore conclude that $w(\rho)=L_{n+l}^{2l+1}(\rho)$, which is the *associated Laguerre polynomial* of order $(2l+1)$ and of degree $(n-l-1)$. Thus the radial function is:

$$R_{nl}(\rho)=Ne^{-\rho/2}\rho^l L_{n+l}^{2l+1}(\rho) \tag{6.27}$$

where N is the normalization constant. An associated Laguerre polynomial is related to a Laguerre polynomial as follows:

$$L_{p-q}^p(x)=(-1)^p\left(\frac{d}{dx}\right)^p L_q(x) \tag{6.28}$$

where:

$$L_q(x)=e^x\left(\frac{d}{dx}\right)^q\left(x^q e^x\right) \tag{6.29}$$

is the Laguerre polynomial. To determine N in Eqn. (6.27), we use the normalization condition:

$$\int\psi_{nlm}^*(r,\theta,\phi)\psi_{nlm}(r,\theta,\phi)d^3r=1$$

$$\Rightarrow\int_0^\infty|R_{nl}(r)|^2r^2dr\int_0^\pi\int_0^{2\pi}|Y_{lm}(\theta,\phi)|^2\sin\theta d\theta d\phi=1 \tag{6.30}$$

Since $\int_0^\pi\int_0^{2\pi}|Y_{lm}(\theta,\phi)|^2\sin\theta d\theta d\phi=1$, we get:

$$\int_0^\infty R_{nl}^2(r)r^2dr=1 \tag{6.31}$$

which on substituting $R_{nl}(\rho)$ from Eqn. (6.27) with $r=\frac{\rho a}{2}$, where a is given by Eqn. (6.23), goes to:

$$\left(\frac{4\pi\varepsilon_0 n\hbar^2}{2\mu e^2}\right)^3|N|^2\int_0^\infty e^{-\rho}\rho^{2l}\left|L_{n+l}^{2l+1}\right|^2\rho^2d\rho=1 \tag{6.32a}$$

The associated Laguerre polynomials satisfy:

$$\int_0^\infty\rho^{2l+2}e^{-\rho}\left|L_{n+l}^{2l+1}(\rho)\right|^2d\rho=2n\frac{\{(n+l)!\}^3}{(n-l-1)!} \tag{6.32b}$$

Therefore:

$$\left(\frac{4\pi\varepsilon_0 n\hbar^2}{2\mu e^2}\right)^3 |N|^2 \frac{2n\left[(n+l)!\right]^3}{(n-l-1)!} = 1 \tag{6.33a}$$

which yields:

$$N = \pm\left[\left(\frac{2\mu e^2}{4\pi\varepsilon_0 n\hbar^2}\right)^3 \frac{(n-l-1)!}{2n\left\{(n+l)!\right\}^3}\right]^{1/2} \tag{6.33b}$$

and the normalized radial wave functions are given by:

$$R_{nl}(r) = \left[\left(\frac{2\mu e^2}{4\pi\varepsilon_0 n\hbar^2}\right)^3 \frac{(n-l-1)!}{2n\left\{(n+l)!\right\}^3}\right]^{1/2} e^{-\rho/2}\rho^l L_{n+l}^{2l+1}(\rho) \tag{6.34}$$

The normalized wave function for the hydrogen atom then is:

$$\begin{aligned}\psi_{nlm}(r,\theta,\phi) &= \left[\left(\frac{2\mu e^2}{4\pi\varepsilon_0 n\hbar^2}\right)^3 \frac{(n-l-1)!}{2n\left\{(n+l)!\right\}^3}\right]^{1/2} e^{-\rho/2}\rho^l L_{n+l}^{2l+1}(\rho) Y_{lm}(\theta,\varphi) \\ &= \left[\left(\frac{2}{na_0}\right)^3 \frac{(n-l-1)!}{2n\left\{(n+l)!\right\}^3}\right]^{1/2} e^{-(r/na_0)}\left(\frac{2r}{na_0}\right)^l L_{n+l}^{2l+1}\left(\frac{2r}{na_0}\right) Y_{lm}(\theta,\varphi)\end{aligned} \tag{6.35}$$

The $Y_l^m(\theta,\varphi)$ are spherical harmonics given by Eqn. (5.45). The ground state of hydrogen belongs to $n=1$, $l=0$, and $m=0$. Thus, the ground state of the hydrogen atom is a spherically symmetric, zero angular momentum state. The ground state energy eigenvalue is:

$$E_1 = -\frac{\mu e^4}{2(4\pi\varepsilon_0)^2\hbar^2} = -13.6 \text{ electron volts.} \tag{6.36}$$

and the corresponding eigenvector is $|\psi_{100}\rangle or |1,0,0\rangle$. The ground state is thus a non-degenerate state. The next energy state corresponds to $n=2$. The other quantum numbers take the values $l=0; m=0$ and $l=1; m=-1,0,1$. Thus, there are 3-states with non-zero angular momentum for $n=2$. Note that the energy levels given by Eqn. (6.22) are independent of the quantum numbers l and m, though radial eigenfunctions given by Eqn. (6.34) depend on l. *This is a special property of a* $1/r$ *Coulomb potential.* There are four eigenvectors, $|2,0,0\rangle;|2,1,-1\rangle;|2,1,0\rangle$ and $|2,1,1\rangle$, which correspond to the same energy eigenvalue $E_2 = -\frac{\mu e^4}{8(4\pi\varepsilon_0)^2\hbar^2}$. Thus, there is four-fold degeneracy for $n=2$. For n = 3, there are 9 states all together: l = 0 gives one, l = 1 gives 3 and l = 2 yields 5 different m values. In fact there are n^2 degenerate states for principal quantum number n. The n^2 is the sum of the first n odd integers:

$$\sum_{l=0}^{n-1}(2l+1) = 2\sum_{l=0}^{n-1} l + n = n(n-1) + n = n^2 \tag{6.37}$$

The wave functions for some of the lower values of n, calculated with the use of Eqns. (6.35) and (5.45), are:

$$\psi_{100}(r) = \frac{1}{a_0}\left(\frac{1}{\pi a_0}\right)^{1/2} e^{-r/a_0}.$$

$$\psi_{200}(r) = \frac{1}{4a_0}\left(\frac{1}{2\pi a_0}\right)^{1/2}\left(2 - \frac{r}{a_0}\right)e^{-r/2a_0}.$$

$$\psi_{210}(r,\theta,\phi) = \frac{1}{4a_0}\left(\frac{1}{2\pi a_0}\right)^{1/2}\left(\frac{r}{a_0}\right)e^{-r/2a_0}\cos\theta$$

$$\psi_{21,\pm1}(r,\theta,\varphi) = \frac{1}{4a_0}\left(\frac{1}{\pi a_0^3}\right)^{1/2} re^{-r/2a_0}(\sin\theta)e^{\pm i\varphi}$$

$$\psi_{300}(r,\theta,\varphi) = \frac{2}{3a_0}\left(\frac{1}{12\pi a_0}\right)^{1/2}\left(1 - \frac{3r}{3na_0} + \frac{2r^2}{27a_0^2}\right)e^{-r/3a_0} \tag{6.38}$$

where a_0 is the Bohr radius, defined above.

6.4 Radial Probability Density

The probability of finding an electron in the volume element of d^3r is given by:

$$|\psi_{nlm}(r,\theta,\phi)|^2 d^3r = |R_{nl}(r)|^2 r^2 dr |Y_{lm}(\theta,\phi)|^2 \sin\theta d\theta d\phi \tag{6.39}$$

The probability of finding the electron in a thin spherical shell having radii between r and $r + dr$ is then given by:

$$\begin{aligned} P(r)dr &= |R_{nl}(r)|^2 r^2 dr \int_0^{\pi}\int_0^{2\pi} |Y_{lm}(\theta,\phi)|^2 \sin\theta d\theta d\phi \\ &= |R_{nl}(r)|^2 r^2 dr \end{aligned} \tag{6.40}$$

where we used the property of normalization of spherical harmonics. Radial probability $P(r) = |R_{nl}(r)|^2 r^2$ is defined as the probability of finding an electron at distance r from nucleus.

The maximum probability density for the ground state of a hydrogen atom exists at a distance where $\frac{dP_{10}}{dr} = 0$. Hence:

$$\begin{aligned} &\frac{d}{dr}\left(|R_{10}|^2 r^2\right) = \frac{4}{a_0^3}\frac{d}{dr}\left(r^2 e^{-2r/a_0}\right) = 0 \\ &\Rightarrow \left(2r - \frac{2r^2}{a_0}\right)e^{-2r/a_0} = 0 \end{aligned} \tag{6.41}$$

which gives $r = a_0$. Thus, the maximum probability of finding an electron in the ground state occurs at a distance equal to the Bohr radius from the nucleus. The plot of $P(r)$ versus r has spherical distribution, which shows a maximum at $r = a_0$. Similarly, $P(r)$ of other states in the hydrogen atom can be analyzed.

6.5 Free Particle Motion

In Section 2.4.1 we discussed the motion of a free particle in one dimension. In the three-dimensional case, the Schrödinger equation for the free particle is:

$$\begin{aligned}&\left(-\frac{\hbar^2}{2m}\nabla^2 + V_0\right)|\psi\rangle = E|\psi\rangle \\ &\Rightarrow \nabla^2\psi(\mathbf{r}) + \frac{2m}{\hbar^2}(E - V_0)\psi(\mathbf{r}) = 0\end{aligned} \tag{6.42}$$

where V_0 is a constant potential. We can choose $V_0 = 0$ as well. By taking $k^2 = \frac{2m}{\hbar^2}(E - V_0) \geq 0$, we find that $\psi(\mathbf{r}) = Ce^{i\mathbf{k}.\mathbf{r}}$ is one of the possible solutions of Eqn. (6.42). By restricting the domain of $\psi(\mathbf{r})$ to an arbitrarily large but finite cubic volume of side L, box normalization yields: $\psi(\mathbf{r}) = \frac{1}{(L)^{3/2}}e^{i\mathbf{k}.\mathbf{r}}$. The values of $\mathbf{k}$ are then discretized: $k_i = \frac{2n_i\pi}{L}$, where $n_i (i = x, y, z)$ takes integer values $1, 2, 3, \ldots\ldots$ Equation (6.42) can be expressed in spherical polar coordinates (r, θ, ϕ) too. Following the procedure displayed in Section 6.2, we get the radial equation; see Eqn. (6.8):

$$\frac{1}{r^2}\frac{d}{dr}\left(r^2\frac{dR_l}{dr}\right) + \left(k^2 - \frac{l(l+1)}{r^2}\right)R_l(r) = 0 \tag{6.43}$$

which on defining $kr = \rho$ reduces to:

$$\begin{aligned}&\frac{1}{\rho^2}\frac{d}{d\rho}\left(\rho^2\frac{dR_l}{d\rho}\right) + \left(1 - \frac{l(l+1)}{\rho^2}\right)R_l(\rho) = 0 \\ &\Rightarrow \frac{d^2R_l(\rho)}{d\rho^2} + \frac{2}{\rho}\frac{dR_l(\rho)}{d\rho} + \left(1 - \frac{l(l+1)}{\rho^2}\right)R_l(\rho) = 0\end{aligned} \tag{6.44}$$

Taking $R_l(\rho) = \xi(\rho)/\sqrt{\rho}$, we obtain:

$$\frac{d^2\xi(\rho)}{d\rho^2} + \frac{1}{\rho}\frac{d\xi(\rho)}{d\rho} + \left\{1 - \left(\frac{l + 1/2}{\rho}\right)^2\right\}\xi(\rho) = 0 \tag{6.45}$$

which is Bessel's equation. Its solution is written as:

$$\begin{aligned}&\xi(\rho) = C_1 J_{\left(l+\frac{1}{2}\right)}(\rho) + C_2 J_{-\left(l+\frac{1}{2}\right)}(\rho) \\ &\Rightarrow R_l(\rho) = \frac{1}{\sqrt{\rho}}\left[C_1 J_{\left(l+\frac{1}{2}\right)}(\rho) + C_2 J_{-\left(l+\frac{1}{2}\right)}(\rho)\right]\end{aligned} \tag{6.46}$$

Here, C_1 and C_2 are constants and $J_{\pm\left(l+\frac{1}{2}\right)}(\rho)$ are Bessel functions. The spherical Bessel functions, $j_l(\rho)$ and spherical Neumann functions $n_l(\rho)$ are defined in terms of Bessel functions as follows:

$$j_l(\rho) = \left(\frac{\pi}{2\rho}\right)^{1/2} J_{l+\frac{1}{2}}(\rho) \tag{6.47}$$

and:

$$n_l(\rho) = (-1)^{l+1}\left(\frac{\pi}{2\rho}\right)^{1/2} J_{-\left(l+\frac{1}{2}\right)}(\rho) \tag{6.48}$$

Therefore:

$$R_l(\rho) = Aj_l(\rho) + Bn_l(\rho) \tag{6.49}$$

where A and B are new constants.

The behavior of spherical Bessel and Neumann functions for $\rho \to 0$ is approximated by:

$$j_l(\rho) \to \frac{\rho^l}{1.3.5...(2l+1)} \tag{6.50}$$

$$n_l(\rho) \to \frac{1.3.5...(2l-1)}{\rho^{l+1}} \tag{6.51}$$

Equation (6.51) suggests that B has to be zero; otherwise $R_l(\rho)$ will go to ∞, as $\rho \to 0$, which violates the requirement that wave function (probability) should have finite value everywhere. We therefore conclude that $R_l(\rho) = Aj_l(\rho)$. Hence, free particle energy $E = \frac{\hbar^2 k^2}{2m} + V_0$ and the wave function for the angular momentum state of l is:

$$\psi_l(r,\theta,\phi) = A_l j_l(kr) Y_{lm}(\theta,\phi) \tag{6.52}$$

A more generalized solution for the free particle in the energy state E is given by:

$$\psi_k(r,\theta,\phi) = \sum_l A_l j_l(kr) Y_{lm}(\theta,\phi) \tag{6.53}$$

6.6 Spherically Symmetric Potential Well

This type of potential is frequently used to approximate several physics problems like that of nano-sized particles and of nuclear physics, where the exact nature of the potential is not known. It is also known as a 3D potential well, depicted in Fig. 6.2.

The potential is defined by:

$$V(r) = \begin{cases} -V_0 & 0 < r < a \\ 0 & r > a \end{cases} \tag{6.54}$$

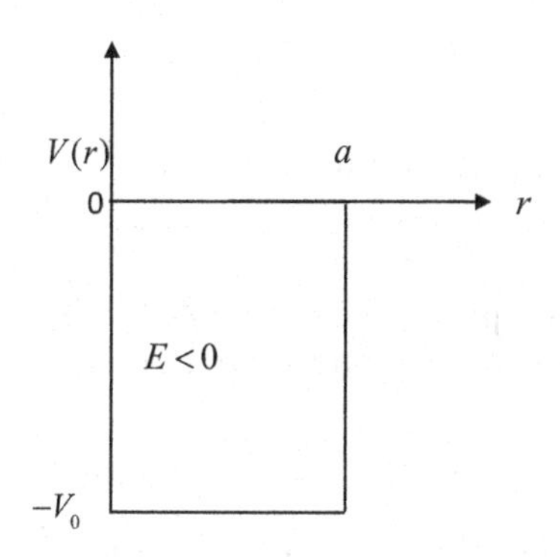

FIGURE 6.2
Spherically symmetric potential well.

for a sphere of radius a. Within the sphere, the potential is given by $-V_0$, while outside the sphere the potential is zero. Since, there is spherical symmetry, we can always look for a solution like: $\psi(r,\theta,\phi)=R_l(r)Y_{lm}(\theta,\phi)$, and the radial equation for $R_l(r)$ is then given by:

$$\frac{1}{r^2}\frac{d}{dr}\left(r^2\frac{dR_l}{dr}\right)+\left(\frac{2\mu}{\hbar^2}(E+V_0)-\frac{l(l+1)}{r^2}\right)R_l(r)=0;\ \text{for}\ 0<r<a \tag{6.55}$$

and:

$$\frac{1}{r^2}\frac{d}{dr}\left(r^2\frac{dR_l}{dr}\right)+\left(\frac{2\mu E}{\hbar^2}-\frac{l(l+1)}{r^2}\right)R_l(r)=0;\text{when}\ r>a \tag{6.56}$$

where μ is the reduced mass of the particle. The above two equations are to be solved separately, and then the two solutions and their first order derivative are to be matched at the boundary which occurs at $r=a$. It is to be noted that energy E can have a negative value ($E<0$, a bound state) as well as a positive value, an unbound state. For both the cases, $E+V_0$ is going to be positive, and hence we take $\alpha^2=\frac{2\mu}{\hbar^2}(E+V_0)$. Then, on using $\alpha r=\rho$, Eqn. (6.55) reduces to Eqn. (6.44). We therefore can conclude that inside the sphere ($0<r<a$):

$$R_l^I(r)=Aj_l(\alpha r) \tag{6.57}$$

We next take the exterior region of the sphere ($r>a$), where Eqn. (6.56) is applicable. There can be two scenarios: (i) $E<0$ and (ii) $E>0$.

Case-I ($E<0$):

Let us take $\beta^2=\frac{2\mu}{\hbar^2}|E|$, which is positive quantity. Then for $E<0$, we can rewrite Eqn. (6.56) as follows:

$$\begin{aligned}&\frac{1}{r^2}\frac{d}{dr}\left(r^2\frac{dR_l}{dr}\right)-\left(\beta^2+\frac{l(l+1)}{r^2}\right)R_l(r)=0\\ \Rightarrow&\frac{1}{\beta^2r^2}\frac{d}{d(\beta r)}\left(\beta^2r^2\frac{dR_l}{d(\beta r)}\right)-\left(1+\frac{l(l+1)}{\beta^2r^2}\right)R_l(r)=0\end{aligned} \tag{6.58}$$

On taking $\beta r = -i\rho$, we get:

$$\frac{d^2R_l(\rho)}{d\rho^2}+\frac{2}{\rho}\frac{dR_l(\rho)}{d\rho}+\left(1-\frac{l(l+1)}{\rho^2}\right)R_l(\rho)=0 \tag{6.59}$$

which is exactly the same as is Eqn. (6.44). Therefore, its solution can be given by Eqn. (6.49) or any other linear combination of $j_l(\rho)$ and $n_l(\rho)$. Let us write the solution as:

$$\begin{aligned}R_l^{II}(\rho)&=(B+C)j_l(\rho)+i(B-C)n_l(\rho)\\&=B\{j_l(\rho)+in_l(\rho)\}+C\{j_l(\rho)-in_l(\rho)\}\\&=Bh_l^{(1)}(\rho)+Ch_l^{(2)}(\rho)\end{aligned} \tag{6.60}$$

where B and C are constants. $h_l^{(1)}(\rho)$ and $h_l^{(2)}(\rho)$ are known as Hankel functions. Also, note that $h_l^{(1)}(\rho)$ and $h_l^{(2)}(\rho)$ are complex conjugates of each other. Writing solution in the form of Eqn. (6.60) is advantageous when asymptotic behavior of solution is discussed. The exterior region of the sphere includes $r\to\infty$, but it does not include $r\to 0$. From the properties of Hankel functions, it is well known that for $\rho\to\infty$:

$$h_l^{(1)}(\rho)\propto\frac{e^{i\rho}}{\rho}\approx C_1\frac{e^{-\beta r}}{r} \tag{6.61a}$$

$$h_l^{(2)}(\rho)\propto\frac{e^{-i\rho}}{\rho}\approx C_2\frac{e^{\beta r}}{r} \tag{6.61b}$$

where C_1 and C_2 are constants. It is obvious from Eqn. (6.61b) that the solution will tend to ∞ for $r\to\infty$, which is unacceptable. Hence, the solution in the exterior region of the sphere should not include $h_l^{(2)}(\rho)$. Therefore, the acceptable solution in the exterior region is:

$$R_l^{II}(r)=Bh_l^{(1)}(i\beta r) \tag{6.62}$$

Thus, for a bound state case ($E<0$), Eqns. (6.57) and (6.62) are acceptable solutions in interior and exterior regions, respectively, of the spherically symmetric potential. Continuity from the interior to exterior or vice versa require that:

$$\begin{aligned}&R_l^{I}(a)=R_l^{II}(a)\\&\Rightarrow Aj_l(\alpha a)=Bh_l^{(1)}(i\beta a)\end{aligned} \tag{6.63a}$$

and:

$$\begin{aligned}&\left.\frac{dR_l^{I}(r)}{dr}\right|_{r=a}=\left.\frac{dR_l^{II}(r)}{dr}\right|_{r=a}\\&\Rightarrow Aj_l'(\alpha a)=Bh_l'^{(1)}(i\beta a)\end{aligned} \tag{6.63b}$$

Division of Eqn. (6.63b) by (6.63a) gives:

$$h_l^{(1)}(i\beta a)j_l'(\alpha a)-j_l(\alpha a)h_l'^{(1)}(i\beta a)=0 \tag{6.64}$$

The α and β are real and are dependent on n and l. Equation (6.64) explicitly determines the energy eigenvalues E_{nl} for existing bound states that correspond to a given value l. The constants A and B can be determined from the normalization condition $\int_0^\infty R_{nl}^2(r)r^2dr = 1$. Hankel functions of the imaginary argument are real apart from the factor i^l. Therefore, radial function:

$$R_l(r) = \begin{cases} Aj_l(\alpha r) & 0 < r < a \\ Bh_l^{(1)}(i\beta r) & r > a \end{cases} \tag{6.65}$$

is real if A and B are chosen to be real. To get more inside the problem, let us consider a simpler case of s-state ($l = 0$), which explains the essential physics of many interesting systems. For $l = 0$: $j_0(\rho) = \frac{\sin\rho}{\rho}$ and $n_0(\rho) = -\frac{\cos\rho}{\rho}$. Hence:

$$\begin{aligned} h_0^{(1)}(\rho) &= \frac{\sin\rho}{\rho} - i\frac{\cos\rho}{\rho} = \frac{e^{i\rho}}{i\rho} \\ \Rightarrow h_0^{(1)}(i\beta r) &= -\frac{e^{-\beta r}}{\beta r} \end{aligned} \tag{6.66}$$

We then have:

$j_0(\alpha a) = \frac{\sin(\alpha a)}{\alpha a}$; $j_0'(\alpha a) = \frac{1}{\alpha a^2}[(\alpha a)\cos(\alpha a) - \sin(\alpha a)]$; $h_0^{(1)}(i\beta a) = -\frac{e^{-\beta a}}{\beta a}$, and $h_0'^{(1)}(i\beta a) = \frac{e^{-\beta a}}{\beta a^2}(\beta a + 1)$. Substitution of these into Eqn. (6.64) yields:

$$\alpha\cot(\alpha a) = -\beta \tag{6.67}$$

which resembles to Eqn. (2.50b) of Section 2.4.3. We thus find that energy levels of the s-state of a spherical potential well are given by odd parity (asymmetric wave function) energy levels of a one-dimensional potential well of depth V_0 and width $2a$, suggesting that it can be solved graphically or numerically. Following the discussions of Section 2.4.3, a solution exists if only if $V_0a^2 > \frac{\pi^2\hbar^2}{8\mu}$ and there would be at least one bound state if:

$$\frac{\pi^2\hbar^2}{8\mu} < V_0a^2 \le \frac{9\pi^2\hbar^2}{8\mu} \tag{6.68}$$

It is to be noted that unlike the case of the 1D potential well, where the bound state exists for any smallest value of V_0, for the spherically symmetric potential well, existence of a bound state is characterized by Eqn. (6.68), which is applicable to the s-state ($l = 0$) only. For bound states with $l > 0$, the width of the well should be even larger. To find bound states for the p-state ($l = 1$) and for the d-state ($l = 2$), we have to evaluate Eqn. (6.64) in terms of the spherical Bessel functions and Hankel functions for $l = 1$ and and $l = 2$, and then solve the resulting equations graphically or numerically.

Case-II ($E > 0$):

For $E > 0$ (non-localized case), we can take $\rho = \beta r$ and then from Eqns. (6.61) we find that both $h_l^{(1)}(\rho)$ and $h_l^{(2)}(\rho)$ can be included in the radial solution for the exterior region $(r > a)$ of a spherical symmetric potential well. Therefore for $r > a$, we have:

$$R_l^{II}(r) = Bh_l^{(1)}(\beta r) + Ch_l^{(2)}(\beta r) \tag{6.69}$$

which means that at $r = a$, we should have:

$$Aj_l(\alpha a) = Bh_l^{(1)}(\beta a) + Ch_l^{(2)}(\beta a) \tag{6.70}$$

and:

$$Aj_l'(\alpha a) = Bh_l'^{(1)}(\beta a) + Ch_l'^{(2)}(\beta a) \tag{6.71a}$$

where:

$$h_l'^{(1)}(\beta a) = \frac{d}{dr}\left\{h_l^{(1)}(\beta r)\right\}\bigg|_{r=a} \text{ and } h_l'^{(2)}(\beta a) = \frac{d}{dr}\left\{h_l^{(2)}(\beta r)\right\}\bigg|_{r=a} \tag{6.71b}$$

$j_l'(\alpha a)$ is defined earlier. Equations (6.70) and (6.71a) can be solved to find B/A and C/A:

$$\frac{B}{A} = \frac{j_l(\alpha a)h_l'^{(2)}(\beta a) - j_l'(\alpha a)h_l^{(2)}(\beta a)}{h_l^{(2)}(\beta a)h_l'^{(1)}(\beta a) - h_l'^{(2)}(\beta a)h_l^{(1)}(\beta a} \tag{6.72a}$$

and:

$$\frac{C}{A} = \frac{j_l(\alpha a)h_l'^{(1)}(\beta a) - j_l'(\alpha a)h_l^{(1)}(\beta a)}{h_l^{(2)}(\beta a)h_l'^{(1)}(\beta a) - h_l'^{(2)}(\beta a)h_l^{(1)}(\beta a} \tag{6.72b}$$

Since $h_l^{(1)}(\beta a)$ and $h_l^{(2)}(\beta a)$ are complex conjugates of each other, B/A and C/A too are complex conjugates of each other, provided A is real. This suggests that Eqn. (6.69) gives a real solution for the exterior region of the spherical symmetric potential well. Equations (6.57) and (6.69) along with (6.72) uniquely define the wave functions for the spherically symmetric potential for $E > V_0$. Further, evaluated values of $\int_0^\infty R_l^2(r)r^2dr$ are equal to some constant but not unity. This clearly establishes that states are non-localized for $E > 0$.

6.7 Electron Confined to a 3D Box

The potential energy can be taken to be zero inside the potential well. It is infinite outside the well. An electron in a conducting solid such as a metal is very well represented by a 3D infinite potential well. Inside the well, we have:

$$-\frac{\hbar^2}{2m}\nabla^2\psi = E\psi$$
$$\Rightarrow -\frac{\hbar^2}{2m}\left(\frac{\partial^2}{\partial x^2}+\frac{\partial^2}{\partial y^2}+\frac{\partial^2}{\partial z^2}\right)\psi(x,y,z) = E\psi(x,y,z) \tag{6.73}$$

Separation of variables allows to write: $\psi(x,y,z)=\psi(x)\psi(y)\psi(z)$. Each of $\psi(x)$, $\psi(y)$, and $\psi(z)$ vanishes at boundaries of the well, which occur at $(x=0, y=0, z=0)$ and at $(x=L, y=L, z=L)$. Further, normalization of wave function demands:

$$\int_0^L\int_0^L\int_0^L \psi^*(x,y,z)\psi(x,y,z)dxdydz = 1$$
$$\Rightarrow \int_0^L |\psi(x)|^2 dx \int_0^L |\psi(y)|^2 dy \int_0^L |\psi(z)|^2 dz = 1 \tag{6.74}$$

This permits us to choose:

$$\psi(x) = \sqrt{\frac{2}{L}}\sin(k_x x);$$
$$\psi(y) = \sqrt{\frac{2}{L}}\sin(k_y y); \tag{6.75}$$
$$\psi(z) = \sqrt{\frac{2}{L}}\sin(k_z z)$$

Next, $\psi(L,y,z)=0$, $\psi(x,L,z)=0$, and $\psi(x,y,L)=0$ force us to take $k_x = \pi n_x/L$, $k_y = \pi n_y/L$, and $k_z = \pi n_z/L$, where n_x, n_y, and n_z take nonzero integer values $1,2,3,\ldots..$

Substituting Eqn. (6.75) in Eqn. (6.73), we obtain:

$$E = \frac{\hbar^2}{2m}\left(k_x^2+k_y^2+k_z^2\right) = \frac{\hbar^2\pi^2}{2mL^2}\left(n_x^2+n_y^2+n_z^2\right) \tag{6.76}$$

The lowest possible energy of the system is:

$$E_1 = \frac{3\hbar^2\pi^2}{2mL^2} \tag{6.77}$$

And, the volume occupied by a single state is:

$$\Delta k = \Delta k_x \Delta k_y \Delta k_z = \left(\frac{\pi}{L}\right)^3 \tag{6.78}$$

Equation (6.76) gives the energy for a set of three quantum numbers(n_x, n_y, n_z) in k-space. Hence, the number of states $N(E)$, the energy of which is less than or equal to E, will be equal to the volume of a sphere of radius of k divided by the volume occupied by a single k-state. Therefore:

$$\begin{aligned} N(E) &= \frac{\left(\frac{4\pi}{3}k^3\right)}{\Delta k} \\ &= \frac{4\pi}{3}\left(\frac{kL}{\pi}\right)^3 \\ &= \frac{4V}{3\pi^2}\left(\frac{2mE}{\hbar^2}\right)^{3/2} \\ &= \frac{4V}{3\pi^2\hbar^3}(2mE)^{3/2} \end{aligned} \tag{6.79}$$

here we have used $V = L^3$. This is a useful relation to compute the number of electrons that occupy the quantum states up to a given energy level in the metal.

The density of states, number of states per unit energy in the range of E to $E + dE$, is obtained as follows:

$$\begin{aligned} D(E) &= \frac{dN}{dE} \\ &= \frac{d}{dE}\left\{\frac{4V}{3\pi^2\hbar^3}(2mE)^{3/2}\right\} \\ &= \frac{2V}{\pi^2\hbar^3}(2m)^{3/2}\sqrt{E} \end{aligned} \tag{6.80}$$

We thus find that the density of the states of a free electron is proportional to $\sqrt{E}$.

6.8 Solved Examples

1. Show the Bohr radius is $a_0 = 0.0529$ nm for both cases of an electron with actual mass and its reduced mass in a hydrogen atom. Also, show that $E_n = -\frac{13.6}{n^2}eV$.

SOLUTION

Let us first calculate:

$$\frac{4\pi\varepsilon_0\hbar^2}{e^2} = \frac{\left(1.054\times10^{-34}\,\text{Js}\right)^2}{\left(8.988\times10^9\,\text{Nm}^2/\text{C}^2\right)\left(1.602\times10^{-19}\,\text{C}\right)^2} = 4.816\times10^{-41}\,\text{Js}/\text{Nm}^2.$$

Hence, for an electron of mass $m_e = 9.11 \times 10^{-31}$ kg:

$$a_0 = \frac{4\pi\varepsilon_0 \hbar^2}{m_e e^2} = \frac{4.816 \times 10^{-41}}{9.11 \times 10^{-31}}$$
$$= 0.5286 \times 10^{-10} \text{ meter}$$
$$= 0.0529 \text{ nm.}$$

The reduced mass of the electron in the hydrogen atom is:

$$\begin{aligned} \mu &= \left(\frac{m_p}{m_e + m_p}\right) m_e \\ &= \left(\frac{1.673 \times 10^{-24}}{9.11 \times 10^{-31} + 1.673 \times 10^{-24}}\right) m_e \\ &= 0.9995 m_e \end{aligned} \tag{6.81}$$

The Bohr radius is:

$$a_0 = \frac{4\pi\varepsilon_0 \hbar^2}{\mu e^2} = \frac{4\pi\varepsilon_0 \hbar^2}{0.9995 m_e e^2} = \frac{0.5286}{0.9995} = 0.5289 \times 10^{-10} \text{m}$$
$$= 0.0529 \text{ nm.}$$

The energy of nth level in the hydrogen atom is given by Eqn. (6.22):

$$\begin{aligned} E_n &= -\frac{\mu e^4}{2(4\pi\varepsilon_0)^2 n^2 \hbar^2} = -\left(\frac{e^2}{4\pi\varepsilon_0}\right)^2 \times \frac{\mu}{2n^2\hbar^2} \\ &= -\left(2.307 \times 10^{-28}\right)^2 \times \frac{0.9995 \times 9.11 \times 10^{-31}}{2n^2 \times (1.054 \times 10^{-34})^2} \text{J} \\ &= \frac{-21.8115 \times 10^{-19}}{n^2} \text{J} \\ &= \frac{-13.6}{n^2} \text{eV} \end{aligned} \tag{6.82}$$

2. Consider an electron confined to the interior of a spherical cavity of radius a, with impenetrable walls (a 3D infinite potential well). The potential seen by the electron inside the cavity is $-V_0$. Calculate the energy and normalized wave function for the ground state of the system.

SOLUTION

The walls of the cavity are impenetrable; hence the wave function of the electron vanishes at the wall $\psi(r = a) = 0$. For inside the spherically symmetric cavity $(0 \le r \le a), V(r) = -V_0$. The radial Eqn. (6.8) for $l = 0$, reduces to:

$$\left[\frac{d^2}{dr^2} + \frac{2}{r}\frac{d}{dr} + \frac{2m}{\hbar^2}(E + V_0)\right] R_0(r) = 0 \tag{6.83}$$

On taking $P(r) = rR_0(r)$, we have:

$$\frac{d^2P(r)}{dr^2} + \frac{2m(E+V_0)}{\hbar^2}P(r) = 0 \tag{6.84}$$

when $r > a$, $R_0(r) = 0$, as the electron cannot be found outside the cavity. The solution of Eqn. (6.84) can be given by:

$$\begin{aligned} P(r) &= A\sin(\beta r) \\ \Rightarrow R_0(r) &= \frac{A}{r}\sin(\beta r) \end{aligned} \tag{6.85}$$

with $\beta = \left(\frac{2m(E+V_0)}{\hbar^2}\right)^{1/2}$, and A is constant. The condition $\psi(r=a) = 0$ yields $\beta a = n\pi$, where n is a positive integer, which gives:

$$\begin{aligned} a\left(\frac{2m(E+V_0)}{\hbar^2}\right)^{1/2} &= n\pi \\ \Rightarrow E_1(n) &= \frac{\hbar^2}{2m}\left(\frac{n\pi}{a}\right)^2 - V_0 \end{aligned} \tag{6.86}$$

Note that the energy levels for an electron in the ground state of a spherical cavity look very similar to those in a 1D potential infinite well. To fulfill the condition of normalization: $\int_0^\infty R_0^2(r)r^2dr = 1$. Let us evaluate:

$$\begin{aligned} \int_0^a R_0^2(r)r^2dr &= A^2\int_0^a \sin^2(\beta r)dr \\ &= A^2\int_0^a \frac{1-\cos(2\beta r)}{2}dr \\ &= \frac{aA^2}{2} \end{aligned} \tag{6.87}$$

which gives $\frac{aA^2}{2} = 1$ or $A = \sqrt{\frac{2}{a}}$. The normalized wave function of the system then is:

$$\psi_{100}(r) = R_0(r)Y_{00} = \frac{1}{\sqrt{2\pi a}}\left(\frac{1}{r}\right)\sin\left\{\frac{n\pi r}{a}\right\} \tag{6.88}$$

3. As an extension of problem 2, consider the case when a particle is constrained to move between two concentric impermeable spheres of radii $r = a$ and $r = b$. Potential in the region between spheres is constant $-V_0$. Find the normalized wave function and the energy for ground state ($l = 0$).

SOLUTION

In this case, Eqn. (6.84) is applicable for $a \le r \le b$. The $R_0(r) = 0$ at both $r = a$ and $r = b$. Hence a normalization condition of the radial wave function is defined by:

$$\int_a^b R_0^2(r) r^2 dr = 1 \tag{6.89}$$

and $R_0(r)$ does not exist when $b < r < a$. To fulfill the requirement that wave function vanishes at $r = a$, we can choose:

$$P(r) = A \sin\{\beta(r-a)\} \tag{6.90}$$

and then we impose the requirement $\beta(b-a) = n\pi$ to make $R_0(r) = 0$ at $r = b$. We thus have:

$$\begin{aligned} &\beta(b-a) = n\pi \\ &\Rightarrow E_1(n) = \frac{\hbar^2}{2m}\left(\frac{n\pi}{b-a}\right)^2 - V_0 \end{aligned} \tag{6.91}$$

We next evaluate:

$$\begin{aligned} 1 &= \int_a^b R_0^2(r) r^2 dr \\ &= A^2 \int_a^b \sin^2\{\beta(r-a)\} dr \\ &= A^2 \int_a^b \frac{1 - \cos\{2\beta(r-a)\}}{2} dr \\ &= \frac{(b-a)A^2}{2} \end{aligned} \tag{6.92}$$

which yields $A = \sqrt{\frac{2}{(b-a)}}$. Therefore, the normalized wave functions and energies for ground state ($l = 0$) are given by:

$$\psi_{100}(r) = R_0(r) Y_{00} = \frac{1}{\sqrt{2\pi(b-a)}}\left(\frac{1}{r}\right)\sin\left\{\frac{n\pi(r-a)}{(b-a)}\right\} \tag{6.93}$$

and:

$$E_1(n) = \frac{n^2\hbar^2\pi^2}{2m(b-a)^2} - V_0 \tag{6.94}$$

4. The wave function ket for the hydrogen atom is given as a linear combination of s-orbital and p-orbitals such as $|\phi\rangle = \frac{1}{\sqrt{15}}\left(2|\psi_{100}\rangle + |\psi_{200}\rangle + 3|\psi_{211}\rangle + |\psi_{21-1}\rangle\right)$. Find the expectation value of the Hamiltonian of the hydrogen atom.

SOLUTION

We know that energy eigenvalues and eigenvectors for the hydrogen atom satisfy:

$$H|\psi_{nlm}\rangle = E_n|\psi_{nlm}\rangle \tag{6.95a}$$

with:

$$E_n = -\frac{\mu e^4}{2(4\pi\varepsilon_0)^2 n^2\hbar^2} \tag{6.95b}$$

where (nlm) are quantum numbers. The expectation value of the Hamiltonian is:

$$E = \langle\phi|H|\phi\rangle$$

$$= \frac{1}{15}\left(2\langle\psi_{100}| + \langle\psi_{2100}| + 3\langle\psi_{211}| + \langle\psi_{21-1}|\right)\left(2H|\psi_{100}\rangle + H|\psi_{2100}\rangle + 3H|\psi_{211}\rangle + |\psi_{21-1}\rangle\right) \tag{6.96a}$$

With the use of Eqns. (6.95) and the orthogonality condition $\langle\psi_{nlm}|\psi_{n'l'm'}\rangle = \delta_{nn'}\delta_{ll'}\delta_{mm'}$, Eqn. (6.96a) simplifies to:

$$E = \frac{1}{15}(4E_1 + E_2 + 9E_2 + E_2) = \frac{1}{15}(4E_1 + 11E_2) \tag{6.96b}$$

From Eqn. (6.82), we have $E_1 = -13.6\text{eV}$ and $E_2 = -3.4\text{eV}$. Hence:

$$E = -\frac{1}{15}(4\times 13.6 + 11\times 3.4) = -6.12 \text{ eV}.$$

5. What is the probability of finding a 1s electron within a distance of a_0 from the proton inside the hydrogen atom?

SOLUTION

The probability of finding the electron in the volume element of d^3r is given by:

$$Pd^3r = |R_{nl}(r)Y_{lm}(\theta,\phi)|^2 r^2 dr \sin\theta d\theta d\phi \tag{6.97}$$

Hence the probability of finding a 1s electron at the distance of a_0 from the nucleus is:

$$\begin{aligned}
P_{1s} &= \int_0^{a_0} |R_{10}(r)|^2 r^2 dr \int_0^{\pi}\int_0^{2\pi} |Y_{00}(\theta,\phi)|^2 \sin\theta d\theta d\phi \\
&= \int_0^{a_0} |R_{10}(r)|^2 r^2 dr \\
&= \frac{4}{(a_0)^3}\int_0^{a_0} e^{-2r/a_0} r^2 dr \\
&= \frac{1}{2}\left[\left(-x^2 - 2x - 2\right)e^{-x}\right]_0^2 = \left(1 - 5e^{-2}\right) \\
&= 0.323
\end{aligned} \tag{6.98}$$

where we used $R_{10}(r) = \frac{2}{(a_0)^{3/2}} e^{-r/a_0}$.

6. Calculate uncertainties $\Delta\mathbf{r} = \sqrt{\langle \mathbf{r}^2 \rangle - \langle \mathbf{r} \rangle^2}$ and $\Delta\mathbf{p} = \sqrt{\langle \mathbf{p}^2 \rangle - \langle \mathbf{p} \rangle^2}$ for the hydrogen atom in the ground state and then evaluate $\Delta\mathbf{r}\Delta\mathbf{p}$.

SOLUTION

$n = 1, l = 0$ and $m = 0$ for the ground state of the hydrogen atom. The, $\psi_{100}(\mathbf{r}) = R_{10}(r)Y_{00}(\theta,\phi) = \left(\frac{1}{\pi a_0^3}\right)^{1/2} e^{-r/a_0} = \psi^*_{100}(\mathbf{r})$. We define the expectation value of an observable A:

$$\langle \mathrm{A} \rangle = \int \psi^* \mathrm{A} \psi d^3 r$$
$$= \int_0^\infty \int_0^\pi \int_0^{2\pi} \left(R_{nl}(r)Y_{lm}(\theta,\phi)\right)^* \mathrm{A}\left(R_{nl}(r)Y_{lm}(\theta,\phi)\right) r^2 dr \sin\theta d\theta d\phi \tag{6.99a}$$

Therefore $\langle \mathbf{r} \rangle = \langle x \rangle \hat{i} + \langle y \rangle \hat{j} + \langle z \rangle \hat{k}$. With the use of $x = r\sin\theta\cos\phi$, we have:

$$\langle x \rangle = \frac{1}{\pi a_0^3} \int_0^\infty r^3 e^{-2r/a_0} dr \int_0^\pi \sin\theta \sin\theta d\theta \int_0^{2\pi} \cos\phi d\phi$$
$$= 0 \tag{6.99b}$$

Similarly, $\langle y \rangle = 0$ and $\langle z \rangle = 0$ giving $\langle \mathbf{r} \rangle = 0$.

Next take:

$$\langle r^2 \rangle = \int \psi^* r^2 \psi d^3 r$$
$$= \frac{4}{a_0^3} \int_0^\infty r^4 e^{-2r/a_0} dr \tag{6.100a}$$
$$= 3a_0^2$$

$$\langle \mathbf{p} \rangle = \langle p_x \rangle \hat{i} + \langle p_y \rangle \hat{j} + \langle p_z \rangle \hat{k} \text{ and,}$$

$$\langle p_x \rangle = -\frac{i\hbar}{\pi a_0^3} \int_0^\infty \int_0^\pi \int_0^{2\pi} e^{-r/a_0} \left(\frac{d}{dx} e^{-r/a_0}\right) r^2 dr \sin\theta d\theta d\phi$$
$$= \frac{i\hbar}{\pi a_0^4} \int_0^\infty \int_0^\pi \int_0^{2\pi} e^{-r/a_0} \left(\frac{x}{r} e^{-r/a_0}\right) r^2 dr \sin\theta d\theta d\phi$$
$$= \frac{i\hbar}{\pi a_0^4} \int_0^\infty e^{-2r/a_0} r^2 dr \int_0^\pi \sin^2\theta d\theta \int_0^{2\pi} \cos\phi d\phi$$
$$= 0$$

Similarly:

$$\langle p_y \rangle = \frac{i\hbar}{\pi a_0^4}\int_0^\infty e^{-2r/a_0} r^2 dr \int_0^\pi \sin^2\theta d\theta \int_0^{2\pi} \sin\phi d\phi = 0$$

and (6.100b)

$$\langle p_z \rangle = \frac{i\hbar}{\pi a_0^4}\int_0^\infty e^{-2r/a_0} r^2 dr \int_0^\pi \cos\theta \sin\theta d\theta \int_0^{2\pi} d\phi = 0$$

Hence, $\langle \mathbf{p} \rangle = 0$. To evaluate $\langle p^2 \rangle$, we take:

$$\begin{aligned}\langle p^2 \rangle &= \int \psi^* p^2 \psi d^3 r \\ &= -\frac{\hbar^2}{\pi a_0^3}\int_0^\infty \int_0^\pi \int_0^{2\pi} e^{-r/a_0}\left(\nabla^2 e^{-r/a_0}\right) r^2 dr \sin\theta d\theta d\phi\end{aligned}$$

with:

$$\nabla^2 = \frac{1}{r^2}\frac{\partial}{\partial r}\left(r^2 \frac{\partial}{\partial r}\right) + \frac{1}{r^2 \sin\theta}\frac{\partial}{\partial \theta}\left(\sin\theta \frac{\partial}{\partial \theta}\right) + \frac{1}{r^2 \sin^2\theta}\frac{\partial^2}{\partial \phi^2}$$

and obtain:

$$\begin{aligned}\langle p^2 \rangle &= -\frac{4\hbar^2}{a_0^3}\left[\frac{2!}{a_0^2 (2/a_0)^3} - \frac{2}{a_0}\frac{1}{(2/a_0)^2}\right] \\ &= \frac{\hbar^2}{a_0^2}\end{aligned} \tag{6.101}$$

We then have:

$$\Delta \mathbf{r} = \sqrt{\langle \mathbf{r}^2 \rangle - \langle \mathbf{r} \rangle^2} = \sqrt{3} a_0, \text{and } \Delta \mathbf{p} = \sqrt{\langle \mathbf{p}^2 \rangle - \langle \mathbf{p} \rangle^2} = \frac{\hbar}{a_0} \tag{6.102}$$

which gives $\Delta \mathbf{r} \Delta \mathbf{p} = \sqrt{3}\hbar$.

7. A particle of mass m oscillates under the spherically symmetric potential given by: $V(r) = \frac{1}{2}kr^2$, where k is some constant. Calculate the energy eigenvalue and eigenfunction of the particle.

SOLUTION

The time-independent Schrödinger equation for the particle is:

$$\left[-\frac{\hbar^2}{2m}\nabla^2 + V(r)\right]\psi(\mathbf{r}) = E\psi(\mathbf{r}) \tag{6.103}$$

which is separable into three equations by writing: $\psi(r) = \psi_1(x)\psi_2(y)\psi_3(z)$, and $E = E_1 + E_2 + E_3$. After taking $\omega = \sqrt{k/m}$, we get:

$$\frac{d^2\psi_1(x)}{dx^2} + \frac{2m}{\hbar^2}\left(E_1 - \frac{1}{2}m\omega^2 x^2\right)\psi_1(x) = 0 \tag{6.104a}$$

$$\frac{d^2\psi_2(y)}{dy^2} + \frac{2m}{\hbar^2}\left(E_2 - \frac{1}{2}m\omega^2 y^2\right)\psi_2(y) = 0 \tag{6.104b}$$

$$\frac{d^2\psi_3(z)}{dz^2} + \frac{2m}{\hbar^2}\left(E_3 - \frac{1}{2}m\omega^2 z^2\right)\psi_3(z) = 0 \tag{6.104c}$$

Each of Eqns. (6.104) is similar to the equation of the 1D harmonic oscillator whose Hamiltonian is defined by Eqn. (3.123). Hence, the method presented for the 1D harmonic oscillator in Section 3.19 applies to solve each the Eqns. (6.104). The particle under consideration is behaving like a 3D harmonic oscillator. We therefore get:

$$E_1 = \left(n_1 + \frac{1}{2}\right)\hbar\omega \tag{6.105a}$$

$$E_2 = \left(n_2 + \frac{1}{2}\right)\hbar\omega \tag{6.105b}$$

$$E_3 = \left(n_3 + \frac{1}{2}\right)\hbar\omega \tag{6.105c}$$

n_1, n_2 and n_3 are integers, which takes values 0,1,2,3 .... The wave functions that satisfy Eqns. (6.104) are then obtained from Eqn. (3.146), which are as follows:

$$\psi_1(\rho) = \frac{1}{\sqrt{(2^{n_1} n_1!)}}\left(\frac{m\omega}{\hbar\pi}\right)^{1/4} e^{-\rho^2/2} H_{n_1}(\rho) \tag{6.106a}$$

$$\psi_2(\sigma) = \frac{1}{\sqrt{(2^{n_2} n_2!)}}\left(\frac{m\omega}{\hbar\pi}\right)^{1/4} e^{-\sigma^2/2} H_{n_2}(\sigma) \tag{6.106b}$$

$$\psi_3(\varepsilon) = \frac{1}{\sqrt{(2^{n_3} n_3!)}}\left(\frac{m\omega}{\hbar\pi}\right)^{1/4} e^{-\varepsilon^2/2} H_{n_3}(\varepsilon) \tag{6.106c}$$

where $\rho = \sqrt{\frac{m\omega}{\hbar}}x$, $\sigma = \sqrt{\frac{m\omega}{\hbar}}y$, and $\varepsilon = \sqrt{\frac{m\omega}{\hbar}}z$. We thus find that the energy and wave function of the nth state of the particle are:

$$E_n = \left(n + \frac{3}{2}\right)\hbar\omega \tag{6.107}$$

and:

$$\psi_n(r) = \frac{1}{\{2^n(n_1!)(n_2!)(n_3!)\}^{1/2}}\left(\frac{m\omega}{\hbar\pi}\right)^{3/4} e^{-(\rho^2+\sigma^2+\varepsilon^2)/2} H_{n1}(\rho)H_{n2}(\sigma)H_{n3}(\varepsilon) \tag{6.108}$$

where $n = n_1 + n_2 + n_3$.

8. A NaCl crystal has certain negative ion vacancies behaving like a free electron, inside a volume having dimensions of the order of a lattice constant. Estimate the longest wavelength of electromagnetic radiation absorbed strongly by these electrons, when the crystal is at room temperature.

SOLUTION

Energy levels for a free electron confined to a cubic box having each side of length L are given by Eqn. (6.76):

$$E = \frac{\hbar^2\pi^2}{2mL^2}\left(n_1^2 + n_1^2 + n_3^2\right) \tag{6.109}$$

where n_1, n_2, and n_3 are positive integers. Taking $L \sim 1$ Angstrom, the ground state energy is:

$$E_{111} = \frac{3\hbar^2\pi^2}{2mL^2} = \frac{3(1.054\times10^{-34})^2(3.14)^2}{2\times9.11\times10^{-31}(10^{-10})^2} = 1.8\times10^{-17}\ \text{Joules.}$$

$$E_{111} = 112.5\text{eV}.$$

Electrons in a crystal at room temperature are almost as in the ground state. Hence, the energy of the first excited states is: $E_{211} = \frac{6\hbar^2\pi^2}{2mL^2} = \frac{3\hbar^2\pi^2}{mL^2}$. The longest wave length for transition from the ground state to the first excited state is given by:

$$\lambda = \frac{c}{\nu} = \frac{ch}{(E_{211} - E_{111})}$$
$$\lambda = \frac{3\times10^8\times6.626\times10^{-34}}{1.8\times10^{-17}} = 1.104\times10^{-8}m \tag{6.110}$$

or $\lambda = 11.04$ nm.

9. A nano-meter sized particle can be modeled as a spherical potential well of radius a, with impenetrable walls. Inside the nano-particle electrons move freely under a constant potential $-V_0$. What is the probability of finding an electron at distance $a/2$ from the center of the nano-particle, for $l = 0$. Calculate the first three energy gaps between the discrete states.

SOLUTION

As discussed in example 2, the wave function of an electron inside the spherical potential well of radius a, with impenetrable walls for $l=0$ is given by:

$$\psi_{100}(r,\theta,\phi)=\sqrt{\frac{2}{a}}\left(\frac{1}{r}\right)\sin\left\{\frac{n\pi r}{a}\right\}Y_{00} \tag{6.111}$$

where $n=1,2,3,\ldots$ is an integer. The probability of finding the electron at distance $a/2$ from the center of the nano-particle is:

$$\begin{aligned}P&=\int_0^{a/2}\int_0^{\pi}\int_0^{2\pi}\left|\psi_{100}(r,\theta,\phi\right|^2 d^3r\\&=\frac{1}{2\pi a}\int_0^{a/2}\int_0^{\pi}\int_0^{2\pi}\frac{1}{r^2}\sin^2\left\{\frac{n\pi r}{a}\right\}r^2dr\sin\theta d\theta d\phi\\&=\frac{2}{a}\int_0^{a/2}\sin^2\left\{\frac{n\pi r}{a}\right\}dr\\&=\frac{1}{a}\int_0^{a/2}\left[1-\cos\left\{\frac{2n\pi r}{a}\right\}\right]dr\\&=\frac{1}{a}\times\frac{a}{2}=\frac{1}{2}\end{aligned} \tag{6.112}$$

As discussed in example 2, the energy eigenvalues of the electron are discrete. We write:

$$E_n=\frac{\hbar^2}{2m}\left(\frac{n\pi}{a}\right)^2-V_0 \tag{6.113}$$

The first three energy gaps between the discrete energy levels are:

$$\begin{aligned}\Delta E_1&=E_2-E_1=\frac{3\hbar^2\pi^2}{2ma^2};\\\Delta E_2&=E_3-E_2=\frac{5\hbar^2\pi^2}{2ma^2};\\\Delta E_3&=E_4-E_3=\frac{7\hbar^2\pi^2}{2ma^2}\end{aligned} \tag{6.114}$$

6.9 Exercises

1. The Hamiltonian for an electron in the ground state of hydrogen-like atoms is given by:

$$\frac{\hbar^2}{2\mu}\left[\frac{d^2}{dr^2}+\frac{2}{r}\frac{d}{dr}\right]\psi(r)+\left(E+\frac{Ze^2}{4\pi\varepsilon_0 r}\right)\psi(r)=0 \tag{6.115}$$

   To choose $\psi(r)=Ae^{-\alpha r}$ as the eigenfunction, what should be the values of A, α and E? Find the expectation values of potential energy $\langle V\rangle$ and of kinetic energy $\langle T\rangle$.
2. Find the most probable radius of the electron in the 1s and 2p states of the hydrogen atom.
3. Calculate $\langle x\rangle,\langle y\rangle,\langle z\rangle,\langle x^2\rangle,\langle y^2\rangle$ and $\langle z^2\rangle$ for the ground state of the hydrogen atom. What do you notice from these calculations?
4. Calculate $\langle p_x\rangle,\langle p_y\rangle,\langle p_z\rangle,\langle p_x^2\rangle,\langle p_y^2\rangle$ and $\langle p_z^2\rangle$ for the ground state of the hydrogen atom. Show that $\Delta x\Delta p_x > \hbar/2$.
5. A particle of mass m is confined to move within a two-dimensional rectangular system of lengths a and b along the x-axis and y-axis, respectively. The confining potential is:

$$V(x,y)=\begin{cases} V_0 & 0<x<a \text{ and } 0<y<b \\ \infty & \text{otherwise} \end{cases}.$$

   Find out the wave function $\psi(x,y)$ and the energy E of the particle.
6. What is the probability of finding the particle described in exercise 5, in the region of $0<x<a/2$ and $b/2<y<b$?
7. A particle of mass m oscillates in the x-y plane under the potential defined by: $V(x,y)=\frac{1}{2}\left(k_1x^2+k_2y^2\right)$, where k_1 and k_2 are constant. Calculate the energy eigenvalue and eigenfunction of the particle.
8. An electron in the 2p-state of the hydrogen atom is described by wave function $\psi(r,\theta,\phi)=Are^{-r/2a_0}$. Use the normalization condition to find the value of A and then calculate the probability of finding the 2p-electron within a distance of $2a_0$ from the proton inside the hydrogen atom.
9. A GaAs-quantum wire can be modeled as a system in which electrons are confined along x-y directions and they move freely along the z-axis. The confining lengths along the x- and y-axes are equal to a, the potential seen by the electron inside the wire is zero, and the effective mass of the electron is $m=0.07m_e$. (a) Calculate the energy of an electron. (b) Find the frequency of the photon released when the electron makes transition from the third excited state to the ground state.

7

Approximation Methods

In Chapters 2 and 6, we discussed some problems for which the Schrödinger equation can be solved exactly either in 1D or in 3D. However, the vast majority of physical and chemical problems cannot be represented by these prototype problems. There exist a large number of applications in physics and chemistry for which the Schrödinger equation cannot be solved exactly, and hence it is to be solved approximately. In this chapter, we will be discussing well known approximation methods developed to solve the Schrödinger equation approximately.

7.1 Perturbation Theory

Perturbation theory is a mathematical technique commonly used in many areas of physics and chemistry, not just in quantum mechanics. In fact, it had been extensively and successfully used as a powerful tool by physicists long before the development of quantum mechanics. One of the procedures was a formulation of perturbation theory utilized by quantum chemists for estimating electron correlation. The basic concept of perturbation theory involves taking a physical system for which it is not possible to obtain exact solutions and separating the Hamiltonian of the system into two parts. The first part can be solved exactly, while the second part is the troublesome part that has no analytic solution. The Hamiltonian of a system is written as $H = H_0 + H_1$, where H_0 is a simple Hamiltonian for which exact eigenvalues and eigenstates are known. An interesting physics is introduced by H_1, but on adding H_1 to H_0 we are unable to find the exact energy eigenvalues and eigenvectors. However, if H_1 can be regarded as being small compared to H_0, we can find the approximate eigenvalues and eigenvectors of the Hamiltonian $H = H_0 + H_1$. Perturbation theory provides a mathematical means to study how the solutions evolve when an exact system undergoes the perturbation represented by H_1.

There exist many approaches; we here follow the Rayleigh-Schrödinger perturbation theory approach. Let H be the Hamiltonian of the system of interest. Energy eigenvalues and eigenfunctions for the state, specified by a composite index$n$ representing a set three quantum numbers (n,l,m) are the solutions to the Schrödinger equation:

$$H|\psi_n\rangle = E_n|\psi_n\rangle \tag{7.1}$$

Further, there are two scenarios: (i) a non-degenerate case, where only one $|\psi_n\rangle$ corresponds to one value of E_n, and (ii) a degenerate case, where more than one $|\psi_n\rangle$ belong to the same E_n value.

7.1.1 Perturbation Theory for Nondegenerate States

Let us further assume that for a system described by:

$$H_0\left|\psi_n^{(0)}\right\rangle = E_n^{(0)}\left|\psi_n^{(0)}\right\rangle \tag{7.2}$$

solutions (eigenvalues and eigenfunctions) are already known or these can easily be obtained. We write the Hamiltonian as:

$$H = H_0 + \lambda H_1 \tag{7.3}$$

where the perturbation parameter is $0 \le \lambda \le 1$. Thus, the Schrödinger equation for the system now is:

$$(H_0 + \lambda H_1)\left|\psi_n\right\rangle = E_n\left|\psi_n\right\rangle \tag{7.4}$$

The eigenfunctions and eigenvalues of H must depend on λ, because H depends on λ. Therefore, $\left|\psi_n\right\rangle$ and E_n can be expanded in a power series with respect to λ, as follows:

$$E_n = E_n^{(0)} + \lambda E_n^{(1)} + \lambda^2 E_n^{(2)} + \ldots\ldots\ldots \tag{7.5}$$

$$\left|\psi_n\right\rangle = \left|\psi_n^{(0)}\right\rangle + \lambda\left|\psi_n^{(1)}\right\rangle + \lambda^2\left|\psi_n^{(2)}\right\rangle + \ldots\ldots\ldots \tag{7.6}$$

where we used the following abbreviations in the expansion:

$$E_n^{(k)} = \frac{1}{k!}\frac{\partial^k E_n}{\partial \lambda^k}\bigg|_{\lambda=0} \tag{7.7}$$

$$\left|\psi_n^{(k)}\right\rangle = \frac{1}{k!}\frac{\partial^k \left|\psi_n\right\rangle}{\partial \lambda^k}\bigg|_{\lambda=0} \tag{7.8}$$

The E_n and $\left|\psi_n\right\rangle$ of a system are the sum of the correction terms. It is obvious that the convergence of these power series is a key issue, because the series must be truncated in practice.

Using Eqns. (7.5) and (7.6) in Eqn. (7.4), we get:

$$\begin{aligned}
&(H_0 + \lambda H_1)\left[\left|\psi_n^{(0)}\right\rangle + \lambda\left|\psi_n^{(1)}\right\rangle + \lambda^2\left|\psi_n^{(2)}\right\rangle + \ldots\ldots\ldots\right] \\
&= \left[E_n^{(0)} + \lambda E_n^{(1)} + \lambda^2 E_n^{(2)} + \ldots\ldots\ldots\right]\left[\left|\psi_n^{(0)}\right\rangle + \lambda\left|\psi_n^{(1)}\right\rangle + \lambda^2\left|\psi_n^{(2)}\right\rangle + \ldots\ldots\ldots\right]
\end{aligned} \tag{7.9}$$

the simplification of which gives terms with various powers of λ on both the sides. If equality is to hold for all $0 \le \lambda \le 1$, then for a given power of λ, all terms on the left-hand side must equal to those on the right-hand side. This gives rise to:

$$H_0\left|\psi_n^{(0)}\right\rangle = E_n^{(0)}\left|\psi_n^{(0)}\right\rangle \tag{7.10}$$

$$H_0\left|\psi_n^{(1)}\right\rangle + H_1\left|\psi_n^{(0)}\right\rangle = E_n^{(0)}\left|\psi_n^{(1)}\right\rangle + E_n^{(1)}\left|\psi_n^{(0)}\right\rangle \tag{7.11}$$

$$H_0\left|\psi_n^{(2)}\right\rangle + H_1\left|\psi_n^{(1)}\right\rangle = E_n^{(0)}\left|\psi_n^{(2)}\right\rangle + E_n^{(1)}\left|\psi_n^{(1)}\right\rangle + E_n^{(2)}\left|\psi_n^{(0)}\right\rangle \tag{7.12}$$

Equations (7.11) and (7.12) can be generalized to:

$$H_0\left|\psi_n^{(k)}\right\rangle + H_1\left|\psi_n^{(k-1)}\right\rangle = \sum_{i=0}^{k} E_n^{(i)}\left|\psi_n^{(k-i)}\right\rangle \tag{7.13}$$

Equation (7.10) provides zeroth order eigenvalues and eigenvectors. We have to find various corrections $E_n^{(k)}$ to energy E_n, and $\left|\psi_n^{(k)}\right\rangle$ to eigenvector $\left|\psi_n\right\rangle$. We assume that eigenvectors of H_0 are orthonormal, and hence $\left\langle\psi_m^{(0)}\middle|\psi_n^{(0)}\right\rangle = \delta_{mn}$. The zeroth order energy is then obtained from:

$$E_n^{(0)} = \left\langle\psi_n^{(0)}\middle|H_0\middle|\psi_n^{(0)}\right\rangle \tag{7.14}$$

7.1.1.1 First Order Corrections to Energy and Wave Function Ket

On operating with $\left\langle\psi_m^{(0)}\right|$ from the left on both the sides of Eqn. (7.11) we get:

$$\begin{aligned} &\left\langle\psi_m^{(0)}\middle|H_0\middle|\psi_n^{(1)}\right\rangle + \left\langle\psi_m^{(0)}\middle|H_1\middle|\psi_n^{(0)}\right\rangle = E_n^{(0)}\left\langle\psi_m^{(0)}\middle|\psi_n^{(1)}\right\rangle + E_n^{(1)}\left\langle\psi_m^{(0)}\middle|\psi_n^{(0)}\right\rangle \\ &\Rightarrow E_m^{0}\left\langle\psi_m^{(0)}\middle|\psi_n^{(1)}\right\rangle + \left\langle\psi_m^{(0)}\middle|H_1\middle|\psi_n^{(0)}\right\rangle = E_n^{(0)}\left\langle\psi_m^{(0)}\middle|\psi_n^{(1)}\right\rangle + E_n^{(1)}\delta_{mn} \end{aligned} \tag{7.15}$$

where we have used $\left\langle\psi_m^{(0)}\middle|\psi_n^{(0)}\right\rangle = \delta_{mn}$ and $\left\langle\psi_m^{(0)}\middle|H_0\middle|\psi_n^{(1)}\right\rangle = E_m^{(0)}\left\langle\psi_m^{(0)}\middle|\psi_n^{(1)}\right\rangle$ because H_0 is a Hermitian operator. We now consider the two possibilities: (i) $m = n$ and (ii) $m \neq n$. For the case of $m = n$ Eqn. (7.15) simplifies to:

$$E_n^{(1)} = \left\langle\psi_n^{(0)}\middle|H_1\middle|\psi_n^{(0)}\right\rangle \tag{7.16}$$

and when $m \neq n$, Eqn. (7.15) is rewritten as:

$$\begin{aligned} &E_m^{(0)}\left\langle\psi_m^{(0)}\middle|\psi_n^{(1)}\right\rangle + \left\langle\psi_m^{(0)}\middle|H_1\middle|\psi_n^{(0)}\right\rangle = E_n^{(0)}\left\langle\psi_m^{(0)}\middle|\psi_n^{(1)}\right\rangle \\ &\Rightarrow \left\langle\psi_m^{(0)}\middle|H_1\middle|\psi_n^{(0)}\right\rangle = \left(E_n^{(0)} - E_m^{(0)}\right)\left\langle\psi_m^{(0)}\middle|\psi_n^{(1)}\right\rangle \end{aligned} \tag{7.17}$$

The Eqn. (7.11) can be rewritten as:

$$\left(H_0 - E_n^{(0)}\right)\left|\psi_n^{(1)}\right\rangle = \left(E_n^{(1)} - H_1\right)\left|\psi_n^{(0)}\right\rangle \tag{7.18}$$

which suggests that $\left|\psi_n^{(1)}\right\rangle$ can be constructed as a linear combination of exact solutions of H_0, which provide a convenient but not a unique choice for a complete orthonormal basis.

The various order corrections $\left|\psi_n^{(k)}\right\rangle$ to the eigenstate can then also be constructed as a linear combination of exact solutions of H_0. We therefore take:

$$\left|\psi_n^{(k)}\right\rangle = \sum_{\mu} C_{n\mu}^{(k)} \left|\psi_\mu^{(0)}\right\rangle \tag{7.19}$$

where expansion coefficients $C_{n\mu}^{(k)}$ are to be determined. To find an expansion coefficient $C_{n\mu}^{(1)}$, we rewrite Eqn. (7.17) with the use of Eqn. (7.19) for $k=1$:

$$\left(E_m^{(0)} - E_n^{(0)}\right) \sum_{\mu} C_{n\mu}^{(1)} \left\langle \psi_m^{(0)} \middle| \psi_\mu^{(0)} \right\rangle = \left\langle \psi_m^{(0)} \middle| H_1 \middle| \psi_n^{(0)} \right\rangle \tag{7.20}$$

which yields:

$$C_{nm}^{(1)} = \frac{\left\langle \psi_m^{(0)} \middle| H_1 \middle| \psi_n^{(0)} \right\rangle}{\left(E_m^{(0)} - E_n^{(0)}\right)} \tag{7.21}$$

It is to be noted that Eqn. (7.20) is valid only for $m \neq n$; it is not valid for $m = n$. We therefore obtain:

$$\left|\psi_n^{(1)}\right\rangle = \sum_{m \neq n} \frac{\left\langle \psi_m^{(0)} \middle| H_1 \middle| \psi_n^{(0)} \right\rangle}{\left(E_m^{(0)} - E_n^{(0)}\right)} \left|\psi_m^{(0)}\right\rangle \tag{7.22}$$

Further, we find that if $\left|\psi_n^{(1)}\right\rangle$ satisfies Eqn. (7.18), then $\left(\left|\psi_n^{(1)}\right\rangle + A\left|\psi_n^{(0)}\right\rangle\right)$ also satisfies it, for any constant A. This provides us freedom to subtract off the $\left|\psi_n^{(0)}\right\rangle$ term, and therefore, there is no need to include the $m = n$ term in the sum for constructing $\left|\psi_n^{(1)}\right\rangle$.

7.1.1.2 Second Order Corrections to Energy and Wave Function Ket

On operating with $\left\langle \psi_m^{(0)} \right|$ from the left on both the sides of Eqn. (7.12) we have:

$$\begin{aligned} &\left\langle \psi_m^{(0)} \middle| H_0 \middle| \psi_n^{(2)} \right\rangle + \left\langle \psi_m^{(0)} \middle| H_1 \middle| \psi_n^{(1)} \right\rangle = \\ &E_n^{(0)} \left\langle \psi_m^{(0)} \middle| \psi_n^{(2)} \right\rangle + E_n^{(1)} \left\langle \psi_m^{(0)} \middle| \psi_n^{(1)} \right\rangle + E_n^{(2)} \left\langle \psi_m^{(0)} \middle| \psi_n^{(0)} \right\rangle \end{aligned} \tag{7.23}$$

When we take $m = n$, the first term on the left-hand side cancels out the first term on the right-hand side, and the second term on the right-hand side vanishes, because $\left\langle \psi_m^{(0)} \middle| \psi_n^{(1)} \right\rangle = 0$ due to non-inclusion of the $m = n$ term in $\left|\psi_n^{(1)}\right\rangle$. We therefore get:

$$E_n^{(2)} = \left\langle \psi_n^{(0)} \middle| H_1 \middle| \psi_n^{(1)} \right\rangle \tag{7.24a}$$

which with the use of Eqn. (7.22) gives:

$$
\begin{aligned}
E_n^{(2)} &= \sum_{m\neq n} \frac{\langle \psi_m^{(0)} | H_1 | \psi_n^{(0)} \rangle \langle \psi_n^{(0)} | H_1 | \psi_m^{(0)} \rangle}{\left(E_m^{(0)} - E_n^{(0)}\right)} \\
&= \sum_{m\neq n} \frac{\left| \langle \psi_m^{(0)} | H_1 | \psi_n^{(0)} \rangle \right|^2}{\left(E_m^{(0)} - E_n^{(0)}\right)}
\end{aligned} \tag{7.24b}
$$

On taking $m \neq n$, Eqn. (7.19) gives:

$$
|\psi_n^{(2)}\rangle = \sum_{\mu} C_{n\mu}^{(2)} |\psi_\mu^{(0)}\rangle \tag{7.25}
$$

Substituting Eqn. (7.25) into Eqn. (7.23), we get for $m \neq n$:

$$
\begin{aligned}
&\left(E_m^{(0)} - E_n^{(0)}\right) \sum_{\mu} C_{n\mu}^{(2)} \langle \psi_m^{(0)} | \psi_\mu^{(0)} \rangle + \sum_{\mu \neq n} C_{n\mu}^{(1)} \langle \psi_m^{(0)} | H_1 | \psi_\mu^{(0)} \rangle \\
&= E_n^{(1)} \sum_{\mu \neq n} C_{n\mu}^{(1)} \langle \psi_m^{(0)} | \psi_\mu^{(0)} \rangle
\end{aligned} \tag{7.26}
$$

which gives:

$$
C_{nm}^{(2)} = \frac{E_n^{(1)} C_{nm}^{(1)} - \sum_{\mu \neq n} C_{n\mu}^{(1)} \langle \psi_m^{(0)} | H_1 | \psi_\mu^{(0)} \rangle}{\left(E_m^{(0)} - E_n^{(0)}\right)} \tag{7.27}
$$

It is to be noted that Eqn. (7.27) cannot be used to calculate $C_{nn}^{(2)}$, because Eqn. (7.27) is obtained for $m \neq n$. The $C_{nn}^{(2)}$ can be calculated by imposing the condition that the perturbed wave function be normalized, that is $\langle \psi_n | \psi_n \rangle = 1$. Retaining the terms up to λ^2 in Eqn. (7.6), we have:

$$
\left[\langle \psi_n^{(0)} | + \lambda \langle \psi_n^{(1)} | + \lambda^2 \langle \psi_n^{(2)} | \right] \left[|\psi_n^{(0)}\rangle + \lambda |\psi_n^{(1)}\rangle + \lambda^2 |\psi_n^{(2)}\rangle \right] = 1 \tag{7.28}
$$

With the use of the orthonormality condition $\langle \psi_\mu^{(0)} | \psi_\nu^{(0)} \rangle = \delta_{\mu\nu}$, Eqn. (7.28) simplifies to:

$$
\lambda\left(C_{nn}^{(1)} + (C_{nn}^{(1)})^*\right) + \lambda^2 \left(C_{nn}^{(2)} + (C_{nn}^{(2)})^* + \sum_{\nu} \left| C_{n\nu}^{(1)} \right|^2 \right) = 0 \tag{7.29}
$$

From discussions presented above in Section 7.1.1.1, we find that $C_{nn}^{(1)} + (C_{nn}^{(1)})^* = 0$. Therefore:

$$
C_{nn}^{(2)} + (C_{nn}^{(2)})^* = -\sum_{\nu} \left| C_{n\nu}^{(1)} \right|^2 \tag{7.30}
$$

The imaginary parts of the left-hand side of Eqn. (7.30) cancel out each other. Therefore, the real part of $C_{n,n}^{(2)}$ is given by:

$$\mathrm{Re}\left(C_{nn}^{(2)}\right) = -\frac{1}{2}\sum_{\nu\neq n}\left|C_{n\nu}^{(1)}\right|^2 \tag{7.31}$$

We thus, have:

$$\begin{aligned}\left|\psi_n^{(2)}\right\rangle = \sum_{\nu\neq n}\sum_{\mu\neq n}&\left[\frac{\left\langle\psi_\mu^{(0)}\middle|H_1\middle|\psi_n^{(0)}\right\rangle\left\langle\psi_\nu^{(0)}\middle|H_1\middle|\psi_\mu^{(0)}\right\rangle}{\left(E_\mu^{(0)}-E_n^{(0)}\right)\left(E_n^{(0)}-E_\nu^{(0)}\right)} - \frac{\left\langle\psi_n^{(0)}\middle|H_1\middle|\psi_n^{(0)}\right\rangle\left\langle\psi_\mu^{(0)}\middle|H_1\middle|\psi_n^{(0)}\right\rangle}{\left(E_\mu^{(0-)}-E_n^{(0)}\right)^2}\right]\left|\psi_\mu^{(0)}\right\rangle \\ &-\frac{1}{2}\sum_{\nu\neq n}\frac{\left|\left\langle\psi_\nu^{(0)}\middle|H_1\middle|\psi_n^{(0)}\right\rangle\right|^2}{\left(E_\nu^{(0)}-E_n^{(0)}\right)^2}\left|\psi_n^{(0)}\right\rangle\end{aligned} \tag{7.32}$$

7.1.1.3 kth Order Corrections to Energy and Wave Function Ket

On operating with $\left\langle\psi_n^{(0)}\right|$ from the left in both sides, Eqn. (7.13) goes to:

$$\begin{aligned}\left\langle\psi_n^{(0)}\middle|H_0\middle|\psi_n^{(k)}\right\rangle + \left\langle\psi_n^{(0)}\middle|H_1\middle|\psi_n^{(k-1)}\right\rangle &= \sum_{i=0}^{k}E_n^{(i)}\left\langle\psi_n^{(0)}\middle|\psi_n^{(k-i)}\right\rangle \\ &= \sum_{i=0}^{k}E_n^{(i)}\delta_{0,k-i}\end{aligned} \tag{7.33}$$

This gives:

$$E_n^{(k)} = \left\langle\psi_n^{(0)}\middle|H_1\middle|\psi_n^{(k-1)}\right\rangle \tag{7.34}$$

which is a generalization of Eqns. (7.16) and (7.24a). We note from Eqn. (7.34) that in general the $(k$ - 1)th order correction to the wave function ket is required to obtain the kth order energy correction.

On operating with $\left\langle\psi_m^{(0)}\right|$ from the left, Eqn. (7.13) becomes for $m\neq n$:

$$\begin{aligned}\sum_{i=0}^{k}E_n^{(i)}\left\langle\psi_m^{(0)}\middle|\psi_n^{(k-i)}\right\rangle &= \left\langle\psi_m^{(0)}\middle|H_0\middle|\psi_n^{(k)}\right\rangle + \left\langle\psi_m^{(0)}\middle|H_1\middle|\psi_n^{(k-1)}\right\rangle \\ &= E_m^0\left\langle\psi_m^{(0)}\middle|\psi_n^{(k)}\right\rangle + \left\langle\psi_m^{(0)}\middle|H_1\middle|\psi_n^{(k-1)}\right\rangle\end{aligned} \tag{7.35}$$

or:

$$\begin{aligned}&\sum_{i=0}^{k}\sum_{\mu\neq n}C_{n\mu}^{(k-i)}E_n^{(i)}\left\langle\psi_m^{(0)}\middle|\psi_\mu^{(0)}\right\rangle = \\ &E_m^{(0)}\sum_{\mu\neq n}C_{n\mu}^{(k)}\left\langle\psi_m^{(0)}\middle|\psi_\mu^{(0)}\right\rangle + \sum_{\mu\neq n}C_{n\mu}^{(k-1)}\left\langle\psi_m^{(0)}\middle|H_1\middle|\psi_\mu^{(0)}\right\rangle\end{aligned} \tag{7.36}$$

which simplifies to:

$$\sum_{i=0}^{k}\sum_{\mu\neq n} C_{n\mu}^{(k-i)} E_n^{(i)} \delta_{m\mu} = E_m^{(0)} \sum_{\mu\neq n} C_{n\mu}^{(k)} \delta_{m\mu} + \sum_{\mu\neq n} C_{n\mu}^{(k-1)} \left\langle \psi_m^{(0)} \middle| H_1 \middle| \psi_\mu^{(0)} \right\rangle \tag{7.37a}$$

giving rise to:

$$C_{nm}^{(k)} = \frac{\sum_{\mu\neq n} C_{n\mu}^{(k-1)} \left\langle \psi_m^{(0)} \middle| H_1 \middle| \psi_\mu^{(0)} \right\rangle - \sum_{i=1}^{k} C_{nm}^{(k-i)} E_n^{(i)}}{\left(E_n^{(0)} - E_m^{(0)}\right)} \tag{7.37b}$$

Note that Eqn. (7.37b) is a generalization of Eqn. (7.27).

7.1.1.4 Anharmonic Oscillator

We come across several realistic problems in physics where the potential energy of an oscillator has both a non-quadradic as well as a quadradic term. Such oscillators are termed anharmonic oscillators. The non-quadradic term can be treated as the perturbative term to the harmonic oscillator, if it is small as compared to the Hamiltonian of the harmonic oscillator. We here consider two such cases of (i) $H_1 = Ax$, and (ii) $H_1 = Bx^4$, where A and B are constants, and calculate first order corrections to the energy and wave function of the *nth* state. The Hamiltonian of the unperturbed harmonic oscillator is $H_0 = \frac{p^2}{2m} + \frac{1}{2}m\omega^2 x^2$. On defining $\alpha = \left(\frac{m\omega}{\hbar}\right)^{1/2}$, $a = \frac{\alpha x}{\sqrt{2}} + \frac{ip}{\hbar\alpha\sqrt{2}}$ and $a^\dagger = \frac{\alpha x}{\sqrt{2}} - \frac{ip}{\hbar\alpha\sqrt{2}}$ we get $H_0 = \left(a^\dagger a + \frac{1}{2}\right)\hbar\omega$, and $x = \frac{1}{\sqrt{2}\alpha}\left(a^\dagger + a\right)$. For the nth eigenstate of the harmonic oscillator, we have $H_0|n\rangle = E_n|n\rangle$, $a|n\rangle = \sqrt{n}|n-1\rangle$, $a^\dagger|n\rangle = \sqrt{n+1}|n+1\rangle$ and $N|n\rangle = a^\dagger a|n\rangle = n|n\rangle$. Also $\langle m|n\rangle = \delta_{mn}$. We then have:

(i)
$$H_1 = \frac{A}{\sqrt{2}\alpha}(a^\dagger + a) \tag{7.38}$$

From Eqn. (7.16), we get:

$$\begin{aligned}
E_n^{(1)} &= \frac{A}{\sqrt{2}\alpha}\langle n|(a^\dagger + a)|n\rangle \\
&= \frac{A}{\sqrt{2}\alpha}\left(\langle n|a^\dagger|n\rangle + \langle n|a|n\rangle\right) \\
&= \frac{A}{\sqrt{2}\alpha}\left(\sqrt{n+1}\langle n|n+1\rangle + \sqrt{n}\langle n|n-1\rangle\right) \\
&= \frac{A}{\sqrt{2}\alpha}\left(\sqrt{n+1}\delta_{n,n+1} + \sqrt{n}\delta_{n,n-1}\right) \\
&= 0
\end{aligned} \tag{7.39a}$$

because n cannot be equal to $n+1$ or $n-1$. The first order correction to the wave function is:

$$\begin{aligned}
\left|\psi_n^{(1)}\right\rangle &= \frac{A}{\sqrt{2}\alpha}\sum_{m\neq n}\frac{\langle m|(a^{\dagger}+a)|n\rangle}{\left(E_m^{(0)}-E_n^{(0)}\right)}|m\rangle \\
&= \frac{A}{\sqrt{2}\alpha}\sum_{m\neq n}\left(\frac{\sqrt{(n+1)}\delta_{m,n+1}+\sqrt{n}\delta_{m,n-1}}{\left(E_m^{(0)}-E_n^{(0)}\right)}\right)|m\rangle \\
&= \frac{A}{\sqrt{2}\alpha}\left(\frac{\sqrt{(n+1)}}{E_{n+1}^{0}-E_n^{(0)}}|n+1\rangle+\frac{\sqrt{n}}{E_{n-1}^{0}-E_n^{(0)}}|n-1\rangle\right) \\
&= \frac{A}{\hbar\omega\sqrt{2}\alpha}\left(\sqrt{(n+1)}|n+1\rangle-\sqrt{n}|n-1\rangle\right)
\end{aligned} \tag{7.39b}$$

(ii)

$$\begin{aligned}
H_1 &= \frac{B}{4\alpha^4}(a^{\dagger}+a)^4 \\
&= \frac{B}{4\alpha^4}\left[(a^{\dagger})^2+a^{\dagger}a+aa^{\dagger}+a^2\right]^2 \\
&= \frac{B}{4\alpha^4}\left[(a^{\dagger})^2+1+2N+a^2\right]^2 \\
&= \frac{B}{4\alpha^4}\left[(a^{\dagger})^4+a^4+(a^{\dagger})^2(1+2N)+(a^{\dagger})^2a^2+(1+2N)(a^{\dagger})^2+(1+2N)^2\right. \\
&\quad \left.+(1+2N)a^2+a^2(a\dagger)^2+a^2(1+2N)\right]
\end{aligned} \tag{7.40a}$$

where we used $a^{\dagger}a = N$ and $\left[a,a^{\dagger}\right]=1$. Hence:

$$\begin{aligned}
E_n^{(1)} &= \frac{B}{4\alpha^4}\langle n|\left[(a^{\dagger})^4+a^4+(a^{\dagger})^2(1+2N)+(a^{\dagger})^2a^2+(1+2N)(a^{\dagger})^2+(1+2N)^2\right. \\
&\quad \left.+(1+2N)a^2+a^2(a\dagger)^2+a^2(1+2N)\right]|n\rangle
\end{aligned} \tag{7.40b}$$

With the argument that is used above in (i), we find:

$$\begin{aligned}
E_n^{(1)} &= \frac{B}{4\alpha^4}\langle n|\left[(a\dagger)^2a^2+a^2(a\dagger)^2+(1+2N)^2\right]|n\rangle \\
&= \frac{B}{4\alpha^4}\left[n(n-1)+(n+1)(n+2)+(1+2n)^2\right] \\
&= \frac{B}{4\alpha^4}(n+1)^2.
\end{aligned}$$

Next we calculate:

$$\left|\psi_n^{(1)}\right\rangle = \sum_{m\neq n}\frac{\langle m|H_1|n\rangle}{\left(E_m^{(0)}-E_n^{(0)}\right)}|m\rangle$$

$$= \frac{B}{4\alpha^4}\left[\sum_{m\neq n}\frac{\langle m|(a^\dagger)^4|n\rangle}{\left(E_m^{(0)}-E_n^{(0)}\right)}|m\rangle + \sum_{m\neq n}\frac{\langle m|a^4|n\rangle}{\left(E_m^{(0)}-E_n^{(0)}\right)}|m\rangle\right.$$

$$+\sum_{m\neq n}\frac{\langle m|(1+2N)^2|n\rangle}{\left(E_m^{(0)}-E_n^{(0)}\right)}|m\rangle + \sum_{m\neq n}\frac{\langle m|a^2(a^\dagger)^2|n\rangle}{\left(E_m^{(0)}-E_n^{(0)}\right)}|m\rangle \tag{7.41a}$$

$$+\sum_{m\neq n}\frac{\langle m|(a^\dagger)^2a^2|n\rangle}{\left(E_m^{(0)}-E_n^{(0)}\right)}|m\rangle + \sum_{m\neq n}\frac{\langle m|a^2(1+2N)|n\rangle}{\left(E_m^{(0)}-E_n^{(0)}\right)}|m\rangle + \sum_{m\neq n}\frac{\langle m|(1+2N)a^2|n\rangle}{\left(E_m^{(0)}-E_n^{(0)}\right)}|m\rangle$$

$$\left.+\sum_{m\neq n}\frac{\langle m|(a^\dagger)^2(1+2N)|n\rangle}{\left(E_m^{(0)}-E_n^{(0)}\right)}|m\rangle + \sum_{m\neq n}\frac{\langle m|(1+2N)(a^\dagger)^2|n\rangle}{\left(E_m^{(0)}-E_n^{(0)}\right)}|m\rangle\right]$$

which simplifies to:

$$\left|\psi_n^{(1)}\right\rangle = \frac{B}{2\sqrt{2}\alpha^3}\left[\sum_{m\neq n}\frac{\langle m|(a^\dagger)^4|n\rangle}{\left(E_m^{(0)}-E_n^{(0)}\right)}|m\rangle + \sum_{m\neq n}\frac{\langle m|a^4|n\rangle}{\left(E_m^{(0)}-E_n^{(0)}\right)}|m\rangle\right.$$

$$+\sum_{m\neq n}\frac{\langle m|a^2(1+2N)|n\rangle}{\left(E_m^{(0)}-E_n^{(0)}\right)}|m\rangle + \sum_{m\neq n}\frac{\langle m|(1+2N)a^2|n\rangle}{\left(E_m^{(0)}-E_n^{(0)}\right)}|m\rangle \tag{7.41b}$$

$$\left.+\sum_{m\neq n}\frac{\langle m|(a^\dagger)^2(1+2N)|n\rangle}{\left(E_m^{(0)}-E_n^{(0)}\right)}|m\rangle + \sum_{m\neq n}\frac{\langle m|(1+2N)(a^\dagger)^2|n\rangle}{\left(E_m^{(0)}-E_n^{(0)}\right)}|m\rangle\right]$$

Further simplification gives:

$$\left|\psi_n^{(1)}\right\rangle = \frac{B}{4\hbar\omega\alpha^4}\left[\frac{1}{4}\left\{((n+1)(n+2)(n+3)(n+4))^{1/2}|n+4\rangle\right.\right.$$

$$\left.-(n(n-1)(n-2)(n-3))^{1/2}|n-4\rangle\right\} + ((n+1)(n+2))^{1/2}(3+2n)|n \tag{7.42}$$

$$\left.-(n(n-1))^{1/2}(2n-1)|n-2\rangle\right]$$

7.1.2 Perturbation Theory for Degenerate States

As stated earlier, when there are more than one eigenvectors, the set of quantum numbers (nlm) takes more than one value, and all belong to same energy eigenvalue, the state is called the degenerate state. As is seen from Eqns. (7.21), (7.27), and (7.37b), the perturbation series converges if $\left\langle\psi_m^{(0)}\right|H_1\left|\psi_n^{(0)}\right\rangle << \left|E_m^{(0)}-E_n^{(0)}\right|$, which means that the mixing matrix element must be small compared to the energy difference between the unperturbed energy levels. We next examine what happens when energy states of H_0 are degenerate. Let us write:

$$H_0\left|\psi_{n\alpha}^{(0)}\right\rangle = E_n^{(0)}\left|\psi_{n\alpha}^{(0)}\right\rangle \tag{7.43}$$

where α takes values 1, 2,3, ….r, for a given value of n. Thus, α-labels represents quantum numbers other than n in a complete set of (nlm). We say that the nth state is r-fold degenerate. In the degenerate case, the mixing matrix element cannot possibly be small, and in principle it couples states of all orders. However, within a degenerate subspace, any linear combination of wave function kets is another wave function ket of H_0, with the same energy eigenvalue. Let us take the linear combination:

$$\left|\phi_n^{(0)}\right\rangle = \sum_{\alpha=1}^{r} D_\alpha \left|\psi_{n\alpha}^{(0)}\right\rangle. \tag{7.44a}$$

which with the use of the above argument gives:

$$H_0\left|\phi_n^{(0)}\right\rangle = E_n^{(0)}\left|\phi_n^{(0)}\right\rangle \tag{7.44b}$$

We further say that the new set of wave function kets satisfies the orthonormality condition $\left\langle\phi_{n\alpha}^{(0)}\middle|\phi_{n'\beta}^{(0)}\right\rangle = \delta_{nn'}\delta_{\alpha\beta}$ because $\left\langle\psi_{n\alpha}^{(0)}\middle|\psi_{n\beta}^{(0)}\right\rangle = \delta_{\alpha\beta}$. Then, on replacing $\left|\psi_n^{(0)}\right\rangle$ by $\left|\phi_n^{(0)}\right\rangle$ in Eqn. (7.11), we get the first order perturbation expansion equation:

$$H_0\left|\psi_n^{(1)}\right\rangle + H_1\left|\phi_n^{(0)}\right\rangle = E_n^{(0)}\left|\psi_n^{(1)}\right\rangle + E_n^{(1)}\left|\phi_n^{(0)}\right\rangle \tag{7.45}$$

Next operate from the left with $\left\langle\psi_{n\beta}^{(0)}\right|$ to get:

$$\left\langle\psi_{n\beta}^{(0)}\middle|H_0\middle|\psi_n^{(1)}\right\rangle + \left\langle\psi_{n\beta}^{(0)}\middle|H_1\middle|\phi_n^{(0)}\right\rangle = E_n^{(0)}\left\langle\psi_{n\beta}^{(0)}\middle|\psi_n^{(1)}\right\rangle + E_n^{(1)}\left\langle\psi_{n\beta}^{(0)}\middle|\phi_n^{(0)}\right\rangle \tag{7.46}$$

Since $\left\langle\psi_{n\beta}^{(0)}\middle|H_0\middle|\psi_n^{(1)}\right\rangle = E_n^{(0)}\left\langle\psi_{n\beta}^{(0)}\middle|\psi_n^{(1)}\right\rangle$, Eqn. (7.46) simplifies to:

$$\sum_{\alpha=1}^{r} D_\alpha\left\langle\psi_{n\beta}^{(0)}\middle|H_1\middle|\psi_{n\alpha}^{(0)}\right\rangle - E_n^{(1)}\sum_{\alpha=1}^{r} D_\alpha\delta_{\beta\alpha} = 0 \tag{7.47}$$

On defining $\left\langle\psi_{n\beta}^{(0)}\middle|H_1\middle|\psi_{n\alpha}^{(0)}\right\rangle = h_{\beta\alpha}$ and then varying both α and β between 1 and r, Eqn. (7.47) expands to a matrix equation:

$$\begin{bmatrix} h_{11}-E_n^{(1)} & h_{12} & \cdots & h_{1r} \\ h_{21} & h_{22}-E_n^{(1)} & \cdots & h_{2r} \\ \vdots & \vdots & \vdots & \vdots \\ h_{r1} & h_{r2} & \cdots & h_{rr}-E_n^{(1)} \end{bmatrix} \begin{bmatrix} D_1 \\ D_2 \\ \vdots \\ D_r \end{bmatrix} = \begin{bmatrix} 0 \\ 0 \\ \vdots \\ 0 \end{bmatrix} \tag{7.48}$$

A non-trivial solution of Eqn. (7.48) demands:

$$\begin{vmatrix} h_{11}-E_n^{(1)} & h_{12} & \cdots & h_{1r} \\ h_{21} & h_{22}-E_n^{(1)} & \cdots & h_{2r} \\ \vdots & \vdots & \vdots & \vdots \\ h_{r1} & h_{r2} & \cdots & h_{rr}-E_n^{(1)} \end{vmatrix} = 0 \tag{7.49}$$

Simplification of the determinant gives rise to a polynomial of order r in $E_n^{(1)}$. Solution of the polynomial yields r-values of $E_n^{(1)}$. Let us term these $E_{n1}^{(1)}, E_{n2}^{(1)}, E_{n3}^{(1)} ....E_{nr}^{(1)}$. All of $E_{n1}^{(1)}, E_{n2}^{(1)}, E_{n3}^{(1)} ....E_{nr}^{(1)}$ can be nonzero and distinct, or some of these could be zero or identical to each other. When all the r-values are non-zero and distinct, we have r-nonzero and distinct values of $E_n = E_n^{(0)} + \lambda E_n^{(1)}$, suggesting that the degeneracy is completely removed by perturbation. For the case when some of the r-values of $E_n^{(1)}$ are zero or identical, degeneracy is partially removed. The expansion coefficients $D_1, D_2, D_3 .....D_r$ are determined by solving Eqn. (7.48) along with the normalizing condition of the wave function, which leads to $D_1^2 + D_2^2 + D_3^2 + ...D_r^2 = 1$. To illustrate the use of degenerate perturbation theory, we take an example of the effect of the application of an electric field to the first excited state in a hydrogen-like atom. The phenomenon is also referred as the linear Stark effect.

7.1.2.1 Effect of an Electric Field on the First Excited State in a Hydrogen Atom (Linear Stark Effect)

The first excited state $(n = 2)$ is a four-fold degenerate state; there are one 2s-state $(l = 0, m = 0)$and three p-states $(l = 1; m = 1, 0, -1)$. Let us suppose that the electric field **E** is applied along the positive z-axis. The electric field interacts with the electron dipole moment $-e'\mathbf{r}$ and gives rise to an additional potential energy $-e'\mathbf{E.r}$ for the electron, where $e' = \frac{e}{\sqrt{4\pi\varepsilon_0}}$, e is the charge on the electron. Since the field is along the z-axis, additional potential energy is $H_1 = -e'\mathrm{E}\hat{z}.\mathbf{r} = -e'\mathrm{E}r\cos\theta$, where θ is an angle between the z-axis and **r** and $\mathrm{E} = |\mathbf{E}|$. In this case, $|\psi_{n\alpha}^{(0)}\rangle$ takes the following 4-values:

$$\begin{aligned} |\psi_{21}^{(0)}\rangle &= |\psi_{200}\rangle, \\ |\psi_{22}^{(0)}\rangle &= |\psi_{210}\rangle, \\ |\psi_{23}^{(0)}\rangle &= |\psi_{211}\rangle, \\ |\psi_{24}^{(0)}\rangle &= |\psi_{21,-1}\rangle \end{aligned} \tag{7.50}$$

As each of α and β has 4-values, the matrix in Eqn. (7.48) is of the order of 4×4. The $h_{\alpha\beta} = -e'\mathrm{E}\langle nlm|r\cos\theta|n'l'm'\rangle$, therefore, has 16-values, obtained from:

$$h_{\alpha\beta} = -e'\mathrm{E}\int_0^\infty\int_0^\pi\int_0^{2\pi} \psi_{nlm}^*(r,\theta,\phi)(r\cos\theta)\psi_{n'l'm'}(r,\theta,\phi)r^2 dr \sin\theta d\theta d\phi \tag{7.51a}$$

Wave functions for the hydrogen atom from Eqn. (6.38) are:

$$\psi_{200}(r) = \frac{1}{4a_0}\left(\frac{1}{2\pi a_0}\right)^{1/2}\left(2-\frac{r}{a_0}\right)e^{-r/2a_0},$$

$$\psi_{210}(r,\theta,\phi) = \frac{1}{4a_0}\left(\frac{1}{2\pi a_0}\right)^{1/2}\left(\frac{r}{a_0}\right)e^{-r/2a_0}\cos\theta, \text{ and} \tag{7.51b}$$

$$\psi_{21,\pm1}(r,\theta,\varphi) = \frac{1}{4a_0}\left(\frac{1}{\pi a_0^3}\right)^{1/2} re^{-r/2a_0}(\sin\theta)e^{\pm i\varphi}$$

We here notice that the 10-values of $h_{\alpha\beta}$, consisting of $\int_0^{2\pi} e^{i(m-m')\phi}d\phi$, are zero because of $m \neq m'$. These are $\langle 200|H_1|211\rangle$; $\langle 200|H_1|21-1\rangle$; $\langle 210|H_1|211\rangle$; $\langle 210|H_1|21,-1\rangle$; $\langle 211|H_1|21,-1\rangle$; $\langle 211|H_1|200\rangle$; $\langle 21-1|H_1|200\rangle$; $\langle 211|H_1|210\rangle$; $\langle 21,-1|H_1|21,0\rangle$, and $\langle 21,-1|H_1|211\rangle$. The diagonal matrix elements $\langle 200|H_1|200\rangle$, $\langle 210|H_1|210\rangle$, $\langle 211|H_1|211\rangle$, and $\langle 21,-1|H_1|21,-1\rangle$ are also zero, because these involve the integral $\int_0^{\pi}(\cos\theta)^q \sin\theta\, d\theta = 0$, since q is an odd integer. The remaining two non-zero matrix elements are $\langle 200|H_1|210\rangle$ and $\langle 210|H_1|200\rangle$:

$$\begin{aligned} h_{12} &= \langle 200|H_1|210\rangle \\ &= -\frac{e'\mathrm{E}}{32\pi a_0^3}\int_0^{\infty}\frac{r}{a_0}\left(2-\frac{r}{a_0}\right)r^2e^{-r/a_0}dr\int_0^{\pi}\cos^2\theta\sin\theta\, d\theta\int_0^{2\pi}d\phi \\ &= -\frac{e'\mathrm{E}}{24a_0^3}\int_0^{\infty}\frac{r}{a_0}\left(2-\frac{r}{a_0}\right)r^2e^{-r/a_0}dr \\ &= -3e'\mathrm{E}a_0 \end{aligned} \tag{7.52}$$

Since $\langle 200|H_1|210\rangle$ is real, it is equal to $\langle 210|H_1|200\rangle$. We thus find that Eqn. (7.49) in the present case reduces to:

$$\begin{vmatrix} -E_2^{(1)} & -3e'\mathrm{E}a_0 & 0 & 0 \\ -3e'\mathrm{E}a_0 & -E_2^{(1)} & 0 & 0 \\ 0 & 0 & -E_2^{(1)} & 0 \\ 0 & 0 & 0 & -E_2^{(1)} \end{vmatrix} = 0 \tag{7.53}$$

On simplification, the above equation gives four roots $3e'\mathrm{E}a_0$; $-3e'\mathrm{E}a_0$; 0, and 0. We thus find that the energy of state n = 2 of the hydrogen atom has four values: $\left\{-\frac{(e')^2}{8a_0}+3e'\mathrm{E}a_0\right\}$; $\left\{-\frac{(e')^2}{8a_0}-3e'\mathrm{E}a\right\}$; $\left\{-\frac{(e')^2}{8a_0}\right\}$ and $\left\{-\frac{(e')^2}{8a_0}\right\}$. The states $|200\rangle$ and $|210\rangle$ are affected on application of the electric field, while the states $|211\rangle$ and $|21,-1\rangle$ remain unaffected. It can therefore be said that degeneracy has partially been lifted on the application field to the 2*s* and 2*p* states of the hydrogen atom. To evaluate eigenvectors, we solve:

$$\begin{bmatrix} -E_2^{(1)} & -3e'\mathrm{E}a_0 & 0 & 0 \\ -3e'\mathrm{E}a_0 & -E_2^{(1)} & 0 & 0 \\ 0 & 0 & -E_2^{(1)} & 0 \\ 0 & 0 & 0 & -E_2^{(1)} \end{bmatrix} \begin{bmatrix} D_1 \\ D_2 \\ D_3 \\ D_4 \end{bmatrix} = \begin{bmatrix} 0 \\ 0 \\ 0 \\ 0 \end{bmatrix} \tag{7.54}$$

along with the condition of normalization $D_1^2 + D_2^2 + D_3^2 + D_4^2 = 1$. Eqn. (7.54) gives:

$$\begin{aligned} &-E_2^{(1)} D_1 - 3e'\mathrm{E}a_0 D_2 = 0; \\ &-3e'\mathrm{E}a_0 D_1 - E_2^{(1)} D_2 = 0; \\ &-E_2^{(1)} D_3 = 0; \\ &-E_2^{(1)} D_4 = 0 \end{aligned} \tag{7.55}$$

This suggests that $D_1 = \pm D_2$ and $D_3 = D_4 = 0$. We take $D_1 = \frac{1}{\sqrt{2}}$ and $D_2 = \mp\frac{1}{\sqrt{2}}$. Thus, there are four states: (i) $\frac{1}{\sqrt{2}}(|200\rangle + |210\rangle)$, which corresponds to the energy eigenvalue $E_2^{(0)} - 3e'Ea_0$; (ii) $\frac{1}{\sqrt{2}}(|200\rangle - |210\rangle)$ belonging to the energy eigenvalue $E_2^{(0)} + 3e'Ea_0$, and (iii) $|211\rangle$ and (iv) $|21,-1\rangle$ both belonging to energy eigenvalue $E_2^{(0)}$. The first excited state of the hydrogen atom in the presence of an electric field applied along the z-axis presents a permanent electric dipole moment of magnitude $3e'Ea_0$ with three orientations: one state parallel to the electric field, one state antiparallel to the electric field, and two states with zero interaction with the electric field, in the first order linear Stark effect.

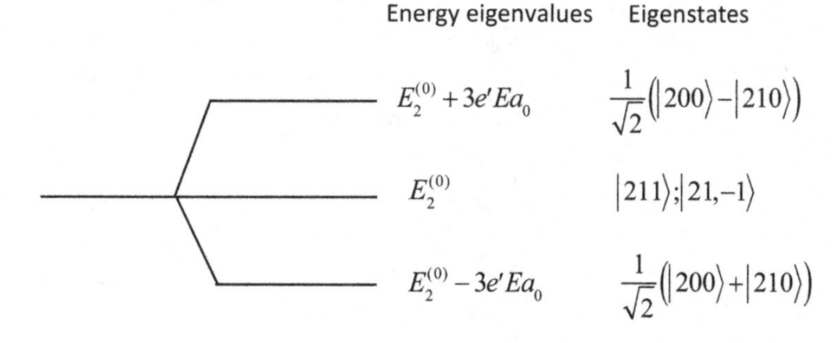

FIGURE 7.1
Schematic diagram of linear Stark effect as an example of degenerate perturbation theory.

7.2 Variation Method

Another approximation method used in quantum mechanics is the variational method. In solving the problems where it is hard to determine a good unperturbed Hamiltonian, to make the perturbation small and solvable, the variational method is more robust in comparison with the perturbation theory. While in cases where a good unperturbed Hamiltonian can be chosen, perturbation theory can be more efficient than the variational method. To guess a trial wave function for the problem in terms of some adjustable parameters, which are termed as variational parameters, is the main idea behind the variational method. The parameters are adjusted until the energy with the use of the trial wave function is minimized. The variational method approximations to the exact wave function and energy of the ground state are represented by the resulting trial wave function and its corresponding energy. We want to calculate ground state energy E_g for a system whose Hamiltonian H is known, but its Schrödinger equation cannot be solved. The variational theorem states that the expectation value of the Hamiltonian $\langle H\rangle$ calculated with the use of a normalized trial wave function ket $|\phi\rangle$ is always greater than or equal to E_g:

$$E_g \leq \langle H\rangle = \langle\phi|H|\phi\rangle \tag{7.56}$$

where

$$\langle\phi|\phi\rangle = 1.$$

To prove the variational theorem, the $|\phi\rangle$ can formally be expanded as a linear combination of the exact wave function kets $|\psi_n\rangle$ of the system. Of course, in practice, we don't know what are the $|\psi_n\rangle$, because we are going to apply the variational method to a problem that cannot be solved analytically. However, this does not prevent us from using the exact wave function kets in our proof, since they certainly do exist and form a complete set. Therefore $|\phi\rangle$ can be written as:

$$|\phi\rangle = \sum_n c_n|\psi_n\rangle \tag{7.57}$$

with $H|\psi_n\rangle = E_n|\psi_n\rangle$ and $\langle\psi_m|\psi_n\rangle = \delta_{mn}$. We take:

$$\begin{aligned} 1 &= \langle\phi|\phi\rangle \\ &= \sum_m c_m^*\langle\psi_m|\sum_n c_n|\psi_n\rangle \\ &= \sum_m\sum_n c_m^* c_n\langle\psi_m|\psi_n\rangle \\ &= \sum_m\sum_n c_m^* c_n\delta_{mn} \\ &= \sum_n |c_n|^2 \end{aligned} \tag{7.58}$$

and:

$$\begin{aligned}\langle H\rangle &= \left(\sum_m c_m^* \langle \psi_m|\right) H \left(\sum_n c_n |\psi_n\rangle\right) \\ &= \sum_m \sum_n c_m^* c_n E_n \langle \psi_m|\psi_n\rangle \\ &= \sum_n E_n |c_n|^2\end{aligned} \tag{7.59}$$

Let us write:

$$\langle H\rangle - E_g = \sum_n (E_n - E_g)|c_n|^2 \tag{7.60}$$

since $E_n \geq E_g$, the right-hand side of Eqn. (7.60) is always greater than or equal to zero, proving that $E_g \leq \langle H\rangle$, which means that the exact ground state energy is equal to or less than the energy computed with the use of an approximate wave function. Any variations in the trial function which lower the computed energy necessarily bring the approximate energy closer to E_g. An example of the above laid down procedure is the calculation of the ground state energy of the hydrogen atom with the use of the variational method by taking the wave function $\phi(r) = Ae^{-\alpha r}$ as a trial wave function, where α is the variational parameter. From Eqn. (6.11), the ground state ($l = 0$) Hamiltonian for the hydrogen atom is given by:

$$H = -\frac{\hbar^2}{2\mu}\left[\frac{d^2}{dr^2} + \frac{2}{r}\frac{d}{dr}\right] - \frac{e^2}{4\pi\varepsilon_0 r} \tag{7.61}$$

To find normalized $\phi(r)$, we take:

$$\begin{aligned}1 &= \langle \phi|\phi\rangle \\ &= A^2 \int_0^\infty \int_0^\pi \int_0^{2\pi} e^{-2\alpha r} r^2 dr \sin\theta d\theta d\phi \\ &= 4\pi A^2 \int_0^\infty e^{-2\alpha r} r^2 dr \\ &= 4\pi A^2 \frac{2!}{(2\alpha)^3} \\ &= A^2 \frac{\pi}{\alpha^3} \\ \Rightarrow A &= \frac{\alpha^{3/2}}{\sqrt{\pi}}\end{aligned} \tag{7.62}$$

Thus $\phi(r)=\frac{\alpha^{3/2}}{\sqrt{\pi}}e^{-\alpha r}$. Then:

$$\langle H\rangle=-\frac{\alpha^3}{\pi}\int_0^\infty\int_0^\pi\int_0^{2\pi}e^{-\alpha r}\left[\frac{\hbar^2}{2\mu}\left(\frac{d^2}{dr^2}+\frac{2}{r}\frac{d}{dr}\right)+\frac{e^2}{4\pi\varepsilon_0 r}\right]\left(e^{-\alpha r}\right)r^2dr\sin\theta d\theta d\phi$$

$$=-4\alpha^3\int_0^\infty e^{-\alpha r}\left[\frac{\hbar^2}{2\mu}\left(\frac{d^2}{dr^2}+\frac{2}{r}\frac{d}{dr}\right)+\frac{e^2}{4\pi\varepsilon_0 r}\right]\left(e^{-\alpha r}\right)r^2dr$$

$$=-4\alpha^3\left[\frac{\hbar^2}{2\mu}\int_0^\infty\alpha^2r^2e^{-2\alpha r}\,dr+\left(\frac{e^2}{4\pi\varepsilon_0}-\frac{\hbar^2\alpha}{\mu}\right)\int_0^\infty re^{-2\alpha r}\,dr\right] \tag{7.63}$$

$$=-4\alpha^3\left[\frac{\hbar^2}{2\mu}\frac{\alpha^2(2!)}{(2\alpha)^3}+\left(\frac{e^2}{4\pi\varepsilon_0}-\frac{\hbar^2\alpha}{\mu}\right)\frac{1}{(2\alpha)^2}\right]$$

$$=\left[\frac{\hbar^2\alpha^2}{2\mu}-\frac{e^2\alpha}{4\pi\varepsilon_0}\right]$$

$\frac{\partial\langle H\rangle}{\partial\alpha}=0$ gives a minimum value of α required to get the minimum value of $\langle H\rangle$. We get:

$$\alpha_{\min}=\frac{\mu e^2}{4\pi\varepsilon_0\hbar^2} \tag{7.64a}$$

and:

$$E_g=\langle H\rangle_{\min}=-\frac{\mu e^4}{2(4\pi\varepsilon_0)^2\hbar^2} \tag{7.64b}$$

which matches with the answer obtained by solving the Schrödinger equation.

7.2.1 The Ground State of the Helium Atom

The helium atom consists of two electrons revolving in an orbit around a nucleus having charge $2e$. Taking the origin of the coordinate system at the nucleus and the two electrons at the distances of $\mathbf{r}_1$ and $\mathbf{r}_2$ from the nucleus, as is shown in Fig. 7.2, the Hamiltonian for the system is:

$$H=-\frac{\hbar^2}{2\mu}\nabla_1^2-\frac{2e^2}{4\pi\varepsilon_0 r_1}-\frac{\hbar^2}{2\mu}\nabla_2^2-\frac{2e^2}{4\pi\varepsilon_0 r_2}+\frac{e^2}{4\pi\varepsilon_0|\mathbf{r}_1-\mathbf{r}_2|} \tag{7.65}$$

We want to calculate the ground state energy for this system. Though it is a simple problem, it does not have an exact solution. We would like to compute the approximate ground

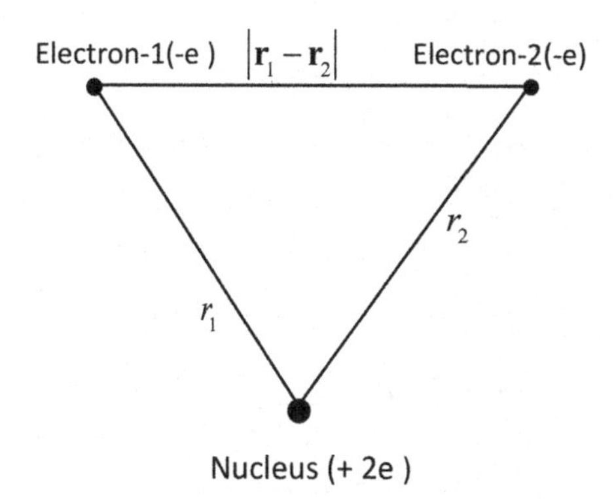

FIGURE 7.2
Schematic diagram for the helium atom.

state energy E_g that could be as close as possible with the experimentally measured value of -78.975 eV. The question is what would be the ground state wave function to compute the E_g. If the electron-electron interaction:

$$V_{ee} = \frac{e^2}{4\pi\varepsilon_0 |\mathbf{r}_1 - \mathbf{r}_2|} \tag{7.66}$$

is ignorable, then the Hamiltonian Eqn. (7.65) simply breaks into two hydrogen-like Hamiltonians, and the ground state energy from Eqn. (6.22) is $2\left\{-\frac{\mu(2e^2)^2}{2(4\pi\varepsilon_0)^2\hbar^2} = -\frac{2e^2}{(4\pi\varepsilon_0)a_0}\right\} = 8E_1$, and the ground state wave function will be the product of two hydrogen-like wave functions:

$$\psi(r) = \frac{8}{\pi a_0^3} e^{-2(r_1+r_2)/a_0} \tag{7.67}$$

However, the electron-electron interaction given by Eqn. (7.66) is not ignorable because it is roughly equal in size to the Hamiltonian of two hydrogen atoms. We hence solve the problem with the use of the variational method to get E_g. We would like to have a realistic trial wave function to apply the variational method. $\psi(r)$, which is given by Eqn. (7.67), is the wave function for the situation when we think that two electrons are not interacting, but in a realistic situation they do interact. On average, each electron represents a cloud of negative charge that partially screens the nucleus, so that another electron sees an effective charge on the nucleus, which should be somewhat less than $2e$. This suggests that a more realistic trial wave function can be taken as:

$$\phi(r) = \frac{Z^3}{\pi a_0^3} e^{-Z(r_1+r_2)/a_0} = \frac{1}{\sqrt{\pi}}\left(\frac{Z}{a_0}\right)^{3/2} e^{-Zr_1/a_0} \times \frac{1}{\sqrt{\pi}}\left(\frac{Z}{a_0}\right)^{3/2} e^{-Zr_2/a_0} = \phi_1(r_1)\phi_2(r_2) \tag{7.68}$$

where Z is the variational parameter to be determined. We then rewrite the Eqn. (7.65) as:

$$H = -\frac{\hbar^2}{2\mu}\nabla_1^2 - \frac{Ze^2}{4\pi\varepsilon_0 r_1} - -\frac{\hbar^2}{2\mu}\nabla_2^2 - \frac{Ze^2}{4\pi\varepsilon_0 r_2} + \frac{e^2}{4\pi\varepsilon_0|\mathbf{r}_1 - \mathbf{r}_2|} + \frac{(Z-2)e^2}{4\pi\varepsilon_0 r_1} + \frac{(Z-2)e^2}{4\pi\varepsilon_0 r_2} \tag{7.69}$$

The expectation value of the Hamiltonian is:

$$\langle H\rangle = 2Z^2E_1 + (Z-2)\frac{e^2}{4\pi\varepsilon_0}\left\langle\frac{1}{r_1}\right\rangle + (Z-2)\frac{e^2}{4\pi\varepsilon_0}\left\langle\frac{1}{r_2}\right\rangle + \frac{e^2}{4\pi\varepsilon_0}\left\langle\frac{1}{|\mathbf{r}_1-\mathbf{r}_2|}\right\rangle \tag{7.70}$$

where:

$$\begin{aligned}\left\langle\frac{1}{r_1}\right\rangle &= \langle\phi|\frac{1}{r_1}|\phi\rangle \\ &= \frac{Z^3}{\pi a_0^3}\int_0^\infty\int_0^\pi\int_0^{2\pi} e^{-2Zr_1/a_0} r_1 dr_1 \sin\theta d\theta d\phi \\ &= \frac{4Z^3}{a_0^3}\int_0^\infty e^{-2Zr_1/a_0} r_1 dr_1 \\ &= \frac{4Z^3}{a_0^3}\frac{a_0^2}{4Z^2} \\ &= \frac{Z}{a_0}\end{aligned} \tag{7.71}$$

Similarly calculation yields $\left\langle\frac{1}{r_2}\right\rangle = \frac{Z}{a_0}$. We next evaluate:

$$\left\langle\frac{1}{|\mathbf{r}_1-\mathbf{r}_2|}\right\rangle = \left(\frac{Z^3}{\pi a_0^3}\right)^2 \int \frac{e^{-2Z(r_1+r_2)/a_0}}{|\mathbf{r}_1-\mathbf{r}_2|} d^3r_1 d^3r_2 \tag{7.72}$$

To evaluate these integrals, we first fix $\mathbf{r}_1$ and then choose the coordinate system for $\mathbf{r}_2$ in such a manner that the polar axis of it is along $\mathbf{r}_1$; see Fig. 7.3.

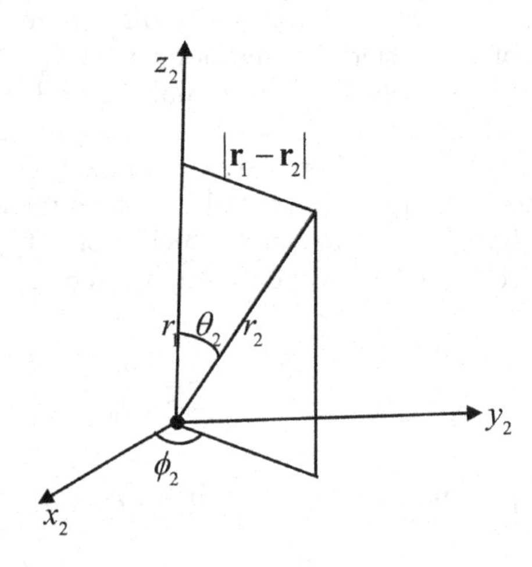

FIGURE 7.3
System of coordinates to evaluate the integral I.

Thus $|\mathbf{r}_1 - \mathbf{r}_2| = \sqrt{\left(r_1^2 + r_2^2 - 2r_1r_2\cos\theta_2\right)}$. Let us take:

$$I = \int \frac{e^{-2Zr_2/a_0}}{|\mathbf{r}_1 - \mathbf{r}_2|} d^3r_2$$

$$= \int\int_0^{\pi}\int_0^{2\pi} \frac{1}{\sqrt{\left(r_1^2 + r_2^2 - 2r_1r_2\cos\theta_2\right)}} e^{-2Zr_2/a_0} r_2^2\, dr_2 \sin\theta_2\, d\phi_2 \tag{7.73a}$$

$$= 2\pi \int e^{-2Zr_2/a_0} r_2^2\, dr_2 \int_0^{\pi} \frac{\sin\theta_2 d\theta_2}{\sqrt{\left(r_1^2 + r_2^2 - 2r_1r_2\cos\theta_2\right)}}$$

The integration over θ_2 is:

$$\int_0^{\pi} \frac{\sin\theta_2 d\theta_2}{\sqrt{\left(r_1^2 + r_2^2 - 2r_1r_2\cos\theta_2\right)}} = \frac{1}{r_1r_2}\left\{\sqrt{r_1^2 + r_2^2 + 2r_1r_2} - \sqrt{r_1^2 + r_2^2 - 2r_1r_2}\right\}$$

$$= \begin{cases} \dfrac{2}{r_1} & \text{for } r_2 < r_1 \\[2ex] \dfrac{2}{r_2} & \text{for } r_2 < r_1 \end{cases} \tag{7.73b}$$

which yields:

$$I = 4\pi\left[\int_0^{r_1} \frac{e^{-2Zr_2/a_0}}{r_1} r_2^2 dr_2 + \int_{r_1}^{\infty} \frac{e^{-2Zr_2/a_0}}{r_2} r_2^2 dr\right]$$

$$= \frac{\pi a_0^3}{Z^3 r_1}\left[1 - \left(1 + \frac{Zr_1}{a_0}\right)e^{-2Zr_1/a_0}\right] \tag{7.73c}$$

Thus:

$$\left\langle \frac{1}{|\mathbf{r}_1 - \mathbf{r}_2|} \right\rangle = \frac{Z^3}{\pi a_0^3} \int_0^{\infty}\int_0^{\pi}\int_0^{2\pi} \left[1 - \left(1 + \frac{Zr_1}{a_0}\right)e^{-2Zr_1/a_0}\right] e^{-2Zr_1/a_0} r_1 dr_1 \sin\theta_1 d\theta_1 d\phi_1$$

$$= \frac{4Z^3}{a_0^3} \int_0^{\infty} \left[1 - \left(1 + \frac{Zr_1}{a_0}\right)e^{-2Zr_1/a_0}\right] e^{-2Zr_1/a_0} r_1\, dr_1$$

$$= \left(\frac{4Z^3}{a_0^3}\right)\left[\int_0^{\infty} e^{-2Zr_1/a_0} r_1\, dr_1 - \int_0^{\infty} e^{-4Zr_1/a_0} r_1\, dr_1 - \frac{Z}{a_0}\int_0^{\infty} e^{-4Zr_1/a_0} r_1^2\, dr_1\right] \tag{7.74}$$

$$= \left(\frac{4Z^3}{a_0^3}\right)\left[\frac{a_0^2}{4Z^2} - \frac{a_0^2}{16Z^2} - \frac{a_0^2}{32Z^2}\right] = \frac{5Z}{8a_0}$$

Therefore:

$$\begin{aligned}\langle H\rangle &= 2Z^2E_1 + (Z-2)\frac{e^2}{4\pi\varepsilon_0}\frac{Z}{a_0} + (Z-2)\frac{e^2}{4\pi\varepsilon_0}\frac{Z}{a_0} + \frac{e^2}{4\pi\varepsilon_0}\frac{5Z}{8a_0} \\ &= 2Z^2E_1 + \frac{e^2}{4\pi\varepsilon_0}\left(2Z-4+\frac{5}{8}\right)\frac{Z}{a_0} \\ &= 2Z^2E_1 - \left(4Z^2 - \frac{27}{4}Z\right)E_1 \\ &= \left(-2Z^2 + \frac{27}{4}Z\right)E_1\end{aligned} \tag{7.75}$$

Application of the variational method requires that $\langle H\rangle$ be minimized with respect to Z:

$$\begin{aligned}&\frac{\partial\langle H\rangle}{\partial Z} = 0 \\ &\Rightarrow \left(-4Z + \frac{27}{4}\right)E_1 = 0 \\ &\Rightarrow Z_{\min} = \frac{27}{16} = 1.69\end{aligned} \tag{7.76}$$

which tells that the other electron screens the nucleus and reduces its charge from $2e$ to $1.69e$. Thus, the ground state energy for helium is:

$$\begin{aligned}E_g &= \left(-2(1.69)^2 + \frac{27}{4}\times 1.69\right)E_1 \\ &= 5.696\times E_1 \\ &= -77.46 \text{ eV}\end{aligned} \tag{7.77}$$

which is a reasonable result when compared with the experimental value –78.975 eV, because the computed result differs from the experimental result by only within 2%. The ground state energy of helium had been calculated with the use of a more complicated trial wave function having a larger number of adjustable parameters and $E_g = -78.7$ eV was achieved.

7.2.2 Rayleigh-Ritz Variational Method

A more convenient method would be to write the trial wave function ket as a linear combination of known fixed basis vectors and then treat the expansion coefficients as a variational parameter. Let us assume that the basis vectors form a complete set, and the trial wave function ket is $|\phi\rangle = \sum_{i=1} c_i|\psi_i\rangle$, where $c_i (i = 1,2,3,......)$ are real or complex variational parameters. We then have:

$$\langle\phi|\phi\rangle = \sum_i c_i^*\langle\psi_i|\sum_j c_j|\psi_j\rangle = \sum_i\sum_j c_i^* c_j\langle\psi_i|\psi_j\rangle \tag{7.78a}$$

and:

$$\langle\phi|H|\phi\rangle = \left(\sum_i c_i^* \langle\psi_i|\right) H \left(\sum_j c_j |\psi_j\rangle\right) = \sum_i \sum_j c_i^* c_i \langle\psi_i|H|\psi_j\rangle \tag{7.78b}$$

This gives:

$$\begin{aligned}\langle H\rangle &= \frac{\langle\phi|H|\phi\rangle}{\langle\phi|\phi\rangle} \\ &= \frac{\sum_i \sum_j c_i^* c_j \langle\psi_i|H|\psi_j\rangle}{\sum_i \sum_j c_i^* c_j \langle\psi_i|\psi_j\rangle} \\ &= \frac{\sum_i \sum_j c_i^* c_j h_{ij}}{\sum_i \sum_j c_i^* c_j s_{ij}}\end{aligned} \tag{7.79}$$

where we defined $h_{ij} = \langle\psi_i|H|\psi_j\rangle$ and $s_{ij} = \langle\psi_i|\psi_j\rangle$. To find ground state energy, we minimize $\langle H\rangle$ with respect to the variational parameter c_k^*; $\frac{\partial\langle H\rangle}{\partial c_k^*} = 0$, where $k = 1, 2, 3.....$ We take:

$$\begin{aligned}&\frac{\partial}{\partial c_k^*}\left[\langle H\rangle \sum_i \sum_j c_i^* c_j s_{ij} - \sum_i \sum_j c_i^* c_j h_{ij}\right] \\ &= \frac{\partial\langle H\rangle}{\partial c_k^*} \sum_i \sum_j c_i^* c_j s_{ij} + \langle H\rangle \sum_j c_j s_{kj} - \sum_j c_j h_{kj} \\ &\Rightarrow \frac{\partial\langle H\rangle}{\partial c_k^*} = \frac{\sum_j c_j h_{kj} - \langle H\rangle \sum_j c_j s_{kj}}{\sum_i \sum_j c_i^* c_j s_{ij}}\end{aligned} \tag{7.80}$$

The $\frac{\partial\langle H\rangle}{\partial c_k^*} = 0$ implies $\sum_j c_j h_{kj} - \langle H\rangle \sum_j c_j s_{kj} = \sum_j \left(h_{kj} - \langle H\rangle s_{kj}\right) c_j = 0$, which in expanded form is:

$$\begin{bmatrix} h_{11} - \langle H\rangle s_{11} & h_{12} - \langle H\rangle s_{12} & \ldots & h_{1n} - \langle H\rangle s_{1n} \\ h_{21} - \langle H\rangle s_{21} & h_{22} - \langle H\rangle s_{22} & \ldots & h_{2n} - \langle H\rangle s_{2n} \\ \vdots & \vdots & \vdots & \vdots \\ h_{n1} - \langle H\rangle s_{n1} & h_{n2} - \langle H\rangle s_{n2} & \ldots & h_{nn} - \langle H\rangle s_{nn} \end{bmatrix} \begin{bmatrix} c_1 \\ c_2 \\ \vdots \\ c_n \end{bmatrix} = \begin{bmatrix} 0 \\ 0 \\ \vdots \\ 0 \end{bmatrix} \tag{7.81a}$$

the non-trivial solution of which demands:

$$\begin{vmatrix} h_{11}-\langle H\rangle s_{11} & h_{12}-\langle H\rangle s_{12} & \dots & h_{1n}-\langle H\rangle s_{1n} \\ h_{21}-\langle H\rangle s_{21} & h_{22}-\langle H\rangle s_{22} & \dots & h_{2n}-\langle H\rangle s_{2n} \\ \vdots & \vdots & \vdots & \vdots \\ h_{n1}-\langle H\rangle s_{n1} & h_{n2}-\langle H\rangle s_{n2} & \dots & h_{nn}-\langle H\rangle s_{nn} \end{vmatrix} = 0 \tag{7.81b}$$

The determinant simplifies to a polynomial of degree n having n-roots. We thus get n-values of $\langle H\rangle$. The smallest value of $\langle H\rangle$ is E_g, the ground state energy for a selected trial wave function.

7.2.3 The Hydrogen Molecule Ion

Another classic example of the variational method is calculation of the ground state energy of the hydrogen molecule ion, which consists of one electron in the field of two protons separated from each other by a distance of **R**, as is shown in Fig. 7.4.

This problem can be solved with the use of the Rayleigh-Ritz variational method. The Hamiltonian for the system is:

$$H = -\frac{\hbar^2}{2m}\nabla^2 - \frac{e^2}{4\pi\varepsilon_0}\frac{1}{r_1} - \frac{e^2}{4\pi\varepsilon_0}\frac{1}{r_2} + \frac{e^2}{4\pi\varepsilon_0 R} \tag{7.82}$$

where r_1 and r_2 are the distances of the electron from proton-1 and proton-2. The electron has equal probability of being associated with either of two protons; we take the following trial wave function:

$$\phi(r) = c_1\psi_1(r_1) + c_2\psi_2(r_2) \tag{7.83}$$

Here $\psi_1(r_1)$ is the atomic orbital when the electron is in the neighborhood of proton-1, and $\psi_2(r_2)$ is the atomic orbital when it is in the neighborhood of proton-2. The c_1 and c_2 are the variational parameters, and Eqn. (7.83) represents the linear combination of atomic orbitals. We take $\psi_1(r_1)$ and $\psi_2(r_2)$ as the ground state wave functions for the hydrogen atom:

$$\begin{aligned} \psi_1(r_1) &= \frac{1}{\sqrt{\pi a_0^3}} e^{-r_1/a_0} \\ \psi_2(r_2) &= \frac{1}{\sqrt{\pi a_0^3}} e^{-r_2/a_0} \end{aligned} \tag{7.84}$$

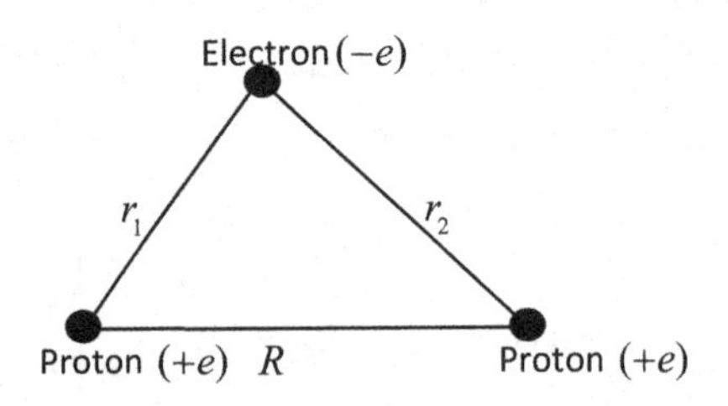

FIGURE 7.4
Schematic diagram for hydrogen molecule H_2^+.

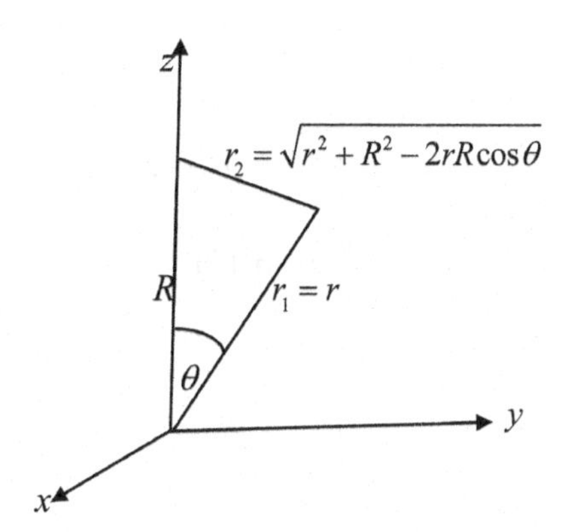

FIGURE 7.5
System of coordinates to evaluate integrals S and I.

The expectation value of the Hamiltonian is:

$$\langle H \rangle = \frac{\langle \phi | H | \phi \rangle}{\langle \phi | \phi \rangle} = \frac{c_1^2 \langle \psi_1 | H | \psi_1 \rangle + c_2^2 \langle \psi_2 | H | \psi_2 \rangle + c_1 c_2 \langle \psi_1 | H | \psi_2 \rangle + c_2 c_1 \langle \psi_2 | H | \psi_1 \rangle}{c_1^2 + c_2^2 + c_1 c_2 \langle \psi_1 | \psi_2 \rangle + c_2 c_1 \langle \psi_2 | \psi_1 \rangle} \tag{7.85}$$

where we have used $\langle \psi_1 | \psi_1 \rangle = \langle \psi_2 | \psi_2 \rangle = 1$. The evaluation of $S = \langle \psi_1 | \psi_2 \rangle = \langle \psi_1 | \psi_2 \rangle = \frac{1}{\pi a_0^3} \int d^3 r e^{-(r_1 + r_2)/a_0}$ is not straight forward; it is a little complicated. As is shown in Fig. 7.5, we choose $r_1 = r$; then $r_2 = |\mathbf{r} - \mathbf{R}| = \sqrt{\left(r^2 + R^2 - 2rR\cos\theta\right)}$, θ is the angle between $\mathbf{r}_1$ and $\mathbf{R}$.
Then:

$$\begin{aligned} S &= \frac{1}{\pi a_0^3} \int d^3 r e^{-(r_1 + r_2)/a_0} \\ &= \frac{1}{\pi a_0^3} \int_0^\infty e^{-r/a_0} \left(\int_0^\pi e^{-\sqrt{\left(r^2 + R^2 - 2rR\cos\theta\right)}/a_0} \sin\theta \, d\theta \right) r^2 dr \int_0^{2\pi} d\phi \\ &= \frac{2}{a_0^3} \int_0^\infty e^{-r/a_0} \left(\int_0^\pi e^{-\sqrt{\left(r^2 + R^2 - 2rR\cos\theta\right)}/a_0} \sin\theta \, d\theta \right) r^2 dr \end{aligned} \tag{7.86}$$

To evaluate:

$$I = \int_0^\pi e^{-\sqrt{\left(r^2 + R^2 - 2rR\cos\theta\right)}/a_0} \sin\theta \, d\theta \tag{7.87a}$$

we choose:

$$y^2 = r^2 + R^2 - 2rR\cos\theta$$
$$\Rightarrow -\frac{ydy}{2rR} = \sin\theta d\theta \tag{7.87b}$$

giving rise to:

$$I = \frac{1}{rR}\int_{|r-R|}^{r+R} e^{-y/a_0} y dy$$
$$= -\frac{a_0}{rR}\left[(r+R+a_0)e^{-(r+R)/a_0} - (|r-R|+a_0)e^{-|r-R|/a_0}\right] \tag{7.88}$$

Then:

$$S = -\frac{2}{a_0^2 R}\int_0^\infty e^{-r/a_0}\left[(r+R+a_0)e^{-(r+R)/a_0} - \left(|r-R|+a_0\right)e^{-|r-R|/a_0}\right] rdr$$
$$= -\frac{2}{a_0^2 R}\left[e^{-R/a_0}\left\{\int_0^\infty r^2 e^{-2r/a_0}dr + (R+a_0)\int_0^\infty re^{-2r/a_0}dr\right\} - e^{-R/a_0}\left\{\int_0^R (R-r+a_0)rdr\right\}\right.$$
$$\left. - e^{R/a_0}\left\{\int_R^\infty r^2 e^{-2r/a_0}dr - (R-a_0)\int_R^\infty re^{-2r/a_0}dr\right\}\right] \tag{7.89}$$
$$= e^{-R/a_0}\left[1 + \left(\frac{R}{a_0}\right) + \frac{1}{3}\left(\frac{R}{a_0}\right)^2\right]$$

Let us define $\langle\psi_1|H|\psi_1\rangle = \langle\psi_2|H|\psi_2\rangle = \alpha$, and $\langle\psi_1|H|\psi_2\rangle = \langle\psi_2|H|\psi_1\rangle = \beta$. Since R does not depend on r_1 and r_2, we have:

$$\alpha = \langle\psi_1|H|\psi_1\rangle$$
$$= \langle\psi_1|\left(-\frac{\hbar^2}{2m}\nabla^2 - \frac{e^2}{4\pi\varepsilon_0 r_1}\right)|\psi_1\rangle - \langle\psi_1|\left(\frac{e^2}{4\pi\varepsilon_0 r_2}\right)|\psi_1\rangle + \frac{e^2}{4\pi\varepsilon_0 R}\langle\psi_1|\psi_1\rangle \tag{7.90}$$
$$= E_1 + \frac{e^2}{4\pi\varepsilon_0 R} - D$$

with:

$$
\begin{aligned}
D &= \langle \psi_1 | \left(\frac{e^2}{4\pi\varepsilon_0 r_2} \right) | \psi_1 \rangle \\
&= \frac{e^2}{4\pi\varepsilon_0} \left(\frac{1}{\pi a_0^3} \right) \int_0^\infty r_1^2 dr_1 e^{-2r_1/a_0} \int_0^\pi \frac{\sin\theta d\theta}{\sqrt{(r_1^2 + R^2 - 2r_1 R\cos\theta)}} \int_0^{2\pi} d\phi \\
&= \frac{e^2}{\varepsilon_0} \left(\frac{1}{\pi a_0^3} \right) \left[\int_0^R r_1^3 dr_1 e^{-2r_1/a_0} + \int_R^\infty r_1^2 dr_1 e^{-2r_1/a_0} \right] \\
&= \frac{e^2}{4\pi\varepsilon_0} \left[\frac{1}{R} - \left(\frac{1}{a_0} + \frac{1}{R} \right) e^{-2R/a_0} \right]
\end{aligned}
\tag{7.91}
$$

Let us next evaluate:

$$
\begin{aligned}
\beta &= \langle \psi_1 | H | \psi_2 \rangle \\
&= \langle \psi_1 | \left(-\frac{\hbar^2}{2m}\nabla^2 - \frac{e^2}{4\pi\varepsilon_0 r_2} \right) | \psi_2 \rangle - \langle \psi_1 | \left(\frac{e^2}{4\pi\varepsilon_0 r_1} \right) | \psi_2 \rangle + \frac{e^2}{4\pi\varepsilon_0 R} \langle \psi_1 | \psi_2 \rangle \\
&= \left(E_1 + \frac{e^2}{4\pi\varepsilon_0 R} \right) \langle \psi_1 | \psi_2 \rangle - X \\
&= \left(E_1 + \frac{e^2}{4\pi\varepsilon_0 R} \right) S - X
\end{aligned}
\tag{7.92}
$$

where:

$$
X = \langle \psi_1 | \left(\frac{e^2}{4\pi\varepsilon_0 r_1} \right) | \psi_2 \rangle \tag{7.93}
$$

evaluation of which by adopting the procedure that was used to evaluate S yields:

$$
\begin{aligned}
X &= \frac{e^2}{4\pi\varepsilon_0 a_0^2} \left[(a_0 + R) e^{-R/a_0} \right] \\
&= -2E_1 \left(1 + \frac{R}{a_0} \right) e^{-R/a_0}
\end{aligned}
\tag{7.94}
$$

The optimum value of c_1 and c_2 is then obtained by minimizing $\langle H \rangle$ with respect to c_1 and c_2. With the use of Eqn. (7.81a), we get:

$$
\begin{bmatrix} \alpha - \langle H \rangle & \beta - \langle H \rangle S \\ \beta - \langle H \rangle S & \alpha - \langle H \rangle \end{bmatrix} \begin{bmatrix} c_1 \\ c_2 \end{bmatrix} = \begin{bmatrix} 0 \\ 0 \end{bmatrix} \tag{7.95}
$$

the non-trivial solution of which demands:

$$\begin{vmatrix} \alpha - \langle H \rangle & \beta - \langle H \rangle S \\ \beta - \langle H \rangle S & \alpha - \langle H \rangle \end{vmatrix} = 0 \tag{7.96}$$

Two roots of the equation are:

$$\langle H \rangle^{\pm} = \frac{\alpha \pm \beta}{1 \pm S} \tag{7.97}$$

with:

$$\begin{aligned} \langle H \rangle^{+} &= \frac{\alpha + \beta}{1 + S} \\ &= E_1 \left\{ 1 - \frac{2a_0}{R} + \frac{2}{(1+S)} \left[\frac{a_0}{R} - \left(1 + \frac{a_0}{R} \right) e^{-2R/a_0} + \left(1 + \frac{R}{a_0} \right) e^{-R/a_0} \right] \right\} \end{aligned} \tag{7.98}$$

and:

$$\begin{aligned} \langle H \rangle^{-} &= \frac{\alpha - \beta}{1 - S} \\ &= E_1 \left\{ 1 - \frac{2a_0}{R} + \frac{2}{(1-S)} \left[\frac{a_0}{R} - \left(1 + \frac{a_0}{R} \right) e^{-2R/a_0} - \left(1 + \frac{R}{a_0} \right) e^{-R/a_0} \right] \right\} \end{aligned} \tag{7.99}$$

Equation (7.95) suggests that there are two possible cases: (i) $c_1 = c_2$, which corresponds to $\langle H \rangle^{+}$ and (ii) $c_1 = -c_2$ belonging to $\langle H \rangle^{-}$. We thus have:

$$\begin{aligned} \phi^{+}(r) &= c_1 \left(\psi_1(r) + \psi_2(r) \right) \\ \phi^{-}(r) &= c_1 \left(\psi_1(r) - \psi_2(r) \right) \end{aligned} \tag{7.100}$$

As is seen from Eqns. (7.98) and (7.99) $\langle H \rangle^{+}$ is smaller than $\langle H \rangle^{-}$. Thus $\phi^{+}(r)$ represents a bonding orbital, while $\phi^{-}(r)$ corresponds to an anti-bonding orbital. The normalization condition $\langle \phi^{+} | \phi^{+} \rangle = 1$ is used to determine c_1:

$$c_1 = \left[\frac{1}{2(1+S)} \right]^{1/2} \tag{7.101}$$

7.2.4 Variational Method for Excited States

If we denote the wave function ket of the ground state of a system by $|\phi_0\rangle$ then the energy of the first excited state can be given by:

$$E_1 = \langle \phi_1 | H | \phi_1 \rangle_{\text{min}} \tag{7.102}$$

under the conditions that $\langle \phi_1 | \phi_1 \rangle = 1$ and $\langle \phi_1 | \phi_0 \rangle = 0$.

Similar to the case of the ground state, $|\phi_1\rangle$ can be expressed in terms of exact but unknown states of the system:

$$|\phi_1\rangle = \sum_{n=0}^{\infty} a_n |\psi_n\rangle \tag{7.103}$$

Then:

$$\begin{aligned} \langle\phi_1|\phi_0\rangle &= \sum_{n=0}^{\infty} a_n^* \langle\psi_n|\phi_0\rangle \\ \Rightarrow 0 &= \sum_{n=0}^{\infty} a_n^* \delta_{n0} \end{aligned} \tag{7.104}$$

which suggests that the coefficient $a_0 = 0$, and hence $|\phi_1\rangle = \sum_{n=1}^{\infty} a_n |\psi_n\rangle$ with the condition $\sum_{n=1}^{\infty} |a_n|^2 = 1$. In a similar manner, the energy of the second excited state will be given by:

$$E_2 = \langle\phi_2|H|\phi_2\rangle_{\text{min}} \tag{7.105}$$

under the conditions $\langle\phi_2|\phi_2\rangle = 1$ and $\langle\phi_2|\phi_1\rangle = 0 = \langle\phi_2|\phi_0\rangle$. We can proceed to compute energy and wave function for the third and further higher state in the same way with additional conditions. However, as can be realized, the variational method becomes increasingly complicated for evaluated energy and wave functions for higher excited states.

7.2.5 Application of Variational Method to the Excited State of a 1D Harmonic Oscillator

The Hamiltonian of a 1D harmonic oscillator is:

$$H = -\frac{\hbar^2}{2m}\frac{d^2}{dx^2} + \frac{1}{2}m\omega^2 x^2 \tag{7.106a}$$

We choose the ground state wave function by keeping in mind that it vanishes at $\pm\infty$ and should not have any nodes. Therefore we take $\phi_0(x) = Ae^{-\alpha x^2}$. Now the trial wave function for the first excited state $\phi_1(x)$ must be orthogonal to $\phi_0(x)$. A right choice can be $\phi_1(x) = Bxe^{-\beta x^2}$, because:

$$\int_{-\infty}^{\infty} xe^{-(\alpha+\beta)x^2} dx = 0 \tag{7.106b}$$

The normalization condition $\langle\phi_1|\phi_1\rangle = 1$ then gives:

$$
\begin{aligned}
1 &= B^2 \int_{-\infty}^{\infty} x^2 e^{-2\beta x^2} dx \\
&= 2B^2 \int_{0}^{\infty} x^2 e^{-2\beta x^2} dx \\
&= 2B^2 \frac{\sqrt{\pi}}{4} \frac{1}{(2\beta)^{3/2}} \\
\Rightarrow B &= 2\left(\frac{2}{\pi}\right)^{1/4} \beta^{3/4}
\end{aligned}
\tag{7.106c}
$$

Next, to calculate $\langle\phi_1|H|\phi_1\rangle$, we evaluate:

$$
\begin{aligned}
H\phi_1 &= B\left[-\frac{\hbar^2}{2m}\frac{d^2}{dx^2} + \frac{1}{2}m\omega^2 x^2\right] x e^{-\beta x^2} \\
&= B\left[\frac{3\hbar^2\beta}{m}x + \left(\frac{1}{2}m\omega^2 - \frac{2\hbar^2\beta^2}{m}\right)x^3\right] e^{-\beta x^2}
\end{aligned}
\tag{7.107}
$$

Then:

$$
\begin{aligned}
\langle\phi_1|H|\phi_1\rangle &= B^2\left[\frac{3\hbar^2\beta}{m}\int_{-\infty}^{\infty} x^2 e^{-2\beta x^2} dx + \left(\frac{1}{2}m\omega^2 - \frac{2\hbar^2\beta^2}{m}\right)\int_{-\infty}^{\infty} x^4 e^{-2\beta x^2} dx\right] \\
&= 2B^2\left[\frac{3\hbar^2\beta}{m}\int_{0}^{\infty} x^2 e^{-2\beta x^2} dx + \left(\frac{1}{2}m\omega^2 - \frac{2\hbar^2\beta^2}{m}\right)\int_{0}^{\infty} x^4 e^{-2\beta x^2} dx\right]
\end{aligned}
\tag{7.108}
$$

On evaluating integrals and then substituting the value of B^2 we get:

$$
\langle\phi_1|H|\phi_1\rangle = \frac{3\hbar^2\beta}{2m} + \frac{3m\omega^2}{8\beta} \tag{7.109}
$$

Minimization of $\langle\phi_1|H|\phi_1\rangle$ with respect to β, gives $\beta_{\min} = \frac{m\omega}{2\hbar}$, substitution of which in Eqn. (7.109) yields:

$$
E_1 = \langle\phi_1|H|\phi_1\rangle_{\min} = \frac{3}{2}\hbar\omega \tag{7.110a}
$$

Then:

$$\phi_1 = 2\left(\frac{2}{\pi}\right)^{1/4}\left(\frac{m\omega}{2\hbar}\right)^{3/4} xe^{-\beta x^2} \tag{7.110b}$$

which are the correct answers obtained by applying the variational method for the excited state.

7.3 The W K B Approximations

The Wentzel, Kramers, and Brillouin (WKB) approximation is applied to obtain an approximate solution of a time-independent 1D Schrödinger equation. The idea can also be applied to many other differential equations and to the radial Schrödinger equation. Here, we confine ourselves to the applications of WKB approximations in calculating bound-state energies and tunneling rates through potential barriers, which involve classical turning points, the points where total energy E is equal to potential energy V; the kinetic energy is zero. Turning points mark the boundaries between two regions: (i) the region where motion is allowed classically and (ii) the region in which motion is not permitted classically, known as the tunneling region, as is shown in Fig. 7.6.

For $E > V(x)$, a particle is allowed to move freely if $V(x)$ is constant, and the solution of the 1D Schrödinger equation is given by $\psi(x) = Ae^{\pm ikx}$, with $k = \sqrt{2m(E-V)}/\hbar$. The plus and minus sign indicate the motion of the particle along the +x-axis and the -x-axis, respectively. The wave function is oscillatory with a constant wave length $\lambda = 2\pi / k$ and amplitude A. When $V(x)$ varies very slowly (not constant) on a distance scale of λ, it is reasonable to assume that $\psi(x)$ remains practically sinusoidal, except that the λ and A change slowly with x on the scale of λ. *This is known as the WKB approximation.* By similar arguments, it can be said that for regions where $E < V(x)$, the solution to the Schrödinger equation for the constant V is given by $\psi(x) = Ae^{\pm Kx}$, with $K = \sqrt{2m(V-E)}/\hbar$. In these regions, a particle would not be allowed classically, but it is said to tunnel quantum mechanically. If potential is not constant but varies slowly with x, in comparison with $1/K$, then the solution remains practically exponential except that A and K are allowed

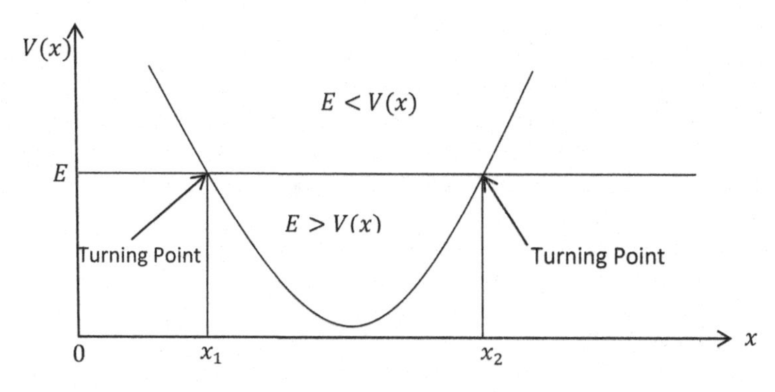

FIGURE 7.6
1D-Potential energy curve: $x_1 \le x \le x_2$ is the classically allowed region.

to vary slowly with x. In between these two types of regions lie the *classical turning points* at which the two wave functions must be properly matched, leading to boundary conditions between the regions. However, in the immediate vicinity the classical turning points $1/k$ tend to infinity and $V(x)$ hardly can be said to vary slowly, and therefore the entire program is bound to fail. Hence, the handling of the turning point is the most difficult aspect of WKB approximation.

7.3.1. The Classical Region

The Schrödinger equation:

$$-\frac{\hbar^2}{2m}\frac{d^2\psi}{dx^2}+V(x)\psi(x)=E\psi(x) \tag{7.111}$$

can be rewritten as:

$$\frac{d^2\psi}{dx^2}=-\frac{p^2}{\hbar^2}\psi \tag{7.112a}$$

where:

$$p(x)=\sqrt{2m[E-V(x)]} \tag{7.112b}$$

is the classical formula for momentum of a particle. In the classical region, $E>V(x)$ and therefore $p(x)$ is real. In general, $\psi(x)$ is a complex function and it is expressed in terms of amplitude $A(x)$ and phase $\phi(x)$; both are real:

$$\psi(x)=A(x)e^{i\phi(x)} \tag{7.113}$$

Differentiating $\psi(x)$ with respect to x, we get:

$$\begin{aligned}\frac{d\psi}{dx}&=\frac{dA(x)}{dx}e^{i\phi(x)}+iA(x)\frac{d\phi(x)}{dx}e^{i\phi(x)}\\&=\left(A'+iA\phi'\right)e^{i\phi(x)}\end{aligned} \tag{7.114a}$$

and:

$$\begin{aligned}\frac{d^2\psi}{dx^2}&=\frac{d}{dx}\left\{\left(A'+iA\phi'\right)e^{i\phi(x)}\right\}\\&=\left(A''+iA'\phi'+iA\phi''\right)e^{i\phi(x)}+i\phi'\left(A'+iA\phi'\right)e^{i\phi(x)}\\&=\left(A''+2iA'\phi'+iA\phi''-A\left(\phi'\right)^2\right)e^{i\phi(x)}\end{aligned} \tag{7.114b}$$

Putting this into Eqn. (7.112a):

$$A''+2iA'\phi'+iA\phi''-A(\phi')^2=\frac{-p^2}{\hbar^2}A \tag{7.115}$$

Then on equating real and imaginary parts of both sides, we have:

$$\begin{aligned} A'' - A(\phi')^2 &= -\frac{p^2}{\hbar^2} A \\ \Rightarrow A'' &= A\left[(\phi')^2 - \frac{p^2}{\hbar^2}\right] \end{aligned} \tag{7.116}$$

and:

$$\begin{aligned} &2A'\phi' + A\phi'' = 0 \\ &\Rightarrow \frac{d}{dx}\left(A^2\phi'\right) = 0 \\ &\Rightarrow A^2\phi' = C^2 \\ &\Rightarrow A = \frac{C}{\sqrt{\phi'}} \end{aligned} \tag{7.117}$$

where C is a real constant. Note that Eqns. (7.116) and (7.117) are equivalent to (7.112a), and no approximation is made by now. We notice that a general solution of Eqn. (7.116) is not possible and therefore here comes the approximation. We assumed that $A(x)$ and $\phi(x)$ vary slowly with x, which suggests that A'' and hence A''/A is negligibly small as compared to $(\phi')^2 - \frac{p^2}{\hbar^2}$, and it can be treated as zero. We then have:

$$\begin{aligned} &(\phi')^2 - \frac{p^2}{\hbar^2} = 0 \\ &\Rightarrow \phi' = \pm\frac{p(x)}{\hbar} \\ &\Rightarrow \phi(x) = \pm\frac{1}{\hbar}\int p(x)dx + C' \end{aligned} \tag{7.118}$$

where C' is a constant of integration. We thus obtain:

$$\psi(x) \simeq \frac{C_0}{\sqrt{p(x)}} e^{\pm\frac{i}{\hbar}\int p(x)dx} \tag{7.119}$$

Here C_0 is a new constant, which is complex because it involve both C and $e^{iC'}$. A general (approximate) solution is:

$$\psi(x) \simeq \frac{1}{\sqrt{p(x)}}\left[C^+ e^{i\frac{1}{\hbar}\int p(x)dx} + C^- e^{-i\frac{1}{\hbar}\int p(x)dx}\right] \tag{7.120a}$$

or more conventionally:

$$\psi(x) \simeq \frac{1}{\sqrt{p(x)}}\left[C_1 \cos\phi(x) + C_2 \sin\phi(x)\right] \tag{7.120b}$$

Equation (7.119) yields:

$$|\psi(x)|^2 = \frac{|C_0|^2}{p(x)} \tag{7.121}$$

which states that the probability of finding a particle at point x is inversely proportional to its classical momentum or its velocity at that point, *which means that the particle does not stay for a longer time where it moves rapidly and it spends more time in the region where it moves slowly. This is in fact the essential idea behind the WKB approximation.*

7.3.2 Alternative Derivation of the WKB Formula

An alternative derivation uses an expansion in powers of $\hbar$. The free particle solution for the constant potential case suggests that for slowly varying potential, wave function can be written as:

$$\psi(x) = e^{if(x)/\hbar} \tag{7.122}$$

without loss of generality. On substituting this into Eqn. (7.112a), we get:

$$i\hbar f'' - (f')^2 + p^2 = 0 \tag{7.123}$$

We expand $f(x)$ in powers of $\hbar$ as follows:

$$f(x) = f_0(x) + \hbar f_1(x) + \hbar^2 f_2(x) + \dots\dots\dots \tag{7.124}$$

Putting this into Eqn. (7.123) and then collecting like powers of $\hbar$, we obtain:

$$(f_0')^2 = p^2 \tag{7.125a}$$

$$if_0'' = 2f_0'f_1' \tag{7.125b}$$

$$if_1'' = 2f_0'f_2' + (f_1')^2 \tag{7.125c}$$

etc. The solution of Eqn. (7.125a) gives:

$f_0(x) = \pm\int p(x)dx + C$, where C is the constant of integration. Equation (7.125b) with the use of (7.125a) yields:

$$\begin{aligned} i\frac{dp}{dx} &= 2p\frac{df_1}{dx} \\ \Rightarrow df_1 &= \frac{i}{2}\frac{dp}{p} \\ \Rightarrow f_1 &= \frac{i}{2}\ln p \\ \Rightarrow f_1 &= i\ln p^{1/2} \\ \Rightarrow p^{-1/2} &= e^{if_1} \end{aligned} \tag{7.126}$$

Now, if we confine ourselves only up to the first two terms in expansion (Eqn. 7.124), then we have:

$$\psi(x) \simeq \frac{B}{\sqrt{p(x)}} e^{\pm \frac{i}{\hbar}\int p(x)dx} \tag{7.127}$$

where B is the complex constant. Note that the Eqns. (7.127) and (7.119) are identical. In obtaining the solution given by Eqn. (7.127) we have ignored Eqn. (7.125c) by neglecting terms beyond f_1 in Eqn. (7.124), which is possible if:

$$\begin{aligned}
&\frac{\hbar f_1(x)}{f_0(x)} \ll 1 \\
&\Rightarrow \hbar\left(\frac{df_1 / dx}{df_0 / dx}\right) \ll 1 \\
&\Rightarrow \hbar\left|\frac{dp / dx}{2p^2}\right| \ll 1 \\
&\Rightarrow \frac{1}{4\pi}\left|\frac{d}{dx}\left(\frac{h}{p}\right)\right| \ll 1 \\
&\Rightarrow \frac{1}{4\pi}\left|\frac{d\lambda}{dx}\right| \ll 1
\end{aligned} \tag{7.128}$$

Thus, the WKB solutions are applicable in the regions where the wave length of the particle changes slowly with x. At the turning points, λ varies rapidly or $V(x)$ has steep behavior; hence WKB approximations fail in the vicinity of turning points.

7.3.3 Non-classical or Tunneling Region

In the non-classical region $E < V(x)$, Eqns. (7.112a) and (7.112b) are still valid but $p(x)$ becomes imaginary. We can write:

$$p(x) = \sqrt{-2m[V(x) - E]} = i\gamma(x) \tag{7.129}$$

and then:

$$\psi(x) = \frac{B}{\sqrt{\gamma(x)}} e^{\pm \frac{1}{\hbar}\int \gamma(x)dx} \tag{7.130}$$

A general approximate solution would be:

$$\psi(x) = \frac{1}{\sqrt{\gamma(x)}}\left[B_1 e^{\frac{1}{\hbar}\int \gamma(x)dx} + B_2 e^{-\frac{1}{\hbar}\int \gamma(x)dx}\right] \tag{7.131}$$

where B_1 and B_2 are constants. A region of finite extent has both exponentially increasing and decreasing terms and therefore both terms are generally retained. As is seen from both the Eqns. (7.119) and (7.130) $\psi(x)$ becomes infinite at the classical turning points $(E = V(x))$ and the approximation fails at the turning points. The WKB approximation is only valid for the situations where the distance over which the potential changes is larger than the wavelength of the perturbing field. However, we need to match wave functions at the turning points for determining bound state energies. One adopts the strategy of connecting formulae to overcome the limitation of the WKB wave functions at the turning points.

7.3.4 Connecting Formulae

We need connecting formulae to deal with two types of problems: (i) the bound state case as shown in Fig. 7.6, and (ii) the scattering case exhibited in Fig. 7.7.

In the regions near the turning point at $x = x_1$, $\frac{dV}{dx} < 0$ for the bound state case, and $\frac{dV}{dx} > 0$ for the scattering case, while in regions near the turning point at $x = x_2$, $\frac{dV}{dx} > 0$ for the bound state case, and $\frac{dV}{dx} < 0$ for the scattering case. Let us consider the bound state case to apply the WKB approximations. The potential in the regions near the turning points can be considered varying linearly with x, as is shown in Fig. 7.8.

We first derive connection formulae for the turning point at $x = x_2$. To solve the Schrödinger equation exactly near the turning point at $x = x_2$, we write: $V(x) \approx V(x_2) + (x - x_2)\left.\frac{dV}{dx}\right|_{x=x_2}$.

On the right-hand side to the turning point, $E < V(x)$ and $x > x_2$:

$$\frac{d^2\psi}{dx^2} + \frac{2m}{\hbar^2}\left[E - V(x_2) - (x - x_2)\left.\frac{dV}{dx}\right|_{x=x_2}\right]\psi(x) = 0 \tag{7.132}$$

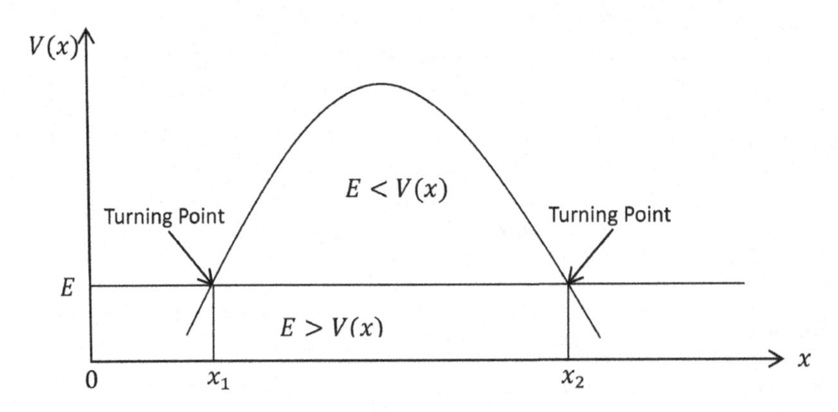

FIGURE 7.7
1D Potential barrier: $x_1 \le x \le x_2$ is a classically forbidden region.

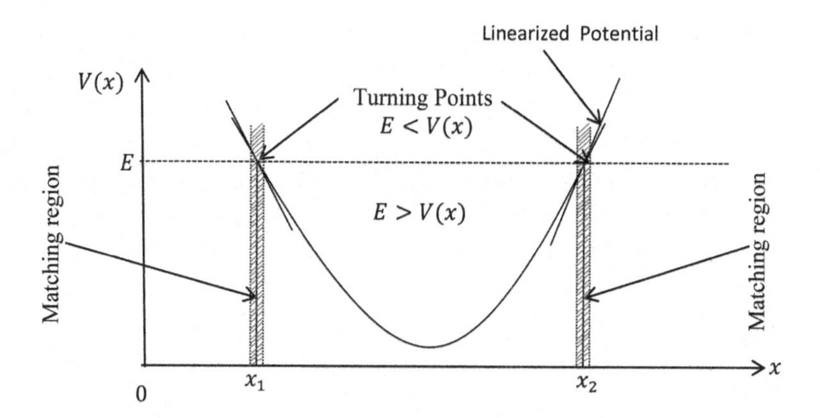

FIGURE 7.8
1D-Potential for bound state case: $x_1 \le x \le x_2$ is classically allowed.

In the vicinity of the turning point, $E \approx V(x_2)$, and therefore in regions near the turning point, we have:

$$\frac{d^2\psi}{dx^2} = \alpha^3 (x - x_2)\psi(x) \tag{7.133}$$

where, we have taken $\alpha^3 = \frac{2m}{\hbar^2}\frac{dV}{dx}\bigg|_{x=x_2}$. Let us define a dimensionless parameter $z = \alpha(x - x_2)$ to obtain:

$$\frac{d^2\psi(z)}{dz^2} = z\psi(z) \tag{7.134}$$

which is Airy's equation having the general solution:

$$\psi_p = aAi(z) + bBi(z) \tag{7.135}$$

Since the sole purpose of $\psi_p(z)$ is to patch together the WKB wave functions on each side of the turning point, it is termed the patching wave function. The asymptotic forms of the Airy function are:

$$\left.\begin{aligned} Ai(z) &\sim \frac{1}{2\sqrt{\pi}z^{1/4}} e^{-\frac{2}{3}z^{3/2}} \\ Bi(z) &\sim \frac{1}{\sqrt{\pi}z^{1/4}} e^{\frac{2}{3}z^{3/2}} \end{aligned}\right\} z \gg 0; \qquad \left.\begin{aligned} Ai(z) &\sim \frac{1}{\sqrt{\pi}(-z)^{1/4}} \sin\left[\frac{2}{3}(-z)^{3/2} + \frac{\pi}{4}\right] \\ Bi(z) &\sim \frac{1}{\sqrt{\pi}(-z)^{1/4}} \cos\left[\frac{2}{3}(-z)^{3/2} + \frac{\pi}{4}\right] \end{aligned}\right\} z \ll 0 \tag{7.136}$$

WKB solutions are not valid in the close vicinity of turning points; however, these can be extended to the linearized potential region to overlap with ψ_p. These overlap zones, on both sides, are close enough to the turning point, as is shown in Fig. 7.8. To match the

WKB solution with ψ_p in the overlap region on the left side of the turning point $(x < x_2)$, we have:

$$p(x) = \left[2m\left(E - V(x_2) - (x - x_2)\frac{dV}{dx}\bigg|_{x=x_2}\right)\right]^{1/2} = \hbar\alpha^{3/2}\sqrt{(x_2 - x)}$$
$$= \hbar\alpha(-z)^{1/2} \tag{7.137}$$

and:

$$\phi(x) = \frac{1}{\hbar}\int_x^{x_2} p(x')dx'$$
$$= \alpha^{3/2}\int_x^{x_2}\sqrt{(x_2 - x')}dx' \tag{7.138}$$
$$= \frac{2}{3}\alpha^{3/2}(x_2 - x)^{3/2}$$

use of which in Eqn. (7.120a) yields:

$$\psi_I = \frac{C^+}{\sqrt{p(x)}}e^{\frac{2}{3}i\alpha^{3/2}(x_2 - x)^{3/2}} + \frac{C^-}{\sqrt{p(x)}}e^{-\frac{2}{3}i\alpha^{3/2}(x_2 - x)^{3/2}} \tag{7.139}$$

The asymptotic form of Eqn. (7.135) for $x < x_2$ is:

$$\psi_p \sim \sqrt{\frac{\hbar\alpha}{\pi}}\frac{1}{2\sqrt{p(x)}}\left[(b - ia)e^{i\pi/4}e^{\frac{2}{3}i\alpha^{3/2}(x_2 - x)^{3/2}} + (b + ia)e^{-i\pi/4}e^{-\frac{2}{3}i\alpha^{3/2}(x_2 - x)^{3/2}}\right] \tag{7.140}$$

Both the Eqns. (7.139) and (7.140) represent the same solution in the overlap region on the left side of the turning point, $x < x_2$ $(z \ll 0)$.Therefore:

$$C^+ = \sqrt{\frac{\hbar\alpha}{\pi}}\frac{(b - ia)}{2}e^{i\pi/4} \text{ and } C^- = \sqrt{\frac{\hbar\alpha}{\pi}}\frac{(b + ia)}{2}e^{-i\pi/4} \tag{7.141a}$$

which gives:

$$a = \sqrt{\frac{\pi}{\hbar\alpha}}\left(C^+e^{i\pi/4} + C^-e^{-i\pi/4}\right),$$
$$b = \sqrt{\frac{\pi}{\hbar\alpha}}\left(C^+e^{-i\pi/4} + C^-e^{i\pi/4}\right) \tag{7.141b}$$

On the right-hand side of the turning point ($x > x_2$) or $z \gg 0$, we have from Eqn. (7.135):

$$\psi_p \sim \frac{a}{2\sqrt{\pi} z^{1/4}} e^{-\frac{2}{3}z^{3/2}} + \frac{b}{\sqrt{\pi} z^{1/4}} e^{\frac{2}{3}z^{3/2}} \tag{7.142}$$

with

$$z = \alpha(x - x_2).$$

Also, from Eqn. (7.129) we have:

$$\gamma(x) = \left[2m\left(V(x_2) + (x - x_2)\frac{dV}{dx}\bigg|_{x=x_2} - E\right)\right]^{1/2} = \hbar\alpha^{3/2}\sqrt{(x - x_2)}$$
$$= \hbar\alpha(z)^{1/2} \tag{7.143a}$$

and:

$$\frac{1}{\hbar}\int_{x_2}^{x} \gamma(x')dx = \alpha^{3/2}\int_{x_2}^{x}(x' - x_2)^{1/2}dx' = \frac{2}{3}\alpha^{3/2}(x - x_2)^{3/2} \tag{7.143b}$$

which on substituting in Eqn. (7.131) yields:

$$\psi_{II}(x) = \frac{B_1}{\sqrt{\gamma(x)}} e^{\frac{2}{3}\alpha^{3/2}(x-x_2)^{3/2}} + \frac{B_2}{\sqrt{\gamma(x)}} e^{-\frac{2}{3}\alpha^{3/2}(x-x_2)^{3/2}}$$
$$= \frac{B_1}{(\hbar\alpha)^{1/2} z^{1/4}} e^{\frac{2}{3}z^{3/2}} + \frac{B_2}{(\hbar\alpha)^{1/2} z^{1/4}} e^{-\frac{2}{3}z^{3/2}} \tag{7.144}$$

On equating Eqns. (7.142) to (7.144), we obtain:

$$\frac{B_1}{(\hbar\alpha)^{1/2}} = \frac{b}{2\sqrt{\pi}}$$
$$\Rightarrow b = 2B_1\sqrt{\frac{\pi}{\hbar\alpha}};$$
$$\frac{B_2}{(\hbar\alpha)^{1/2}} = \frac{a}{\sqrt{\pi}}$$
$$\Rightarrow a = B_2\sqrt{\frac{\pi}{\hbar\alpha}} \tag{7.145}$$

Substituting values of a and b from Eqn. (7.141b) we get the connection formulae on the right of the turning point $(x > x_2)$:

$$B_1 = \frac{1}{2}\left(C^+ e^{-i\pi/4} + C^- e^{i\pi/4}\right)$$
$$B_2 = C^+ e^{i\pi/4} + C^- e^{-i\pi/4} \tag{7.146}$$

To get connection formulae on the left side of the turning point $(x < x_2)$, we substitute values of a and b from Eqn. (7.145) into Eqn. (7.141a) and obtain:

$$C^+ = \frac{1}{2} B_2 e^{-i\pi/4} + B_1 e^{i\pi/4}$$
$$C^- = \frac{1}{2} B_2 e^{i\pi/4} + B_1 e^{-i\pi/4} \tag{7.147}$$

We next consider another classical turning point occurring at $x = x_1$. Left of the turning point $(x < x_1)$, $E < V(x)$, while $E > V(x)$ on the right of the turning point $(x > x_1)$. The results can be derived with the use of a similar procedure. We find that the connection formulae on the left of the turning point are:

$$B_1 = C^+ e^{i\pi/4} + C^- e^{-i\pi/4}, \text{and } B_2 = \frac{1}{2}\left(C^+ e^{-i\pi/4} + C^- e^{i\pi/4}\right) \tag{7.148a}$$

while on the right of the turning point, we have:

$$C^+ = \frac{1}{2} B_2 e^{i\pi/4} + B_1 e^{-i\pi/4}, \text{and } C^- = \frac{1}{2} B_2 e^{-i\pi/4} + B_1 e^{i\pi/4} \tag{7.148b}$$

It is to be noted that in the connection formulae there is no mention of the linearized potential and Airy functions. The procedure served only as a mechanism to relate the constants C^+ and C^- on the left of the turning point to B_1 and B_2 on the right of the turning point occurring at $x = x_2$ and vice versa for the turning point at $x = x_1$. The patching wave function is no longer needed. We thus know how to deal with a turning point where $\frac{dV}{dx} > 0$ and the turning points with $\frac{dV}{dx} < 0$.

7.3.5 Quantum Condition for Bound State

After knowing how to match WKB wave functions at the turning points, we can derive the quantization condition. Let us take the case of the bound state, Fig. 7.8, having potential similar to the harmonic oscillator with $\frac{dV}{dx} < 0$ in the region close to the turning point at $x = x_1$, and $\frac{dV}{dx} > 0$ near the turning point at $x = x_2$. We call the region having $x < x_1$ region I, and that of $x > x_2$ as region III. Both regions I and III are classically forbidden. With the use of Eqn. (7.131), we can write:

$$\psi_I(x) = \frac{B_1}{\sqrt{\gamma}} e^{\frac{1}{\hbar}\int_x^{x_1} \gamma(x')dx'} + \frac{B_2}{\sqrt{\gamma}} e^{-\frac{1}{\hbar}\int_x^{x_1} \gamma(x')dx'} \tag{7.149}$$

with $\gamma(x)=\sqrt{2m(V(x)-E)}$. In region II ($x_1 < x < x_2$). For $x > x_1$, we have:

$$\psi_{II}(x)=\frac{C^+}{\sqrt{p}}e^{\frac{i}{\hbar}\int_{x1}^{x}p(x')dx'}+\frac{C^-}{\sqrt{p}}e^{-\frac{i}{\hbar}\int_{x1}^{x}p(x')dx'} \tag{7.150}$$

Region I extends up to $-\infty$ and therefore to have a physically acceptable wave function $B_2=0$. Hence, from Eqn. (7.148a) we have $C^-e^{i\pi/4}+C^+e^{-i\pi/4}=0$, which implies $C^+=-iC^-$. Therefore:

$$\begin{aligned}\psi_{II}(x)&=\frac{-iC^-}{\sqrt{p}}e^{\frac{i}{\hbar}\int_{x1}^{x}p(x')dx'}+\frac{C^-}{\sqrt{p}}e^{-\frac{i}{\hbar}\int_{x1}^{x}p(x')dx'}\\&=\frac{C^-e^{-i\pi/2}}{\sqrt{p}}e^{\frac{i}{\hbar}\int_{x1}^{x}p(x')dx'}+\frac{C^-}{\sqrt{p}}e^{-\frac{i}{\hbar}\int_{x1}^{x}p(x')dx'}\end{aligned} \tag{7.151a}$$

and from Eqn. (7.148b):

$$C^-=B_1e^{i\pi/4} \tag{7.151b}$$

We thus have:

$$\psi_{II}(x)=\frac{B_1}{\sqrt{p}}e^{\frac{i}{\hbar}\int_{x1}^{x}p(x')dx'-\frac{i\pi}{4}}+\frac{B_1}{\sqrt{p}}e^{-\frac{i}{\hbar}\int_{x1}^{x}p(x')dx'+\frac{i\pi}{4}} \tag{7.151c}$$

which can be rewritten as:

$$\begin{aligned}\psi_{II}(x)&=\frac{B_1}{\sqrt{p}}e^{\frac{i}{\hbar}\int_{x1}^{x2}p(x')dx'-\frac{i\pi}{4}}e^{-\frac{i}{\hbar}\int_{x}^{x2}p(x')dx'}+\frac{B_1}{\sqrt{p}}e^{-\frac{i}{\hbar}\int_{x1}^{x2}p(x')dx'+\frac{i\pi}{4}}e^{+\frac{i}{\hbar}\int_{x}^{x2}p(x')dx'}\\&=\frac{B'}{\sqrt{p}}e^{\frac{i}{\hbar}\int_{x}^{x2}p(x')dx'}+\frac{B''}{\sqrt{p}}e^{-\frac{i}{\hbar}\int_{x}^{x2}p(x')dx'}\end{aligned} \tag{7.152}$$

where we have defined:

$$\begin{aligned}B'&=B_1e^{-\frac{i}{\hbar}\int_{x1}^{x2}p(x')dx'+\frac{i\pi}{4}}\\B''&=B_1e^{\frac{i}{\hbar}\int_{x1}^{x2}p(x')dx'-\frac{i\pi}{4}}\end{aligned} \tag{7.153}$$

Across the turning point at $x = x_2$, we also should have solutions like:

$$\psi_{III}(x) = \frac{D_1}{\sqrt{\gamma}} e^{\frac{1}{\hbar}\int_{x_2}^{x} \gamma(x')dx'} + \frac{D_2}{\sqrt{\gamma}} e^{-\frac{1}{\hbar}\int_{x_2}^{x} \gamma(x')dx'} \tag{7.154a}$$

$$\psi_{II}(x) = \frac{A'}{\sqrt{p}} e^{\frac{i}{\hbar}\int_{x}^{x_2} p(x')dx'} + \frac{A''}{\sqrt{p}} e^{-\frac{i}{\hbar}\int_{x}^{x_2} p(x')dx'} \tag{7.154b}$$

And D_1 must be zero to have an acceptable solution. This, from Eqn. (7.146), implies that $A'e^{-i\pi/4} = -A''e^{i\pi/4}$. Also, the Eqns. (7.152) and (7.154b) represent the same solution and hence:

$$\begin{aligned} A' &= B_1 e^{-\frac{i}{\hbar}\int_{x_1}^{x_2} p(x')dx' + \frac{i\pi}{4}} \\ A'' &= B_1 e^{\frac{i}{\hbar}\int_{x_1}^{x_2} p(x')dx' - \frac{i\pi}{4}} \end{aligned} \tag{7.155}$$

giving rise to:

$$\begin{aligned} & B_1 e^{-\frac{i}{\hbar}\int_{x_1}^{x_2} p(x')dx' + \frac{i\pi}{4}} e^{-i\pi/4} = -B_1 e^{\frac{i}{\hbar}\int_{x_1}^{x_2} p(x')dx' - \frac{i\pi}{4}} e^{i\pi/4} \\ & \Rightarrow e^{\frac{i}{\hbar}\int_{x_1}^{x_2} p(x')dx'} + e^{-\frac{i}{\hbar}\int_{x_1}^{x_2} p(x')dx'} = 0 \\ & \Rightarrow \cos\left(\frac{1}{\hbar}\int_{x_1}^{x_2} p(x')dx'\right) = 0 = \cos\left(n + \frac{1}{2}\right)\pi \end{aligned} \tag{7.156}$$

which gives:

$$\frac{1}{\hbar}\int_{x_1}^{x_2} p(x')dx' = \left(n + \frac{1}{2}\right)\pi \tag{7.157}$$

where n takes values $0, 1, 2, 3, \ldots\ldots$ Equation (7.157) determines the allowed bound state energies of the system. The left-hand side of Eqn. (7.157) has an integral over the complete cycle of motion (from x_1 to x_2); hence, we write:

$$\oint p(x)dx = \left(n + \frac{1}{2}\right)h \tag{7.158}$$

which is simply the Sommerfeld quantization condition with a half integer.

7.4 Solved Examples

1. A delta function bump $H_1 = \alpha\delta(x - a/2)$ is introduced at the center of the infinite potential well of width a, with walls at $x = 0$ and $x = a$. Calculate the first order correction to the allowed energies and wave function.

SOLUTION

The first order corrections to the allowed energies are given by $E_n^{(1)} = \langle \psi_n^{(0)} | H_1 | \psi_n^{(0)} \rangle$.

Since $\psi_n^{(0)}(x) = 0$ at $x = 0$ and at $x = a$, we take $\psi_n^{(0)}(x) = \sqrt{\frac{2}{a}} \sin\left(\frac{n\pi}{a} x\right)$.

Therefore:

$$\begin{aligned} E_n^{(1)} &= \frac{2}{a}\int_0^a \left(\sin\left(\frac{n\pi}{a}x\right)\right)^2 \alpha\delta(x - a/2)dx \\ &= \frac{2\alpha}{a}\left(\sin\left(\frac{n\pi}{2}\right)\right)^2 \end{aligned} \tag{7.159}$$

Thus $E_n^{(1)}$ is zero for even values of n and it is $\frac{2\alpha}{a}$ for odd values of n. Then:

$|\psi_n^{(1)}\rangle = \sum_{m \neq n} \frac{\langle \psi_m^{(0)} | H_1 | \psi_n^{(0)} \rangle}{\left(E_m^{(0)} - E_n^{(0)}\right)} |\psi_m^{(0)}\rangle$, with:

$$\begin{aligned} \langle \psi_m^{(0)} | H_1 | \psi_n^{(0)} \rangle &= \frac{2}{a}\int_0^a \sin\left(\frac{m\pi}{a}x\right)\alpha\delta(x - a/2)\sin\left(\frac{n\pi}{a}x\right)dx \\ &= \frac{2\alpha}{a}\sin\left(\frac{m\pi}{2}\right)\sin\left(\frac{n\pi}{2}\right) \end{aligned} \tag{7.160a}$$

which is equal to $\frac{2\alpha}{a}(-1)^{m+n}$ for odd values of m and n, and vanishes for even values of m and n. Therefore, for odd values of m and n we get $\frac{\langle \psi_m^{(0)} | H_1 | \psi_n^{(0)} \rangle}{\left(E_m^{(0)} - E_n^{(0)}\right)} = \frac{4m^* a\alpha}{\hbar^2\pi^2} \frac{(-1)^{m+n}}{\left(m^2 - n^2\right)}$. Here, we have used $E_n^{(0)} = \frac{\hbar^2 n^2 \pi^2}{2m^* a^2}$, m^* being the mass of the particle. Hence:

$$|\psi_n^{(1)}\rangle = \frac{4m^*\alpha a}{\hbar^2\pi^2}\sum_{m \neq n} \frac{(-1)^{m+n}}{\left(m^2 - n^2\right)} |\psi_m^{(0)}\rangle, \text{when } m \text{ and } n \text{ take odd values.} \tag{7.160b}$$

2. A particle of mass μ and charge e moves under a central potential defined by:

$$V(r)=\begin{cases} -\dfrac{e^2}{4\pi\varepsilon_0 r} & r<a \\ -\dfrac{e^2}{4\pi\varepsilon_0 r}e^{-\lambda r} & r>a \end{cases} \tag{7.161}$$

Apply the perturbation theory to calculate the approximate energy of the ground state.

SOLUTION

The Hamiltonian of the particle can be written as:

$$\begin{aligned} H &= H_0 + H_1 \\ &= \frac{p^2}{2\mu} - \frac{e^2}{4\pi\varepsilon_0 r} + \frac{e^2}{4\pi\varepsilon_0 r}\left[1-e^{-\lambda r}\right]\Theta(r-a) \end{aligned} \tag{7.162a}$$

We then take:

$$\begin{aligned} H_0 &= \frac{p^2}{2\mu} - \frac{e^2}{4\pi\varepsilon_0 r}; \\ H_1 &= \frac{e^2}{4\pi\varepsilon_0 r}\left[1-e^{-\lambda r}\right]\Theta(r-a) \end{aligned} \tag{7.162b}$$

where Θ is the step function. Since H_0 is the Hamiltonian of the hydrogen atom, we have: $H_0\psi_0^{(0)}(r)=E_0^{(0)}\psi_0^{(0)}(r)$, and the approximate ground state energy is: $E_0 \approx E_0^{(0)} + E_0^{(1)}$, with $E_0^{(0)} = -\dfrac{1}{4\pi\varepsilon_0}\dfrac{e^2}{2a_0}$ and $\psi_0^{(0)} = \dfrac{1}{\sqrt{\pi a_0^3}}e^{-r/a_0}$. The first order correction to energy $E_0^{(1)}$ is:

$$\begin{aligned} E_0^{(1)} &= \left\langle \psi_0^{(0)} \middle| H_1 \middle| \psi_0^{(0)} \right\rangle \\ &= \frac{e^2}{4\pi\varepsilon_0}\frac{1}{\pi a_0^3}\int e^{-r/a_0}\frac{1}{r}\left[1-e^{-\lambda r}\right]\Theta(r-a)e^{-r/a_0}d^3r \\ &= \frac{e^2}{4\pi\varepsilon_0}\frac{1}{\pi a_0^3}\int_a^\infty\int_0^\pi\int_0^{2\pi}\frac{1}{r}\left[1-e^{-\lambda r}\right]e^{-2r/a_0}r^2 dr\sin\theta d\theta d\phi \\ &= \left(\frac{1}{4\pi\varepsilon_0}\right)\left(\frac{4e^2}{a_0^3}\right)\int_a^\infty\left(1-e^{-\lambda r}\right)e^{-2r/a_0}r dr \\ &= \left(\frac{e^2}{4\pi\varepsilon_0 a_0}\right)e^{-2a/a_0}\left[\left(1+\frac{2a}{a_0}\right)-\frac{e^{-\lambda a}}{\left(1+\frac{\lambda a_0}{2}\right)^2}\left\{1+a\lambda+\frac{2a}{a_0}\right\}\right] \end{aligned} \tag{7.163a}$$

which vanishes for $\lambda \to 0$. Thus, the approximate energy is:

$$E_0 \approx -\left(\frac{1}{4\pi\varepsilon_0}\right)\frac{e^2}{2a_0}\left[1-2e^{-2a/a_0}\left\{\left(1+\frac{2a}{a_0}\right)-\frac{e^{-\lambda a}}{\left(1+\frac{\lambda a_0}{2}\right)^2}\left\{1+a\lambda+\frac{2a}{a_0}\right\}\right\}\right] \quad (7.163b)$$

3. A quantum mechanical rigid rotor has moment of inertia I about its axis of rotation and the in-plane electric dipole moment μ is constrained to rotate in a weak uniform electric field **E** in the plane of rotation. Apply the perturbation theory to find the first non-zero corrections to the energy levels and wave function of the rotator.

SOLUTION

We take the x-y plane as the plane of rotation and the z-axis parallel to electric field $\mathbf{E}$. In the absence of $\mathbf{E}$, the Hamiltonian of the rotator is $H_0 = -\frac{\hbar^2}{2I}\frac{\partial^2}{\partial\phi^2}$ and the equation of motion is:

$$-\frac{\hbar^2}{2I}\frac{\partial^2\psi_m^{(0)}(\phi)}{\partial\phi^2} = E\psi_m^{(0)}(\phi) \quad (7.164)$$

whose solution is:

$$\psi_m^{(0)}(\phi) = \frac{1}{\sqrt{2\pi}}e^{im\phi}, \text{ and } E_m^{(0)} = \frac{\hbar^2 m^2}{2I}, \text{with } m = \pm 1, \pm 2, \pm 3, \ldots.. \quad (7.165)$$

In the presence of the electric field $H = H_0 + H_1$ with $H_1 = -\mu.\mathrm{E} = -\mu\mathrm{E}\cos\phi$.

The first order correction to the energy is:

$$E_m^{(1)} = \left\langle\psi_m^{(0)}\middle|H_1\middle|\psi_m^{(0)}\right\rangle = -\frac{\mu\mathrm{E}}{2\pi}\int_0^{2\pi} e^{-im\phi}\cos\phi e^{im\phi}d\phi = 0 \quad (7.166)$$

The first order correction to the energy is zero. We then calculate the *second order correction to the energy,* which is given by:

$$E_m^{(2)} = \sum_{n\neq m}\frac{\left|\left\langle\psi_m^{(0)}\middle|H_1\middle|\psi_n^{(0)}\right\rangle\right|^2}{E_m^{(0)} - E_n^{(0)}} \quad (7.167a)$$

where:

$$\begin{aligned}\left\langle\psi_m^{(0)}\middle|H_1\middle|\psi_n^{(0)}\right\rangle &= -\frac{\mu\mathrm{E}}{2\pi}\int_0^{2\pi} e^{-im\phi}\cos\phi e^{in\phi}d\phi \\ &= -\frac{\mu\mathrm{E}}{4\pi}\left[\int_0^{2\pi} e^{i(n-m+1)\phi}d\phi + \int_0^{2\pi} e^{i(n-m-1)\phi}d\right] \\ &= -\frac{\mu\mathrm{E}}{2}\left(\delta_{n,m-1} + \delta_{n,m+1}\right)\end{aligned} \quad (7.167b)$$

Therefore:

$$E_m^{(2)} = \left(\frac{\mu \mathrm{E}}{2}\right)^2 \frac{2I}{\hbar^2}\left[\frac{1}{m^2-(m-1)^2}+\frac{1}{m^2-(m+1)^2}\right] \tag{7.168a}$$

Hence, the first non-zero correction to the energy is:

$$E_m^{(2)} = \left(\frac{\mu \mathrm{E}}{\hbar}\right)^2 \frac{I}{(4m^2-1)} \tag{7.168b}$$

The first order correction to the wave function is:

$$\begin{aligned} \left|\psi_m^{(1)}\right\rangle &= \sum_{n\neq m} \frac{\left\langle \psi_n^{(0)}\right| H_1 \left|\psi_m^{(0)}\right\rangle}{\left(E_n^{(0)}-E_m^{(0)}\right)}\left|\psi_n^{(0)}\right\rangle \\ &= -\frac{\mu \mathrm{E} I}{\hbar^2}\sum_{n\neq m}\left[\frac{\delta_{n,m+1}}{n^2-m^2}\left|\psi_n^{(0)}\right\rangle + \frac{\delta_{n,m-1}}{n^2-m^2}\left|\psi_n^{(0)}\right\rangle\right] \end{aligned} \tag{7.169a}$$

Therefore the non-zero correction to the wave function is:

$$\left|\psi_m^{(1)}\right\rangle = -\frac{\mu \mathrm{E} I}{\hbar^2}\left[\frac{\left|\psi_{m+1}^{(0)}\right\rangle}{2m+1} - \frac{\left|\psi_{m-1}^{(0)}\right\rangle}{2m-1}\right] = -\frac{\mu E I}{\hbar^2\sqrt{2\pi}} e^{im\phi}\left[\frac{e^{i\phi}}{2m+1} - \frac{e^{-i\phi}}{2m-1}\right] \tag{7.169b}$$

4. A massless rod of length l attached to a mass m at one end and to a pivot at the other end swings in a vertical plane under gravity. Use perturbation theory to calculate the approximate ground state energy of the system.
(Similar Problem No. 5014 in "Problems & Solutions on Quantum Mechanics," edited by Yung-Kuo Lim.)

SOLUTION

By taking the equilibrium position of the point mass as the zero point of potential energy, the Hamiltonian of the system at a deflection of θ, is given by:

$$H = \frac{1}{2} m l^2 \dot{\theta}^2 + mgl(1-\cos\theta) \tag{7.170a}$$

For small values of θ, we write:

$$H \simeq \frac{1}{2} m l^2 \dot{\theta}^2 + \frac{1}{2} mgl\theta^2 - \frac{1}{24} mgl\theta^4 \tag{7.170b}$$

On taking $H_0 = \frac{1}{2} m l^2 \dot{\theta}^2 + \frac{1}{2} mgl\theta^2$ and $H_1 = -\frac{1}{24} mgl\theta^4$, we have:

$$E_n^{(0)} = \left(n + \frac{1}{2}\right)\hbar\omega \tag{7.171a}$$

and:

$$\psi_0^{(0)}(x) = \left(\frac{\alpha}{\sqrt{\pi}}\right)^{1/2} e^{-\frac{1}{2}\alpha^2 x^2} \tag{7.171b}$$

with $\omega = \sqrt{\frac{g}{l}}$ and $\alpha = \left(\frac{m\omega}{\hbar}\right)^{1/2}$. The perturbation part of the Hamiltonian is: $H_1 = -\frac{1}{24} mgl\theta^4 = -\frac{1}{24}\frac{mg}{l^3}x^4$. Then the first order correction to the energy is:

$$\begin{aligned} E_0^{(1)} &= \left\langle \psi_0^{(0)} \middle| H_1 \middle| \psi_0^{(0)} \right\rangle \\ &= -\frac{mg\alpha}{24l^3\sqrt{\pi}} \int_{-\infty}^{\infty} x^4 e^{-\alpha^2 x^2} dx = -\frac{mg}{24l^3} \times \frac{3}{4\alpha^4} \\ &= -\frac{\hbar^2}{32ml^2} \end{aligned} \tag{7.171c}$$

Therefore, the approximate ground state energy of the system is:

$$E_0 \simeq \frac{\hbar}{2}\sqrt{\frac{g}{l}} - \frac{\hbar^2}{32ml^2} \tag{7.172}$$

5. The energy levels of the electron in the hydrogen atom split up when the atom is placed in a magnetic field. The phenomenon is known as the Zeeman effect. Consider a homogeneous magnetic field $\mathbf{B} = B_0\hat{z}$ applied along the z-axis. Calculate the first order corrections to the energy of the H-atom.

SOLUTION

In presence of a magnetic field, the Hamiltonian of the electron in the H-atom is $H = H_0 + H_1$, where H_0 is the unperturbed part defined by Eqn. (6.11). Energy eigenvalues and eigenwave functions are given by Eqns. (6.22) and (6.35), respectively. The perturbative part is the scalar product of the electron dipole moment μ and $\mathbf{B}$. It is given by:

$$\begin{aligned} H_1 &= -\mu.\mathbf{B} \\ &= \frac{e}{2m}(\mathbf{L} + g\mathbf{S}).\mathbf{B} \\ &= \frac{e}{2m}(\mathbf{L} + g\mathbf{S}).\hat{z}B_0 \\ &= \frac{e}{2m}(L_z + gS_z)B_0 \end{aligned} \tag{7.173}$$

for electron $g \simeq 2$.

Let us further assume that the magnetic field is sufficiently strong not to have coupling of **L** and **S**. Thus, the unperturbed states are product states of the form $\left|\psi_{nlm}^{(0)}\right\rangle\left|\uparrow\right\rangle$ and $\left|\psi_{nlm}^{(0)}\right\rangle\left|\downarrow\right\rangle$. The first order correction to the energy is obtained from:

$$E_n^{(1)} = \left\langle \updownarrow \right|\left\langle \psi_{nlm}^{(0)} \right| H_1 \left| \psi_{nlm}^{(0)} \right\rangle \left| \updownarrow \right\rangle = \frac{e\hbar B_0}{2m}(m+2s_z) \tag{7.174}$$

where $-l<m<l$ and $s_z=\pm 1$. We thus find that every l-level splits up by an energy correction term that depends on the spin orientation. The Zeeman effect can be observed experimentally. In absence of a magnetic field, two ground states $\left|\psi_{100}^{(0)}\right\rangle\left|\uparrow\right\rangle$ and $\left|\psi_{100}^{(0)}\right\rangle\left|\downarrow\right\rangle$ would have the same energy $E_1=-13.6$ eV. But, on application of a magnetic field they have different energies: $E_{100\uparrow}=E_1+\frac{e\hbar}{2m}B_0$ and $E_{100\downarrow}=E_1-\frac{e\hbar}{2m}B_0$. The single spectral line belonging to eight $\left|\psi_{2lm}^{(0)}\right\rangle\left|\updownarrow\right\rangle$ states will similarly split into five closely spaced but separate lines, corresponding to the five possible values $-2,-1,0,1,2$ of the factor $m + 2s_z$. It is to be noted that the energy change due to an extremely strong magnetic field of 100 Tesla is only 0.006 eV.

6. A particle of mass m oscillates under the influence of potential $V(x)=\frac{cx^3}{2}-\frac{3}{2}\left(ac-\frac{b}{2}\right)x^2+\frac{3a}{2}(ac-b)x+\frac{a^2}{2}\left(\frac{3b}{2}-ac\right)$ about the position of minimum of $V(x)$ occurring at x_0. Find the x_0 and express the potential in terms of x_0. Calculate the ground state energy with use of the variation method by taking the trial wave function $\psi(x)=\left(\frac{2\beta}{\pi}\right)^{1/4} e^{\ \beta(x\ \ x_0)^2}$ where β is the variational parameter.

SOLUTION

At the minima, $\frac{dV(x)}{dx}=0$ and $\frac{d^2V}{dx^2}>0$. We thus have:

$$\begin{aligned}\frac{dV}{dx} &= \frac{d}{dx}\left\{\frac{cx^3}{2}-\frac{3}{2}\left(ac-\frac{b}{2}\right)x^2+\frac{3a}{2}(ac-b)x+\frac{a^2}{2}\left(\frac{3b}{2}-ac\right)\right\}\\ &= \frac{3cx^2}{2}-3\left(ac-\frac{b}{2}\right)x+\frac{3a}{2}(ac-b)\\ &= \frac{3}{2}(x-a)[b+c(x-a)]\end{aligned} \tag{7.175a}$$

and:

$$\frac{d^2V}{dx^2}=\frac{3b}{2}+3c(x-a) \tag{7.175b}$$

$\frac{dV}{dx}$ goes to zero at $x=a$ and $x=a-\frac{b}{c}$. At $x=a$, $\frac{d^2V}{dx^2}=\frac{3b}{2}>0$, while at $x=a-\frac{b}{c}$, $\frac{d^2V}{dx^2}=-\frac{3b}{2}<0$. Thus, the minimum of $V(x)$ occurs at $x=a$, not at $x=a-\frac{b}{c}$, and therefore $a=x_0$. In terms of $a=x_0$, $V(x)$ can be expressed as:

$$\begin{aligned} V(x) &= \frac{cx^3}{2}-\frac{3}{2}\left(ac-\frac{b}{2}\right)x^2+\frac{3a}{2}(ac-b)x+\frac{a^2}{2}\left(\frac{3b}{2}-ac\right) \\ &= \frac{3b}{4}(x-x_0)^2+\frac{c}{2}(x-x_0)^3 \end{aligned} \tag{7.176}$$

The trial wave function therefore is $\psi(x)=\left(\frac{2\beta}{\pi}\right)^{1/4}e^{-\beta(x-x_0)^2}$. To compute the expectation value of the Hamiltonian $\langle H\rangle=\frac{\langle\psi|\left(\frac{p^2}{2m}+V(x)\right)|\psi\rangle}{\langle\psi|\psi\rangle}$, we evaluate:

$$\langle\psi|\psi\rangle=\left(\frac{2\beta}{\pi}\right)^{1/2}\int_{-\infty}^{\infty}e^{-2\beta(x-x_0)^2}dx=\left(\frac{2\beta}{\pi}\right)^{1/2}\left(\frac{\pi}{2\beta}\right)^{1/2}=1 \tag{7.177a}$$

and:

$$\langle\psi|\left(\frac{p^2}{2m}+V(x)\right)|\psi\rangle=-\frac{\hbar^2}{2m}\langle\psi|\frac{d^2}{dx^2}|\psi\rangle+\langle\psi|\frac{3b}{4}(x-x_0)^2+\frac{c}{2}(x-x_0)^3|\psi\rangle \tag{7.177b}$$

Take:

$$\begin{aligned} \langle\psi|\frac{d^2}{dx^2}|\psi\rangle &= \left(\frac{2\beta}{\pi}\right)^{1/2}\left[4\beta^2\int_{-\infty}^{\infty}(x-x_0)^2e^{-2\beta(x-x_0)^2}dx-2\beta\int_{-\infty}^{\infty}e^{-2\beta(x-x_0)^2}dx\right] \\ &= \left(\frac{2\beta}{\pi}\right)^{1/2}\left[\sqrt{\frac{\pi\beta}{2}}-\sqrt{2\beta\pi}\right] \\ &= -\beta \end{aligned} \tag{7.178a}$$

and, hence:

$$\langle\psi|\frac{p^2}{2m}|\psi\rangle=\frac{\hbar^2\beta}{2m}$$

Next, we calculate:

$$\langle\psi|V|\psi\rangle = \frac{c}{2}\langle\psi|(x-x_0)^3|\psi\rangle + \frac{3b}{4}\langle\psi|(x-x_0)^2|\psi\rangle$$

$$= \left(\frac{2\beta}{\pi}\right)^{1/2}\left[\frac{c}{2}\int_{-\infty}^{\infty}(x-x_0)^3 e^{-2\beta(x-x_0)^2}dx + \frac{3b}{4}\int_{-\infty}^{\infty}(x-x_0)^2 e^{-2\beta(x-x_0)^2}dx\right]$$

$$= \left(\frac{2\beta}{\pi}\right)^{1/2}\left[\frac{c}{2}\left\{\int_{-\infty}^{\infty}y^3 e^{-2\beta y^2}dy\right\} + \frac{3b}{4}\left\{\int_{-\infty}^{\infty}y^2 e^{-2\beta y^2}dy\right\}\right] \tag{7.179a}$$

$$= \left(\frac{2\beta}{\pi}\right)^{1/2}\left[\frac{3b}{4}\frac{\sqrt{\pi}}{2}\left(\frac{1}{2\beta}\right)^{3/2}\right]$$

$$= \frac{3b}{16\beta}$$

Therefore:

$$\langle H\rangle = \frac{\hbar^2\beta}{2m} + \frac{3b}{16\beta} \tag{7.179b}$$

which gives $\beta_{\min} = \left(\frac{3bm}{8\hbar^2}\right)^{1/2}$ and hence ground state energy is:

$$E_g = \frac{\hbar}{2}\left(\frac{3b}{2m}\right)^{1/2} \tag{7.180}$$

If we consider the case of small oscillations and identify $\frac{1}{2}m\omega^2 = \frac{3b}{4}$ in Eqn. (7.176), then $\omega = \sqrt{\frac{3b}{2m}}$ giving:

$$E_g = \frac{1}{2}\hbar\omega \tag{7.181}$$

which is the energy of the ground state of the 1D harmonic oscillator. Note that if we use the non-degenerate perturbation theory and calculate first order corrections to the ground state energy of the 1D harmonic oscillator, we will get the same answer because of the zero contribution coming from perturbative terms having odd powers of $(x-a)$, as has been shown in Section 7.1.1.4.

7. Take $\psi(r) = \left(\frac{\alpha^3}{\pi}\right)^{1/2} e^{-\alpha r}$ as a trial wave function to calculate the expectation value of the Hamiltonian of the electron in the ground state of a hydrogen atom when

coulomb potential is replaced by screened potential defined by $V(r) = \frac{e^2}{4\pi\varepsilon_0 r} e^{-\lambda r}$ where $1/\lambda$ is known as the screening length and α is the variational parameter. If parameter λ is written as $\lambda = y\alpha$ where $y \ll 1$, what is approximate ground state energy?

SOLUTION

The Hamiltonian for the electron in the ground state ($l = 0$) of the hydrogen atom is:

$$H = -\frac{\hbar^2}{2\mu}\left[\frac{d^2}{dr^2} + \frac{2}{r}\frac{d}{dr}\right] - \frac{e^2}{4\pi\varepsilon_0 r} e^{-\lambda r} \tag{7.182a}$$

We notice that $\langle\psi|\psi\rangle = 1$ and:

$$\langle H\rangle = \langle\psi|\left[-\frac{\hbar^2}{2\mu}\left(\frac{d^2}{dr^2} + \frac{2}{r}\frac{d}{dr}\right)\right]|\psi\rangle + \langle\psi|V|\psi\rangle \tag{7.182b}$$

As has been shown by Eqn. (7.63), the expectation value is:

$$\langle\psi|\left[-\frac{\hbar^2}{2\mu}\left(\frac{d^2}{dr^2} + \frac{2}{r}\frac{d}{dr}\right)\right]|\psi\rangle = \frac{\hbar^2\alpha 2}{2\mu} \tag{7.183}$$

We next calculate the expectation value of $V(r)$:

$$\begin{aligned}
\langle\psi|V|\psi\rangle &= -\frac{e^2}{4\pi\varepsilon_0}\left(\frac{\alpha^3}{\pi}\right)\int_0^\infty e^{-\lambda r} r e^{-2\alpha r}\int_0^\pi \sin\theta\, d\theta \int_0^{2\pi} d\phi \\
&= -\frac{e^2\alpha^3}{\pi\varepsilon_0}\int_0^\infty e^{-\lambda r} r e^{-2\alpha r} \\
&= -\frac{e^2\alpha^3}{\pi\varepsilon_0}\frac{1}{(\lambda + 2\alpha)^2} \\
&= \left(-\frac{e^2}{4\pi\varepsilon_0}\right)\frac{4\alpha^3}{(\lambda + 2\alpha)^2}
\end{aligned} \tag{7.184}$$

Therefore:

$$\begin{aligned}
\langle H\rangle &= \frac{\hbar^2\alpha^2}{2\mu} - \left(\frac{e^2}{4\pi\varepsilon_0}\right)\frac{4\alpha^3}{(\lambda + 2\alpha)^2} \\
&= \frac{\hbar^2}{\mu}\left[\frac{\alpha^2}{2} - \frac{1}{a_0}\left\{\frac{4\alpha^3}{(\lambda + 2\alpha)^2}\right\}\right]
\end{aligned} \tag{7.185a}$$

Taking $\lambda = y\alpha$, we have:

$$\langle H\rangle = \frac{\hbar^2}{\mu}\left[\frac{\alpha^2}{2} - \frac{\alpha}{a_0(y/2-1)^2}\right] \tag{7.185b}$$

The condition of minimization of $\langle H\rangle$ with respect to α gives:

$$\alpha_{\min} = \frac{1}{a_0(y/2+1)^2} \tag{7.186}$$

Note that $\alpha_{\min} \sim 1/a_0$. The ground state energy is:

$$\begin{aligned} E_g &= -\frac{\hbar^2}{2\mu a_0^2}\frac{1}{(1+y/2)^4} \\ \Rightarrow E_g &\approx -\frac{\hbar^2}{2\mu a_0^2} + \frac{\hbar^2}{2\mu a_0^2}\frac{y(1-3y)}{2} = E_1 - E_1\frac{y(1-3y)}{2} \end{aligned} \tag{7.187}$$

We thus find that the magnitude of ground state energy of hydrogen atom is slightly reduced if the electron is considered moving under the influence of screened electrostatic field rather than coulombic field.

8. A system is described by the Hamiltonian:

$$H = -\frac{\hbar^2}{2m}\frac{d^2}{dx^2} - \left(\frac{\hbar^2}{ma_0}\right)\delta(x-\alpha/2) \tag{7.188a}$$

Use the triangular trail wave function defined by:

$$\psi(x) = \begin{cases} Ax & \text{if}(0 \le x \le \alpha/2) \\ A(\alpha - x) & \text{if}(\alpha/2 \le x \le \alpha) \\ 0 & \text{otherwise} \end{cases} \tag{7.188b}$$

as the trial wave function and calculate the ground state energy. A is to be determined by the normalization condition and α is the variation parameter.

SOLUTION

From the normalization condition, we have:

$$\begin{aligned} 1 &= |A|^2\left[\int_0^{\alpha/2} x^2 dx + \int_{\alpha/2}^{\alpha} (\alpha - x)^2 dx\right] \\ &= |A|^2\frac{\alpha^3}{12} \\ \Rightarrow A &= \frac{2}{\alpha}\sqrt{\frac{3}{\alpha}} \end{aligned} \tag{7.189}$$

We want to calculate $\langle H\rangle = -\langle\psi|\frac{\hbar^2}{m}\left[\frac{1}{2}\frac{d^2}{dx^2}+\frac{1}{a_0}\delta(x-\alpha/2)\right]|\psi\rangle$.

$$\frac{d\psi}{dx}=\begin{cases} A & \text{if}(0\le x\le\alpha/2) \\ -A & \text{if}(\alpha/2\le x\le\alpha) \\ 0 & \text{otherwise}\end{cases} \tag{7.190a}$$

which is the step function. The derivative of the step function is the delta function. Hence

$$\frac{d^2\psi}{dx^2}=A\delta(x)-2A\delta(x-\alpha/2)+A\delta(x-\alpha) \tag{7.190b}$$

Therefore:

$$\begin{aligned}\langle\psi|\frac{p^2}{2m}|\psi\rangle &= -A\frac{\hbar^2}{2m}\int[\delta(x)-2\delta(x-\alpha/2)+\delta(x-\alpha)]\psi(x)dx \\ &= -A\frac{\hbar^2}{2m}[\psi(0)-2\psi(\alpha/2)+\psi(\alpha)] \\ &= \frac{\hbar^2A^2\alpha}{2m}=\frac{6\hbar^2}{m\alpha^2}\end{aligned} \tag{7.191a}$$

and:

$$\begin{aligned}\langle\psi|V|\psi\rangle &= -\left(\frac{\hbar^2}{ma_0}\right)\int\psi^*(x)\delta(x-\alpha/2)\psi(x)dx \\ &= -\left(\frac{\hbar^2}{ma_0}\right)[\psi(\alpha/2)]^2=-\left(\frac{\hbar^2}{ma_0}\right)A^2\left(\frac{\alpha}{2}\right)^2 \\ &= -\frac{3}{\alpha}\left(\frac{\hbar^2}{ma_0}\right)\end{aligned} \tag{7.191b}$$

Thus:

$\langle H\rangle=\frac{6\hbar^2}{m\alpha^2}-\frac{3}{\alpha}\left(\frac{\hbar^2}{ma_0}\right)$ and $\frac{\partial\langle H\rangle}{\partial\alpha}=-\frac{12\hbar^2}{m\alpha^3}+\frac{3}{\alpha^2}\left(\frac{\hbar^2}{ma_0}\right)$, which gives $\alpha_{\min}=4a_0$.

We get:

$$E_g=-\frac{3\hbar^2}{8ma_0^2} \tag{7.192}$$

9. Use the WKB method to calculate the energy of a particle of mass m confined to an infinite potential well of width a and a constant potential energy $V_0 < E$.

SOLUTION

Inside the potential well $V_0 < E$ and therefore, we have the solution:

$$\psi(x) \approx \frac{1}{\sqrt{p(x)}}\left[C^+ e^{i\phi(x)} + C^- e^{-i\phi(x)}\right] \tag{7.193a}$$

The more convenient form is:

$$\psi(x) \approx \frac{1}{\sqrt{p(x)}}\left[C_1 \cos\phi(x) + C_2 \sin\phi(x)\right] \tag{7.193b}$$

At the boundaries of the well, the wave function vanishes. Hence $\psi(0) = \psi(a) = 0$.

This suggests that $C_1 = 0$ and $\phi(a) = n\pi$, where $n = 1, 2, 3....$ The $\phi(x) = \frac{1}{\hbar}\int\limits_0^x p(x')dx'$ gives:

$$n\pi = \frac{1}{\hbar}\int\limits_0^a p(x)dx \tag{7.194}$$

which is the quantization condition and it gives:

$$\begin{aligned} &\int\limits_0^a \sqrt{2m(E - V_0)}dx = n\pi\hbar \\ &\Rightarrow E_n = \frac{\hbar^2 n^2 \pi^2}{2ma^2} + V_0 \end{aligned} \tag{7.195}$$

We thus find the WKB method yields the exact result for an infinite potential well.

10. Apply the WKB approximation to obtain the energy eigenvalues of a harmonic oscillator.

SOLUTION

The classical turning points are those where $p = 0$ or $E = V(x) = \frac{1}{2}m\omega^2$. Therefore, we have two classical turning points, $x_1 = -\left(\frac{2E}{m\omega^2}\right)^{1/2}$ and $x_2 = \left(\frac{2E}{m\omega^2}\right)^{1/2}$. The energy eigenvalues for a bound state problem can be determined by Eqn. (7.157). In this case, we have:

$$\frac{1}{\hbar}\int_{x_1}^{x_2} p(x)dx = \left(n+\frac{1}{2}\right)\pi$$
$$\Rightarrow \frac{\sqrt{2mE}}{\hbar}\int_{x_1}^{x_2}\left(1-\frac{m\omega^2}{2E}x^2\right)^{1/2} dx = \left(n+\frac{1}{2}\right)\pi \tag{7.196}$$

To evaluate the integral, we take $\sqrt{\frac{m\omega^2}{2E}}x = \sin\theta$.

$$\int_{x_1}^{x_2}\left(1-\frac{m\omega^2}{2E}x^2\right)^{1/2} dx = \sqrt{\frac{2E}{m\omega^2}}\int_{-\pi/2}^{\pi/2}\cos^2\theta\, d\theta = \frac{\pi}{2}\sqrt{\frac{2E}{m\omega^2}},$$ which yields:

$$\frac{2E}{\hbar\omega}\frac{\pi}{2} = \left(n+\frac{1}{2}\right)\pi$$
$$\Rightarrow E_n = \left(n+\frac{1}{2}\right)\hbar\omega \tag{7.197}$$

11. Explain the emission of an alpha particle using the WKB approximation.

SOLUTION

Spontaneous emission of an alpha particle ((two protons and two neutrons) by certain radioactive nuclei) can be explained using WKB approximations. Since the alpha particle carries a positive charge ($2e$), it will be repelled by the left-over nucleus of the charge (Ze) as soon as it gets far enough away to escape from the nucleus binding force. The potential energy curve for the alpha particle emission can be represented by the finite square well potential extended from r_1 (radius of the nucleus) to the repulsive coulomb tail, as is shown in Fig. 7.9. The outer turning point r_2 can be taken where $\frac{1}{4\pi\varepsilon_0}\frac{2Ze^2}{r_2} = E$.

In the repulsive region, we can write:

$$\psi_I(r) = \frac{B_1}{\sqrt{\gamma}}e^{\frac{1}{\hbar}\int_r^{r_2}\gamma(r')dr'} + \frac{B_2}{\sqrt{\gamma}}e^{-\frac{1}{\hbar}\int_r^{r_2}\gamma(r')dr'} \tag{2.208}$$

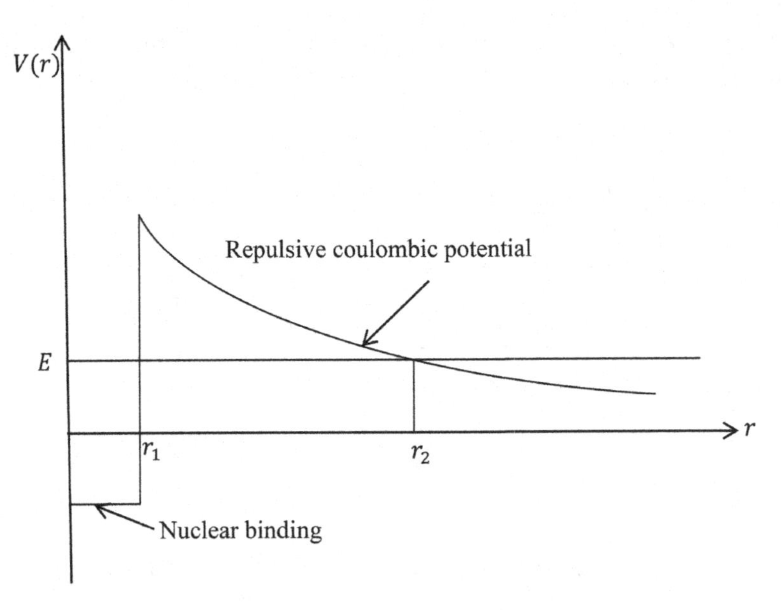

FIGURE 7.9
Model potential energy curve for alpha particle emission.

but the barrier is very high and wide and hence the chances of having an exponential increasing term are almost zero. Therefore, we have chosen $B_1 = 0$. The exponent of the decaying term has:

$$\begin{aligned}\int_{r_1}^{r_2} \frac{\gamma(r)}{\hbar} dr &= \frac{1}{\hbar}\int_{r_1}^{r_2} \sqrt{2m\left(\frac{2Ze^2}{4\pi\varepsilon_0 r} - E\right)} dr \\ &= \sqrt{\frac{2mE}{\hbar^2}} \int_{r_1}^{r_2} \sqrt{2m\left(\frac{r_2}{r} - 1\right)} dr \\ &= \sqrt{\frac{2mE}{\hbar^2}} \left[r_2 \cos^{-1}(\sqrt{r_1/r_2} - \sqrt{r_1(r_2 - r_1)} \right]\end{aligned} \tag{7.198}$$

Generally, $r_1 \ll r_2$, and hence the result is simplified to:

$$\begin{aligned}&\sqrt{\frac{2mE}{\hbar^2}} \left[r_2 \cos^{-1}(\sqrt{r_1/r_2} - \sqrt{r_1(r_2 - r_1)} \right] \\ &\cong \sqrt{\frac{2mE}{\hbar^2}} \left(\frac{\pi}{2} r_2 - 2\sqrt{r_1 r_2} \right) \\ &\cong \frac{K_1 Z}{\sqrt{E}} - K_2\sqrt{Z r_1}\end{aligned} \tag{7.199}$$

where $K_1 = \left(\frac{e^2}{4\pi\varepsilon_0}\right)\frac{\pi\sqrt{2m}}{\hbar}$ and $K_2 = \left(\frac{e^2}{4\pi\varepsilon_0}\right)^{1/2}\frac{4\sqrt{m}}{\hbar}$. The transmission coefficient T and decay constant λ are then given by $T = e^{-2\int_{r_1}^{r_2}\gamma(r)dr} = e^{-2K_1Z/\sqrt{E}}e^{2K_2\sqrt{Zr_1}}$ and $\lambda = 10^{21}T$. If we imagine that the alpha particle is rattling around the nucleus with an average speed v with the probability of escape at each collision T then the probability of emission per unit time is $\left(\frac{v}{2r_1}\right)T$ and hence, the life time of the parent nucleus is $\tau = \frac{2r_1T}{v}$. For the alpha particle $v \approx 10^7\ \text{ms}^{-1}$ and the radius of the heavy nucleus is of the order of 10^{-14}m, which gives $\lambda = \frac{1}{\tau} = 10^{21}T$.

12. A particle moving in a potential well has one infinite high vertical side at $x = 0$ and one sloping side given by $V(x) = Ax$ for $x > 0$:

$$V(x) = \begin{cases} \infty & x = 0 \\ Ax & x > 0 \end{cases} \tag{7.200}$$

where A is constant. Apply the WKB approximation to find the energy of the particle.

SOLUTION

We have $p(x) = \sqrt{2m\{E - V(x)\}} = \sqrt{2m\{E - Ax\}}$. The turning point will appear at $x_2 = \frac{E}{A}$. The WKB wave function of the particle for $x > x_2$ is given by Eqn. (7.131). The right-hand side of potential $(x > 0)$ can be extended up to ∞. Hence to have a well-defined wave function, we should have $B_1 = 0$ in Eqn. (7.131). Then, with the use of Eqns. (7.146) and (7.147), we write:

$$\begin{aligned}\psi(x) &= \frac{B_2}{2}e^{\left(\frac{i}{\hbar}\int_x^{x_2}p(x')dx' - \frac{i\pi}{4}\right)} + \frac{B_2}{2}e^{-\left(\frac{i}{\hbar}\int_x^{x_2}p(x')dx' - \frac{i\pi}{4}\right)} \\ &= B_2\cos\left(\frac{1}{\hbar}\int_x^{x_2}p(x')dx' - \frac{\pi}{4}\right) \\ &= B_2\sin\left(\frac{1}{\hbar}\int_x^{x_2}p(x')dx' + \frac{\pi}{4}\right)\end{aligned} \tag{7.201a}$$

Since the wave function vanishes at $x = 0$, $\psi(0) = 0$, we have:

$$\frac{1}{\hbar}\int_0^{x_2}p(x')dx' + \frac{\pi}{4} = n\pi, \text{where}\, n = 1, 2, 3..... \tag{7.201b}$$

which yields:

$$\frac{1}{\hbar}\int_0^{x_2}\sqrt{2m\{E-Ax'\}}dx'+\frac{\pi}{4}=n\pi$$
$$\Rightarrow\int_0^{x_2}\sqrt{\{x_2-x'\}}dx'=\sqrt{\frac{1}{2mA}}\left(n-\frac{1}{4}\right)\pi\hbar \tag{7.202}$$
$$\Rightarrow\frac{2}{3}x_2^{3/2}=\sqrt{\frac{1}{2mA}}\left(n-\frac{1}{4}\right)\pi\hbar$$

the evaluation of which gives:

$$E_n=\frac{1}{(2m)^{1/3}}\left\{\frac{3A(4n-1)h}{16}\right\}^{2/3} \tag{7.203}$$

7.5 Exercises

1. Take $H_1=\lambda x^2$ as a perturbative term in the Hamiltonian of a one dimensional harmonic oscillator. (a) Use $\psi_0^{(0)}(x)=Ae^{-\beta x^2}$ as the ground state wave function and then calculate first the order correction to the ground state energy. (b) Determine the values of A and β to obtain $E_0^{(0)}=\frac{1}{2}\hbar\omega$.
2. In the nonrelativistic limit, the Hamiltonian of a one dimensional harmonic oscillator is: $H=\frac{p^2}{2m}+\frac{1}{2}m\omega^2x^2$. If the relativistic corrections are applied, the approximate Hamiltonian can be written as: $H=\frac{p^2}{2m}+\frac{1}{2}m\omega^2x^2-\frac{p^4}{8m^3c^2}$. Treat $H_1=-\frac{p^4}{8m^3c^2}$ as a perturbative term and calculate the first order correction to the ground state energy.
3. A particle moves in a one dimensional box of width a, for which potential is defined by:

$$V(x)=\begin{cases}\infty & \text{for } 0>x>a\\ V_0 & \text{for } 0<x<a/2.\\ 0 & \text{for } a/2<x<a\end{cases}$$

Find the ground state energy using first order perturbation theory.

4. If the proton at the nucleus of a hydrogen atom is treated as a sphere of radius R with charge e, the perturbation to the Hamiltonian is given by:

$$H_1 = \begin{cases} \dfrac{e^2}{4\pi\varepsilon_0 r} - \dfrac{e^2}{4\pi\varepsilon_0 R} & \text{for } 0 \le r \le R \\ 0 & R \le r \le \infty \end{cases}.$$

Compute first order corrections to the ground state energy $E_0^{(1)} = \langle \psi_{100} | H_1 | \psi_{100} \rangle$ by taking $R \ll a_0$.

5. A particle of mass m oscillating with frequency ω along the x-axis is subjected to the perturbation, $H_1 = Bx$. Calculate first and second order energy corrections to the nth state of the system.

6. Take the trial wave function $\psi(x) = Ae^{-\alpha x^2}$. (a) Use the normalization condition to calculate the value of A. (b) Find the expectation values of kinetic energy $\langle T \rangle$ and of potential energy $\langle V \rangle$ for a 1D harmonic oscillator. Apply the variational principle to calculate ground state energy.

7. The Hamiltonian of a particle of mass m is given by $-\dfrac{\hbar^2}{2m}\dfrac{d^2}{dx^2} - b\delta(x - x_0)$, where b is a positive constant. Take the trial wave function $\psi(x) = \left(\dfrac{2\alpha}{\pi}\right)^{1/4} e^{-\alpha(x-x_0)^2}$ to calculate the ground state energy of a particle, using the variational principle.

8. Estimate the ground state energy of a 1D harmonic oscillator using $\psi(x) = Ae^{-\beta|x|}$, with β as the variational parameter. Estimate the deviation from the exact value of the ground state energy.

9. Estimate the ground state energy for an electron in a hydrogen atom using the trial wave function $\psi(r) = A(1 + \alpha r)e^{-\alpha r}$, with α as the variational parameter.

10. A particle of mass m and positive charge q, moving along the x-axis is perturbed by $qE|x|$, where E is a uniform electric field. Apply the WKB approximation to calculate the energy levels of the particle.

8

Quantum Theory of Scattering

A major part of our knowledge on physics at the microscopic level originates from scattering experiments. For example, the concept of the atom was an outcome of the Rutherford experiment of α-particles scattering. In a typical scattering experiment, particles of an incident beam collide with a fixed target containing scattering centers, such as atoms, molecules, or nuclei. As a result of interactions between the particles in the incident beam and those in the target, some of the particles of the incoming beam are deflected and emerge from the target, and travel along a direction (θ,ϕ) with respect to the original beam direction, while others are left un-scattered and emerge out to the other side without any deflection, which is known as forward scattering. The number of deflected particles is counted in a detector of some sort; see Fig. 8.1.

In general, this type of analysis of interactions and the general scattering can be quite complicated. However, we discuss, in this chapter, the scattering of incident particles from the target under the following conditions:

1. The incident beam has point particles which are idealized spinless and structureless.
2. The interaction of the incident particles with the scattering centers is elastic, the energy of an incident particle is the same before and after scattering, the internal structure of the scattered particle does not change (particles are not excited and there is no particle production), the potential seen by the scattered particle does not change during the scattering event.
3. Each incident particle interacts with at most one scattering center (no multiple scattering). This condition can be obtained with sufficiently thin or dilute targets.
4. The interaction between an incident particle and a scattering center is described by the potential $V(\mathbf{r}_1 - \mathbf{r}_2)$ that depends only on relative positions of particles. We thus can work in the relative coordinate and reduced mass of the system.

8.1 Scattering Cross-Section and Frame of Reference

We imagine a beam of monoenergetic particles is scattered by a target located at the origin of the coordinate system $\mathbf{r} = 0$. Further, we think that the detector covers a solid angle $d\Omega$ in the direction (θ,ϕ) from the scattering center. We then choose a polar coordinate system (r,θ,ϕ), where the incoming beam of the particle travels along the z -direction $\mathbf{k} = \frac{\sqrt{2\mu E}}{\hbar}\hat{z}$, as is shown in Fig. 8.1. Let us assume that the $Nd\Omega$ particles are entering into the detector per unit time, and F is the number of particles per unit time crossing a unit area placed normal to the direction of incidence (the particle flux); then the differential scattering

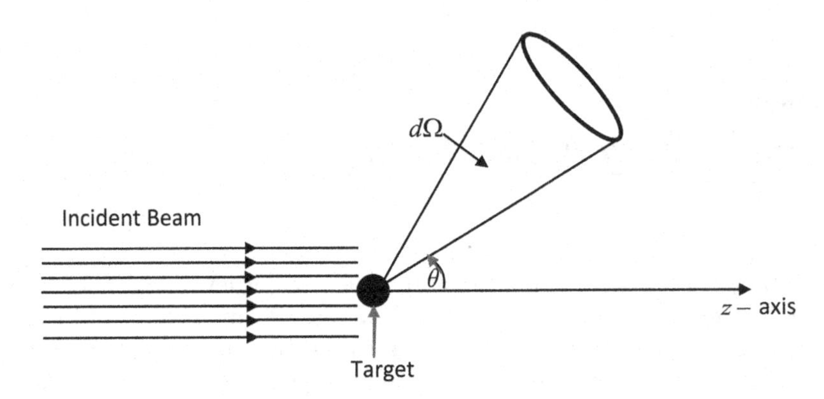

FIGURE 8.1
Schematic diagram for scattering, solid angle, $d\Omega$ in the direction of (θ,ϕ) and angle of scattering, θ.

cross-section is given by $\sigma(\theta,\phi)=\frac{d\sigma}{d\Omega}=\frac{N}{F}$. Thus, the differential scattering cross-section is the ratio of the number of particles scattered into the direction (θ,ϕ) per unit time, per unit solid angle, divided by the incident particle flux. The total scattering cross-section is given by the integral of the differential scattering cross-section over all solid angles:

$$\sigma_{tot}=\int\left(\frac{d\sigma}{d\Omega}\right)d\Omega=\int_0^{\pi}\int_0^{2\pi}\sigma(\theta,\phi)\sin\theta\,d\theta\,d\phi \tag{8.1}$$

Both the differential and the total scattering cross-sections have the dimension of the area. For spherically symmetric scattering potential, $\sigma(\theta,\phi)$ is independent of ϕ and therefore:

$$\sigma_{tot}=2\pi\int_0^{\pi}\sigma(\theta)\sin\theta\,d\theta \tag{8.2}$$

In the center of mass system, the kinematics of the reduced 1-body problem are given by the reduced mass $\mu=\frac{m_1m_2}{m_1+m_2}$ and momentum $\mathbf{p}=\frac{\mathbf{p}_1m_2-\mathbf{p}_2m_1}{m_1+m_2}$, where $\mathbf{p}_1(m_1)$ and $\mathbf{p}_2(m_2)$ are the momentum (mass) of incoming and target particles, respectively.

8.2 Asymptotic Expansion and Scattering Amplitude

We consider the scattering of a beam of particles by a fixed scattering center. Let us say that μ is reduced mass and $\mathbf{r}$ is the relative coordinate. When the beam of particles is switched on for a longer time as compared to the time one particle needs to cross the interaction area, steady-state conditions are achieved and we work out stationary solutions of the time independent Schrödinger equation:

$$\left[-\frac{\hbar^2}{2\mu}\nabla^2+V(\mathbf{r})\right]\psi(\mathbf{r})=E\psi(\mathbf{r}) \tag{8.3a}$$

The energy eigenvalue E is related to the momentum $\mathbf{p}$ and wave vector $\mathbf{k}$ of the incident particle by $E = \frac{p^2}{2\mu} = \frac{\hbar^2 k^2}{2\mu}$. Therefore:

$$\left[\nabla^2 + k^2 - \frac{2\mu}{\hbar^2} V(\mathbf{r})\right]\psi(\mathbf{r}) = 0 \tag{8.3b}$$

For potentials that asymptotically decrease faster than r^{-1} when $r \to \infty$, we can neglect $\frac{2\mu}{\hbar^2} V(\mathbf{r})$ at larger distances from the target, and then we have:

$$\left[\nabla^2 + k^2\right]\psi_{as}(\mathbf{r}) = 0 \tag{8.4}$$

which is the Helmholtz equation for a free particle. For larger $\mathbf{r}$, the wave function is decomposed into two parts: (a) ψ_{in} that represents the incoming beam and (b) ψ_{sc}, which represents the scattered beam: $\psi(\mathbf{r}) \to \psi_{in}(\mathbf{r}) + \psi_{sc}(\mathbf{r})$ for $\mathbf{r} \to \infty$. Since we have taken the z-axis as the direction of incidence, we write $\psi_{in}(\mathbf{r}) = e^{i\mathbf{k}.\mathbf{r}} = e^{ikz}$. The scattered wave function represents an outward radial flow of particles at distances far away from the scattering center. Hence $\psi_{sc}(\mathbf{r})$ can be parametrized in terms of the scattering amplitude $f(k,\theta,\phi)$ as follows:

$$\psi_{sc}(\mathbf{r}) = f(k,\theta,\phi)\frac{e^{ikr}}{r} + O\left(\frac{1}{r^{\alpha}}\right) \tag{8.5}$$

The polar coordinates (r,θ,ϕ) represent the position of a scattered particle. Thus, the asymptotic form of the wave function is:

$$\psi_{as}(\mathbf{r}) = \left(e^{i\mathbf{k}.\mathbf{r}}\right)_{as} + f(k,\theta,\phi)\frac{e^{ikr}}{r} \tag{8.6}$$

To relate the scattering amplitude to the differential cross-section, we calculate the probability current $\mathbf{S} = \frac{\hbar}{2i\mu}\left[\psi^*\nabla\psi - \psi\nabla\psi^*\right]$. The incident flux ($F$) related to the incoming plane wave part e^{ikz} is:

$$F = S_{in} = \frac{\hbar}{2i\mu}\left[e^{-ikz}\frac{\partial e^{ikz}}{\partial z} - e^{ikz}\frac{\partial e^{-ikz}}{\partial z}\right] = \frac{\hbar k}{\mu} \tag{8.7a}$$

Similarly, the scattered flux, which is related to $\psi_{sc}(r)$, is given by:

$$\begin{aligned}\mathbf{S}_{sc} &= \frac{\hbar}{2i\mu}\left[\psi_{sc}^*\nabla\psi_{sc} - \psi_{sc}\nabla\psi_{sc}^*\right] \\ &= \frac{\hbar}{2i\mu}\left[\psi_{sc}^*\left(\hat{r}\frac{\partial}{\partial r} + \hat{\theta}\frac{1}{r}\frac{\partial}{\partial\theta} + \hat{\phi}\frac{1}{r\sin\theta}\frac{\partial}{\partial\phi}\right)\psi_{sc} - \psi_{sc}\left(\hat{r}\frac{\partial}{\partial r} + \hat{\theta}\frac{1}{r}\frac{\partial}{\partial\theta} + \hat{\phi}\frac{1}{r\sin\theta}\frac{\partial}{\partial\phi}\right)\psi_{sc}^*\right]\end{aligned} \tag{8.7b}$$

where we expressed the gradient operator in terms of the spherical polar coordinates. Writing $\mathbf{S}_{sc} = \hat{r}S_r + \hat{\theta}S_\theta + \hat{\phi}S_\phi$ and then taking $\psi_{sc}(\mathbf{r}) = f(k,\theta,\phi)\frac{e^{ikr}}{r}$, we get:

$$S_r = \frac{\hbar k}{\mu r^2}\left|f(k,\theta,\phi)\right|^2 \tag{8.8a}$$

$$S_\theta = \frac{\hbar}{\mu r^3}\text{Re}\left[\frac{1}{i}f^*\frac{\partial f}{\partial \theta}\right] \tag{8.8b}$$

$$S_\phi = \frac{\hbar}{\mu r^3}\frac{1}{\sin\theta}\text{Re}\left[\frac{1}{i}f^*\frac{\partial f}{\partial \phi}\right] \tag{8.8c}$$

We find that angular components S_θ and S_ϕ become negligibly small as compared to S_r, for $r \to \infty$. Hence, we approximate $\mathbf{S}_{sc} = \hat{r}S_r$. The area of the detector is $r^2 d\Omega$ and therefore the number of particles entering into the detector is $S_r r^2 d\Omega$, which is also equal to $N d\Omega$. Therefore:

$$N d\Omega = \frac{\hbar k}{\mu r^2}|f(k,\theta,\phi)|^2 r^2 d\Omega = \frac{\hbar k}{\mu}|f(k,\theta,\phi)|^2 d\Omega \tag{8.9}$$

We thus have:

$$\sigma(k,\theta,\phi) = \frac{N}{F} = |f(k,\theta,\phi)|^2 \tag{8.10}$$

which states that the scattering amplitude is related to the experimentally observable differential scattering cross-section, which has the dimensions of $(\text{length})^2$. Thus, the scattering amplitude has the dimension of the length.

The two commonly used theoretical techniques employed to evaluate scattering amplitude are (i) partial wave analysis and (ii) Born approximation.

8.3 Partial Wave Analysis

For a spherically symmetric central potential, which is independent of θ and ϕ, the solution of the Schrödinger equation is given by $\psi_{nlm}(\mathbf{r}) = R_l(r)Y_{lm}(\theta,\phi)$, where $Y_{lm}(\theta,\phi)$ are spherical harmonics. As discussed in Section 6.1, the radial wave function $R_l(r)$ satisfies the equation:

$$-\frac{\hbar^2}{2\mu}\left[\frac{1}{r^2}\frac{d}{dr}\left(\frac{r^2 dR_l}{dr}\right) - \frac{l(l+1)R_l}{r^2}\right] + [V(r) - E]R_l(r) = 0 \tag{8.11}$$

which suggests that a general solution of Eqn. (8.3b) should be:

$$\psi(\mathbf{r}) = \sum_{l=0}^{\infty}\sum_{m=-l}^{l} R_l(r)Y_{lm}(\theta,\phi) \tag{8.12}$$

Since we have chosen the z-axis as the incoming direction, the z-component of angular momentum must be zero and hence $m = 0$. Also, $Y_{lm}(\theta,\phi) \propto P_l(\cos\theta)e^{im\phi}$, Eqn. (8.12) can be chosen to have:

$$\psi(\mathbf{r}) = \sum_{l=0}^{\infty} R_l(r)P_l(\cos\theta) \tag{8.13}$$

where $P_l(\cos\theta)$ is the Legendre polynomial of order l. Each value of l represents a partial wave in the series (Eqn. 8.13). On introducing $E = \frac{\hbar^2 k^2}{2\mu}$ and $U(r) = \frac{2\mu}{\hbar^2}V(r)$, Eqn. (8.11) reduces to:

$$\left[\frac{d^2}{dr^2} + \frac{2}{r}\frac{d}{dr} - \frac{l(l+1)}{r^2} - U(r) + k^2\right]R_l(k,r) = 0 \tag{8.14}$$

8.3.1 Free Particle and Asymptotic Solutions

We solve Eqn. (8.14) for $V(r) = 0$, which represents two scenarios: (i) a free particle solution at any value of $0 \le r < \infty$ and (ii) an asymptotic form of solution for $r \to \infty$, when potentials are of finite range. On introducing $\rho = kr$, a dimensionless variable, Eqn. (8.14) becomes:

$$\left[\frac{d^2}{d\rho^2} + \frac{2}{\rho}\frac{d}{d\rho} + \left(1 - \frac{l(l+1)}{\rho^2}\right)\right]R_l(\rho) = 0 \tag{8.15}$$

which is a spherical Bessel differential equation. As was shown in Section 6.5, the independent solutions of Eqn. (8.15) are spherical Bessel functions:

$$j_l(\rho) = (-\rho)^l \left(\frac{1}{\rho}\frac{d}{d\rho}\right)^l \frac{\sin\rho}{\rho} \tag{8.16}$$

and, spherical Neumann functions:

$$n_l(\rho) = -(-\rho)^l \left(\frac{1}{\rho}\frac{d}{d\rho}\right)^l \frac{\cos\rho}{\rho} \tag{8.17}$$

A generalized solution of Eqn. (8.15) is:

$$R_l(\rho) = A_l j_l(\rho) + B_l n_l(\rho) \tag{8.18}$$

The two limiting cases are (a) $\rho \to 0$, and (b) $\rho \to \infty$. For $\rho \to 0$, $j_l(\rho)$ and $n_l(\rho)$ are given by Eqns. (6.50) and (6.51).

Therefore at the origin ($\rho = 0$), the spherical Bessel function is the regular solution with a zero of order l, while the spherical Neumann function has a pole of order $(l+1)$ and hence it is an irregular solution, which means that the radial part of the wave function of a free particle should not include the spherical Neumann function and hence $R_l(\rho) = A_l j_l(\rho)$. Then, the complete wave function for the free particle from Eqn. (8.13) is:

$$\psi(\mathbf{r}) = \sum_{l=0}^{\infty} A_l j_l(kr) P_l(\cos\theta) \tag{8.19}$$

We have assumed that the incoming beam of free particles travels along the z-axis, and it is represented by $e^{ikz} = e^{ikr\cos\theta}$. We therefore write: $e^{ikr\cos\theta} = \sum_{l=0}^{\infty} A_l j_l(kr) P_l(\cos\theta)$ or:

$$e^{ikr\cos\theta} P_{l'}(\cos\theta)\sin\theta = \sum_{l=0}^{\infty} A_l j_l(kr) P_l(\cos\theta) P_{l'}(\cos\theta)\sin\theta \tag{8.20}$$

after multiplying both sides with $P_{l'}(\cos\theta)\sin\theta$. On integrating both sides over θ in the range of $0-\pi$, we have:

$$\int_0^{\pi} e^{ikr\cos\theta} P_{l'}(\cos\theta)\sin\theta d\theta = \sum_{l=0}^{\infty} A_l j_l(kr) \int_0^{\pi} P_l(\cos\theta)P_{l'}(\cos\theta)\sin\theta\, d\theta \tag{8.21}$$

With the use of orthogonality relations for Legendre polynomials, the right-hand side simplifies to:

$$\sum_{l=0}^{\infty} A_l j_l(kr) \int_0^{\pi} P_l(\cos\theta)P_{l'}(\cos\theta)\sin\theta\, d\theta = \sum_{l=0}^{\infty} A_l j_l(kr) \frac{2}{(2l+1)} \delta_{ll'} \tag{8.22a}$$

To simplify the left-hand side of Eqn. (8.21), we use:

$$j_0(\rho) = \frac{\sin(\rho)}{\rho} = \frac{1}{2}\int_{-1}^{1} e^{i\rho t} dt \tag{8.22b}$$

in Eqn. (8.16) to obtain:

$$j_l(\rho) = \frac{1}{2}(-\rho)^l \left(\frac{1}{\rho}\frac{d}{d\rho}\right)^l \int_{-1}^{1} e^{i\rho t} dt \tag{8.22c}$$

which on simplification with the use of the Rodrigues formula for Legendre polynomials, gives:

$$j_l(\rho) = \frac{(-i)^l}{2} \int_{-1}^{1} P_l(t) e^{i\rho t} dt \tag{8.22d}$$

the use of which in Eqn. (8.21) gives:

$$2i^{l'} j_{l'}(kr) = \sum_{l=0}^{\infty} A_l j_l(kr) \frac{2}{(2l+1)} \delta_{ll'} \tag{8.23a}$$

or,

$$A_l = i^l(2l+1) \text{ and:}$$

$$e^{ikz} = \sum_{l=0}^{\infty} i^l (2l+1) j_l(kr) P_l(\cos\theta) \tag{8.23b}$$

With the use of the theorem on spherical harmonics:

$$\left(\frac{2l+1}{4\pi}\right) P_l(cos\theta) = \sum_{m=-l}^{l} Y_{lm}^*(\theta_1,\phi_1) Y_{lm}(\theta_2,\phi_2) \tag{8.24a}$$

where θ is taken as the angle between the directions defined by (θ_1, ϕ_1) and (θ_2, ϕ_2), a more generalized expression is:

$$e^{ikz} = 4\pi \sum_{l=0}^{\infty} \sum_{m=-l}^{l} i^l j_l(kr) Y_{lm}^*(\theta_1, \phi_1) Y_{lm}(\theta_2, \phi_2) \tag{8.24b}$$

In this equation, if $\mathbf{k}$ is taken along (θ_1, ϕ_1) then $\mathbf{r}$ is along (θ_2, ϕ_2).

8.3.2 Scattering Amplitude and Phase Shift

We first analyze the asymptotic form of the radial solution given by Eqn. (8.18), in order to find out how to extract and interpret the scattering data. For $r \to \infty$:

$$\begin{aligned} j_l(kr) &\underset{r\to\infty}{\to} \frac{1}{kr} \sin\left(kr - \frac{l\pi}{2}\right) \\ n_l(kr) &\underset{r\to\infty}{\to} -\frac{1}{kr} \cos\left(kr - \frac{l\pi}{2}\right) \end{aligned} \tag{8.25}$$

Hence:

$$\begin{aligned} R_l(kr) &\underset{r\to\infty}{\to} \frac{1}{kr}\left[A_l \sin\left(kr - \frac{l\pi}{2}\right) - B_l \cos\left(kr - \frac{l\pi}{2}\right)\right] \\ \Rightarrow R_l(kr) &\underset{r\to\infty}{\to} \frac{C_l}{kr} \sin\left(kr - \frac{l\pi}{2} + \delta_l\right) \end{aligned} \tag{8.26}$$

where:

$$C_l = \sqrt{A_l^2 + B_l^2} \text{ and } \delta_l = -\tan^{-1}\left(\frac{B_l}{A_l}\right) \tag{8.27}$$

The $\delta_l(k)$ are known as phase shifts, which are real functions of k, and they completely characterize the strength of the scattering of the lth partial wave by the potential at the energy E. From Eqn. (8.13), we then have:

$$\psi(\mathbf{r}) \underset{r\to\infty}{\to} \sum_{l=0}^{\infty} \frac{C_l}{kr} \sin\left(kr - \frac{l\pi}{2} + \delta_l\right) P_l(\cos\theta) \tag{8.28}$$

To find the relation between phase shifts and scattering amplitude, we take the asymptotic form of Eqn. (8.23b) for $r \to \infty$:

$$e^{ikz} \underset{r\to\infty}{\to} \frac{1}{kr} \sum_{l=0}^{\infty} i^l (2l+1) \sin\left(kr - \frac{l\pi}{2}\right) P_l(\cos\theta) \tag{8.29}$$

and then substitute it in Eqn. (8.6) to obtain:

$$\psi_{as}(\mathbf{r}) = \frac{1}{kr}\sum_{l=0}^{\infty} i^l (2l+1)\sin\left(kr - \frac{l\pi}{2}\right)P_l(\cos\theta) + f(k,\theta,\phi)\frac{e^{ikr}}{r} \tag{8.30}$$

Since Eqns. (8.28) and (8.30) represent the same wave, we have:

$$\begin{aligned}&\frac{1}{kr}\sum_{l=0}^{\infty} i^l (2l+1)\sin\left(kr - \frac{l\pi}{2}\right)P_l(\cos\theta) + f(k,\theta,\phi)\frac{e^{ikr}}{r}\\ &= \frac{1}{kr}\sum_{l=0}^{\infty} C_l \sin\left(kr - \frac{l\pi}{2} + \delta_l\right)P_l(\cos\theta)\end{aligned} \tag{8.31}$$

Using $\sin x = \dfrac{e^{ix} - e^{-ix}}{2i}$ and comparing coefficients of e^{-ikr} on both sides of Eqn. (8.31), we obtain: $i^l(2l+1)e^{il\pi/2} = C_l e^{il\pi/2}e^{-i\delta_l}$, which gives:

$$C_l = i^l(2l+1)e^{i\delta_l} \tag{8.32}$$

Comparison of coefficients of e^{ikr} on both sides of Eqn. (8.31) gives:

$$\sum_{l=0}^{\infty} i^l \frac{(2l+1)}{2ikr} e^{-il\pi/2} P_l(\cos\theta) + \frac{f(k,\theta,\phi)}{r} = \sum_{l=0}^{\infty} \frac{C_l}{2ikr} e^{-il\pi/2} e^{i\delta_l} P_l(\cos\theta) \tag{8.33a}$$

which with the use of Eqn. (8.32) yields:

$$f(k,\theta,\phi) = \frac{1}{2ik}\sum_{l=0}^{\infty} i^l (2l+1) e^{-il\pi/2}\left(e^{2i\delta_l} - 1\right)P_l(\cos\theta) \tag{8.33b}$$

which is independent of ϕ. We write $i^l = (e^{i\pi/2})^l = e^{i\pi l/2}$ to obtain:

$$\begin{aligned}f(k,\theta) &= \frac{1}{k}\sum_{l=0}^{\infty}(2l+1)e^{i\delta_l}\left(\frac{e^{i\delta_l} - e^{-i\delta_l}}{2i}\right)P_l(\cos\theta)\\ &= \frac{1}{k}\sum_{l=0}^{\infty}(2l+1)e^{i\delta_l}\sin(\delta_l)P_l(\cos\theta)\end{aligned} \tag{8.34}$$

The differential scattering cross-section defined by Eqn. (8.10) is then given by:

$$\begin{aligned}\sigma(k,\theta) &= |f(k,\theta)|^2\\ &= \frac{1}{k^2}\sum_{l=0}^{\infty}(2l+1)e^{i\delta_l}\sin(\delta_l)P_l(\cos\theta)\sum_{l'=0}^{\infty}(2l'+1)e^{-i\delta_{l'}}\sin(\delta_{l'})P_{l'}(\cos\theta)\end{aligned} \tag{8.35}$$

Therefore, the total scattering cross-section, Eqn. (8.2), is:

$$\begin{aligned}\sigma_{tot}(k) &= \frac{2\pi}{k^2}\sum_{l=0}^{\infty}(2l+1)e^{i\delta_l}\sin(\delta_l)\sum_{l'=0}^{\infty}(2l'+1)e^{-i\delta_{l'}}\sin(\delta_{l'})\int_0^{\pi}P_l(\cos\theta)P_{l'}(\cos\theta)\sin\theta d\theta \\ &= \frac{2\pi}{k^2}\sum_{l=0}^{\infty}(2l+1)e^{i\delta_l}\sin(\delta_l)\sum_{l'=0}^{\infty}(2l'+1)e^{-i\delta_{l'}}\sin(\delta_{l'})\frac{2}{(2l'+1)}\delta_{ll'} \\ &= \frac{4\pi}{k^2}\sum_{l=0}^{\infty}(2l+1)\sin^2(\delta_l)\end{aligned} \tag{8.36}$$

If we write $\sigma_{tot}(k) = \sum_{l=0}^{\infty}\sigma_l(k)$ then:

$$\sigma_l(k) = \frac{4\pi}{k^2}(2l+1)\sin^2(\delta_l) \tag{8.37}$$

It is to be noted from Eqn. (8.35) that the differential scattering cross-section has interference terms belonging to $l \neq l'$, which vanish in total scattering because of the orthogonality properties of Legendre polynomials.

8.3.3 Optical Theorem

For $\theta = 0$, $P_l(1) = 1$ and Eqn. (8.34) give:

$$\begin{aligned}f(k,0) &= \frac{1}{k}\sum_{l=0}^{\infty}(2l+1)e^{i\delta_l}\sin(\delta_l) \\ \Rightarrow Im\{f(k,0)\} &= \frac{1}{k}\sum_{l=0}^{\infty}(2l+1)\sin^2(\delta_l)\end{aligned} \tag{8.38a}$$

which suggest that:

$$\sigma_{tot}(k) = \frac{4\pi}{k}Im\{f(k,0)\} \tag{8.38b}$$

In analogy with the relation between the absorption coefficient and the imaginary part of the complex refraction index in optics, Eqn. (8.38b) is known as the optical theorem.

8.3.4 Scattering Length

As is seen from Eqn. (8.34), $f(k,\theta)$ has the dimensions of the length. We write:

$$f(k,\theta) = \sum_{l=0}^{\infty}f_l(k,\theta) \tag{8.39a}$$

with:

$$f_l(k,\theta) = \frac{1}{k}(2l+1)e^{i\delta_l}\sin(\delta_l)P_l(\cos\theta) \tag{8.39b}$$

Thus, the scattering amplitude is contributed by different partial waves corresponding to $l = 0,1,2,3......$ The most prominent contribution comes from the s-wave ($l = 0$). When energy E is low and the range of the scattering potential is also small, one neglects all contributions to $f(k,\theta)$, except that coming from $l = 0$. In that case:

$$f_0(k,\theta) = \frac{1}{k}e^{i\delta_0}\sin(\delta_0) \tag{8.40a}$$

The energy for which Eqn. (8.40a) holds good is known as zero energy and the limiting value of $-f_0(k,\theta)$ for $E \to 0$ or $k \to 0$ is called the scattering length:

$$a_s = \lim_{k\to 0}\left(-\frac{1}{k}e^{i\delta_0}\sin\delta_0\right) \tag{8.40b}$$

Since δ_0 is very small, $e^{i\delta_0} \approx 1$ and $\sin\delta_0 \approx \delta_0$, we therefore write $a_s = \lim_{k\to 0}\left(-\frac{\delta_0}{k}\right)$. Thus the zero energy scattering cross-section, involving the s-wave scattering only, is $\sigma_0 = 4\pi a_s^2 = \frac{4\pi}{k^2}\delta_0^2$ and $\delta_0 = -ka_s$.

8.3.5 Scattering by a Square Well Potential

We consider a centrally symmetric square well potential for which the phase shifts can be calculated by analytical methods:

$$V(r) = \begin{cases} -V_0 & r \le a \\ 0 & r > a \end{cases} \tag{8.41}$$

Inside the potential well, radial Eqn. (8.14) becomes:

$$\left[\frac{d^2}{dr^2} + \frac{2}{r}\frac{d}{dr} - \frac{l(l+1)}{r^2} + U_0 + k^2\right]R_l(r) = 0 \tag{8.42}$$

where $U_0 = \frac{2\mu V_0}{\hbar^2}$. Since $r \to 0$ is in region I, by taking $K = \sqrt{k^2 + U_0}$, the solution of Eqn. (8.42) is given by:

$$R_l^I(Kr) = A_l j_l(Kr) \tag{8.43}.$$

For $(r > a)$, region II, the radial equation is:

$$\left[\frac{d^2}{dr^2} + \frac{2}{r}\frac{d}{dr} - \frac{l(l+1)}{r^2} + k^2\right]R_l(r) = 0 \tag{8.44}$$

for which the generalized solution is:

$$R_l^{II}(kr) = B_l j_l(kr) + C_l n_l(kr) \tag{8.45}$$

For continuity of the radial part of the wave function and its derivative, we should have:

$$\begin{aligned} R_l^I(Ka) = R_l^{II}(ka) \quad &\Rightarrow \quad A_l j_l(Ka) = B_l j_l(ka) + C_l n_l(ka) \\ \left.\frac{dR_l^I(Kr)}{dr}\right|_{r=a} = \left.\frac{dR_l^{II}(kr)}{dr}\right|_{r=a} \quad &\Rightarrow \quad A_l K j_l'(Ka) = k\left[B_l j_l'(ka) + C_l n_l'(ka)\right] \end{aligned} \tag{8.46}$$

which yields:

$$\frac{K j_l'(Ka)}{j_l(Ka)} = \frac{k\left[j_l'(ka) + \frac{C_l}{B_l} n_l'(ka)\right]}{j_l(ka) + \frac{C_l}{B_l} n_l(ka)} \tag{8.47}$$

Now, by taking $\tan(\delta_l) = -\frac{C_l}{B_l}$ we get:

$$\tan\delta_l = \frac{k j_l'(ka) j_l(Ka) - K j_l'(Ka) j_l(ka)}{k n_l'(ka) j_l(Ka) - K j_l'(Ka) n_l(ka)} \tag{8.48}$$

For the low energy scattering case ($ka \ll 1$), $j_l(\rho) \propto \rho^l; j_l'(\rho) \propto \rho^{l-1}$ and $n_l(\rho) \propto \rho^{-l-1}$; $n_l'(\rho) \propto \rho^{-l-2}$. We thus find that $E \to 0$, $\tan\delta_l$ goes to zero as a constant time k^{2l+1}.

As stated earlier, the most prominent contribution to the scattering cross-section comes from the s-wave ($l = 0$). For $l = 0$, the Eqns. (8.43) and (8.45) reduce to:

$$\begin{aligned} R_0^I(Kr) &= A_0 \frac{\sin(Kr)}{Kr} \\ &\Rightarrow rR_0^I(Kr) = A\sin(Kr) \end{aligned} \tag{8.49a}$$

and:

$$\begin{aligned} R_0^{II}(kr) &= B_0 \frac{\sin(kr)}{kr} - C_0 \frac{\cos(kr)}{kr} \\ &= \frac{B_0}{kr}[\sin(kr) + \tan\delta_0 \cos(k \\ &\Rightarrow rR_0^{II}(kr) = B\sin(kr + \delta_0) \end{aligned} \tag{8.49b}$$

Matching of two solutions at $r = a$ gives:

$$A\sin(Ka) = B\sin(ka + \delta_0) \tag{8.50a}$$

The matching of first-order derivatives of $R_0^I(Kr)$ and $R_0^{II}(kr)$ at $r = a$ yields $KA\cos(Ka) - A\sin(Ka) = kB\cos(kr + \delta_0) - B\sin(ka + \delta_0)$, which with the use of (8.50a) simplifies to:

$$KA\cos(Ka) = kB\cos(kr + \delta_0) \qquad (8.50b).$$

Dividing Eqn. (8.50b) by (8.50a), we get:

$$k\tan(Ka) = K\tan(ka + \delta_0) \qquad (8.51).$$

Therefore:

$$\delta_0 = \tan^{-1}\left[\frac{k}{K}\tan(Ka)\right] - ka \qquad (8.52)$$

or:

$$\tan\delta_0 = \frac{k\tan(Ka) - K\tan(ka)}{K + k\tan(Ka)\tan(ka)} \qquad (8.53)$$

for very low energy scattering $ka \ll 1$ and $\frac{k}{K} \ll 1$. We then write:

$$\tan\delta_0 \approx \left\{\frac{\frac{k}{K}\tan(Ka) - ka}{1 + \frac{k^2 a}{K}\tan(Ka)}\right\} \approx \frac{k}{K}\tan(Ka) - ka$$
$$\Rightarrow \tan\delta_0 \approx ka\left[\frac{\tan(Ka)}{Ka} - 1\right] \qquad (8.54)$$

which suggests that $\tan\delta_0 \propto k$ for $k \to 0$. Thus, for a square well:

$$a_s = -\lim_{k\to 0}\frac{\delta_0}{k} = a\left[1 - \frac{\tan\left(\sqrt{U_0}\,a\right)}{\sqrt{U_0}\,a}\right] \qquad (8.55)$$

Equation (8.37) for $l \to 0$ gives:

$$\sigma_0(k) = \frac{4\pi}{k^2}\sin^2(\delta_0) = \frac{4\pi}{k^2}\left(\frac{1}{1 + \cot^2\delta_0}\right) \qquad (8.56)$$

and:

$$\sigma_0(k) \underset{k\to 0}{\rightarrow} \frac{4\pi}{k^2}\left(\frac{1}{1 + 1/\delta_0^2}\right) \approx 4\pi\frac{\delta_0^2}{k^2} = 4\pi a_s^2 \qquad (8.57)$$

To define the effective range r_0 of potential $k\cot\delta_0$ is expanded as:

$$k\cot\delta_0 \approx \frac{k}{\delta_0} - \frac{k\delta_0}{3} - \frac{k\delta_0^3}{45} + \ldots.. \tag{8.58a}$$

which with the use of Eqn. (8.55) is written as:

$$k\cot\delta_0 \approx -\frac{1}{a_s} + \frac{1}{2}r_0k^2 + \ldots\ldots. \tag{8.58b}$$

The short-range potential scattering can then be expressed in terms of a_s and r_0.

8.3.6 Scattering by a Hard Sphere Potential

The problem of hard sphere potential can also be solved exactly. The hard sphere potential is defined as:

$$V(r) = \begin{cases} \infty & 0 \le r \le a \\ 0 & r > a \end{cases} \tag{8.59}$$

Since a particle cannot penetrate into the region of $0 \le r \le a$, wave function in this region has to be zero. Let us study the problem in two parts.

a. **Low Energy Limit**
For low energy, most of the contribution comes from s-waves ($l = 0$), the solution of the radial equation for $r > a$ can be given by Eqn. (8.49b):

$$\chi_0(r) = A\sin(kr + \delta_0) \tag{8.60}$$

where we have used $\chi_0 = rR_0$. Matching of wave function at the surface of the sphere requires that $\chi_0(r = a) = 0$. Hence, we have $A\sin(ka + \delta_0) = 0$, which implies that $ka + \delta_0 = n\pi$, where n is an integer. Since we are ignoring the contributions from the partial wave having $l \neq 0$, $\sigma_{tot} = \sigma_0$ and therefore:

$$\begin{aligned} \sigma_{tot} &= \frac{4\pi}{k^2}\sin^2\delta_0 = \frac{4\pi}{k^2}\sin^2(n\pi - ka) \\ &= \frac{4\pi}{k^2}\sin^2(ka) \end{aligned} \tag{8.61}$$

Since $ka \ll 1$, we take $\sin(x) \approx x$ and obtain:

$$\sigma_{tot} = 4\pi a^2 \tag{8.62}$$

We see that a_s coincides with a for $k \to 0$, and the scattering cross-section is four times its classical value.

b. **High Energy Limit**
In this limit $ka \gg 1$ and therefore the partial wave other than $l = 0$ would also contribute. Also, the incident particles will be scattered if $l \le ka$. The magnitude of

angular momentum $|\mathbf{L}|$ of the particle in the incident beam is related to its linear momentum and impact parameter b (perpendicular distance between the scattering center and the incident direction) and hence one can write $|\mathbf{L}| = |\mathbf{p}|b$. Further, $b \le a$ where a is the range of scattering potential. Therefore, if we represent $|\mathbf{L}|$ by $\sqrt{l(l+1)}\hbar \sim l\hbar$ and $|\mathbf{p}|$ by $\hbar k$ then $l \le ka$. We thus have:

$$\sigma_{tot}(k) = \frac{4\pi}{k^2}\sum_{l=0}^{ka}(2l+1)\sin^2(\delta_l) \tag{8.63}$$

The radial part of the Schrödinger equation for $r > a$ is given by Eqn. (8.44), for which the asymptotic solution is given by Eqn. (8.26), which can be written as:

$$\chi_l(r) = B_l \sin\left(kr - \frac{l\pi}{2} + \delta_l\right) \tag{8.64}$$

The matching of the wave function at $r = a$ demands $\sin\left(ka - \frac{l\pi}{2} + \delta_l\right) = 0$. One of the solutions is $\delta_l = l\pi/2 - ka$. Therefore:

$$\sigma_{tot}(k) = \frac{4\pi}{k^2}\sum_{l=0}^{ka}(2l+1)\sin^2(ka - l\pi/2) \tag{8.65}$$

Since $ka \gg 1$, summation over l is replaced by an integration:

$$\sigma_{tot}(k) = \frac{4\pi}{k^2}\int_0^{ka}(2l+1)\sin^2(ka - l\pi/2)dl \tag{8.66}$$

The integral is simplified further on replacing $\sin^2(ka - l\pi/2)$ by its average value 1/2 to get:

$$\sigma_{tot}(k) = \frac{2\pi}{k^2}\int_0^{ka}(2l+1)dl \tag{8.67a}$$

which simplifies to:

$$\sigma_{tot} \approx 2\pi a^2 \tag{8.67b}$$

When $ka \gg 1$, the wave lengths of the scattered particles tend to 0 and one expects to observe that classical area $\pi a^2 = \sigma_{tot}$. We however find the quantum mechanical value is two times the classical value, which is in accordance with refraction phenomena observed in optics. It is attributed to interference between the incoming and the scattered beam near the forward direction, which is known as *shadow scattering*.

8.3.7 Interpretation of the Phase Shift

When scattering potential is weak and it varies slowly, we say that phase shift is arising from the change in the effective wave vector $k = \sqrt{2\mu(E - V(r)}/\hbar$ due to the presence of the potential. Thus, we expect an advanced oscillation and a positive phase shift ($\delta_l > 0$) for an attractive potential, while we expect that a repulsive potential should lead to retarded oscillation and a negative phase ($\delta_l < 0$). In the case of the square well potential discussed above, we find that Eqn. (8.55), on using $\tan(x) \approx x + \frac{1}{3}x^3$ for small values of U_0, gives $a_s = -\frac{1}{3}a^3 U_0$, which is negative. And Eqn. (8.54) yields a positive value of δ_0 for the attractive potential ($U_0 > 0$). For ($U_0 < 0$), Eqns. (8.54) and (8.55) give a positive value of a_s and negative value of the phase shift, which are in accordance with our expectations. Small angular momenta dominate the scattering at low energies and therefore $\sigma_l(k)$ is negligible when $l > ka$, where a is the range of the potential. The partial scattering cross-section is:

$$\sigma_l(k) = \frac{4\pi}{k^2}(2l+1)\sin^2(\delta_l) = \frac{4\pi}{k^2}(2l+1)\left(\frac{1}{1+\cot^2\delta_l}\right) \tag{8.68}$$

which goes through the maximum and zero for δ_l becoming an odd and even multiple of $\frac{\pi}{2}$, respectively, on increasing the energy. For small energies, σ_0 dominates so that we can get minima, where the target becomes almost transparent. This is called the **Ramsauer-Townsend effect**. If there is a bound state as the energy tends to zero and the phase shift goes through $\frac{\pi}{2}$, the cross-section attains maximum value. This is known as the **resonance effect**. We next investigate the cross-section near the resonance. Let us say that the resonance occurs at energy E_r. The phase shift will be a function of E_r. Since phase shift at resonance is $\pi/2$, we have $\sin\delta_l(E_r) = 1$ and $\cos\delta_l(E_r) = 0$. The Taylor series expansion near E_r gives:

$$\begin{aligned}
\sin\delta_l(E) &\simeq \sin\delta_l(E_r) + \left(\frac{\partial \sin\delta_l}{\partial E}\right)_{E=E_r}(E-E_r) + \ldots\ldots \\
&\simeq \sin\delta_l(E_r) + \left(\cos\delta_l(E)\frac{\partial\delta_l}{\partial E}\right)_{E=E_r}(E-E_r) + \ldots\ldots \\
&\simeq \sin\delta_l(E_r) + \cos\delta_l(E_r)\left(\frac{\partial\delta_l}{\partial E}\right)_{E=E_r}(E-E_r) + \ldots\ldots
\end{aligned} \tag{8.69a}$$

and:

$$\begin{aligned}
\cos\delta_l(E) &\simeq \cos\delta_l(E_r) + \left(\frac{\partial \cos\delta_l}{\partial E}\right)_{E=E_r}(E-E_r) + \ldots\ldots \\
&\simeq \cos\delta_l(E_r) - \left(\sin\delta_l(E)\frac{\partial\delta_l}{\partial E}\right)_{E=E_r}(E-E_r) + \ldots\ldots \\
&\simeq \cos\delta_l(E_r) - \sin\delta_l(E_r)\left(\frac{\partial\delta_l}{\partial E}\right)_{E=E_r}(E-E_r) + \ldots\ldots
\end{aligned} \tag{8.69b}$$

Let us define:

$$\frac{2}{\Gamma} = \left.\frac{\partial\delta_l(E)}{\partial E}\right|_{E=E_r} \tag{8.70a}$$

Then putting $\sin\delta_l(E_r)=1$ and $\cos\delta_l(E_r)=0$ in Eqns. (8.69), we obtain:

$$\cot\delta_l(E) \simeq -\frac{2}{\Gamma}(E-E_r) \tag{8.70b}$$

which on substituting in Eqn. (8.68) gives:

$$\sigma_l(k) = \frac{4\pi}{k^2}(2l+1)\left(\frac{\Gamma^2/4}{\Gamma^2/4+(E-E_r)^2}\right) \tag{8.71}$$

This is the **Breit-Wigner formula** describing the cross-section for resonance scattering. The Γ represents the width of the resonance curve.

The dwelling time of the scattered particles in the interaction region is related to the inverse width Γ^{-1}. It is to be noted that the maximum of σ_l at a resonance is determined by the momentum of the scattered particles, and it is not the properties of the target. Resonances can also be interpreted as poles in the scattering amplitudes, which occur near the real axis and the imaginary part related to the lifetime. On the other hand, poles on the positive imaginary axis correspond to bound states for the potential. The phase shift also consists of the information on the number of such bound states.

8.4 Expression for Phase Shift

The radial Schrödinger Eqn. (8.14) that describes the scattering can be rewritten as:

$$\frac{d^2\chi_l(r)}{dr^2} + \left[k^2 - U(r) - \frac{l(l+1)}{r^2}\right]\chi_l(r) = 0 \tag{8.72}$$

where $\chi_l(r) = rR_l(r)$ and $k^2 = \frac{2\mu E}{\hbar^2}$. In the incident region $U(r)=0$ and we get:

$$\frac{d^2\chi_l^{in}(r)}{dr^2} + \left[k^2 - \frac{l(l+1)}{r^2}\right]\chi_l^{in}(r) = 0 \tag{8.73}$$

for which the solution is:

$$\chi_l^{in}(r) = krj_l(kr) \tag{8.74}$$

As discussed in Section 8.3.2, the asymptotic form of it is:

$$\chi_l^{in}(r) \underset{r\to\infty}{\longrightarrow} \sin(kr - l\pi/2) \tag{8.75}$$

Similarly, the asymptotic solution of Eqn. (8.72) for $U(r) \neq 0$ is:

$$\chi_l^{sc}(r) \underset{r\to\infty}{\rightarrow} \sin(kr - l\pi/2 + \delta_l) \tag{8.76}$$

Multiply Eqn. (8.73) by $\chi_l^{sc}(r)$ and Eqn. (8.72) by $\chi_l^{in}(r)$, after replacing $\chi_l(r)$ by $\chi_l^{sc}(r)$. Then subtracting Eqn. (8.73) from Eqn. (8.72), gives:

$$\chi_l^{in}(r)\frac{d^2\chi_l^{sc}(r)}{dr^2} - \chi_l^{sc}(r)\frac{d^2\chi_l^{in}(r)}{dr^2} = \chi_l^{in}(r)U(r)\chi_l^{sc}(r) \tag{8.77}$$

Integrate both sides in between the limits of 0 to r to get:

$$\chi_l^{in}(r)\frac{d\chi_l^{sc}(r)}{dr} - \chi_l^{sc}(r)\frac{d\chi_l^{in}(r)}{dr} = \int_0^r \chi_l^{in}(r)U(r)\chi_l^{sc}(r)dr \tag{8.78}$$

In the asymptotic limit of $r \to \infty$, substitute Eqns. (8.75) and (8.76) into Eqn. (8.78) to obtain:

$$k\left[\sin(kr - l\pi/2)\cos(kr - l\pi/2 + \delta_l) - \sin(kr - l\pi/2 + \delta_l)\cos(kr - l\pi/2)\right]$$

$$= \int_0^\infty \chi_l^{in}(r)U(r)\chi_l^{sc}(r)dr$$

or:

$$\sin\delta_l(k) = -\frac{1}{k}\int_0^\infty \chi_l^{in}(r)U(r)\chi_l^{sc}(r)dr \tag{8.79}$$

which is an exact expression for phase shift. At high energies and weak potential, phase shift is small and therefore one uses $\chi_l^{in}(kr) \cong \chi_l^{sc}(kr) \cong krj_l(kr)$. Then:

$$\sin\delta_l(k) \approx \delta_l(k) = -\frac{2\mu k}{\hbar^2}\int_0^\infty V(r)\left(j_l(kr)\right)^2 r^2 dr \tag{8.80}$$

Equation (8.80) is known as the *Born approximation for phase shift*, which is valid only for weak potentials. It is obvious from this equation that phase shift is positive for attractive potential and negative for repulsive potential.

8.5 Integral Equation

Equation (8.3b), which describes the scattering, can also be solved by the method of Green functions. The Green function $G_0(k,\mathbf{r},\mathbf{r}')$ is the solution of the Helmholtz equation:

$$\left(\nabla^2 + k^2\right)G_0(k,\mathbf{r},\mathbf{r}') = \delta(\mathbf{r} - \mathbf{r}') \tag{8.81}$$

We want the solution $\psi(\mathbf{r})$ of Eqn. (8.3b) to be such that in the asymptotic limit $\psi(\mathbf{r}) \to e^{i\mathbf{k}.\mathbf{r}} + \psi_{sc}(\mathbf{r})$, the substitution for which in Eqn. (8.3b) gives:

$$\left(\nabla^2 + k^2\right)\left(e^{i\mathbf{k}.\mathbf{r}} + \psi_{sc}(\mathbf{r})\right) = U(\mathbf{r})\psi(\mathbf{r}) \tag{8.82}$$

Since $\left(\nabla^2 + k^2\right)e^{i\mathbf{k}.\mathbf{r}} = 0$, we obtain:

$$\left(\nabla^2 + k^2\right)\psi_{sc}(\mathbf{r}) = U(\mathbf{r})\psi(\mathbf{r}) \tag{8.83}$$

which is an inhomogeneous equation, where the inhomogeneous term itself depends on $\psi(\mathbf{r})$. A formal solution of (8.83) is:

$$\psi_{sc}(\mathbf{r}) = \int G_0(k, \mathbf{r}, \mathbf{r}')U(\mathbf{r}')\psi(\mathbf{r}')d^3r' \tag{8.84}$$

Since, there is translation invariance, we write $G_0(k, \mathbf{r}, \mathbf{r}') = G_0(k, \mathbf{r} - \mathbf{r}') = G_0(k, \mathbf{x})$, where we have defined $\mathbf{x} = \mathbf{r} - \mathbf{r}'$. Equation (8.81) is a linear equation and therefore one determines the Green function with the use of Fourier transforms:

$$G_0(k, \mathbf{x}) = \frac{1}{(2\pi)^3}\int g_0(k, k')e^{i\mathbf{k}'.\mathbf{X}}d^3k' \tag{8.85a}$$

and:

$$\delta(\mathbf{x}) = \frac{1}{(2\pi)^3}\int e^{i\mathbf{k}'.\mathbf{X}}d^3k' \tag{8.85b}$$

On substituting Eqns. (8.85) into Eqn. (8.81), we get:

$$\begin{aligned}&\left(\nabla^2 + k^2\right)\int g_0(k, k')e^{i\mathbf{k}'.\mathbf{X}}d^3k' = \int e^{i\mathbf{k}'.\mathbf{X}}d^3k' \\ &\Rightarrow \int\left\{(k^2 - k'^2)g_0(k, k') - 1\right\}e^{i\mathbf{k}'.\mathbf{X}}d^3k' = 0\end{aligned} \tag{8.86}$$

which yields:

$$g_0(k, k') = \frac{1}{(k^2 - k'^2)} \tag{8.87}$$

From Eqn. (8.85a) we then have:

$$G_0(k, \mathbf{x}) = \frac{1}{(2\pi)^3}\int \frac{e^{i\mathbf{k}'.\mathbf{X}}}{(k^2 - k'^2)}d^3k' \tag{8.88}$$

To simplify this, we choose spherical polar coordinates where the polar axis is along-**x**.

$$G_0(k,\mathbf{x}) = \frac{1}{(2\pi)^3}\int_0^\infty\int_0^\pi\int_0^{2\pi}\frac{e^{ik'|\mathbf{x}|\cos\theta}}{(k^2-k'^2)}k'^2dk'\sin\theta d\theta d\phi$$

$$= \frac{1}{(2\pi)^2}\int_0^\infty\int_0^\pi\frac{e^{ik'|\mathbf{x}|\cos\theta}}{(k^2-k'^2)}k'^2dk'\sin\theta d\theta$$

$$= \frac{1}{(2\pi)^2}\int_0^\infty\int_{-1}^1\frac{e^{ik'|\mathbf{x}|t}}{(k^2-k'^2)}k'^2dk'dt \qquad (8.89)$$

$$= \frac{1}{(2\pi)^2|\mathbf{x}|}\int_0^\infty\frac{\left(e^{ik'|\mathbf{x}|}-e^{ik'|\mathbf{x}|}\right)}{ik'(k^2-k'^2)}k'^2dk'$$

$$= \frac{2}{(2\pi)^2|\mathbf{x}|}\int_0^\infty\frac{\sin(k'|\mathbf{x}|)}{(k^2-k'^2)}k'dk'$$

Since, the integrand is an even function of k', we write:

$$G_0(k,\mathbf{x}) = \frac{1}{(2\pi)^2|\mathbf{x}|}\int_{-\infty}^{\infty}\frac{\sin(k'|\mathbf{x}|)}{(k^2-k'^2)}k'dk'$$

$$= \frac{1}{2(2\pi)^2|\mathbf{x}|}\int_{-\infty}^{\infty}\sin(k'|\mathbf{x}|)\left[\frac{1}{k-k'}-\frac{1}{k+k'}\right]dk' \qquad (8.90)$$

$$= -\frac{1}{4i(2\pi)^2|\mathbf{x}|}\left[\int_{-\infty}^{\infty}e^{ik'|\mathbf{x}|}\left\{\frac{1}{k'+k}+\frac{1}{k'-k}\right\}dk' - \int_{-\infty}^{\infty}e^{-ik'|\mathbf{x}|}\left\{\frac{1}{k'+k}+\frac{1}{k'-k}\right\}dk'\right]$$

The integrands of two integrals have poles on the real axis at $k' = \pm k$. The integrals are performed with the use of the Cauchy integral formula:

$$\frac{1}{2\pi i}\oint\frac{f(z)dz}{z-a} = f(a) \qquad (8.91a)$$

The requirement to use the Cauchy formula is that the contour for integration be closed in such a manner that it encloses $f(z)$. Addition of an infinitesimally small imaginary part $i\alpha$ to k makes it clear that the contour which encloses $e^{ik'|\mathbf{X}|}$ has the pole or singularity at $k' = k$, while, the contour for the other integral with the integrand having $e^{-ik'|\mathbf{X}|}$ encloses the pole at $k' = -k$. Therefore:

$$\int_{-\infty}^{\infty}e^{ik'|\mathbf{x}|}\left\{\frac{1}{k'+k}+\frac{1}{k'-k}\right\}dk' = 2\pi i e^{ik|\mathbf{x}|} \qquad (8.91b)$$

and:

$$\int_{-\infty}^{\infty} e^{-ik'|\mathbf{x}|}\left\{\frac{1}{k'+k}+\frac{1}{k'-k}\right\}dk' = -2\pi i e^{ik|\mathbf{x}|} \tag{8.91c}$$

which yield:

$$\begin{aligned} G_0(k,\mathbf{r},\mathbf{r}') &= -\frac{1}{4i(2\pi)^2|\mathbf{x}|}\left(2\pi i e^{ik|\mathbf{x}|}+2\pi i e^{ik|\mathbf{x}|}\right) \\ &= -\frac{e^{ik|\mathbf{r}-\mathbf{r}'||\mathbf{r}-\mathbf{r}'|}}{4\pi|\mathbf{r}-\mathbf{r}'|} \end{aligned} \tag{8.92a}$$

and:

$$\psi_{sc}(\mathbf{r}) = -\frac{1}{4\pi}\int \frac{e^{ik|\mathbf{r}-\mathbf{r}'|}}{|\mathbf{r}-\mathbf{r}'|}U(\mathbf{r}')\psi(\mathbf{r}')d^3r' \tag{8.92b}$$

We thus obtain:

$$\psi(\mathbf{r}) = e^{i\mathbf{k}.\mathbf{r}} - \frac{1}{4\pi}\int \frac{e^{ik|\mathbf{r}-\mathbf{r}'|}}{|\mathbf{r}-\mathbf{r}'|}U(\mathbf{r}')\psi(\mathbf{r}')d^3r' \tag{8.93}$$

The integral Eqn. (8.93) is also known as the **Lippmann-Schwinger equation** for potential scattering, which is equivalent to Eqn. (8.6). Hence, we can find an expression for scattering amplitude by considering the situation where the distance of the detector $r \to \infty$, is much larger than the range of the potential to which the integration variable ($\mathbf{r}'$) is essentially confined so that $r' \ll r$. Therefore:

$$|\mathbf{r}-\mathbf{r}'| = \sqrt{r^2 - 2\mathbf{r}.\mathbf{r}' + r'^2} \approx r - \frac{\mathbf{r}.\mathbf{r}'}{r} + \frac{1}{2}\left(\frac{r'}{r}\right)^2 + \ldots\ldots \tag{8.94}$$

Let $\hat{n}$ be the unit vector in the direction of r ($\mathbf{r} = \hat{n}r$); therefore $|\mathbf{r}-\mathbf{r}'| \approx r - \hat{n}.\mathbf{r}'$, after neglecting the terms of the order of $1/r^2$. Further, for elastic scattering, the momentum of the incident particle is the same as that of scattered particle; we take $k|\mathbf{r}-\mathbf{r}'| \approx kr - \mathbf{k}'.\mathbf{r}'$ and:

$$\begin{aligned} \frac{1}{|\mathbf{r}-\mathbf{r}'|} &= \frac{1}{r}\left(1 - 2\frac{\hat{n}.\mathbf{r}'}{r} + \frac{r'^2}{r^2}\right)^{-1/2} \approx \frac{1}{r} + \frac{\hat{n}.\mathbf{r}'}{r^2} - \frac{r'^2}{2r^3} + \ldots\ldots \\ &\approx \frac{1}{r} \end{aligned} \tag{8.95a}$$

on neglecting the terms of the order of $1/r^2$ or smaller. Hence:

$$\frac{e^{ik|\mathbf{r}-\mathbf{r}'|}}{|\mathbf{r}-\mathbf{r}'|} \underset{r\to\infty}{\to} \frac{e^{ikr}}{r}e^{-i\mathbf{k}'.\mathbf{r}'} + \ldots\ldots.. \tag{8.95b}$$

which yields:

$$\psi(\mathbf{r}) \underset{r\to\infty}{\to} e^{i\mathbf{k}.\mathbf{r}} - \frac{1}{4\pi}\frac{e^{ikr}}{r}\int e^{-i\mathbf{k}'.\mathbf{r}'}U(\mathbf{r}')\psi(\mathbf{r}')d^3r' \tag{8.96}$$

Comparing with Eqn. (8.6), we get:

$$\begin{aligned} f(k,\theta,\phi) &= -\frac{1}{4\pi}\int e^{-i\mathbf{k}'.\mathbf{r}'}U(\mathbf{r}')\psi(\mathbf{r}')d^3r' \\ &= -\frac{\mu}{2\pi\hbar^2}\int e^{-i\mathbf{k}'.\mathbf{r}'}V(\mathbf{r}')\psi(\mathbf{r}')d^3r' \end{aligned} \tag{8.97}$$

8.6 The Born Approximation

The Lippmann-Schwinger Eqn. (8.93) is an integral transform of the differential Eqn. (8.3b) with the help of Green functions. In this equation, evaluation of $\psi(\mathbf{r})$ requires $\psi(\mathbf{r}')$. Born used the iterative method to get $\psi(\mathbf{r})$. In the first Born approximation, $\psi(\mathbf{r}')$ can be replaced by the incoming wave function form $e^{i\mathbf{k}.\mathbf{r}'}$ in Eqn. (8.93). This will lead to an improved value of $\psi(\mathbf{r})$ that will differ from $e^{i\mathbf{k}.\mathbf{r}}$. This improved value will again be used in Eqn. (8.93), after replacing $\mathbf{r}$ by $\mathbf{r}'$ and so on. The iterative method should continue until input and out wave functions are almost identical. As is obvious, the procedure gets complicated on moving to higher order approximations. On confining to the first order only, we write:

$$\begin{aligned} f(k,\theta,\phi) &= -\frac{\mu}{2\pi\hbar^2}\int e^{-i\mathbf{k}'.\mathbf{r}'}V(\mathbf{r}')e^{i\mathbf{k}.\mathbf{r}'}d^3r' \\ &= -\frac{\mu}{2\pi\hbar^2}\int e^{i(\mathbf{k}-\mathbf{k}').\mathbf{r}'}V(\mathbf{r}')d^3r' \end{aligned} \tag{8.98}$$

The $\mathbf{k}$ and $\mathbf{k}'$ represent the wave vectors of incident and scattered waves. Let us take $\mathbf{q} = \mathbf{k} - \mathbf{k}'$. $\hbar\mathbf{q}$ is interpreted as momentum transferred to an incoming particle in the process of its scattering from potential. If θ is scattering angle, the directions of $\mathbf{k}$ and $\mathbf{k}'$ are separated θ, then $q = \sqrt{(k^2 - 2kk'\cos\theta + k'^2)} = 2k\sin(\theta/2)$, since $|\mathbf{k}| = |\mathbf{k}'|$. We find that the scattering amplitude is related to the Fourier transform of potential in the first Born approximation. Thus, the differential scattering cross-section in the first Born approximation is:

$$\sigma(k,\theta,\phi) = |f(k,\theta,\phi)|^2 = \left|\frac{\mu}{2\pi\hbar^2}\int e^{i(\mathbf{k}-\mathbf{k}').\mathbf{r}'}V(\mathbf{r}')d^3r'\right|^2 \tag{8.99}$$

We evaluate $f(k,\theta)$ for central potential, by introducing spherical polar coordinates (θ',ϕ') and by taking q along the polar axis:

$$\begin{aligned} f(k,\theta) &= -\frac{\mu}{2\pi\hbar^2}\int e^{i\mathbf{q}.\mathbf{r}'}V(r')d^3r' \\ &= -\frac{\mu}{2\pi\hbar^2}\int_0^\infty\int_0^\pi\int_0^{2\pi} e^{iqr'\cos\theta'}V(r')r'^2dr'\sin\theta'd\theta'd\phi' \\ &= -\frac{\mu}{\hbar^2}\int_0^\infty\int_0^\pi e^{iqr'\cos\theta'}V(r')r'^2dr'\sin\theta'd\theta' \\ &= -\frac{2\mu}{\hbar^2 q}\int_0^\infty r'V(r')\sin(qr')dr' \end{aligned} \tag{8.100a}$$

On substituting the value of q we obtain:

$$f(k,\theta) = -\frac{\mu}{k\hbar^2\sin(\theta/2)}\int_0^\infty r'V(r')\sin(qr')dr' \tag{8.100b}$$

and:

$$\sigma(k,\theta) = \left(\frac{\mu}{\hbar^2 k\sin(\theta/2)}\right)^2\left|\int_0^\infty r'V(r')\sin(qr')dr'\right|^2 \tag{8.101}$$

We again notice that the scattering amplitude and differential scattering cross-section are independent of ϕ.

8.6.1 Scattering by Screened Coulomb Potential

We apply the Born approximation to calculate scattering amplitude for scattering by a screened Coulomb potential, defined as:

$$V(r) = -\frac{ZZ'e^2}{4\pi\varepsilon_0 r}e^{-\alpha r} \tag{8.102}$$

where an incoming particle having charge Ze interacts with a target particle of charge $Z'e$. α is a parameter having dimensions of the inverse of the length. On substituting Eqn. (8.102) into (8.100b), we get:

$$\begin{aligned} f(k,\theta) &= \frac{ZZ'\mu e^2}{4\pi\varepsilon_0 k\hbar^2\sin(\theta/2)}\int_0^\infty e^{-\alpha r}\sin(qr)dr \\ &= \frac{ZZ'\mu e^2}{4\pi\varepsilon_0 k\hbar^2\sin(\theta/2)}\left[\frac{1}{2i}\left\{\int_0^\infty e^{-(\alpha-iq)r}dr - \int_0^\infty e^{-(\alpha+iq)r}dr\right\}\right] \\ &= \frac{ZZ'\mu e^2}{4\pi\varepsilon_0 k\hbar^2\sin(\theta/2)}\left(\frac{q}{\alpha^2+q^2}\right) \end{aligned} \tag{8.103a}$$

which on substituting the value of q yields:

$$f(k,\theta)=\left(\frac{ZZ'\mu e^2}{4\pi\varepsilon_0\hbar^2}\right)\frac{2}{\alpha^2+(2k\sin(\theta/2))^2} \tag{8.103b}$$

and:

$$\sigma(k,\theta)=\left(\frac{ZZ'\mu e^2}{4\pi\varepsilon_0\hbar^2}\right)^2\frac{4}{\left\{\alpha^2+(2k\sin(\theta/2))^2\right\}^2} \tag{8.104}$$

For screening length $\frac{1}{\alpha}\to\infty$, Eqn. (8.102) gives the bare Coulomb potential. Then, in limit $\alpha\to 0$, Eqn. (8.104) reduces to:

$$\sigma(k,\theta)=\left(\frac{ZZ'\mu e^2}{4\pi\varepsilon_0\hbar^2}\right)^2\frac{1}{4k^4\sin^4(\theta/2)} \tag{8.105a}$$

or:

$$\sigma(E,\theta)=\left(\frac{ZZ'e^2}{4\pi\varepsilon_0}\right)^2\frac{1}{16E^2\sin^4(\theta/2)} \tag{8.105b}$$

The result given by Eqn. (8.105a) for the differential cross-section for scattering by a Coulomb potential is identical to the formula given by Rutherford in 1911 by using classical mechanics. It has been shown that the exact quantum mechanical treatment (not the Born approximation) of scattering by the Coulomb potential yields the same result for the differential cross-section. However, scattering amplitude differs by a phase factor. It is to be noted that the Rutherford differential cross-section varies as $1/E^2$ at all angles and hence the angular distribution is independent of the energy.

8.6.2 Validity of the Born Approximation

In the first-order Born approximation, scattering amplitude has been evaluated using $\psi(\mathbf{r}')=e^{i\mathbf{k}.\mathbf{r}'}$, a plane wave, which is possible if $\psi_{sc}(\mathbf{r}')$ is negligibly small as compared to the incident plane wave; see Eqn. (8.6). The $\psi_{sc}(\mathbf{r}')$ has its maximum value in the interaction region, where $r\cong 0$. On putting $r\cong 0$ and $\psi(\mathbf{r}')=e^{i\mathbf{k}.\mathbf{r}'}$, Eqn. (8.92b) reduces to:

$$\psi_{sc}(\mathbf{r}\to 0)=-\frac{1}{4\pi}\int\frac{e^{ikr'}}{r'}U(\mathbf{r}')e^{i\mathbf{k}.\mathbf{r}'}d^3r' \tag{8.106}$$

and then the requirement of $|\psi_{sc}|\ll 1$ results to:

$$\left|\frac{1}{4\pi}\int\frac{e^{ikr'}}{r'}U(\mathbf{r}')e^{ik\mathbf{k}.\mathbf{r}'}d^3r'\right|\ll 1 \tag{8.107}$$

On introducing spherical polar coordinates $\mathbf{r}' \equiv (r', \theta', \phi')$ and taking k along polar axis, we have for the central potential:

$$\left| \frac{\mu}{2\pi\hbar^2} \int_0^{\infty}\int_0^{\pi}\int_0^{2\pi} \frac{e^{ikr'}}{r'} V(r') e^{ikr'\cos\theta'} r'^2 dr' \sin\theta' d\theta' d\phi' \right| \ll 1 \tag{8.108}$$

On carrying out integrals over θ' and ϕ', we obtain:

$$\begin{aligned} &\left| \frac{2\mu}{k\hbar^2} \int_0^{\infty} e^{ikr'} V(r') \sin(kr') dr' \right| \ll 1 \\ \Rightarrow &\left| \frac{\mu}{k\hbar^2} \int_0^{\infty} (e^{2ikr'} - 1) V(r') dr' \right| \ll 1 \end{aligned} \tag{8.109}$$

We notice that to fulfill the above requirement, $\frac{2\mu}{k\hbar^2}$ and the value of the integral should be small. This demands that (i) energy should be high so that k is high and $\sin(kr')$ varies rapidly and (ii) $V(r')$ should be weak. It can therefore be inferred that the Born approximation is valid for high energies and weak scattering potential.

For the case of the square well potential:

$$V(r) = \begin{cases} V_0 & 0 \le r \le a \\ 0 & r > a \end{cases},$$

Equation (8.109) goes to:

$$\left| \frac{\mu V_0}{k\hbar^2} \int_0^{a} (e^{2ikr'} - 1) dr' \right| \ll 1 \tag{8.110}$$

which simplifies to:

$$\begin{aligned} &\left| \frac{\mu V_0}{k\hbar^2} \left(\frac{e^{2ika} - 1}{2ik} - a \right) \right| \ll 1 \\ \Rightarrow &\left| \frac{\mu V_0}{k^2\hbar^2} \left(e^{ika} \sin(ka) - ka \right) \right| \ll 1 \\ \Rightarrow &\frac{\mu V_0}{k^2\hbar^2} \left\{ \sin^2(ka) + (ka)^2 - 2ka \sin(ka)\cos(ka) \right\}^{1/2} \ll 1 \end{aligned} \tag{8.111}$$

For low energy scattering, $ka \ll 1$, we write $e^{ika}\sin(ka) - ka \approx i(ka)^2$ to give:

$$\frac{\mu V_0 a^2}{\hbar^2} \ll 1 \tag{8.112}$$

which states that the potential should be weak enough not to bind the particle for scattering from the square well potential in the low energy regime.

8.7 Transformation from the Center of Mass Coordinate System to the Laboratory Coordinate System

The theory of scattering is usually developed in the center of mass (CM) coordinate system, while experiments to measure the scattering cross-section are performed in the laboratory coordinate system. Therefore, a relation connecting the differential cross-section in CM and the laboratory frames is needed. The CM coordinate system is more convenient for making calculations, because there the two-body problem with 6-degrees of freedom reduces to two one-body problems with 3-degrees of freedom. In the laboratory system, the target or center of scattering is initially at rest, whereas in the CM system, the center of mass of two interacting particles (incident and target) is always at rest.

To get the relation between two coordinate systems, we consider a particle of mass m_1 moving with velocity $\mathbf{v}_1$, let us say along x-axis, that collides with another particle of mass m_2 at rest at the origin in the laboratory system. After collision, the incident particle is scattered with velocity $\mathbf{v}_1'$ in a direction of (α, β). The motion in the laboratory system is shown in Fig. 8.2.

The velocity of the center of mass is then given by:

$$\mathbf{v}_{cm} = \frac{m_1 \mathbf{v}_1}{m_1 + m_2} \tag{8.113}$$

Now let us examine the same scattering with respect to an observer located at the center of mass. The observer sees that the particle of mass m_2 is approaching him with velocity $\mathbf{v}_{cm}$ from the left, while the particle of mass m_1 is traveling from the right with velocity:

$$\mathbf{v}_c = \mathbf{v}_1 - \mathbf{v}_{cm} = \mathbf{v}_1 \left(1 - \frac{m_1}{m_1 + m_2} \right) = \mathbf{v}_1 \left(\frac{m_2}{m_1 + m_2} \right) \tag{8.114}$$

After collision, the two particles should scatter in opposite directions, to keep the center of mass at rest in the CM coordinate system. Let us say the direction of scattering in the CM coordinate system is (θ, ϕ). Also, elastic scattering demands that the magnitude of velocity of the two particles be equal. The collision process is illustrated in Fig. 8.3.

Since two systems of coordinates move relative to each other with velocity $\mathbf{v}_{cm}$, $\mathbf{v}_1' = \mathbf{v}_c + \mathbf{v}_{cm}$, therefore:

$$\begin{aligned} v_1' \cos\alpha &= v_c \cos\theta + v_{cm} \\ v_1' \sin\alpha &= v_c \sin\theta \\ \beta &= \phi \end{aligned} \tag{8.115}$$

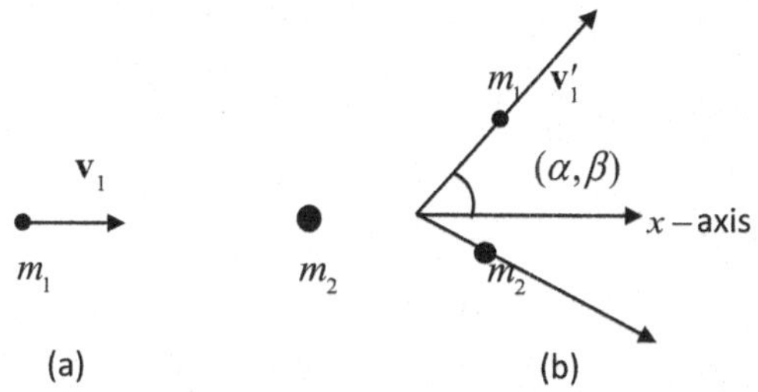

FIGURE 8.2
Motion of a particle along the x-axis in the laboratory coordinate system; (a) before scattering and (b) after scattering.

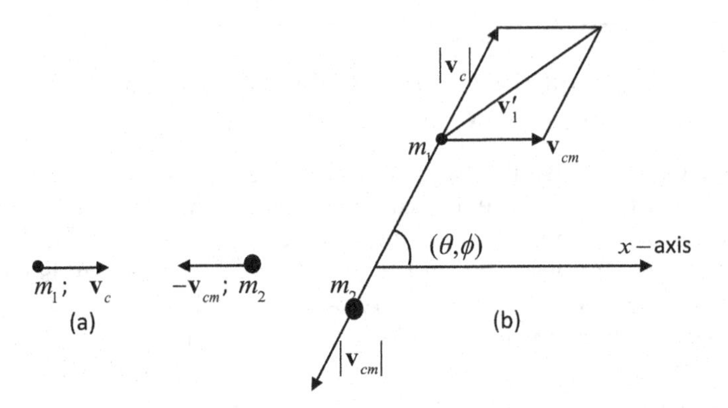

FIGURE 8.3
Motion of a particle along the x-axis in the center of mass coordinate system; (a) before scattering and (b) after scattering.

We thus have:

$$\tan\alpha = \frac{\sin\theta}{\cos\theta + \frac{v_{cm}}{v_c}} = \frac{\sin\theta}{\cos\theta + \gamma} \tag{8.116}$$

From Eqns. (8.113) and (8.114) $\gamma = \frac{v_{cm}}{v_c} = \frac{m_1}{m_2}$. Hence, for $m_2 \gg m_1$, $\theta \simeq \alpha$ and recoil is negligible. In such cases, the target behaves as a fixed system.

The relationship between the differential scattering cross-section in the laboratory system $\sigma^L(\alpha,\beta)$ and in the CM coordinate system $\sigma^C(\theta,\phi)$ is established by using the condition that the number of particles scattered into a given solid angle must be same in both the systems of coordinates:

$$\begin{aligned} &N\sigma^L(\alpha,\beta)d\Omega^L = N\sigma^C(\theta,\phi)d\Omega^C \\ &\Rightarrow \sigma^L(\alpha,\beta)\sin\alpha d\alpha d\beta = \sigma^C(\theta,\phi)\sin d\theta d\phi \end{aligned} \tag{8.117a}$$

$\beta = \phi \Rightarrow d\beta = d\phi$, and the square of both sides of Eqn. (8.116) gives:

$$\begin{aligned} &\frac{\sin^2\alpha}{\cos^2\alpha} + 1 = \frac{\sin^2\theta}{(\cos\theta+\gamma)^2} + 1 \\ &\Rightarrow \cos^2\alpha = \frac{(\cos\theta+\gamma)^2}{(\cos\theta+\gamma)^2 + \sin^2\theta} = \frac{(\cos\theta+\gamma)^2}{1+2\gamma\cos\theta+\gamma^2} \\ &\Rightarrow \cos\alpha = \frac{(\cos\theta+\gamma)}{\left(1+2\gamma\cos\theta+\gamma^2\right)^{1/2}} \end{aligned} \tag{8.117b}$$

Differentiating both the sides, we get:

$$\begin{aligned} -\sin\alpha\, d\alpha &= \frac{-\sin\theta d\theta}{\left(1+2\gamma\cos\theta+\gamma^2\right)^{1/2}} + \frac{(\cos\theta+\gamma)\times\gamma\sin\theta\, d\theta}{\left(1+2\gamma\cos\theta+\gamma^2\right)^{3/2}} \\ \Rightarrow \sin\alpha\, d\alpha &= \frac{(\gamma\cos\theta+1)}{\left(1+2\gamma\cos\theta+\gamma^2\right)^{3/2}}\sin\theta\, d\theta \end{aligned} \tag{8.118}$$

Therefore:

$$\sigma^L(\alpha,\beta) = \frac{\left(1+2\gamma\cos\theta+\gamma^2\right)^{3/2}}{(\gamma\cos\theta+1)}\sigma^C(\theta,\phi) \tag{8.119}$$

As is seen for $\gamma \to 0$, $\sigma^L(\alpha,\beta) \cong \sigma^C(\theta,\phi)$ and for $\gamma = 1$, Eqn. (8.116) yields $\alpha = \theta/2$ and $\sigma^L(\alpha,\beta) = 4\cos(2\alpha)\sigma^C(2\alpha,\beta)$.

8.8 Solved Examples

1. Calculate scattering amplitude and differential scattering cross-section, using the Born approximation, for an attractive potential defined by $V(r) = -\frac{\hbar^2\lambda}{2\mu}\delta(r-a)$, where $\lambda > 0$. Then, evaluate the total scattering cross-section for the low energy and small angle scattering case.

SOLUTION

The scattering amplitude is given by Eqn. (8.100b):

$f(k,\theta) = -\frac{2\mu}{q\hbar^2}\int_0^\infty r'V(r')\sin(qr')dr'$, with $q = 2k\sin(\theta/2)$. Substituting the value of $V(r')$, we have:

$$f(k,\theta) = \frac{\lambda}{q}\int_0^a r'\delta(r'-a)\sin(qr')dr' \tag{8.120a}$$

which simplifies to:

$$f(k,\theta) = \frac{\lambda a\sin(qa)}{q} \tag{8.120b}$$

Therefore:

$$\sigma(k,\theta) = \left(\frac{\lambda a}{q}\right)^2 \sin^2(qa) \tag{8.121}$$

For low energy $k \ll 1/a$, and for small θ $qa \ll 1$. We therefore get: $\sigma(k,\theta) = \left(\frac{\lambda a}{q}\right)^2 \sin^2(qa) \approx (\lambda a^2)^2$, which gives:

$$\sigma_{tot} = \int \sigma(k,\theta) d\Omega = 4\pi a^2 (\lambda a)^2.$$

2. For the potential described in problem 1: (a) Find the minimum value of λ for which there exists a bound state. (b) Use partial wave analysis to calculate the phase shift and scattering cross-section when the particle incident on the potential has a low energy.

   (Similar Problem 10.2 in "Problems & Solutions in Quantum Mechanics," Kyrikos Tamvakis.)

SOLUTION

(a) The bound state of a single particle is described by $\psi(r) = R_l(k,r)Y_{lm}(\theta,\phi)$. Then, the radial equation is:

$$\frac{d^2R_l}{dr^2} + \frac{2}{r}\frac{dR_l}{dr} + \left[\frac{2\mu E}{\hbar^2} - \frac{l(l+1)}{r^2} + \lambda\delta(r-a)\right] R_l(r) = 0 \tag{8.122}$$

for the bound state $E < 0$. We define $\alpha = \sqrt{\frac{-2\mu|E|}{\hbar^2}}$. When $r \neq a$, the equation is an imaginary-variable spherical Bessel equation. The solution of the equation, which is finite at $r = 0$, in the region of $r < a$ is:

$$R_l^{<}(r) = Aj_l(i\alpha r) \tag{8.123a}$$

As discussed in Section 6.6, the solution of Eqn. (8.122) for $r > a$ can be given as:

$$R_l^{>}(r) = Bh_l^{(1)}(i\alpha r) \tag{8.123b}$$

where

$$h_l^{(1)}(i\alpha r) = j_l(i\alpha r) + in_l(i\alpha r).$$

Continuity of wave function at $r = a$, demands:

$$Aj_l(i\alpha a) = Bh_l^{(1)}(i\alpha a)$$
$$\Rightarrow \frac{A}{B} = \frac{h_l^{(1)}(i\alpha a)}{j_l(i\alpha a)} \tag{8.124}$$

Integration of Eqn. (8.122) between $a + \varepsilon \to a - \varepsilon$ and then taking $\varepsilon \to 0$ limit gives the discontinuity in the first-order derivative at the same point:

$$R_l'(a + \varepsilon \to 0) - R_l'(a - \varepsilon \to 0) = -\lambda R_l(a) \tag{8.125}$$

For low energy scattering $l = 0$ case is the most prominent. For $l = 0$:

$$R_0^<(r) = \frac{A\sin(i\alpha r)}{i\alpha r} \tag{8.126a}$$

$$R_0^>(r) = -B\frac{e^{-\alpha r}}{\alpha r} \tag{8.126b}$$

and then:

$$R_0'(a+0) = \frac{Be^{-\alpha a}}{\alpha a}\left(\alpha + \frac{1}{a}\right) \tag{8.127a}$$

$$R_0'(a-0) = \frac{A}{\alpha}\left[\frac{\alpha\cosh(\alpha a)}{a} - \frac{\sinh(\alpha a)}{a^2}\right] \tag{8.127b}$$

On substituting Eqns. (8.127) into Eqn. (8.125) and then using Eqn. (8.124), we get:

$$a\lambda = \frac{2\alpha a}{1 - e^{-2\alpha a}} \tag{8.128}$$

When energy is low, $1 - e^{-2\alpha a} \approx 2\alpha a - 2(\alpha a)^2 + ...$, suggesting that the minimum value of $a\lambda$ is unity. Hence, $\lambda_{\min} = 1/a$.

(b) To calculate phase shift, we take $k = \sqrt{\frac{2\mu E}{\hbar^2}}$ and obtain:

$$\begin{aligned} R_l^<(r) &= Aj_l(kr) \\ R_l^>(r) &= B\left[\cos\delta_l j_l(kr) - \sin\delta_l n_l(kr)\right] \end{aligned} \tag{8.129}$$

Continuity of wave function at $r = a$ then requires:

$$\begin{aligned} &Aj_l(ka) = B\left[\cos\delta_l j_l(ka) - \sin\delta_l n_l(ka)\right] \\ &\Rightarrow \frac{A}{B} = \cos\delta - \sin\delta_l\frac{n_l(ka)}{j_l(ka)} \end{aligned} \tag{8.130}$$

Equation (8.125) gives:

$$\begin{aligned} &kB\left[\cos\delta_l j_l'(ka) - \sin\delta_l n_l'(ka)\right] - Akj_l'(ka) = -\lambda Aj_l(ka) \\ &\Rightarrow \frac{A}{B} = \frac{\cos\delta_l j_l'(ka) - \sin\delta_l n_l'(ka)}{j_l'(ka) - \frac{\lambda}{k}j_l(ka)} \end{aligned} \tag{8.131}$$

We thus have:

$$1-\tan\delta_l\frac{n_l(ka)}{j_l(ka)}=\frac{j_l'(ka)-\tan\delta_l n_l'(ka)}{j_l'(ka)-\frac{\lambda}{k}j_l(ka)}$$

$$\Rightarrow \tan\delta_l\left[\frac{n_l(ka)}{j_l(ka)}-\frac{n_l'(ka)}{j_l'(ka)-\frac{\lambda}{k}j_l(ka)}\right]=1-\frac{j_l'(ka)}{j_l'(ka)-\frac{\lambda}{k}j_l(ka)} \tag{8.132}$$

which yields:

$$\tan\delta_l=\frac{-\frac{\lambda}{k}j_l^2(ka)}{n_l(ka)j_l'(ka)-j_l(ka)\left\{\frac{\lambda}{k}n_l(ka)+n_l'(ka)\right\}} \tag{8.133}$$

which for $l=0$ reduces to:

$$\frac{ka}{\tan(ka+\delta_0)}-\frac{ka}{\tan(ka)}=-\lambda a \tag{8.134a}$$

And for $ka \ll 1$, we obtain:

$$\frac{ka}{\tan\delta_0}-1=-\lambda a$$
$$\Rightarrow \sin\delta_0=\frac{ka}{\left((ka)^2+(1-\lambda a)^2\right)^{1/2}} \tag{8.134b}$$

Hence, the total scattering cross-section is:

$$\sigma_{tot}=\frac{4\pi}{k^2}\sin^2\delta_0=4\pi\left(\frac{a^2}{(ka)^2+(1-\lambda a)^2}\right)$$
$$\approx\frac{4\pi a^2}{(1-\lambda a)^2} \tag{8.135}$$

3. Calculate the energy dependent phase shift δ_0 for an attractive square well potential defined by:

$$V(r)=\begin{cases}-V_0 & r<a\\ 0 & r>a\end{cases},$$

and then show that at high energies $\delta_0\rightarrow\frac{\mu a V_0}{\hbar^2 k}$.

SOLUTION

On taking $\chi(r) = rR(r)$, the radial equation for $l = 0$ is:

$$\chi'' + k'^2\chi = 0 \quad \text{for} \quad r < a \tag{8.136a}$$

and:

$$\chi'' + k^2\chi = 0 \quad \text{when} \quad r > a \tag{8.136b}$$

where $k^2 = \dfrac{2\mu E}{\hbar^2}$ and $k'^2 = \dfrac{2\mu}{\hbar^2}(E + V_0)$. The solution of Eqns. (8.136) can be written as:

$$\chi(r) = \begin{cases} A\sin(k'r) & r < a \\ B\sin(kr + \delta_0) & r > a \end{cases} \tag{8.137}$$

The continuity at $r = a$ demands:

$$A\sin(k'a) = B\sin(ka + \delta_0) \tag{8.138a}$$

and:

$$k'A\cos(k'a) = kB\cos(ka + \delta_0) \tag{8.138b}$$

which yield:

$$k\tan(k'a) = k'\tan(ka + \delta_0) \tag{8.139}$$

Since $k'^2 = k^2\left(1 + \dfrac{V_0}{E}\right)$, at high energies we have:

$$\tan(k'a) = \tan\left(ka\sqrt{(1 + V_0 / E)}\right) \approx \tan\left(ka + \frac{kaV_0}{2E}\right) \tag{8.140}$$

Also $k' \to k$ for $E \to \infty$. Hence:

$$\tan\left(ka + \frac{kaV_0}{2E}\right) \approx \tan(ka + \delta_0) \tag{8.141}$$

which gives:

$$\delta_0 = \frac{kaV_0}{2E} = \frac{\mu a V_0}{\hbar^2 k} \tag{8.141}$$

4. Calculate the total scattering cross-section using the Born approximation for attractive square well potential defined above in problem 3 and compare your results with that obtained using the partial wave method for very low energies.

SOLUTION

From Eqn. (8.100a), we obtain:

$$f(k,\theta)=\frac{2\mu}{\hbar^2 q}\int_0^a r'V_0\sin(qr')dr'$$

$$=\frac{2\mu V_0}{\hbar^2 q}\left[-\frac{r'}{q}\cos(qr')\Big|_0^a+\frac{1}{q^2}\sin(qr')\Big|_0^a\right] \tag{8.142}$$

$$=\frac{2\mu V_0}{\hbar^2 q^2}\left[\frac{1}{q}\sin(qa)-a\cos(qa)\right]$$

which gives:

$$\sigma(k,\theta)=\left(\frac{2\mu V_0}{\hbar^2 q^2}\right)^2\left[\frac{1}{q}\sin(qa)-a\cos(qa)\right]^2 \tag{8.143}$$

where $q=2k\sin(\theta/2)=2\sqrt{\frac{2\mu E}{\hbar^2}}\sin(\theta/2)$. For low energies $k\to 0\Rightarrow qa\ll 1$, we get:

$$\sigma(k,\theta)\approx\left(\frac{2\mu V_0}{\hbar^2 q^2}\right)^2\left[\frac{1}{q}\left(qa-\frac{(qa)^3}{3!}+\ldots\right)-a\left(1-\frac{(qa)^2}{2!}+\ldots.\right)\right]^2$$

$$\Rightarrow\sigma(k,\theta)\approx\left(\frac{2\mu V_0}{\hbar^2 q^2}\right)^2\left[\frac{a(qa)^2}{3}\right]^2\approx\left(\frac{2\mu V_0 a^3}{3\hbar^2}\right)^2 \tag{8.144}$$

Then the total scattering cross-section is:

$$\sigma_{tot}=\int\sigma(k,\theta)d\Omega$$

$$\Rightarrow\sigma_{tot}=\int_0^{\pi}\int_0^{2\pi}\left(\frac{2\mu V_0 a^3}{3\hbar^2}\right)^2\sin\theta\,d\theta d\phi=4\pi\left(\frac{2\mu V_0 a^3}{3\hbar^2}\right)^2 \tag{8.145}$$

From the partial wave analysis method, σ_{tot} is given by:

$$\sigma_{tot}=\frac{4\pi}{k^2}\sum_{l=0}^{\infty}(2l+1)\sin^2\delta_l,$$

which for $l=0$, reduces to:

$$\sigma_{tot}=\frac{4\pi}{k^2}\sin^2\delta_0 \tag{8.146}$$

The δ_0 for the attractive square potential well is given by Eqn. (8.139) $k\tan(k'a)=k'\tan(ka+\delta_0)$. For $k\to 0$, $k'\to k_0=\sqrt{\frac{2\mu V_0}{\hbar^2}}$

where $k_0 = \sqrt{\frac{2\mu V_0}{\hbar^2}}$. We thus have:

$$\frac{\tan(k'a)}{k'} \approx \frac{\tan(k_0 a)}{k_0} \approx \frac{1}{k_0}\left(k_0 a + \frac{(k_0 a)^3}{3} + ...\right) \approx a + \frac{a(k_0 a)^2}{3} + ..., \text{if } k_0 a \ll 1 \qquad (8.147a)$$

Also:

$$\frac{\tan(ka + \delta_0)}{k} \approx a + \frac{\delta_0}{k} \qquad (8.147b)$$

Therefore $\delta_0 \approx \frac{ka(k_0 a)^2}{3}$. Then, Eqn. (8.146) gives:

$$\sigma_{tot} = \frac{4\pi}{k^2}\sin^2\delta_0 \approx \frac{4\pi}{k^2}\delta_0^2$$

$$\Rightarrow \sigma_{tot} \approx 4\pi\frac{a^6 k_0^4}{9} \approx 4\pi\left(\frac{2\mu V_0 a^3}{3\hbar^2}\right)^2 \qquad (8.148)$$

which is the same as is given by Eqn. (8.145).

5. Calculate the phase shift using the Born approximation for $l = 0$ for scattering potential defined by $V(r) = V_0 e^{-r/a}$.

SOLUTION

The Born approximation for phase shift for a potential is given by Eqn. (8.80):

$$\delta_l(k) = -\frac{2\mu k}{\hbar^2}\int_0^\infty V(r)\left(j_l(kr)\right)^2 r^2 dr.$$

For $l = 0$, we get:

$$\delta_0(k) = -\frac{2\mu k V_0}{\hbar^2}\int_0^\infty e^{-r/a}\left(j_0(kr)\right)^2 r^2 dr$$

$$= -\frac{2\mu V_0}{k\hbar^2}\int_0^\infty e^{-r/a}\sin^2 kr dr \qquad (8.149a)$$

$$= -\frac{\mu V_0}{k\hbar^2}\int_0^\infty e^{-r/a}\{1 - \cos(2kr)\}dr$$

$$= -\frac{\mu V_0}{k\hbar^2}\left[a - \frac{a}{4k^2a^2 + 1}\right]$$

which simplifies to:

$$\delta_0 = -\frac{\mu V_0 a}{k\hbar^2}\left(\frac{4k^2a^2}{1+4k^2a^2}\right) \tag{8.149b}$$

The Born approximation is valid for high energies ($ka \gg 1$) and weak potential. Therefore: $\delta_0 \approx -\dfrac{\mu V_0 a}{k\hbar^2}$.

6. A particle of charge e, mass m, and wave vector $\mathbf{k}$ is incident perpendicular to the direction of a dipole made of two electric charges e and $-e$ at a mutual distance of $2a$. Calculate the scattering amplitude using the Born approximation and find the directions in which the scattering cross-section has maximum value.

SOLUTION

On placing charge $-e$ at $-\mathbf{a}$ and e at $\mathbf{a}$, the potential created by the dipole is:

$$V(r) = -\frac{e^2}{4\pi\varepsilon_0|\mathbf{r}+\mathbf{a}|} + \frac{e^2}{4\pi\varepsilon_0|\mathbf{r}-\mathbf{a}|} \tag{8.150}$$

Substituting the above equation into Eqn. (8.98), we have:

$$f(k,\theta) = -\frac{m}{2\pi\hbar^2}\left(\frac{1}{4\pi\varepsilon_0}\right)\int e^{i(\mathbf{k}-\mathbf{k}').\mathbf{r}'}\left[-\frac{e^2}{|\mathbf{r}'+\mathbf{a}|} + \frac{e^2}{|\mathbf{r}'-\mathbf{a}|}\right]d^3r' \tag{8.151}$$

Using $\mathbf{q} = \mathbf{k} - \mathbf{k}'$ and:

$$\frac{1}{|\mathbf{r}'\pm\mathbf{a}|} = \frac{4\pi}{(2\pi)^3}\int \frac{e^{-i\mathbf{k}.(\mathbf{r}'\pm\mathbf{a})}}{K^2}d^3K \tag{8.152}$$

we get:

$$\begin{aligned}
\int e^{i\mathbf{q}.\mathbf{r}'}\frac{1}{|\mathbf{r}'\pm\mathbf{a}|}d^3r' &= \frac{4\pi}{(2\pi)^3}\iint e^{i\mathbf{q}.\mathbf{r}'}\frac{e^{-i\mathbf{k}.\mathbf{r}'}}{K^2}e^{\mp i\mathbf{k}.\mathbf{a}}d^3r'd^3K \\
&= \frac{4\pi}{(2\pi)^3}\iint \frac{e^{i(\mathbf{q}-\mathbf{k}).\mathbf{r}'}}{K^2}e^{\mp i\mathbf{k}.\mathbf{a}}d^3r'd^3K \\
&= \frac{4\pi}{(2\pi)^3}\int \frac{\delta(\mathbf{q}-\mathbf{K})}{K^2}e^{\mp i\mathbf{k}.\mathbf{a}}d^3K \\
&= \frac{4\pi}{q^2}e^{\mp i\mathbf{q}.\mathbf{a}}
\end{aligned} \tag{8.153}$$

Therefore:

$$f(k,\theta) = -\frac{2me^2}{\hbar^2q^2}\left(\frac{e^{i\mathbf{q}.\mathbf{a}} - e^{-i\mathbf{q}.\mathbf{a}}}{4\pi\varepsilon_0}\right) = -\frac{4ime^2}{\hbar^2q^2}\frac{1}{4\pi\varepsilon_0}\sin(\mathbf{q}.\mathbf{a}) \tag{8.154a}$$

On taking $\mathbf{k} = k\hat{z}; \mathbf{k}' = k\sin\theta\hat{x} + k\cos\theta\hat{z}$ and $\mathbf{a} = a\hat{x}$, we get:

$$q^2 = (2k\sin(\theta/2))^2 = 2k^2(1-\cos\theta) \text{ and } \mathbf{q.a} = -ka\sin\theta \tag{8.154b}$$

which yields to:

$$\sigma(k,\theta) = |f(\theta,\phi)|^2 = \left(\frac{2m^2}{\hbar^2 k^2}\right)^2 \left(\frac{1}{4\pi\varepsilon_0}\right)^2 \frac{\sin^2(ka\sin\theta)}{(1-\cos\theta)^2} \tag{8.155}$$

$\sigma(k,\theta)$ becomes maximum when $ak\sin\theta = \frac{\pi}{2}(2n+1)$, where $n = 0,1,2,3........$

7. Consider the scattering by the hard sphere potential, defined by Eqn. (8.59), and show that $\tan\delta_l = \frac{j_l(ka)}{n_l(ka)}$, where a is the range of the potential. Calculate the scattering cross-section. Find out values of the phase shift and scattering cross-section for $l = 0$.

SOLUTION

In the exterior region $r > a$, the solution of the radial equation is:

$$\begin{aligned} R_l(kr) &= A_l j_l(kr) + B_l n_l(kr) \\ &= C_l\left[j_l(kr)\cos\delta_l - \sin\delta_l n_l(kr)\right] \end{aligned} \tag{8.156}$$

where $C_l = \sqrt{A_l^2 + B_l^2}$, $A_l = C_l\cos\delta_l$, and $B_l = -C_l\sin\delta_l$. The particle cannot penetrate into the region of $r < a$; hence $R_l(ka) = 0$. This gives:

$$\begin{aligned} j_l(ka)\cos\delta_l &= \sin\delta_l n_l(ka) \\ \Rightarrow \tan\delta_l &= \frac{j_l(ka)}{n_l(ka)} \end{aligned} \tag{8.157}$$

To calculate the scattering cross-section, we need:

$$\sin^2\delta_l = \frac{\tan^2\delta_l}{1+\tan^2\delta_l} = \frac{j_l^2(ka)}{j_l^2(ka) + n_l^2(ka)} \tag{8.158}$$

Thus:

$$\sigma_{tot} = \frac{4\pi}{k^2}\sum_{l=0}^{\infty}(2l+1)\sin^2\delta_l = \frac{4\pi}{k^2}\sum_{l=0}^{\infty}(2l+1)\frac{j_l^2(ka)}{j_l^2(ka)+n_l^2(ka)} \tag{8.159}$$

For $l = 0$:

$$\begin{aligned} \tan\delta_0 &= \frac{j_0(ka)}{n_0(ka)} = -\frac{\sin(ka)}{\cos(ka)} = -\tan(ka) \\ \Rightarrow \delta_0 &= -ka \end{aligned} \tag{8.160a}$$

and:

$$\sigma_{tot} = \frac{4\pi}{k^2}\sin^2(ka) \tag{8.160b}$$

For $ka \ll 1$, the above equation reduces to $\sigma_{tot} \approx 4\pi a^2$, while Eqn. (8.159) goes to:

$$\sigma_{tot} \approx 4\pi a^2 \sum_{l=0}^{\infty} \frac{1}{(2l+1)} \left[\frac{2^l (l!)}{(2l)!} \right]^4 (ka)^{4l} \tag{8.160c}$$

8.9 Exercises

1. Use the Born approximation, Eqn. (8.100a), to calculate the scattering amplitude $f(k,\theta)$ and differential scattering cross-section $\sigma(k,\theta)$ for the scattering of a particle of reduced mass μ from the potential described by $V(r) = -\frac{\beta}{r^2 + a^2}$, where β and a are constants. Use $\int_0^{\infty} \frac{x \sin(bx)}{x^2 + a^2} dx = \frac{\pi}{2} e^{-ab}$. Analyze your result for $a \to 0$.
2. Consider the case of the scattering of a particle from an attractive potential $V(r) = -\frac{\beta}{r^2}$. (a) Calculate the phase shift for the s-state (δ_0) using the Born approximation, Eqn. (8.80). Does δ_0 depend on the energy of the incident particle? (b) Calculate the scattering cross-section and compare your results with that obtained above in exercise 1 for $a \to 0$. (c) Compute δ_0 by taking μ equal to the mass of the electron and $\beta = 3.82 \text{meV}.(nm)^2$. Use $\int_0^{\infty} \frac{\sin^2(ax)}{x^2} dx = \frac{\pi a}{2}$.
3. For a particle of mass m, which is scattered from a potential $V(r)$, the wave function can be written as $\psi(r) = e^{ikz} + \phi(r)$. $k^2 = \frac{2mE}{\hbar^2}$, where E is the energy of the particle. Derive the differential equation for $\phi(r)$ within the first Born approximation.
4. Show that for $r \neq 0$, $\phi(r) = -\frac{e^{ikr}}{r}$ satisfies the equation $\left(\nabla^2 + k^2\right)\phi(r) = 0$.
5. Evaluate the validity condition for the Born approximation, Eqn. (8.109), for the attractive delta function potential $V(r) = -\frac{\hbar^2 \lambda}{2\mu} \delta(r-a)$, where $\lambda > 0$. What should be the relation between λ and a to use $ka \ll 1$?
6. Find the Green function for the 1D Schrodinger equation $G_0(k, x, x_0)$ and then find the 1D form of the integral (Lippmann-Schwinger) equation.
7. Simplify the 1D Schrodinger integral equation obtained in the above exercise for the asymptotic region ($x \gg x'$) and use $k' = \frac{|x|}{x}$. (a) Deduce the expressions for scattering amplitude and for differential scattering cross-section. (b) Show that the reflection coefficient can be expressed as $R \simeq \left| \frac{m}{\hbar^2 k} \right|^2 \left| \int_{-\infty}^{\infty} e^{2ik'x'} V(x') dx' \right|^2$.
8. For exercise 7, estimate the differential scattering cross-section and reflection coefficient for $V(x) = b\delta(x)$, where b is a constant. Explain why your estimated value of the scattering cross-section differs from its exact value $\sigma(k) = \frac{m^2 b^2}{m^2 b^2 + \hbar^4 k^2}$.

9

Quantum Theory of Many Particle Systems

There are no restrictions in applying the general postulates of quantum mechanics to any kind of system. Though until now we have mainly discussed the quantum theory of single particle systems, the postulates are applicable to all quantum systems, including those containing many particles. Experimental information of considerable interest exists on systems of many particles. In this chapter, we will be discussing the quantum mechanical treatment of many particle systems.

9.1 System of Indistinguishable Particles

The particles that can be interchanged without any change in the physical properties of a system are called indistinguishable or identical particles. For example, in a system of N-electrons one can assign N-position coordinates and N-momenta, but it is impossible to tell which electron occupies which position and momentum state. The Hamiltonian of the system remains unchanged on interchanging the position and momentum of any two electrons. Any two electrons in a many electron system are identical to one another and there exists no experiment that could possibly distinguish one from the other. A Hamiltonian for an N-particles system can be expressed as:

$$H(1,2,3.....,N)=\sum_{i=1}^{N}\frac{p_i^2}{2m}+V(1,2,3,....,N) \tag{9.1}$$

where the numbers $1,2,3,....,N$ represent composite coordinates (both position and spin) for particles $1,2,3,....,N$, respectively. There exists no way (experimental or theoretical) to tell which particle out of N-particles is to be called particle 1 or 2 or so on. The above equation is invariant under the interchange of any two numbers. A quantum state representing the system at a given time t may in general be defined by $|\psi(1,2,3,.....,N,t)\rangle$. For brevity, let us consider a system of two particles. The Schrödinger equation is:

$$H(1,2)|\psi(1,2\rangle=E|\psi(1,2\rangle \tag{9.2a}$$

On interchanging 1 and 2 we have:

$$H(2,1)|\psi(2,1\rangle=E|\psi(2,1\rangle \tag{9.2b}$$

Next, let us define the particle exchange operator by P_{12}, such that $P_{12}f(1,2)=f(2,1)$. We then have:

$$\begin{aligned} P_{12}H(1,2)|\psi(1,2)\rangle &= P_{12}E|\psi(1,2)\rangle \\ &= E|\psi(2,1)\rangle \\ &= H(2,1)|\psi(2,1)\rangle \\ &= H(2,1)P_{12}|\psi(1,2)\rangle \end{aligned} \tag{9.3a}$$

which implies:

$$\{P_{12}H(1,2)-H(2,1)P_{12}\}|\psi(1,2)\rangle=0 \tag{9.3b}$$

From Eqn. (9.1), we notice that $H(2,1)$ can be replaced by $H(1,2)$, as the interchanging of 1 and 2 does not change the Hamiltonian. Therefore:

$$[P_{12},H(1,2)]|\psi(1,2)\rangle=0 \tag{9.4}$$

Since $|\psi(1,2)\rangle\neq 0$, we have $[P_{12},H(1,2)]=0$ saying that the exchange operator commutes with the Hamiltonian. This means that $|\psi(1,2)\rangle$ is an eigenstate of P_{12} as well, and hence, we write:

$$P_{12}|\psi(1,2)\rangle=\lambda|\psi(1,2)\rangle \tag{9.5}$$

Operating both sides with P_{12} from the left, we obtain:

$$\begin{aligned} &P_{12}^2|\psi(1,2)\rangle=\lambda P_{12}|\psi(1,2)\rangle=\lambda^2|\psi(1,2)\rangle \\ &\Rightarrow|\psi(1,2)\rangle=\lambda^2|\psi(1,2)\rangle \\ &\Rightarrow(\lambda^2-1)|\psi(1,2)\rangle=0 \end{aligned} \tag{9.6}$$

where we used $P_{12}^2|\psi(1,2)\rangle=P_{12}|\psi(2,1)\rangle=|\psi(1,2)\rangle$. Therefore, $\lambda=\pm 1$ and $\psi(2,1)=\pm\psi(1,2)$. Hence, the eigenvalue of the particle exchange operator is either +1 or –1. Thus, there can exist only two types of many particle systems: (a) whose wave function is symmetric $\psi(2,1)=\psi(1,2)$, and (b) which has an antisymmetric wave function $\psi(2,1)=-\psi(1,2)$.

The above conclusions can be generalized for a system of an N-particle, whose quantum state at a given time t is represented by the wave function ket $|\psi(1,2,3,..,\alpha,..,\beta,...,N,t)\rangle$. An operator $P_{\alpha\beta}$, whose operation on $|\psi(1,2,3,..,\alpha,..,\beta,...,N,t)\rangle$ interchanges α and β, will commute with $H(1,2,3,.....,N)$ and hence:

$$\begin{aligned} &P_{\alpha\beta}|\psi(1,2,3,..,\alpha,..,\beta,...,N,t)\rangle=\lambda|\psi(1,2,3,..,\alpha,..,\beta,...,N,t)\rangle \\ &\Rightarrow|\psi(1,2,3,..,\alpha,..,\beta,...,N,t)\rangle=\lambda^2|\psi(1,2,3,..,\alpha,..,\beta,...,N,t)\rangle \end{aligned} \tag{9.7}$$

which again yields $\lambda=\pm 1$. The α and β can be assigned any two values from 1 to N. Also, more than one exchange operator can be applied simultaneously to a wave function. Each operation will give rise to an eigenvalue λ, and the multiple of all values of λ will be either +1 or −1. Hence, $|\psi(1,2,3,..,\alpha,..,\beta,...,N,t)\rangle$ is either symmetric or antisymmetric. Further, in the time-dependent Schrödinger equation:

$$i\hbar\frac{\partial|\psi(1,2,...,N,t)\rangle}{\partial t}=H(1,2,....,N)|\psi(1,2,...,N,t)\rangle \tag{9.8}$$

Since the Hamiltonian is always symmetric, the right-hand side is either symmetric or antisymmetric, depending on the wave function, which means that at a given time, $\frac{\partial|\psi\rangle}{\partial t}$ is symmetric if the wave function is symmetric. This implies that $\frac{\partial^2|\psi\rangle}{\partial t^2}$ will also be symmetric, because $\frac{\partial|\psi\rangle}{\partial t}$ is symmetric. We thus find that a symmetric will remain symmetric at all the times. A similar argument applies to an antisymmetric wave function. Hence, the symmetry of wave function is independent of time.

9.1.1 Non-interacting System of Particles

In the Schrödinger representation, the Hamiltonian is time-independent, which allows us to take: $|\psi(1,2,3,.....,N,t)\rangle=|\psi(1,2,3,.....,N)\rangle e^{-iEt/\hbar}$. This gives:

$$H(1,2,....,N)|\psi(1,2,...,N)\rangle=E|\psi(1,2,...,N)\rangle \tag{9.9}$$

As is obvious, one can generate $N!$ wave function kets by applying $P_{\alpha\beta}$ to $|\psi(1,2,...,N)\rangle$. And, all of these wave function kets will also be the solution of Eqn. (9.9), for the same energy eigenvalue E. This kind of degeneracy, which arises due to the interchanging of an indistinguishable particle, is known as *exchange degeneracy*. Any linear combination of all these solutions will also be the solution of Eqn. (9.9). For brevity, let us again consider a two particle system. As discussed above, both $\psi(1,2)$ and $\psi(2,1)$ are eigenfunctions of $H(1,2)$ with the eigenvalue E. Then, $\psi(1,2)\pm\psi(2,1)$ too will be an eigenfunction of $H(1,2)$ for energy eigenvalue E.

The Hamiltonian $H(1,2)$ can be expressed as $H(1,2)=H(1)+H(2)+\delta H$, where $H(1)\{H(2)\}$ is the Hamiltonian for particle 1(2) in the absence of particle 2(1). The δH is interaction energy between the particles when both are present.

Let us consider the case of $\delta H=0$, a non-interacting particle system. For the non-interacting case, we write $H(1)|\phi_1(1)\rangle=E_1|\phi_1(1)\rangle$ and $H(2)|\phi_2(2)\rangle=E_2|\phi_2(2)\rangle$ with $H(1,2)=H(1)+H(2)$ and $E=E_1+E_2$, where suffixes 1(2) to ϕ represent a set of quantum numbers (n,l,m,m_s), specifying the quantum states of particle 1(2). The $\psi(1,2)$ and $\psi(2,1)$ are given by $\phi_1(1)\phi_2(2)$ and $\phi_1(2)\phi_2(1)$, respectively. If $|\phi_1(1)\rangle$ and $|\phi_2(2)\rangle$ are normalized kets, $\langle\phi_1(1)|\phi_1(1)\rangle=1=\langle\phi_2(2)|\phi_2(2)\rangle$, then normalized linear combinations of $|\psi(1,2)\rangle$ and $|\psi(2,1)\rangle$ are:

$$\psi_s(1,2)=\frac{1}{\sqrt{2}}\{\phi_1(1)\phi_2(2)+\phi_1(2)\phi_2(1)\} \tag{9.10a}$$

and:

$$\begin{aligned}\psi_a(1,2)&=\frac{1}{\sqrt{2}}\{\phi_1(1)\phi_2(2)-\phi_1(2)\phi_2(1)\}\\&=\frac{1}{\sqrt{2}}\sum_{n_1=1}^{2}\sum_{n_2=1}^{2}\epsilon_{n_1n_2}\,\phi_{n_1}(1)\phi_{n_2}(2)\\&=\frac{1}{\sqrt{2}}\begin{vmatrix}\phi_1(1)&\phi_1(2)\\\phi_2(1)&\phi_2(2)\end{vmatrix}\end{aligned} \tag{9.10b}$$

where $\epsilon_{n_1n_2} = -\epsilon_{n_2n_1}$ and its value is one when n_1n_2 take values in cyclic order such as 12 and 21, zero otherwise. Equations (9.10) give wave function kets of $H(1,2)$ for symmetric $(\lambda = 1)$ and antisymmetric $(\lambda = -1)$ cases. A generalized form of Eqn. (9.10b) for a system of N non-interacting particles is:

$$\psi_a(1,2,....N) = \frac{1}{\sqrt{N!}} \begin{vmatrix} \phi_1(1) & \phi_1(2) & \phi_1(3) & \cdots & \phi_1(N) \\ \phi_2(1) & \phi_2(2) & \phi_2(3) & \cdots & \phi_2(N) \\ \phi_3(1) & \phi_3(2) & \phi_3(3) & \cdots & \phi_3(N) \\ \vdots & \vdots & \cdots & \cdots & \vdots \\ \phi_N(1) & \phi_N(2) & \phi_N(3) & \cdots & \phi_N(N) \end{vmatrix} \tag{9.11a}$$

The determinant is known as the *Slater determinant*. The determinant expends as a linear combination of $N!$ terms, with each term having a multiplication of N-single particle wave functions, as follows:

$$\psi_a(1,2,....N) = \frac{1}{\sqrt{N!}} \sum_{n_1=1}^{N} \sum_{n_2=1}^{N} \cdots \sum_{n_N=1}^{N} \epsilon_{n_1n_2....n_N}\, \phi_{n_1}(1)\phi_{n_2}(2)....\phi_{n_N}(N) \tag{9.11b}$$

where $\epsilon_{n_1n_2....n_N}$ takes one of the values $1,-1,0$ at a time, depending upon in which order $n_1n_2...n_N$ are assigned the numbers.

As is seen from Eqns. (9.10), $\psi_s(1,2)$ is obtainable from $\psi_a(1,2)$ by replacing the minus sign with a plus sign. In the same manner, $\psi_s(1,2,3,......N)$ for a system of non-interacting particles is obtained by replacing all the minus signs with plus signs in a linear form of $\psi_a(1,2,3,......N)$. The total energy of the N-particles system is $E = E_1 + E_2 + ... + E_N$.

The properties of a determinant remind us of the following:

1. If two rows of a determinant are identical, its value is zero, which implies that the probability $|\psi_a|^2$ of finding more than one particle in a quantum state is zero. This is exactly the statement of Pauli's exclusion principle. We thus see that a system represented by an antisymmetric wave function ket consists of Fermions, which obey the Fermi-Dirac statistics and have half odd integer spins $\left\{\frac{1}{2}, \frac{3}{2}, ...\right\}$.
2. The determinant also goes to zero if any two columns are identical. This means that no two Fermions can occupy the same position.

On the other hand, $|\psi_s|^2$ is nonzero even if more than one particle are allowed to occupy the same quantum state or the same position. The particles which are represented by a symmetric wave function ket are known as Bosons; they follow Bose-Einstein statistics and have integer spins $(1,2,3,....)$. The treatment of indistinguishable particles tells us all elementary particles found in nature fall into two categories, one that is described by a symmetric wave function ket and another which is represented by an antisymmetric wave function ket. Though the analysis presented in this section has been made for non-interacting particles, the conclusions drawn are of a general nature and these are applicable to a system of N interacting particles, too.

9.1.2 Space and Spin Parts of Wave Function

In the above section we said that numbers compositely represent both space and spin coordinates. The wave function of a single particle consists of both space and spin dependent parts. If the Hamiltonian $H(1,2,3,.....,N)$ does not have a space-spin interaction term, then the single particle wave function ket is expressible as the product of the space-dependent and spin-dependent wave function $|\phi_\alpha\rangle = |\varphi_\alpha(\mathbf{r})\rangle|\chi_{m_s}\rangle$. Here, composite suffix α represents a set of quantum numbers n,l,m. The spin part of the wave function ket $|\chi_{m_s}\rangle$ describes the orientation of spin of the particle. Unlike position coordinates, a particle with spin s can only have $(2s+1)$ spin orientations. As discussed in Section 5.5, $|\chi_{m_s}\rangle$ are represented by column matrices having $(2s+1)$ rows with zeros at all places except one place. Two spin wave function kets of s_z for a spin $\left(\pm\frac{1}{2}\right)$ system are:

$$|\chi_\uparrow\rangle = \begin{pmatrix} 1 \\ 0 \end{pmatrix} = \uparrow; \quad m_s = \frac{1}{2}, \quad \text{and} \quad |\chi_\downarrow\rangle = \begin{pmatrix} 0 \\ 1 \end{pmatrix} = \downarrow; \quad m_s = -\frac{1}{2} \tag{9.12}$$

In case of two electron system, symmetric and antisymmetric wave function kets are written as follows:

$$|\psi_s(1,2)\rangle = \begin{cases} |\varphi_s(\mathbf{r}_1,\mathbf{r}_2)\rangle|\chi_s(1,2)\rangle \\ |\varphi_a(\mathbf{r}_1,\mathbf{r}_2)\rangle|\chi_a(1,2)\rangle \end{cases} \tag{9.13a}$$

and:

$$|\psi_a(1,2)\rangle = \begin{cases} |\varphi_a(\mathbf{r}_1,\mathbf{r}_2)\rangle|\chi_s(1,2)\rangle \\ |\varphi_s(\mathbf{r}_1,\mathbf{r}_2)\rangle|\chi_a(1,2)\rangle \end{cases} \tag{9.13b}$$

where, spatial parts are given by:

$$\varphi_s(\mathbf{r}_1,\mathbf{r}_2) = \frac{1}{\sqrt{2!}}\{\phi_\alpha(\mathbf{r}_1)\phi_\beta(\mathbf{r}_2) + \phi_\alpha(\mathbf{r}_2)\phi_\beta(\mathbf{r}_1)\} \tag{9.14a}$$

and:

$$\varphi_a(\mathbf{r}_1,\mathbf{r}_2) = \frac{1}{\sqrt{2!}}\{\phi_\alpha(\mathbf{r}_1)\phi_\beta(\mathbf{r}_2) - \phi_\alpha(\mathbf{r}_2)\phi_\beta(\mathbf{r}_1)\} \tag{9.14b}$$

The spin parts of wave function kets are:

$$\chi_a = \frac{1}{\sqrt{2}}\{\uparrow_1\downarrow_2 - \uparrow_2\downarrow_1\} m_s = 0;\ S = 0, \text{ singlet state} \tag{9.15a}$$

$$\chi_s = \begin{bmatrix} \uparrow_1\uparrow_2 & m_s = 1 \\ \frac{1}{\sqrt{2}}\{\uparrow_1\downarrow_2 + \uparrow_2\downarrow_1\} & m_s = 0 \\ \downarrow_1\downarrow_2 & m_s = -1 \end{bmatrix};\ S = 1, \text{ triplet state} \tag{9.15b}$$

The spin part of the wave function ket is an example of addition of angular momenta. Symmetric and antisymmetric wave function kets for a system of more than two electrons are constructed in a similar manner by taking the appropriate products of spatial and spin wave functions.

9.2 The Helium Atom

The helium atom is the simplest many particle system. Calculation of the ground state energy of the helium atom with the use of the variation method is presented in Section 7.2.1, without taking into consideration the spin symmetry effect and exchange degeneracy. It has distinct sets of quantum numbers where two electrons can have an antiparallel spins state or a parallel spins state. However, Pauli's exclusion principle does not permit parallel spin state in the ground state ($1s^2$) of helium. In the ground state, $n=1$, $l=0$ and $m=0$ for both the electrons; therefore two electrons must have antiparallel spins to obey the Pauli's exclusion principle. This is depicted in Fig. 9.1.

Let us rewrite the Hamiltonian of the helium atom given by Eqn. (7.65) as follows:

$$H = H^0 + H';$$
$$H^0 = -\frac{\hbar^2}{2\mu}\nabla_1^2 - \frac{2e^2}{4\pi\varepsilon_0 r_1} - \frac{\hbar^2}{2\mu}\nabla_2^2 - \frac{2e^2}{4\pi\varepsilon_0 r_2} \tag{9.16}$$
$$H' = \frac{e^2}{4\pi\varepsilon_0 |\mathbf{r}_1 - \mathbf{r}_2|}$$

where μ is the reduced mass of the electron. Since, helium has two identical electrons, which obey Puali's exclusion principle, the wave function ket described by Eqn. (9.13b) has to be antisymmetric. As is seen, the Hamiltonian does not have any spin-dependent term, and hence the single particle wave function ket can be taken as a product of space-dependent and spin-dependent wave function kets. We aim to evaluate $H|\psi_a(1,2)\rangle = E|\psi_a(1,2)\rangle$, where $|\psi_a(1,2)\rangle$ is a twofold degenerate state, as has been described above by Eqn. (9.13b). We use the first order perturbation theory to estimate the E by treating H' as the perturbating part of the Hamiltonian. Following the procedure discussed in Section 7.1.2, we write:

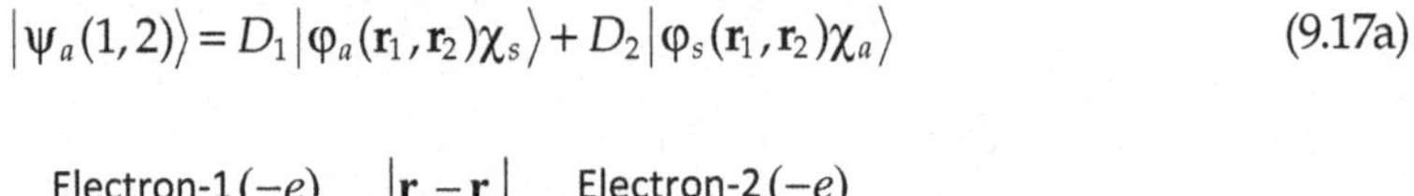

$$|\psi_a(1,2)\rangle = D_1|\varphi_a(\mathbf{r}_1,\mathbf{r}_2)\chi_s\rangle + D_2|\varphi_s(\mathbf{r}_1,\mathbf{r}_2)\chi_a\rangle \tag{9.17a}$$

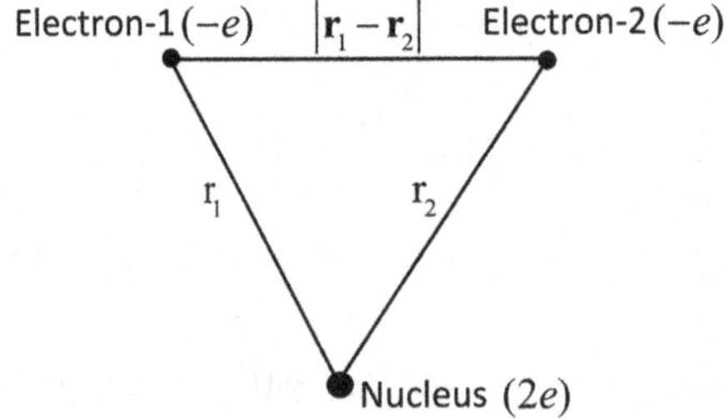

FIGURE 9.1
Schematic diagram of the helium atom.

and:

$$H^0|\psi_a(1,2)\rangle = E^0|\psi_a(1,2)\rangle \tag{9.17b}$$

It is to be noted here that if both the electrons occupy the same state (1*s*-state), $\alpha=\beta$ then the antisymmetric wave function is simply given by $\psi_a(1,2)=\phi_\alpha(1)\phi_\alpha(2)\chi_a(1,2)$.

As is seen from Eqn. (9.16), H^0 is the sum of two Hamiltonians for hydrogen-like atoms. Therefore, we represent $\phi_\alpha(\mathbf{r}_1)$ and $\phi_\beta(\mathbf{r}_2)$ by hydrogen-like wave functions given by Eqn. (6.35). We have:

$$\left.\begin{aligned} H(1)|\phi_i(\mathbf{r}_1)\rangle &= \left(-\frac{\hbar^2}{2\mu}\nabla_1^2 - \frac{2e^2}{4\pi\varepsilon_0 r_1}\right)|\phi_i(\mathbf{r}_1)\rangle = E_i|\phi_i(\mathbf{r}_1)\rangle; \\ H(2)|\phi_i(\mathbf{r}_2)\rangle &= \left(-\frac{\hbar^2}{2\mu}\nabla_2^2 - \frac{2e^2}{4\pi\varepsilon_0 r_2}\right)|\phi_i(\mathbf{r}_2)\rangle = E_i|\phi_i(\mathbf{r}_2)\rangle \end{aligned}\right\}, \text{where } i\equiv(\alpha,\beta) \tag{9.18}$$

Let us consider:

$$\begin{aligned} &H^0\{\phi_\alpha(\mathbf{r}_1)\phi_\beta(\mathbf{r}_2)\pm\phi_\alpha(\mathbf{r}_2)\phi_\beta(\mathbf{r}_1)\} \\ &=\phi_\beta(\mathbf{r}_2)H(1)\phi_\alpha(\mathbf{r}_1)+\phi_\alpha(\mathbf{r}_1)H(2)\phi_\beta(\mathbf{r}_2)\pm\phi_\alpha(\mathbf{r}_2)H(1)\phi_\beta(\mathbf{r}_1)\pm\phi_\beta(\mathbf{r}_1)H(2)\phi_\alpha(\mathbf{r}_2) \\ &=(E_\alpha+E_\beta)\phi_\alpha(\mathbf{r}_1)\phi_\beta(\mathbf{r}_2)\pm(E_\beta+E_\alpha)\phi_\alpha(\mathbf{r}_2)\phi_\beta(\mathbf{r}_1) \\ &=(E_\alpha+E_\beta)\{\phi_\alpha(\mathbf{r}_1)\phi_\beta(\mathbf{r}_2)\pm\phi_\alpha(\mathbf{r}_2)\phi_\beta(\mathbf{r}_1)\} \end{aligned} \tag{9.19}$$

which implies:

$$\begin{aligned} H^0|\psi_a(1,2)\rangle &= D_1H^0|\varphi_a(\mathbf{r}_1,\mathbf{r}_2)\chi_s\rangle + D_2H^0|\varphi_s(\mathbf{r}_1,\mathbf{r}_2)\chi_a\rangle \\ &= (E_\alpha+E_\beta)|\psi_a(1,2)\rangle \end{aligned} \tag{9.20}$$

Therefore:

$$E^0 = E_\alpha + E_\beta \tag{9.21}$$

where each of E_α and E_β is the energy of a state in the hydrogen atom with charge $2e$ at the nucleus. Therefore:

$$E^0 = -\frac{4\mu e^4}{2(4\pi\varepsilon_0)^2\hbar^2}\left[\frac{1}{n_\alpha^2}+\frac{1}{n_\beta^2}\right] \tag{9.22}$$

where both n_α and n_β take values 1,2,3,.....

The first order correction to energy E', which is obtained from Eqn. (7.49) for the degenerate system, is:

$$\begin{vmatrix} h_{aa}-E' & h_{as} \\ h_{sa} & h_{ss}-E' \end{vmatrix} = 0 \tag{9.23}$$

where:

$$h_{aa} = \langle \varphi_a(\mathbf{r}_1,\mathbf{r}_2)|H'|\varphi_a(\mathbf{r}_1,\mathbf{r}_2)\rangle\langle \chi_s|\chi_s\rangle = \langle \varphi_a(\mathbf{r}_1,\mathbf{r}_2)|H'|\varphi_a(\mathbf{r}_1,\mathbf{r}_2)\rangle \tag{9.24a}$$

$$h_{ss} = \langle \varphi_s(\mathbf{r}_1,\mathbf{r}_2)|H'|\varphi_s(\mathbf{r}_1,\mathbf{r}_2)\rangle\langle \chi_a|\chi_a\rangle = \langle \varphi_s(\mathbf{r}_1,\mathbf{r}_2)|H'|\varphi_s(\mathbf{r}_1,\mathbf{r}_2)\rangle \tag{9.24b}$$

and:

$$h_{as} = h_{sa}^* = \langle \varphi_a(\mathbf{r}_1,\mathbf{r}_2)|H'|\varphi_s(\mathbf{r}_1,\mathbf{r}_2)\rangle\langle \chi_a|\chi_s\rangle \tag{9.24c}$$

Since $\langle \chi_a|\chi_s\rangle = 0 = \langle \chi_s|\chi_a\rangle$, $h_{as} = 0 = h_{sa}$. Therefore, two values of E' are:

$$E'_- = h_{aa} = \langle \varphi_a(\mathbf{r}_1,\mathbf{r}_2)|\frac{e^2}{4\pi\varepsilon_0|\mathbf{r}_1-\mathbf{r}_2|}|\varphi_a(\mathbf{r}_1,\mathbf{r}_2)\rangle \tag{9.25a}$$

and:

$$E'_+ = h_{ss} = \langle \varphi_s(\mathbf{r}_1,\mathbf{r}_2)|\frac{e^2}{4\pi\varepsilon_0|\mathbf{r}_1-\mathbf{r}_2|}|\varphi_s(\mathbf{r}_1,\mathbf{r}_2)\rangle \tag{9.25b}$$

After substituting Eqns. (9.14) into Eqn. (9.25), we rearrange as follows:

$$\begin{aligned} E'_\pm = \frac{e^2}{4\pi\varepsilon_0}\Bigg[&\frac{1}{2}\int\int|\phi_\alpha(\mathbf{r}_1)|^2|\phi_\beta(\mathbf{r}_2)|^2\frac{1}{|\mathbf{r}_1-\mathbf{r}_2|}d^3r_1d^3r_2 \\ &+\frac{1}{2}\int\int|\phi_\alpha(\mathbf{r}_2)|^2|\phi_\beta(\mathbf{r}_1)|^2\frac{1}{|\mathbf{r}_1-\mathbf{r}_2|}d^3r_1d^3r_2 \\ &\pm\frac{1}{2}\int\int\phi^*_\alpha(\mathbf{r}_1)\phi^*_\beta(\mathbf{r}_2)\frac{1}{|\mathbf{r}_1-\mathbf{r}_2|}\phi_\beta(\mathbf{r}_1)\phi_\alpha(\mathbf{r}_2)d^3r_1d^3r_2 \\ &\pm\frac{1}{2}\int\int\phi^*_\alpha(\mathbf{r}_2)\phi^*_\beta(\mathbf{r}_1)\frac{1}{|\mathbf{r}_1-\mathbf{r}_2|}\phi_\beta(\mathbf{r}_2)\phi_\alpha(\mathbf{r}_1)d^3r_1d^3r_2\Bigg] \end{aligned} \tag{9.26}$$

As is seen, integration is done over both $\mathbf{r}_1$ and $\mathbf{r}_2$; hence, interchanging of $\mathbf{r}_1$ and $\mathbf{r}_2$ does not affect the value of any term. However, interchanging of $\mathbf{r}_1$ and $\mathbf{r}_2$ in the second and fourth terms makes these equal to first and third terms, respectively. We therefore obtain $E'_\pm = J_{\alpha\beta} \pm K_{\alpha\beta}$, where:

$$J_{\alpha\beta} = \int\int|\phi_\alpha(\mathbf{r}_1)|^2|\phi_\beta(\mathbf{r}_2)|^2\frac{e^2}{4\pi\varepsilon_0|\mathbf{r}_1-\mathbf{r}_2|}d^3r_1d^3r_2 \tag{9.27a}$$

and:

$$K_{\alpha\beta} = \int\int\phi^*_\alpha(\mathbf{r}_1)\phi^*_\beta(\mathbf{r}_2)\frac{e^2}{4\pi\varepsilon_0|\mathbf{r}_1-\mathbf{r}_2|}\phi_\beta(\mathbf{r}_1)\phi_\alpha(\mathbf{r}_2)d^3r_1d^3r_2 \tag{9.27b}$$

The $J_{\alpha\beta}$ is a direct electron-electron interaction contribution, whereas $K_{\alpha\beta}$ is an exchange electron-electron contribution to the energy of the helium atom. Thus the two energies of helium atoms are: $E_+ = E^0 + J_{\alpha\beta} + K_{\alpha\beta}$; $E_- = E^0 + J_{\alpha\beta} - K_{\alpha\beta}$, and their difference:

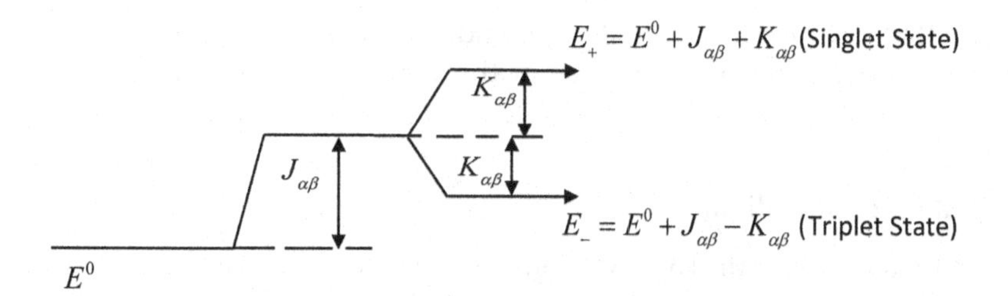

FIGURE 9.2
Schematic diagram for energy level splitting for the $|100\rangle$ to $|nlm\rangle$ transition in the helium atom.

$\Delta E = E_+ - E_- = 2K_{\alpha\beta}$. Note in the case of $\alpha = \beta$, there is no exchange electron-electron interaction and hence $K_{\alpha\beta} = 0$.

In Eqn. (9.27a), $-e|\phi_\alpha(\mathbf{r}_1)|^2 d^3r_1$ and $-e|\phi_\beta(\mathbf{r}_2)|^2 d^3r_2$ represent the charge distribution in infinitesimally small volumes d^3r_1 and d^3r_2, respectively. Therefore, $J_{\alpha\beta}$ is the quantum analogue of classical electrostatic interaction between two charges. However, there is no classical equivalence of $K_{\alpha\beta}$; it is purely a quantum mechanical correction to the Coulomb interaction energy. Calculation of $J_{\alpha\beta}$ and $K_{\alpha\beta}$ with the use of hydrogen-like wave functions shows that both are positive. Hence, $E_+ > E_-$. Looking at Eqns. (9.13) to (9.15), we notice that (i) E_+ corresponds to $|\varphi_s\rangle$ and $|\chi_a\rangle$, which represents the singlet state, and (ii) E_- belongs to $|\varphi_a\rangle$ and $|\chi_s\rangle$, which corresponds to the triplet state. Therefore, the singlet state has higher energy as compared to the triplet state, as is shown in Fig. 9.2.

The physical interpretation is as follows: The singlet state is represented by the symmetric space-coordinate dependent part of the wave function, and therefore electrons have the tendency to come closer, making the electrostatic repulsion less serious, which results in higher energy, whereas, the triplet state involves the antisymmetric space-dependent part of the wave function, giving rise to stronger repulsive interaction and hence lower energy. Helium in the singlet state is known as **parahelium,** while in the triplet state it is termed **orthohelium**. In the ground state helium is *parahelium* only. It has been verified experimentally that the singlet state has more energy than the triplet state.

9.2.1 Ground State of Helium

In the ground state both the electrons are in the $|100\rangle$ state ($\alpha = \beta$); therefore, $K_{\alpha\beta} = 0$ and:

$$\begin{aligned} J_{100} &= \frac{e^2}{4\pi\varepsilon_0}\int\int|\psi_{100}(\mathbf{r}_1)|^2|\psi_{100}(\mathbf{r}_2)|^2\frac{1}{|\mathbf{r}_1-\mathbf{r}_2|}d^3r_1d^3r_2 \\ &= \left(\frac{e^2}{4\pi\varepsilon_0}\right)\times\frac{5}{4a_0} \\ &= -\frac{5}{2}E_1 \end{aligned} \tag{9.28}$$

where, we have used the results reported in Section 7.2.1 (see Eqn. 7.74 with $Z = 2$). The E_1 (= −13.6 eV) is the ground state energy of hydrogen atom; Eqn. (6.22). The ground state energy of the helium atom then is:

$$E_g = 8E_1 - \frac{5}{2}E_1 = \frac{11}{2}E_1 = -\frac{11}{2}\times 13.6 = -74.8\text{eV}. \tag{9.29}$$

We thus see that the perturbation theory underestimates the ground state energy as compared to that evaluated with the use of the variational method ($E_g = -77.6$ eV); see Eqn. (7.77).

9.2.2 Excited State of Helium

In the excited state of helium two elections are in two different states. Let us consider one electron in $\alpha \equiv |100\rangle$ ($1s$) and the other electron in $\beta \equiv |nlm\rangle$, which could be any of $2s, 2p$, etc. In the excited state, both $J_{\alpha\beta}$ and $K_{\alpha\beta}$ will be nonzero, and therefore both *para* and *ortho* forms exist. As is seen from Eqns. (9.14a) and (9.14b), computation of $|\varphi_a(\mathbf{r}_1, \mathbf{r}_2)|^2$ and $|\varphi_s(\mathbf{r}_1, \mathbf{r}_2)|^2$ as a function of $|\mathbf{r}_1 - \mathbf{r}_2|$, gives $|\varphi_a(\mathbf{r}_1, \mathbf{r}_2)|^2 \to 0$ and $|\varphi_s(\mathbf{r}_1, \mathbf{r}_2)|^2$ equal to a maximum value at $r_1 = r_2$. This too means that two electrons have the tendency to avoid each other when represented by $|\varphi_a\rangle$, whereas they like to reach closest to each other when represented by $|\varphi_s\rangle$. Hence, in the excited state, parahelium has more energy than orthohelium. The dip in $|\varphi_a(\mathbf{r}_1, \mathbf{r}_2)|^2$ around $r_1 = r_2$ is called the **Fermi hole**, while an increase in $|\varphi_s(\mathbf{r}_1, \mathbf{r}_2)|^2$ is termed a **Fermi heap**. Each configuration, $1s-2s$, $1s-2p$, etc., will split into the *para state* and the *ortho state*, where the para state lies higher on the energy scale. It is important to note that though the Hamiltonian described by Eqn. (9.16) is independent of spin variables, yet there is a spin-dependent effect on energy, which arises from the Fermi-Dirac statistics. This can be understood from the Heisenberg theory of ferromagnetism. The idea of the alignment of spins has been extended over microscopic distances in the helium atom to give a physical interpretation of spin-dependent energy.

The $J_{\alpha\beta}$ and $K_{\alpha\beta}$ have been evaluated for $1s-2s$ and $1s-2p$ configurations. The values for $1s-2s$ configuration are: $J_{1s,2s} = -\frac{68}{81}E_1 = 11.42$ eV and $K_{1s,2s} = -\left(\frac{2}{3}\right)^6 E_1 = 1.20$ eV; for $1s-2p$ configuration: $J_{1s,2p} = -\frac{236}{243}E_1 = 13.22$ eV and $K_{1s,2p} = -\frac{7\times 2^6}{3^8}E_1 = 0.94$ eV. The singlet and triplet states are separated by 2.4 eV in $1s-2s$ configuration and by 1.88 eV in $1s-2p$ configuration.

9.3 Systems of N-Electrons

The Hamiltonian for an N-electron system such as a metal or a large atom is the generalization of Eqn. (9.16):

$$H = \sum_{i=1}^{N}\left(-\frac{\hbar^2}{2m}\nabla_i^2 + V_{ext}(r_i) + \frac{1}{2}\sum_{j\neq i}\frac{e^2}{4\pi\varepsilon_0|\mathbf{r}_i - \mathbf{r}_j|}\right) \tag{9.30}$$

where single electron potential at the position $\mathbf{r}_i$ is $V_{ext}(r_i) = -\frac{Ze^2}{4\pi\varepsilon_0}\sum_R \frac{1}{|\mathbf{r}_i - \mathbf{R}|}$ for a metal, with $\mathbf{R}$ a position vector of a bare nuclei at a Bravais lattice point, and $V_{ext}(r_i) = -\frac{Ze^2}{4\pi\varepsilon_0 r_i}$ for an atom. In the case of a metal, the Hamiltonian may also be written to have an additional constant term arising from nuclei-nuclei interactions. Since this does not affect the physics of electrons, it is dropped here. The N-electrons wave function $\psi(\mathbf{r}_1 s_1, \mathbf{r}_2 s_2, \mathbf{r}_3 s_3, ...\mathbf{r}_N s_N)$, $\mathbf{r}_i s_i \equiv i$ representing both position and spin coordinates, satisfies the Schrödinger equation:

$$H\psi = E\psi \tag{9.31}$$

The exact solution of Eqn. (9.31) is not possible, and hence one looks for an approximate solution that is most reasonable. An electron interacts with N-1 electrons, in addition to its interaction with nuclei.

The methods based on the variational principle have been found most successful and reasonable approximations to solve the N-electrons Schrödinger equation. The variational principle assumes that the equation:

$$\frac{\delta}{\delta|\psi\rangle}\{\langle\psi|H|\psi\rangle - E\langle\psi|\psi\rangle\} = 0 \tag{9.32}$$

is equivalent to the Schrödinger equation. In other words, $|\psi\rangle$ that satisfies above equation is a solution of the Schrödinger equation or is an eigenfunction. More explicitly, if we take the functional derivative of:

$$\langle H - E\rangle = \int \psi^*(\mathbf{r})H\psi(\mathbf{r})d^3r - E\int \psi^*(\mathbf{r})\psi(\mathbf{r})d^3r \tag{9.33a}$$

with respect to $\psi^*(\mathbf{r})$, we get:

$$\delta\langle H - E\rangle = \int \{H\psi(\mathbf{r}) - E\psi(\mathbf{r})\}\delta\psi^*(\mathbf{r})d^3r \tag{9.33b}$$

To make this stationary, the variation must be zero for all possible forms of $\delta\psi^*(\mathbf{r})$, and hence we should have $H\psi(\mathbf{r}) - E\psi(\mathbf{r}) = 0$, which is the Schrödinger equation.

9.3.1 Hartree Approximation

We rewrite Eqn. (9.30) as follows:

$$H = \sum_{i=1}^{N}\left(H_i + \frac{1}{2}\sum_{j\neq i} V_{ij}\right) \tag{9.34}$$

with:

$$H_i = -\frac{\hbar^2}{2m}\nabla_i^2 + V_{ext}(\mathbf{r}_i) \tag{9.35a}$$

and:

$$V_{ij} = \frac{e^2}{4\pi\varepsilon_0|\mathbf{r}_i - \mathbf{r}_j|} \tag{9.35b}$$

We take the trial wave function:

$$\psi(1,2,3,.....N) = \phi_1(1_1)\phi_2(2)\phi_3(3).........\phi_N(N) \tag{9.36}$$

with $\langle \phi_i(\mathbf{r}_i s_i) | \phi_j(\mathbf{r}_j s_j) \rangle = \delta_{ij}$. We then have:

$$\begin{aligned} &\langle \psi | H | \psi \rangle \\ &= \sum_{i=1}^{N} \int d^3 r_1 \int d^3 r_2 ... \int d^3 r_i .... \int d^3 r_N \phi_1^*(\mathbf{r}_1)\phi_2^*(\mathbf{r}_2)...\phi_i^*(\mathbf{r}_i)...\phi_N^*(\mathbf{r}_N) \\ &\times \left\{ H_i + \sum_{j \neq i} V_{ij} \right\} \phi_1(\mathbf{r}_1)\phi_2(\mathbf{r}_2)...\phi_i(\mathbf{r}_i)...\phi_N(\mathbf{r}_N) \end{aligned} \tag{9.37}$$

Here,

$$\begin{aligned} &\int d^3 r_1 \int d^3 r_2 ... \int d^3 r_i .... \int d^3 r_N \phi_1^*(\mathbf{r}_1)\phi_2^*(\mathbf{r}_2)...\phi_i^*(\mathbf{r}_i)...\phi_N^*(\mathbf{r}_N) H_i \phi_1(\mathbf{r}_1)\phi_2(\mathbf{r}_2)...\phi_i(\mathbf{r}_i)...\phi_N(\mathbf{r}_N) \\ &= \int d^3 r_i \phi_i^*(\mathbf{r}_i) H_i \phi_i(\mathbf{r}_i) \end{aligned} \tag{9.38a}$$

Note that: because H_i operates only on $\phi_i(\mathbf{r}_i)$ and integration over each of the other coordinates is equal to one. The electron-electron interaction term simplifies to $\int\int d^3 r_i d^3 r_j \phi_i^*(\mathbf{r}_i)\phi_j^*(\mathbf{r}_j) V_{ij} \phi_i(\mathbf{r}_i)\phi_j(\mathbf{r}_j)$. We then have:

$$\langle \psi | H | \psi \rangle = \sum_{i=1}^{N} \int d^3 r_i \phi_i^*(\mathbf{r}_i) \left\{ H_i \phi_i(\mathbf{r}_i) + \frac{1}{2} \sum_{j(\neq i)=1}^{N} \int d^3 r_j \phi_j^*(\mathbf{r}_j) V_{ij} \phi_j(\mathbf{r}_j) \phi_i(\mathbf{r}_i) \right\} \tag{9.38b}$$

To minimize $\langle \psi | H | \psi \rangle$, we take the functional derivative with respect to $\phi_k^*(\mathbf{r}_k)$ and obtain:

$$\int d^3 r_k \delta\phi_k^*(\mathbf{r}_k) \left\{ H_k \phi_k(\mathbf{r}_k) + \sum_{j(\neq k)=1}^{N} \int d^3 r_j \phi_j^*(\mathbf{r}_j) V_{kj} \phi_j(\mathbf{r}_j) \phi_k(\mathbf{r}_k) \right\} = 0 \tag{9.39a}$$

It is to be noted that the factor of $\frac{1}{2}$ before interaction term drops out because two equal terms, one for $i = k$ and the other for $j = k$, appear when we take the derivative with respect to $\phi_k^*(\mathbf{r}_k)$.

Further notice that $\langle \phi_k^* | \phi_k \rangle = 1$, and therefore:

$$\varepsilon_k \int d^3 r_k \delta\phi_k^*(\mathbf{r}_k) \phi_k(\mathbf{r}_k) = 0 \text{ with } k = 1, 2, 3, ...N \tag{9.39b}$$

Here, ε_k is a multiplying constant. Subtracting Eqn. (9.39b) from Eqn. (9.39a) we get:

$$\int d^3 r_k \, \delta\phi_k^*(\mathbf{r}_k) \left\{ H_k \phi_k(\mathbf{r}_k) + \sum_{j(\neq k)=1}^{N} \int d^3 r_j \phi_j^*(\mathbf{r}_j) V_{kj} \phi_j(\mathbf{r}_j) \phi_k(\mathbf{r}_k) - \varepsilon_k \phi_k(\mathbf{r}_k) \right\} = 0 \tag{9.40}$$

To make it stationary, the variations are zero for all possible forms of $\delta\phi_k^*(\mathbf{r}_k)$; hence we should have:

$$H_k\phi_k(\mathbf{r}_k) + \sum_{j(\neq k)=1}^{N} \int d^3r_j \phi_j^*(\mathbf{r}_j) V_{kj} \phi_j(\mathbf{r}_j) \phi_k(\mathbf{r}_k) = \varepsilon_k \phi_k(\mathbf{r}_k) \tag{9.41a}$$

More explicitly:

$$\left\{ -\frac{\hbar^2}{2m}\nabla_i^2 + V_{ext}(\mathbf{r}_i) + \frac{e^2}{4\pi\varepsilon_0} \sum_{j(\neq i)=1}^{N} \int d^3r_j \phi_j^*(\mathbf{r}_j) \frac{1}{|\mathbf{r}_i - \mathbf{r}_j|} \phi_j(\mathbf{r}_j) \right\} \phi_i(\mathbf{r}_i) = \varepsilon_i \phi_i(\mathbf{r}_i) \tag{9.41b}$$

which is an *integro-differential equation,* commonly known as the Hartree equation. It is an eigenvalue equation for an electron located at position $\mathbf{r}_i$ and moving under the influence of an effective potential:

$$V_{eff}(\mathbf{r}_i) = V_{ext}(\mathbf{r}_i) + V_{ee}(\mathbf{r}_i) \tag{9.42a}$$

with:

$$V_{ee}(\mathbf{r}_i) = \frac{e^2}{4\pi\varepsilon_0} \sum_{j\neq i} \int d^3r_j \frac{|\phi_j(\mathbf{r}_j)|^2}{|\mathbf{r}_i - \mathbf{r}_j|} \tag{9.42b}$$

The simple interpretation of this is as follows: The ith electron interacts with both nuclei and the remaining N-1 electrons. The remaining electrons are treated as a smooth distribution of negative charge with charge density $\rho(\mathbf{r}_j)$ around the position $\mathbf{r}_j$; then the potential energy due to the interaction between the ith electron and the charge in infinitesimally small volume d^3r_j is: $-\frac{e}{4\pi\varepsilon_0}\frac{\rho(\mathbf{r}_j)}{|\mathbf{r}_i - \mathbf{r}_j|}d^3r_j$. The contribution to charge density by a single electron and by the N-1 electron would be $-e|\phi_j(\mathbf{r}_j)|^2$ and $-e\sum_{j\neq i}|\phi_j(\mathbf{r}_j)|^2$, respectively. Hence, Eqn. (9.42b) represents the potential energy due to the interaction of the ith electron with the remaining N-1 electrons.

We thus find that the Hartree approximation converts the N-particle problem into a set of N-single particle equations that can be solved. It is to be noted that solving Eqn. (9.41b) for state $|\phi_i(\mathbf{r}_i)\rangle$ and eigenvalue ε_i requires the prior knowledge of $|\phi_j(\mathbf{r}_j)\rangle$. Therefore, the equation is to be solved by iterative methods. One starts with a guessed single particle state to compute the Eqn. (9.42b). The computed $V_{eff}(r_i)$ is then used to compute a new state. If the computed state differs from that used to compute $V_{ee}(r_i)$, repeat the procedure until a *self-consistency* is not achieved.

9.3.2 Hartree-Fock Approximation

Equation (9.41b) is a single electron equation that uses a potential obtained by averaging over the positions of the remaining N-1 electrons. It does not represent the way in which a particular electron is affected by the configuration of other N-1 electrons. In other words, Eqn. (9.41b) does not take care of the exchange effect, which demands that the wave function for a system of N-Fermions must be anti-symmetric. As discussed before, an anti-symmetric wave function is given by Eqn. (9.11b) not by Eqn. (9.36). However, the Hamiltonian of the system of N-electrons is still given by Eqn. (9.34). To derive the single electron equation within the Hartree-Fock approximation, we follow the procedure outlined for the Hartree approximation, with the use of the wave function given by Eqn. (9.11b), which is:

$$\psi_a(\mathbf{r}_1,\mathbf{r}_2,....\mathbf{r}_N)=\frac{1}{\sqrt{N!}}\sum_{n_1=1}^{N}\sum_{n_2=1}^{N}...\sum_{n_N=1}^{N}\epsilon_{n_1n_2....n_N}\ \phi_{n_1}(\mathbf{r}_1)\phi_{n_2}(\mathbf{r}_2)....\phi_{n_N}(\mathbf{r}_N) \tag{9.43}$$

Let us first evaluate the matrix element of H_i, defined in Eqn. (9.35a):

$$\begin{aligned}&\langle\psi_a|H_i|\psi_a\rangle\\&=\frac{1}{N!}\int d^3r_1d^3r_2......d^3r_N\left[\sum_{n_1'=1}^{N}\sum_{n_2'=1}^{N}...\sum_{n_N'=1}^{N}\epsilon_{n_1'n_2'....n_N'}\left\{\phi_{n_1'}(\mathbf{r}_1)\phi_{n_2'}(\mathbf{r}_2)....\phi_{n_N'}(\mathbf{r}_N)\right\}^*\right]\times\\&\quad H_i\left[\sum_{n_1=1}^{N}\sum_{n_2=1}^{N}...\sum_{n_N=1}^{N}\epsilon_{n_1n_2....n_N}\ \phi_{n_1}(\mathbf{r}_1)\phi_{n_2}(\mathbf{r}_2)....\phi_{n_N}(\mathbf{r}_N)\right]\end{aligned} \tag{9.44}$$

There will be an integral over $(N!)^2$ terms, where each term involves a product of the N wave functions and an equal number of their complex conjugates. To understand the simplification of this equation, let us consider a case of two electrons situated at $\mathbf{r}_i$ and $\mathbf{r}_j$.

$$\begin{aligned}&\langle\psi_a|H_i|\psi_a\rangle\\&=\frac{1}{2!}\int d^3r_id^3r_j\left[\sum_{n_1'=1}^{2}\sum_{n_2'=1}^{2}\epsilon_{n_1'n_2'}\left\{\phi_{n_1'}(\mathbf{r}_i)\phi_{n_2'}(\mathbf{r}_j)\right\}^*\right]H_i\left[\sum_{n_1=1}^{2}\sum_{n_2=1}^{2}\epsilon_{n_1n_2}\ \phi_{n_1}(\mathbf{r}_i)\phi_{n_2}(\mathbf{r}_j)\right]\end{aligned} \tag{9.45a}$$

Using $\epsilon_{11}=\epsilon_{22}=0$ and $\epsilon_{12}=-\epsilon_{21}=1$, the equation reduces to:

$$\begin{aligned}&\langle\psi_a|H_i|\psi_a\rangle\\&=\frac{1}{2!}\int d^3r_id^3r_j\left[\phi_1^*(\mathbf{r}_i)\phi_2^*(\mathbf{r}_j)-\phi_2^*(\mathbf{r}_i)\phi_1^*(\mathbf{r}_j)\right]H_i\left[\phi_1(\mathbf{r}_i)\phi_2(\mathbf{r}_j)-\phi_2(\mathbf{r}_i)\phi_1(\mathbf{r}_j)\right]\end{aligned} \tag{9.45b}$$

which expands to:

$$\begin{aligned}&\langle\psi_a|H_i|\psi_a\rangle\\&=\frac{1}{2!}\left[\int d^3r_i\int d^3r_j|\phi_2(\mathbf{r}_j)|^2\,\phi_1^*(\mathbf{r}_i)H_i\phi_1(\mathbf{r}_i)+\int d^3r_i\int d^3r_j|\phi_1(\mathbf{r}_j)|^2\,\phi_2^*(\mathbf{r}_i)H_i\phi_2(\mathbf{r}_i)\right]\\&-\frac{1}{2!}\left[\int d^3r_i\left\{\int d^3r_j\phi_2^*(\mathbf{r}_j)\phi_1(\mathbf{r}_j)\right\}\phi_1^*(\mathbf{r}_i)H_i\phi_2(\mathbf{r}_i)+\int d^3r_i\left\{\int d^3r_j\phi_1^*(\mathbf{r}_j)\phi_2(\mathbf{r}_j)\right\}\phi_2^*(\mathbf{r}_i)H_i\phi_1(\mathbf{r}_i)\right]\end{aligned} \tag{9.45c}$$

The third and fourth terms are zero because of the orthonormality of $|\phi_1\rangle$ and $|\phi_2\rangle$. The second term becomes identical to the first term on interchanging integrals over $\mathbf{r}_i$ and $\mathbf{r}_j$. We therefore get:

$$\langle\psi_a|H_i|\psi_a\rangle = \int d^3r_i\phi_1^*(\mathbf{r}_i)H_i\phi_1(\mathbf{r}_i) \tag{9.46a}$$

The answer remains same, even when the two-electrons antisymmetric wave function is replaced by an N-electrons wave function to evaluate $\langle\psi_a|H_i|\psi_a\rangle$. However, in place of suffix 1, we can use a more generalized suffix n_i to write:

$$\langle\psi_a|H_i|\psi_a\rangle = \int d^3r_i\phi_{n_i}^*(\mathbf{r}_i)H_i\phi_{n_i}(\mathbf{r}_i) \tag{9.46b}$$

We next take the evaluation of $\langle\psi_a|V_{ij}|\psi_a\rangle$. Following the procedure of Eqns. (9.26) and (9.27), we find that for a two-electrons system:

$$\begin{aligned}\langle\psi_a|V_{ij}|\psi_a\rangle = &\frac{e^2}{4\pi\varepsilon_0}\iint|\phi_1(\mathbf{r}_i)|^2|\phi_2(\mathbf{r}_j)|^2\frac{1}{|\mathbf{r}_i-\mathbf{r}_j|}d^3r_id^3r_j \\ &-\frac{e^2}{4\pi\varepsilon_0}\iint\phi_1^*(\mathbf{r}_i)\phi_2^*(\mathbf{r}_j)\frac{1}{|\mathbf{r}_i-\mathbf{r}_j|}\phi_2(\mathbf{r}_i)\phi_1(\mathbf{r}_j)d^3r_id^3r_j\end{aligned} \tag{9.47a}$$

A generalization for N-electrons can made by representing quantum states by n_i and n_j in place of 1 and 2:

$$\begin{aligned}\langle\psi_a|V_{ij}|\psi_a\rangle = &\frac{e^2}{4\pi\varepsilon_0}\iint|\phi_{n_i}(\mathbf{r}_i)|^2|\phi_{n_j}(\mathbf{r}_j)|^2\frac{1}{|\mathbf{r}_i-\mathbf{r}_j|}d^3r_id^3r_j \\ &-\frac{e^2}{4\pi\varepsilon_0}\iint\phi_{n_i}^*(\mathbf{r}_i)\phi_{n_j}^*(\mathbf{r}_j)\frac{1}{|\mathbf{r}_i-\mathbf{r}_j|}\phi_{n_j}(\mathbf{r}_i)\phi_{n_i}(\mathbf{r}_j)d^3r_id^3r_j\end{aligned} \tag{9.47b}$$

From Eqn. (9.34), we then have:

$$\begin{aligned}\langle\psi_a|H|\psi_a\rangle = \sum_{i=1}^{N}\Bigg[&\int\phi_{n_i}^*(\mathbf{r}_i)\left\{-\frac{\hbar^2}{2m}\nabla_i^2+V_{ext}(\mathbf{r}_i)\right\}\phi_{n_i}(\mathbf{r}_i)d^3r_i \\ &+\frac{1}{2}\sum_{i\neq j}\iint|\phi_{n_i}(\mathbf{r}_i)|^2|\phi_{n_j}(\mathbf{r}_j)|^2\frac{e^2}{4\pi\varepsilon_0|\mathbf{r}_i-\mathbf{r}_j|}d^3r_id^3r_j \\ &-\frac{1}{2}\sum_{i\neq j}\iint\phi_{n_i}^*(\mathbf{r}_i)\phi_{n_j}^*(\mathbf{r}_j)\frac{e^2}{4\pi\varepsilon_0|\mathbf{r}_i-\mathbf{r}_j|}\phi_{n_j}(\mathbf{r}_i)\phi_{n_i}(\mathbf{r}_j)d^3r_id^3r_j\Bigg]\end{aligned} \tag{9.48}$$

On the right-hand side, the third term consists of $\phi^*_{n_i}(\mathbf{r}_i)\phi_{n_i}(\mathbf{r}_j)$, while the second term has $|\phi_{n_i}(\mathbf{r}_i)|^2$. Minimization with respect to $\phi^*_{n_i}(\mathbf{r}_i)$ following the procedure described for Eqns. (9.39) to (9.41) gives us:

$$\left\{-\frac{\hbar^2}{2m}\nabla_i^2+V_{ext}(\mathbf{r}_i)\right\}\phi_{n_i}(\mathbf{r}_i)+\frac{e^2}{4\pi\varepsilon_0}\left\{\sum_{j\neq i}\int|\phi_{n_j}(\mathbf{r}_j)|^2\frac{1}{|\mathbf{r}_i-\mathbf{r}_j|}d^3r_j\right\}\phi_{n_i}(\mathbf{r}_i)$$
$$-\frac{e^2}{4\pi\varepsilon_0}\left\{\sum_{j\neq i}\int\phi^*_{n_j}(\mathbf{r}_j)\frac{1}{|\mathbf{r}_i-\mathbf{r}_j|}\phi_{n_i}(\mathbf{r}_j)d^3r_j\phi_{n_j}(\mathbf{r}_i)\right\}=\varepsilon_{n_i}\phi_{n_i}(\mathbf{r}_i) \tag{9.49}$$

This equation is known as the Hartree-Fock equation. On comparing Eqn. (9.49) with Eqn. (9.41b), we find that the third term on the left is originated from exchange interactions. Unlike the direct interactions term (second term), the exchange interactions term (third term) is nonlinear having the structure $V(\mathbf{r})\phi(\mathbf{r})=\int U(\mathbf{r}',\mathbf{r})\phi(\mathbf{r}')d^3r'$, which is an integral operator

9.3.3 Thomas-Fermi Theory

For a system of N-electrons in a stationary state, this theory tries to avoid the complicated many-electron wave function by using electron density $n(\mathbf{r})$, which is physically observable, measured, calculated, and easily visualized. $n(\mathbf{r})$ is simply a probability density to find the particle near some position $\mathbf{r}$ and it is defined by $n(\mathbf{r})=\phi^*(\mathbf{r})\phi(\mathbf{r})$, if there is one particle only. For a system of N-particles, electron density to find a particle at or around $\mathbf{r}$ is given by:

$$n(\mathbf{r})=N\int d^3r_2\int d^3r_3\ldots\ldots\int d^3r_N\psi^*(\mathbf{r},\mathbf{r}_2,\mathbf{r}_3,\ldots\ldots\mathbf{r}_N)\psi(\mathbf{r},\mathbf{r}_2,\mathbf{r}_3,\ldots\ldots\mathbf{r}_N) \tag{9.50}$$

Essentially, the Thomas-Fermi method concentrates on the particle density as a key variable by circumventing completely the discussion on the wave function.

The Thomas-Fermi approach is semiclassical theory where certain ideas are borrowed from quantum mechanics. The condition to apply a semi-classical approach to a system is that the spatial variations of the de Broglie wavelength must be small; mathematically it means:

$$\left|\frac{d\lambda(x)}{dx}\right|\ll 1 \tag{9.51}$$

Thomas-Fermi method borrows two ideas from quantum mechanics: (i) Fermi statistics–all the states up to some maximum energy and momentum, say p_F, which may vary over the space are occupied, and (ii) The principle of uncertainty–every cell of phase space (of volume h^3) can host up to two electrons with opposite spin directions. Thus in the ground state volume occupied by the electrons in the phase space would be $\frac{4\pi V}{3}p_F^3$, where V is the volume of a system in real space. It is assumed that all electrons are accommodated up to the phase space sphere of radius p_F. Therefore, the total number of electrons is:

$$N = \left(\frac{4\pi V p_F^3}{3}\right) / \left(\frac{h^3}{2}\right) = \frac{8\pi V p_F^3}{3h^3} \tag{9.52}$$

and hence:

$$n(\mathbf{r}) = \frac{N}{V} = \frac{8\pi p_F^3}{3h^3} = \frac{p_F^3}{3\pi^2\hbar^3} \tag{9.53}$$

which gives:

$$p_F(\mathbf{r}) = \hbar k_F(\mathbf{r}), \quad \text{with } k_F(\mathbf{r}) = \left\{3\pi^2 n(\mathbf{r})\right\}^{1/3} \tag{9.54}$$

The p_F and k_F are known as Fermi momentum and Fermi wave vector, respectively. It is assumed here that p_F and hence $n(\mathbf{r})$ vary with space coordinates over the region in which the condition, Eqn. (9.51), is fulfilled.

Assuming that all electrons move as classical particles under the influence of common potential, the classical energy of fastest moving electrons is:

$$E_{\max} = \frac{p_F^2(r)}{2m} + V(r) \tag{9.55}$$

where both kinetic and potential energy parts may independently depend on r, and their sum in equilibrium remains constant. We next take total energy $(T+U)$ of the entire electrons distribution. With kinetic energy density $t(\mathbf{r})$, we write $T = \int t(\mathbf{r}) d^3r$. To evaluate $t(\mathbf{r})$, let us calculate the fraction of electrons that have momentum between p and $p+dp$:

$$F(p)dp = \left(\frac{4\pi p^2 dp}{\frac{h^3}{2}}\right) = \frac{p^2 dp}{\pi^2\hbar^3} \quad \text{when } p < p_F, \text{zero otherwise.} \tag{9.56}$$

The $t(\mathbf{r})$ at around $\mathbf{r}$, when the classical expression for the kinetic energy of an electron is used, is:

$$\begin{aligned} t(\mathbf{r}) &= \int_0^{p_F} \frac{p^2}{2m} F(p) dp \\ &= \frac{1}{2m\pi^2\hbar^3} \int_0^{p_F} p^4 dp \\ &= \frac{1}{2m\pi^2\hbar^3} \times \frac{p_F^5}{5} \\ &= \frac{3\hbar^2 (3)^{2/3} \pi^{4/3}}{10m} \{n(\mathbf{r})\}^{5/3} \end{aligned} \tag{9.57a}$$

or:

$$t(\mathbf{r}) = C_k n^{5/3}(\mathbf{r}), \text{ with } C_k = \frac{3\hbar^2 (3)^{2/3} \pi^{4/3}}{10m} \tag{9.57b}$$

The total kinetic energy therefore is:

$$T = C_k \int n^{5/3}(\mathbf{r}) d^3 r \tag{9.58a}$$

The potential energy at $\mathbf{r}$ is due to the interaction with the external field and electrostatic interaction of the electron density with itself. Therefore:

$$U = \int n(\mathbf{r}) V_{ext}(\mathbf{r}) d^3 r + \frac{1}{2} \frac{e^2}{4\pi\varepsilon_0} \iint \frac{n(\mathbf{r}) n(\mathbf{r}')}{|\mathbf{r} - \mathbf{r}'|} d^3 r d^3 r' \tag{9.58b}$$

In the case of an atom with nucleus of charge Ze centered at $\mathbf{r} = 0$, $V_{ext}(r) = -\frac{Ze^2}{4\pi\varepsilon_0 r}$.
The total energy of the system is:

$$E_{tot} = T + U = C_k \int n^{5/3}(\mathbf{r}) d^3 r + \int n(\mathbf{r}) V_{ext}(\mathbf{r}) d^3 r + \frac{1}{2} \frac{e^2}{4\pi\varepsilon_0} \iint \frac{n(\mathbf{r}) n(\mathbf{r}')}{|\mathbf{r} - \mathbf{r}'|} d^3 r d^3 r' \tag{9.59}$$

Next, we search for the electron density, which minimizes the total energy subject to the normalization condition $\int n(\mathbf{r}) d^3 r = N$. On introducing a Lagrange multiplier μ, the method of the variational principle with respect to $n(\mathbf{r})$ gives:

$$\delta(E_{tot} - \mu N) = \int \left[\frac{5}{3} C_k n^{2/3}(\mathbf{r}) + V_{ext}(\mathbf{r}) + \frac{e^2}{4\pi\varepsilon_0} \int \frac{n(\mathbf{r}')}{|\mathbf{r} - \mathbf{r}'|} d^3 r' - \mu \right] \delta n(\mathbf{r}) d^3 r = 0 \tag{9.60}$$

which yields:

$$\mu = \frac{5}{3} C_k \{n(\mathbf{r})\}^{2/3} + V_{ext}(\mathbf{r}) + \frac{e^2}{4\pi\varepsilon_0} \int \frac{n(\mathbf{r}')}{|\mathbf{r} - \mathbf{r}'|} d^3 r' \tag{9.61}$$

This equation is known as the Thomas-Fermi equation for determining the equilibrium distribution of the electron density. The $\mu = \frac{\partial E_{tot}}{\partial N}$ suggests that it is the chemical potential. Also, it is energy of the fastest moving electron, which is generally known as Fermi energy.

9.3.4 Thomas-Fermi Model of Atom

Since the electrons cannot escape from the atom, $\mu = 0$ for an atom in equilibrium. Hence:

$$n(\mathbf{r}) = \left[-\frac{3}{5C_k} V_{eff}(\mathbf{r}) \right]^{3/2} \tag{9.62a}$$

with:

$$V_{eff}(\mathbf{r}) = V_{ext}(\mathbf{r}) + \frac{e^2}{4\pi\varepsilon_0} \int \frac{n(\mathbf{r}')}{|\mathbf{r} - \mathbf{r}'|} d^3 r' \tag{9.62b}$$

Because of spherical symmetry inside the atom, both electron density and $V_{eff}(\mathbf{r})$ would be a function of $r = |\mathbf{r}|$. The electrostatic potential $\frac{-V_{eff}(r)}{e}$ and charge density $\rho = -en(r)$ satisfy the Poisson's equation:

$$\frac{1}{e}\nabla^2 V_{eff}(r) = -\frac{en(r)}{\varepsilon_0}$$
$$\Rightarrow \frac{1}{r^2}\frac{d}{dr}\left(r^2 \frac{dV_{eff}}{dr}\right) = -\frac{e^2 n(r)}{\varepsilon_0} \tag{9.63a}$$

which with the use of Eqn. (9.62a) gives:

$$\frac{1}{r^2}\frac{d}{dr}\left(r^2 \frac{dV_{eff}(r)}{dr}\right) = -\frac{e^2}{\varepsilon_0}\left[-\frac{3}{5C_k}V_{eff}(\mathbf{r})\right]^{3/2} \tag{9.63b}$$

For $r \to 0$ the leading term in $V_{eff}(r)$ is the potential due to the nucleus and hence $V_{eff}(r) \to -\frac{Ze^2}{4\pi\varepsilon_0 r}$. Therefore, it is convenient to introduce $V_{eff}(r) = -\frac{Ze^2}{4\pi\varepsilon_0 r}\chi(r)$. We then have:

$$\frac{d^2\chi(r)}{dr^2} = \frac{e^3}{\varepsilon_0}\sqrt{\frac{Z}{4\pi\varepsilon_0}}\left\{\frac{3}{5C_k}\right\}^{3/2}\frac{\chi^{3/2}}{\sqrt{r}} \tag{9.64}$$

which on choosing $x = \frac{r}{b}$ with $b = \frac{5\varepsilon_0 C_k}{3e^2}\left(\frac{Z}{4\pi}\right)^{1/3}$, reduces to:

$$\frac{d^2\chi(x)}{dx^2} = \frac{\{\chi(x)\}^{3/2}}{\sqrt{x}} \tag{9.65}$$

which is known as the Thomas-Fermi equation for an atom. Note that $\chi(0) \to 1$, while $\chi(\infty) \to 0$.

9.3.5 Density Functional Theory

The Thomas-Fermi theory, which is the oldest theory to describe the electronic energy in terms of electron density distribution, was found quite useful to explain qualitative features of atoms. However, it is not of much use when one wants to study the electronic structures of large molecules and materials. A more generalized theory in terms of electron density distribution was derived by Hohenberg and Kohn, and later by Kohn and Sham, where it is found that the Thomas-Fermi theory is a special case of a generalized density functional theory (DFT). The two *Hohenberg-Kohn theorems*, which are the basis of DFT are as follows:

Theorem-1: The ground state properties of a many-electron system depend only on the electronic density $n(\mathbf{r})$ and the external potential is determined by $n(\mathbf{r})$.

Theorem-2: The correct ground state density for a system is the one that minimizes the total energy through the functional $E[n(\mathbf{r})]$ and yields $\int n(\mathbf{r})d^3r = N$.

As compared to wave function dependent approaches, such the Hartree-Fock or post Hartree-Fock theories, DFT provides an alternative method to reformulate the many-body problem, where the density of electrons rather than a many-electrons wave function plays a central role. The wave function approach involves a large number of orbitals in terms of 3N coordinates of N-electrons, whereas DFT involves only one variable $n(\mathbf{r})$ that depends on 3-coordinates.

The N-electrons Hamiltonian, Eqn. (9.30), is:

$$H = T + V_{ee} + \sum_{i=1}^{N} V_{ext}(r_i) \tag{9.66}$$

where T and V_{ee}, kinetic energy and electron-electron interaction energy operators, are called universal operators, as they are the same for any of the N-electron systems. The third term, potential energy due to the interaction of the electron with an external potential (such as a positively charged nuclei), is system dependent. The N and $V_{ext}(r)$ determine all the properties of the ground state. The functional, which is universal in the sense that it does not refer to any system or external potential $V_{ext}(\mathbf{r})$ is defined as:

$$F[n] = \langle \phi | (T + V_{ee}) | \phi \rangle_{\min} \tag{9.67}$$

The minimum is taken over all single particle wave functions ϕ_i, which constitute the Hartree wave function: $\psi(\mathbf{r}_1, \mathbf{r}_2, \mathbf{r}_3, \ldots\ldots \mathbf{r}_N) = \phi_1(\mathbf{r}_1)\phi_2(\mathbf{r}_2)\phi_3(\mathbf{r}_3)\ldots\ldots\ldots\phi_N(\mathbf{r}_N)$. We then write:

$$E[n] = F[n] + \int V_{ext}(\mathbf{r}) n(\mathbf{r}) d^3 r \geq E_g \tag{9.68}$$

for all N-representable $n(\mathbf{r})$. Here, E_g is ground state energy.

Many of the drawbacks of the Thomas-Fermi theory are due to the approximate treatment of kinetic energy. Use of a different separation introduced by Kohn and Sham greatly simplifies the task of finding good approximations to the energy functional:

$$E[n] = T_0[n] + \int n(\mathbf{r}) \left\{ V_{ext}(\mathbf{r}) + \frac{1}{2} \frac{e^2}{4\pi\varepsilon_0} \int \frac{n(\mathbf{r}')}{|\mathbf{r} - \mathbf{r}'|} d^3 r' \right\} d^3 r + E_{xc}(n) \tag{9.69}$$

Here $T_0[n]$ is the kinetic energy of a system with density $n(\mathbf{r})$ in the absence of electron-electron interactions. The $E_{xc}[n]$ defines the exchange-correlation energy. The exact treatment of T_0 in this approach removes many of the deficiencies faced in the Thomas-Fermi approximation. Though T_0 differs from the true kinetic energy T, magnitudes of two are comparable. All terms except $E_{xc}[n]$ in Eqn. (9.69) are evaluated exactly. Therefore, unavoidable approximations for $E_{xc}[n]$ play a central role in DFT calculations. The exact exchange-correlation energy functionals either are not known or they are so complicated that it is not useful to compute these. Hence, various *approximate* exchange-correlation functionals (S-VWN, B3LYP, etc.) are used.

Application of the variational principle to Eqn. (9.69) yields:

$$\frac{\delta E[n]}{\delta n(\mathbf{r})} = \frac{\delta T_0[n]}{\delta n(\mathbf{r})} + V_{ext}(\mathbf{r}) + \frac{e^2}{4\pi\varepsilon_0} \int \frac{n(\mathbf{r}')}{|\mathbf{r} - \mathbf{r}'|} d^3 r' + \frac{\delta E_{xc}(n)}{\delta n(\mathbf{r})} = \mu \tag{9.70}$$

where μ is the Lagrange multiplier associated with the requirement of the constant particle number N. As is seen, mathematically the problem is identical to that described in the Thomas-Fermi theory, if effective potential is defined by:

$$V_{eff}(\mathbf{r}) = V_{ext}(\mathbf{r}) + \frac{e^2}{4\pi\varepsilon_0}\int \frac{n(\mathbf{r}')}{|\mathbf{r}-\mathbf{r}'|}d^3r' + \frac{\delta E_{ex}(n)}{\delta n(\mathbf{r})} \tag{9.71}$$

in place of that given by Eqn. (9.62b). A major step towards the solution of such an equation was introduced by Kohn and Sham by introducing an orbital method, which exactly evaluates:

$$T_0[n] = \sum_{i=1}^{N} \langle \phi_i | \left(-\frac{\hbar^2}{2m}\nabla_i^2 \right) | \phi_i \rangle \tag{9.72}$$

In order to evaluate the kinetic energy of N-noninteracting particles whose density distribution is known, they simply found the corresponding $V_{eff}(\mathbf{r})$ and used the Schrödinger equation:

$$\left(-\frac{\hbar^2}{2m}\nabla^2 + V_{eff}(\mathbf{r}) \right)\phi_i(\mathbf{r}) = \varepsilon_i \phi_i(\mathbf{r}) \tag{9.73}$$

such that:

$$n(\mathbf{r}) = \sum_{i=1}^{N} |\phi_i(\mathbf{r})|^2 \tag{9.74}$$

The $\phi_i(\mathbf{r})$ here are ordered so that the energies ε_i are non–decreasing, and the spin index is included in suffix i. Equations (9.71) and (9.73) are nonlinear and are to be solved self-consistently subject to the condition (Eqn. 9.74), with the use of the iterative method. Much like the Hartree-Fock equation, exchange-correlation potential is used in Eqn. (9.73). Looking at Eqns. (9.61) and (9.71), it is said that the Thomas-Fermi theory is a special case of DFT.

9.4 Solved Examples

1. Consider a system of two non-interacting particles; both are independently attached to the origin by the force constants k_1 and k_2, respectively. Each particle has mass m. What are the eigenfunction and eigenvalue for the system in the nth state. The single particle coordinates are x_1 and x_2.

   SOLUTION

   For a single particle case, the Schrödinger equation for a harmonic oscillator is:

$$-\frac{\hbar^2}{2m}\frac{d^2\psi}{dx^2} + \frac{1}{2}kx^2\psi = E\psi \tag{9.75}$$

As discussed in Chapter 3, Section 3.19, the solution of Eqn. (9.75) for the nth state is given by: $\psi(y) = N_n H_n(y) e^{-y^2/2}$ and $E_n = (n + \frac{1}{2})\hbar\omega$, where $\omega = \sqrt{k/m}$, and $y = \alpha x$ with $\alpha = \left(\frac{m\omega}{\hbar}\right)^{1/2}$.

For the given system of two particles, we have:

$$\left[-\frac{\hbar^2}{2m}\frac{\partial^2}{\partial x_1^2} + \frac{1}{2}k_1 x_1^2 - \frac{\hbar^2}{2m}\frac{\partial^2}{\partial x_2^2} + \frac{1}{2}k_2 x_2^2\right]\psi(x_1, x_2) = E\psi(x_1, x_2)$$
$$\Rightarrow [H_1(x_1) + H_2(x_2)]\psi(x_1, x_2) = E\psi(x_1, x_2) \quad (9.76)$$

As is seen, the Hamiltonian is the sum of two independent Hamiltonians and hence the solution of Eqn. (9.76) can be taken as $\psi(x_1, x_2) = \phi_1(x_1)\phi_2(x_2)$. We thus obtain:

$$H_1(x_1)\phi_1(x_1)\phi_2(x_2) + \phi_1(x_1)H_2(x_2)\phi_2(x_2) = E\phi_1(x_1)\phi_2(x_2)$$
$$\Rightarrow \frac{1}{\phi_1(x_1)}H_1(x_1)\phi_1(x_1) + \frac{1}{\phi_2(x_2)}H_2(x_2)\phi_2(x_2) = E \quad (9.77a)$$

Writing $E = E_1 + E_2$, we obtain:

$$H_i\phi_i(x_i) = E_i\phi_i(x_i), (i = 1, 2) \quad (9.77b)$$

The solution of Eqn. (9.77b) is:

$$E_{n_i} = \left(n_i + \frac{1}{2}\right)\hbar\omega_i, \text{ and } \phi_i(x_i) = N_{n_i}H_{n_i}(y_i)e^{-y_i^2/2}, \text{with } i = (1, 2) \quad (9.78)$$

where we used $y_1 = \alpha_1 x_1$ and $y_2 = \alpha_2 x_2$, with $\alpha_1^4 = \frac{mk_1}{\hbar^2} = \left(\frac{m\omega_1}{\hbar}\right)^2$ and $\alpha_2^4 = \frac{mk_2}{\hbar^2} = \left(\frac{m\omega_2}{\hbar}\right)^2$. We thus have:

$$E_{n_1, n_2} = \left(n_1 + \frac{1}{2}\right)\hbar\omega_1 + \left(n_2 + \frac{1}{2}\right)\hbar\omega_2 \quad (9.79a)$$

and:

$$\psi_{n_1, n_2}(y_1, y_2) = N_{n_1, n_2}H_{n_1}(y_1)H_{n_2}(y_2)e^{-(y_1^2/2 + y_2^2/2)} \quad (9.79b)$$

The total energy of the system is the sum of the energies of two non-interacting systems, which is in conformity to the statement made in Section 9.1.1.

2. Consider again the system of two oscillating particles, each of mass m. Unlike problem 1, two particles have the same spring constant and they interact with each other. The Hamiltonian of the system is $H = -\frac{\hbar^2}{2m}\frac{\partial^2}{\partial x_1^2} - \frac{\hbar^2}{2m}\frac{\partial^2}{\partial x_2^2} + \frac{1}{2}k(x_1^2 + x_2^2 + x_1 x_2)$. Find out the energy and wave function for the interacting system.

SOLUTION

Let us introduce $x_1 = \frac{1}{\sqrt{2}}(z_1 + z_2)$ and $x_2 = \frac{1}{\sqrt{2}}(z_1 - z_2)$, which gives $x_1^2 + x_2^2 = z_1^2 + z_2^2$ and $x_1 x_2 = \frac{1}{2}\left(z_1^2 - z_2^2\right)$. This allows us to write:

$$\frac{\partial}{\partial x_1} = \frac{\partial}{\partial z_1}\frac{\partial z_1}{\partial x_1} + \frac{\partial}{\partial z_2}\frac{\partial z_2}{\partial x_1} = \frac{1}{\sqrt{2}}\left(\frac{\partial}{\partial z_1} + \frac{\partial}{\partial z_2}\right) \tag{9.80}$$

and:

$$\frac{\partial}{\partial x_2} = \frac{\partial}{\partial z_1}\frac{\partial z_1}{\partial x_2} + \frac{\partial}{\partial z_2}\frac{\partial z_2}{\partial x_2} = \frac{1}{\sqrt{2}}\left(\frac{\partial}{\partial z_1} - \frac{\partial}{\partial z_2}\right) \tag{9.81}$$

which yield:

$$\frac{\partial^2}{\partial x_1^2} = \frac{1}{2}\left(\frac{\partial^2}{\partial z_1^2} + \frac{\partial^2}{\partial z_2^2} + 2\frac{\partial}{\partial z_1}\frac{\partial}{\partial z_2}\right) \tag{9.82a}$$

and:

$$\frac{\partial^2}{\partial x_2^2} = \frac{1}{2}\left(\frac{\partial^2}{\partial z_1^2} + \frac{\partial^2}{\partial z_2^2} - 2\frac{\partial}{\partial z_1}\frac{\partial}{\partial z_2}\right) \tag{9.82b}$$

Therefore:

$$\frac{\partial^2}{\partial x_1^2} + \frac{\partial^2}{\partial x_2^2} = \frac{\partial^2}{\partial z_1^2} + \frac{\partial^2}{\partial z_2^2} \tag{9.83a}$$

The Hamiltonian in terms of z_1 and z_2 is:

$$\begin{aligned} H &= -\frac{\hbar^2}{2m}\frac{\partial^2}{\partial z_1^2} - \frac{\hbar^2}{2m}\frac{\partial^2}{\partial z_2^2} + \frac{k}{2}(z_1^2 + z_2^2) + \frac{k}{4}(z_1^2 - z_2^2) \\ &= \left[-\frac{\hbar^2}{2m}\frac{\partial^2}{\partial z_1^2} + \frac{3k}{4}z_1^2\right] + \left[-\frac{\hbar^2}{2m}\frac{\partial^2}{\partial z_2^2} + \frac{k}{4}z_2^2\right] \end{aligned} \tag{9.83b}$$

Taking $\psi(z_1, z_2) = \phi_1(z_1)\phi_2(z_2)$, the eigenvalue equation $H\psi = E\psi$ is:

$$\begin{aligned} &H_1(z_1)\phi_1(z_1)\phi_2(z_2) + \phi_1(z_1)H_2(z_2)\phi_2(z_2) = E\phi_1(z_1)\phi_2(z_2) \\ &\Rightarrow \frac{1}{\phi_1(z_1)}H_1(z_1)\phi_1(z_1) + \frac{1}{\phi_2(z_2)}H_2(z_2)\phi_2(z_2) = E \end{aligned} \tag{9.84}$$

Equations (9.84) and (9.77a) look similar. Thus, the Hamiltonian of two interacting oscillating particles is transformed to the Hamiltonian of two non-interacting oscillating particles.

Defining $\alpha_1^4 = \frac{3mk}{2\hbar^2}$, and $\alpha_2^4 = \frac{mk}{2\hbar^2}$, a generalized solution of Eqn. (9.84b) is:

$$\psi_{n_1,n_2}(\rho_1,\rho_2) = N_{n_1,n_2} H_{n_1}(\rho_1) H_{n_2}(\rho_2) e^{-(\rho_1^2/2+\rho_2^2/2)} \tag{9.85}$$

and:

$$E_{n_1,n_2} = \left(n_1 + \frac{1}{2}\right)\hbar\omega_1 + \left(n_2 + \frac{1}{2}\right)\hbar\omega_2 \tag{9.86}$$

with: $\rho_1 = \alpha_1 z_1 = \frac{1}{\sqrt{2}}\left(\frac{3mk}{2\hbar^2}\right)^{1/4}(x_1 + x_2)$; $\rho_2 = \alpha_2 z_2 = \frac{1}{\sqrt{2}}\left(\frac{mk}{2\hbar^2}\right)^{1/4}(x_1 - x_2)$; $\omega_1 = \sqrt{\frac{3k}{2m}}$ and $\omega_2 = \sqrt{\frac{k}{2m}}$, with $n_1, n_2 = 0, 1, 2\dots\dots$ On defining $\omega = \sqrt{\frac{k}{m}}$ we get:

$$E_{n_1,n_2} = \frac{1}{\sqrt{2}}\left(\sqrt{3}n_1 + n_2 + \frac{\sqrt{3}+1}{2}\right)\hbar\omega \tag{9.87}$$

Note that Eqn. (9.87) endorses the statement of Section 9.1.1, "analysis and conclusions drawn for non-interacting particles are applicable to a system of N-interacting particles too."

3. A central potential well has three possible single particle states, $|\phi_1\rangle$, $|\phi_2\rangle$, and $|\phi_3\rangle$. If there are two electrons, what would be possible wave functions? What would be the matrix element of two electrons interaction potential $v(1,2)$ when they interact?

(Problem 7009 in "Problems & Solutions in Quantum Mechanics," edited by Yung-Kuo Lim.)

SOLUTION

Electrons obey the Pauli exclusion principle and therefore a many particle wave function is antisymmetric. All the possible antisymmetric wave functions from Eqn. (9.11) are:

$$\begin{aligned}
\psi_{12}(1,2) &= \frac{1}{\sqrt{2}}[\phi_1(1)\phi_2(2) - \phi_1(2)\phi_2(1)]; \\
\psi_{23}(1,2) &= \frac{1}{\sqrt{2}}[\phi_2(1)\phi_3(2) - \phi_2(2)\phi_3(1)]; \\
\psi_{31}(1,2) &= \frac{1}{\sqrt{2}}[\phi_3(1)\phi_1(2) - \phi_3(2)\phi_1(1)]
\end{aligned} \tag{9.88}$$

For a two electron interaction potential $v(1,2)$, the matrix element can be given by:

$$\begin{aligned}
&\langle\psi_{ij}|v(1,2)|\psi_{kl}\rangle \\
&= \frac{1}{2}\langle\phi_i(1)\phi_j(2)|v(1,2)|\phi_k(1)\phi_l(2)\rangle + \frac{1}{2}\langle\phi_i(2)\phi_j(1)|v(1,2)|\phi_k(2)\phi_l(1)\rangle \\
&- \frac{1}{2}\langle\phi_i(1)\phi_j(2)|v(1,2)|\phi_k(2)\phi_l(1)\rangle - \frac{1}{2}\langle\phi_i(2)\phi_j(1)|v(1,2)|\phi_k(1)\phi_l(2)\rangle
\end{aligned} \tag{9.89}$$

where each of the pairs ij and kl can be assigned $12, 23$ and 31. Note that 1 and 2 represent both space and spin coordinates.

4. A system of two particles, each having mass m, is confined to volume V. Particles interact with each other via the potential $V(\mathbf{r}_1, \mathbf{r}_2) = k|\mathbf{r}_1 - \mathbf{r}_2|^2$. Calculate the energy of the system.

SOLUTION

The eigenvalue equation for a system of two interacting particles is:

$$-\frac{\hbar^2}{2m}\left(\nabla_1^2 + \nabla_2^2\right)\psi(\mathbf{r}_1, \mathbf{r}_2) + k|\mathbf{r}_1 - \mathbf{r}_2|^2 \psi(\mathbf{r}_1, \mathbf{r}_2) = E\psi(\mathbf{r}_1, \mathbf{r}_2) \tag{9.90}$$

The Hamiltonian of the system is separable into three 1D Hamiltonians. We write:

$$H = H_x + H_y + H_z$$

with:

$$\begin{aligned} H_x &= -\frac{\hbar^2}{2m}\left(\frac{\partial^2}{\partial x_1^2} + \frac{\partial^2}{\partial x_2^2}\right) + k|x_1 - x_2|^2, \\ H_y &= -\frac{\hbar^2}{2m}\left(\frac{\partial^2}{\partial y_1^2} + \frac{\partial^2}{\partial y_2^2}\right) + k|y_1 - y_2|^2, \\ H_z &= -\frac{\hbar^2}{2m}\left(\frac{\partial^2}{\partial z_1^2} + \frac{\partial^2}{\partial z_2^2}\right) + k|z_1 - z_2|^2 \end{aligned} \tag{9.91}$$

which allows us to write $\psi(\mathbf{r}_1, \mathbf{r}_2) = \phi(x_1, x_2)\varphi(y_1, y_2)\xi(z_1, z_2)$. Substitution of these into Eqn. (9.90) yields the following three equations:

$$-\frac{\hbar^2}{2m}\left(\frac{\partial^2}{\partial x_1^2} + \frac{\partial^2}{\partial x_2^2}\right)\phi(x_1, x_2) + k(x_1 - x_2)^2 \phi(x_1, x_2) = E_1\phi(x_1, x_2) \tag{9.92a}$$

$$-\frac{\hbar^2}{2m}\left(\frac{\partial^2}{\partial y_1^2} + \frac{\partial^2}{\partial y_2^2}\right)\varphi(y_1, y_2) + k(y_1 - y_2)^2 \varphi(y_1, y_2) = E_2\varphi(y_1, y_2) \tag{9.92b}$$

$$-\frac{\hbar^2}{2m}\left(\frac{\partial^2}{\partial z_1^2} + \frac{\partial^2}{\partial z_2^2}\right)\xi(z_1, z_2) + k(z_1 - z_2)^2 \xi(z_1, z_2) = E_1\xi(z_1, z_2) \tag{9.92c}$$

with $E = E_1 + E_2 + E_3$. The three equations are similar and hence we need to solve only one of these. Let us solve Eqn. (9.92a). On defining $s = \frac{1}{\sqrt{2}}(x_1 + x_2)$ and $t = \frac{1}{\sqrt{2}}(x_1 - x_2)$, and then adopting the procedure of example 2, we get:

$$-\frac{\hbar^2}{2m}\left(\frac{\partial^2}{\partial s^2} + \frac{\partial^2}{\partial t^2}\right)\phi(s, t) + \frac{1}{2}kt^2\phi(s, t) = E_1\phi(s, t) \tag{9.93a}$$

Next, taking $\phi(s,t)=\phi_1(s)\phi_2(t)$ we obtain:

$$-\frac{\hbar^2}{2m}\frac{\partial^2\phi_1}{\partial s^2}\phi_2(t)+\left\{-\frac{\hbar^2}{2m}\frac{\partial^2\phi_2}{\partial t^2}+\frac{1}{2}kt^2\phi_2(t)\right\}\phi_1(s)=E_1\phi_1(s)\phi_2(t) \tag{9.93b}$$

On writing:

$$-\frac{\hbar^2}{2m}\frac{\partial^2\phi_1(s)}{\partial s^2}=E_{11}\phi_1(s) \tag{9.94a}$$

and:

$$-\frac{\hbar^2}{2m}\frac{\partial^2\phi_2(t)}{\partial t^2}+\frac{1}{2}kt^2\phi_2(t)=E_{12}\phi_2(t) \tag{9.94b}$$

we get $E_1=E_{11}+E_{12}$. Solutions of Eqns. (9.94a) and (9.94b) are:

$$\phi_1(s)=A\sin(k_1 s)\ \text{ and }\ E_{11}=\frac{\hbar^2k_1^2}{2m} \tag{9.95a}$$

and:

$$\phi_2(t)=N_{n_1}H_{n_1}(\alpha t)e^{-\alpha^2t^2/2}\ \text{ and }\ E_{12}=\left(n_1+\frac{1}{2}\right)\hbar\omega,\ n_1=0,1,2,\ldots \tag{9.95b}$$

with $\omega=\sqrt{k/m}$ and $\alpha^4=\dfrac{m\omega}{\hbar^2}$. Here, A and N_{n_1} are constants and H_{n_1} is the Hermite polynomial. Since particles are confined to volume V, $\phi(x_1,x_2)$ must vanish at the boundary, which means that if we take $V=L^3$ then $\phi(L,x_2)=\phi(x_1,L)=0$. This implies that:

$$\begin{aligned}\phi(x_1,x_2)&=\phi_1\left(\frac{x_1+x_2}{\sqrt{2}}\right)\phi_2\left(\frac{x_1-x_2}{\sqrt{2}}\right)\\&=A\sin\left(\frac{k_1(x_1+x_2)}{\sqrt{2}}\right)\phi_2\left(\frac{x_1-x_2}{\sqrt{2}}\right)\end{aligned} \tag{9.96}$$

should go to zero when any of x_1 and x_2 is equal to L. This would be fulfilled if $\dfrac{k_1L}{\sqrt{2}}=l_1\pi$, where $l_1=1,2,3,\ldots$. We thus obtain:

$$E_1=\frac{\hbar^2\pi^2l_1^2}{mL^2}+\left(n_1+\frac{1}{2}\right)\hbar\omega \tag{9.97a}$$

It is then straightforward to write:

$$E_2=\frac{\hbar^2\pi^2l_2^2}{mL^2}+\left(n_2+\frac{1}{2}\right)\hbar\omega \tag{9.97b}$$

and:

$$E_3=\frac{\hbar^2\pi^2l_3^2}{mL^2}+\left(n_3+\frac{1}{2}\right)\hbar\omega \tag{9.97c}$$

Therefore:

$$E_{l,n} = \frac{\hbar^2\pi^2 l^2}{mV^{2/3}} + \left(n + \frac{3}{2}\right)\hbar\omega \tag{9.98}$$

with $l^2 = l_1^2 + l_2^2 + l_3^2$ and $n = n_1 + n_2 + n_3$. The lowest value of energy is:

$$E_{10} = \frac{3\hbar^2\pi^2}{mV^{2/3}} + \frac{3}{2}\hbar\omega \tag{9.99}$$

5. Consider a system of N non-interacting particles whose Hamiltonian is written as: $H = \sum_{i=1}^{N} H_i$ where $H_i|\phi_i\rangle = \varepsilon_i|\phi_i\rangle$. Calculate the ground state energy of the system for both the cases of Bosons (spin = 0) and Fermions (spin = $\frac{1}{2}$). Write explicitly the ground state energy and wave function when $N = 5$.

SOLUTION

In the case of Bosons, all N-particles can occupy the one state. Thus if ε_1 is the energy of lowest energy state, the ground state energy of the system would be $E_0^b = N\varepsilon_1$.

However, for Fermions, the Pauli exclusion principle allows a maximum of two particles of opposite spins to occupy the same state. Therefore $E_0^f = 2(\varepsilon_1 + \varepsilon_2 + \varepsilon_3 + ...\varepsilon_{N/2})$, if N is even, and $E_0^f = 2(\varepsilon_1 + \varepsilon_2 + \varepsilon_3 + ...\varepsilon_{(N-1)/2}) + \varepsilon_N$ when N is odd.

The wave function of five identical Bosons has to be symmetric with respect to interchanges of the particles. Hamiltonian is a sum of N-single particle Hamiltonians; the eigenfunction would be products of single-particle eigenfunctions. Therefore:

$$\psi(1,2,3,4,5) = \phi_1(1)\phi_1(2)\phi_1(3)\phi_1(4)\phi_1(5) \text{ and } E_0^b = 5\varepsilon_1 \tag{9.100}$$

In the case of five identical Fermions, wave function has to be antisymmetric with respect to the interchange of any two particles. In the ground state, two Fermions occupy the lowest energy ε_1 state $|\phi_1\rangle$ with opposite spins, the next two occupy the state $|\phi_2\rangle$ of energy ε_2, and the fifth Fermion occupies the state $|\phi_3\rangle$ having energy eigenvalue ε_3. Hence:

$$\psi(1,2,3,4,5) = \frac{1}{\sqrt{5!}} \begin{vmatrix} \phi_{1\uparrow}(1) & \phi_{1\uparrow}(2) & \phi_{1\uparrow}(3) & \phi_{1\uparrow}(4) & \phi_{1\uparrow}(5) \\ \phi_{1\downarrow}(1) & \phi_{1\downarrow}(2) & \phi_{1\downarrow}(3) & \phi_{1\downarrow}(4) & \phi_{1\downarrow}(5) \\ \phi_{2\uparrow}(1) & \phi_{2\uparrow}(2) & \phi_{2\uparrow}(3) & \phi_{2\uparrow}(4) & \phi_{2\uparrow}(5) \\ \phi_{2\downarrow}(1) & \phi_{2\downarrow}(2) & \phi_{2\downarrow}(3) & \phi_{2\downarrow}(4) & \phi_{2\downarrow}(5) \\ \phi_{3\uparrow}(1) & \phi_{3\uparrow}(2) & \phi_{3\uparrow}(3) & \phi_{3\uparrow}(4) & \phi_{3\uparrow}(5) \end{vmatrix} ; E_0^f = 2(\varepsilon_1 + \varepsilon_2) + \varepsilon_3 \tag{9.101}$$

6. Two non-interacting electrons are confined to a one dimensional infinite potential well:

$$V(x)=\begin{cases} 0 & 0<x<a \\ \infty & 0>x>a \end{cases}.$$

Find the energy eigenvalue and wave function for the ground state.

SOLUTION

Inside the well, each of the electrons obeys the free particle Schrödinger equation. Therefore, the single particle eigenvalue equation for the nth eigenstate is:

$$-\frac{\hbar^2}{2m}\frac{d^2\phi_{n_i}(x_i)}{dx_i^2}=\varepsilon_{n_i}\phi_{n_i}(x_i)\,\text{with}\,i\equiv(1,2)\,\text{for}\,0<x<a \tag{9.102}$$

along with the condition $\phi_{n_i}(0)=\phi_{n_i}(a)=0$. Therefore two solutions are:

$$\begin{aligned}\phi_{n_1}(x_1)&=\sqrt{\frac{2}{a}}\sin\left(\frac{\pi n_1 x_1}{a}\right);\\ \varepsilon_{n_1}&=\frac{\hbar^2\pi^2 n_1^2}{2ma^2}\end{aligned} \tag{9.103}$$

and:

$$\begin{aligned}\phi_{n_2}(x_2)&=\sqrt{\frac{2}{a}}\sin\left(\frac{\pi n_2 x_2}{a}\right);\\ \varepsilon_{n_2}&=\frac{\hbar^2\pi^2 n_2^2}{2ma^2}\end{aligned} \tag{9.104}$$

where n_1 and n_2 are integers and take values 1,2,3,....... Total wave function has to be antisymmetric for a system of electrons. In the ground state both electrons occupy the lowest energy state with opposite spin orientations. Therefore, from Eqns. (9.13) to (9.15), we find that the space-dependent part of wave function should be symmetric and the spin-dependent part is antisymmetric. We thus have:

$$\begin{aligned}\psi_{11}(x_1,x_2)&=\frac{2\times 2}{\sqrt{2}a}\sin\left(\frac{\pi x_1}{a}\right)\sin\left(\frac{\pi x_2}{a}\right)\times\frac{1}{\sqrt{2}}\left[\uparrow_1\downarrow_2-\downarrow_1\uparrow_2\right]\\ &=\frac{2}{a}\left[\cos\left(\frac{\pi}{a}(x_1-x_2)\right)-\cos\left(\frac{\pi}{a}(x_1+x_2)\right)\right]\left[\uparrow_1\downarrow_2-\downarrow_1\uparrow_2\right]\end{aligned} \tag{9.105a}$$

and:

$$E_g=2\varepsilon_1=\frac{\hbar^2\pi^2}{ma^2} \tag{9.105b}$$

7. In some of the metallic systems it is found that $V_{ext}(\mathbf{r}_i)$, which is generated by distribution of positively charged ions, is equal and opposite to the potential created by the distribution of negatively charged electrons. For such systems, Eqn. (9.49) reduces to:

$$-\frac{\hbar^2}{2m}\nabla_i^2\phi_i(\mathbf{r}_i)-\frac{e^2}{4\pi\varepsilon_0}\left\{\sum_{j\neq i}\int\phi_j^*(\mathbf{r}_j)\frac{1}{|\mathbf{r}_i-\mathbf{r}_j|}\phi_i(\mathbf{r}_j)d^3r_j\phi_j(\mathbf{r}_i)\right\}=\varepsilon_i\phi_i(\mathbf{r}_i) \tag{9.106}$$

where $\phi_i(\mathbf{r}_i)$ can be given by a plane wave. Such systems are called free electron systems. Calculate single particle energy for a free electron system.

SOLUTION

For N-free electrons confined to volume V, the single electron wave function can be given by:

$$\phi_i(\mathbf{r}_i)=\frac{1}{\sqrt{V}}e^{i\mathbf{k}_i.\mathbf{r}_i} \tag{9.107}$$

We then have:

$$-\frac{\hbar^2}{2m}\nabla_i^2\phi_i(\mathbf{r}_i)=\frac{\hbar^2k_i^2}{2m}\phi_i(\mathbf{r}_i) \tag{9.108a}$$

To evaluate the exchange interaction energy term, let us write the Coulomb potential in the Fourier transformed form:

$$\begin{aligned}\frac{e^2}{4\pi\varepsilon_0|\mathbf{r}_i-\mathbf{r}_j|}&=\frac{e^2}{\varepsilon_0V}\sum_q\frac{e^{i\mathbf{q}.(\mathbf{r}_i-\mathbf{r}_j)}}{q^2}\\&=\frac{e^2}{\varepsilon_0(2\pi)^3}\int d^3q\frac{e^{i\mathbf{q}.(\mathbf{r}_i-\mathbf{r}_j)}}{q^2}\end{aligned} \tag{9.108b}$$

which yields:

$$\begin{aligned}&\frac{e^2}{4\pi\varepsilon_0}\sum_{j\neq i}\int\phi_j^*(\mathbf{r}_j)\frac{1}{|\mathbf{r}_i-\mathbf{r}_j|}\phi_i(\mathbf{r}_j)d^3r_j\phi_j(\mathbf{r}_i)\\&=\frac{e^2}{\varepsilon_0}\frac{1}{\left(2\pi\sqrt{V}\right)^3}\sum_{k_j}\int\int d^3r_jd^3q\frac{e^{i\mathbf{q}.(\mathbf{r}_i-\mathbf{r}_j)}}{q^2}e^{i(\mathbf{k}_i-\mathbf{k}_j).\mathbf{r}_j}e^{i\mathbf{k}_j.\mathbf{r}_i}\\&=\frac{e^2}{\varepsilon_0}\frac{1}{\left(2\pi\sqrt{V}\right)^3}\sum_{k_j}\int\int d^3r_jd^3q\frac{e^{i\mathbf{q}.\mathbf{r}_i}}{q^2}e^{i(\mathbf{k}_i-\mathbf{k}_j-\mathbf{q}).\mathbf{r}_j}e^{i\mathbf{k}_j.\mathbf{r}_i}\\&=\frac{e^2}{\varepsilon_0}\frac{1}{V^{3/2}}\sum_{k_j}\int d^3q\frac{e^{i(\mathbf{q}+\mathbf{k}_j).\mathbf{r}_i}}{q^2}\delta(\mathbf{k}_i-\mathbf{k}_j-\mathbf{q})\\&=\frac{e^2}{\varepsilon_0}\frac{1}{V^{3/2}}\sum_{k_j}\frac{e^{i\mathbf{k}_i.\mathbf{r}_i}}{|\mathbf{k}_i-\mathbf{k}_j|^2}\\&=\frac{e^2}{\varepsilon_0}\frac{1}{V}\sum_{k_j}\frac{1}{|\mathbf{k}_i-\mathbf{k}_j|^2}\phi_i(\mathbf{r}_i)\end{aligned} \tag{9.109}$$

Substitution of Eqns. (9.108a) and (9.109) into Eqn. (9.106) gives:

$$\left[\frac{\hbar^2 k_i^2}{2m} - \frac{e^2}{\varepsilon_0}\frac{1}{V}\sum_{k_j}\frac{1}{\left|\mathbf{k}_i - \mathbf{k}_j\right|^2}\right]\phi_i(\mathbf{r}_i) = \varepsilon_i\phi_i(\mathbf{r}_i) \tag{9.110}$$

On changing $\mathbf{k}_i \to \mathbf{k}$ and $\mathbf{k}_j \to \mathbf{k}'$, we have:

$$\varepsilon(k) = \frac{\hbar^2 k^2}{2m} - \frac{e^2}{\varepsilon_0}\frac{1}{V}\sum_{k'}\frac{1}{\left|\mathbf{k}-\mathbf{k}'\right|^2} = \frac{\hbar^2 k^2}{2m} - \frac{e^2}{\varepsilon_0}\int\frac{d^3k'}{(2\pi)^3}\frac{1}{\left|\mathbf{k}-\mathbf{k}'\right|^2} \tag{9.111}$$

Since one value of $\mathbf{k}$ can accommodate a maximum of two electrons, distribution of electrons occupies the states having k-values from 0 to k_F. Choosing spherical polar coordinates and taking $\mathbf{k}$ along the polar axis, we obtain:

$$\varepsilon(k) = \frac{\hbar^2 k^2}{2m} - \frac{e^2}{(2\pi)^3\varepsilon_0}\int_0^{k_F} k'^2 dk' \int_0^{\pi}\sin\theta\, d\theta \int_0^{2\pi} d\phi \frac{1}{\left(k^2 + k'^2 - 2kk'\cos\theta\right)} \tag{9.112}$$

Evaluation of integrals gives:

$$\begin{aligned}
I &= \int_0^{k_F} k'^2 dk' \int_0^{\pi}\sin\theta\, d\theta \int_0^{2\pi} d\phi \frac{1}{\left(k^2 + k'^2 - 2kk'\cos\theta\right)} \\
&= 2\pi\int_0^{k_F} k'^2 dk' \int_0^{\pi}\sin\theta d\theta \frac{1}{\left(k^2 + k'^2 - 2kk'\cos\theta\right)} \\
&= 2\pi\int_0^{k_F} k'^2 dk' \left[\frac{1}{2kk'}\ln\left(k^2 + k'^2 - 2kk'\cos\theta\right)\Big|_0^{\pi}\right] \\
&= 2\pi\int_0^{k_F} k'\ln\left\{\frac{k+k'}{|k-k'|}\right\}dk' \\
&= 4\pi\left[\frac{k_F}{2} + \frac{\left(k_F^2 - k^2\right)}{4k}\ln\left|\frac{k+k_F}{k-k_F}\right|\right]
\end{aligned} \tag{9.113}$$

Therefore:

$$\varepsilon(k) = \frac{\hbar^2 k^2}{2m} - \frac{e^2}{2\pi^2\varepsilon_0}\left[\frac{k_F}{2} + \frac{\left(k_F^2 - k^2\right)}{4k}\ln\left|\frac{k+k_F}{k-k_F}\right|\right] \tag{9.114}$$

8. Two spinless identical particles are confined to a one dimensional infinite potential well of width a, described by:

$$V(x) = \begin{cases} 0 & 0 \le x \le a \\ \infty & 0 > x > a \end{cases}$$

If the particles interact with each other via a potential $b\delta(x_1 - x_2 - a/2)$, find the approximate ground state energy of the system with the use of the first order perturbation theory. b is constant.

SOLUTION

The Hamiltonian of the system for $0 \le x \le a$:

$$H = -\frac{\hbar^2}{2m}\left(\frac{\partial^2}{\partial x_1^2} + \frac{\partial^2}{\partial x_2^2}\right) \tag{9.115}$$

The interaction potential can be used as the perturbative term. The unperturbed single particle wave functions for a state of an infinitely deep potential well are:

$$\phi_n(1) = \sqrt{\frac{2}{a}}\sin\left(\frac{n\pi x_1}{a}\right) \text{ and } \phi_l(2) = \sqrt{\frac{2}{a}}\sin\left(\frac{l\pi x_2}{a}\right), \text{where}(n,l) = 1,2,3,\ldots \tag{9.116}$$

Therefore, the unperturbed wave function and energy are as follows:

$$\begin{aligned} \psi(1,2) &= \frac{2}{a}\sin\left(\frac{n\pi x_1}{a}\right)\sin\left(\frac{l\pi x_2}{a}\right); \\ E_{n,l}^{(0)} &= \frac{\hbar^2\pi^2}{2ma^2}\left(n^2 + l^3\right) \end{aligned} \tag{9.117}$$

The first order correction to the energy is:

$$\begin{aligned} E_{n,l}^{(1)} &= b\langle\psi(1,2)|\delta(x_1 - x_2 - a/2)|\psi(1,2)\rangle \\ &= b\left(\frac{2}{a}\right)^2 \int_0^a dx_2 \int_0^a dx_1 \sin^2\left(\frac{n\pi x_1}{a}\right)\delta(x_1 - x_2 - a/2)\sin^2\left(\frac{l\pi x_2}{a}\right) \\ &= b\left(\frac{2}{a}\right)^2 \int_0^a dx_2 \sin^2\left(\frac{n\pi x_2}{a} + \frac{n\pi}{2}\right)\sin^2\left(\frac{l\pi x_2}{a}\right) \end{aligned} \tag{9.118a}$$

For the ground state, $n = 1$ and $l = 1$. Taking $y = \frac{\pi x_2}{a}$, we get:

$$\begin{aligned} E_{11}^{(1)} &= \frac{4b}{\pi a}\int_0^\pi dy \cos^2(y)\sin^2(y) \\ &= \frac{b}{\pi a}\int_0^\pi \cos^2(2y)dy \\ &= \frac{b}{2\pi a}\left[\int_0^\pi dy + \int_0^\pi \cos(4y)dy\right] \\ \Rightarrow E_{11}^{(1)} &= \frac{b}{2a} \end{aligned} \tag{9.118b}$$

Hence, ground state energy is:

$$E_{11} = \frac{\hbar^2\pi^2}{ma^2} + \frac{b}{2a} \tag{9.119}$$

9. Combining Eqns. (9.54) and (9.55), the particles density (number of particles per unit volume) within the Thomas-Fermi theory, is given by: $n(r) = \frac{1}{3\pi^2\hbar^3}\left[2m(\mu - V_{ext}(r))\right]^{3/2}$ for $\mu > V_{ext}(r)$ and zero otherwise. μ is called the chemical potential or Fermi energy. Calculate the total number of particles $N = \int n(r)d^3r$ when $V_{ext}(r) = \alpha r^2$.

SOLUTION

Note that for $r > r_0$, $n(r)$ will become imaginary, which is unphysical. Therefore, the number density must go to zero at $\mu = V_{ext}(r_0) = \alpha r_0^2$ or at $r_0 = \sqrt{\frac{\mu}{\alpha}}$. We thus have:

$$\begin{aligned} N &= \int_0^{r_0} \frac{1}{3\pi^2\hbar^3}\left[2m(\mu - \alpha r^2)\right]^{3/2} r^2 dr \int_0^{\pi} \sin\theta\, d\theta \int_0^{2\pi} d\phi \\ &= \frac{4}{3\pi}\left(\frac{2m\mu}{\hbar^2}\right)^{3/2} \int_0^{r_0} \left(1 - r^2/r_0^2\right)^{3/2} r^2 dr \end{aligned} \tag{9.120}$$

Taking $r/r_0 = \sin x$, we get:

$$\begin{aligned} \int_0^{r_0} \left\{1 - (r/r_0)^2\right\}^{3/2} r^2 dr &= r_0^3 \int_0^{\pi/2} \cos^4(x)\sin^2(x)dx \\ &= \frac{r_0^3}{16}\int_0^{\pi} (1 + \cos(y))\sin^2(y)dy \\ &= \frac{\pi r_0^3}{32} = \frac{\pi}{32}\left(\frac{\mu}{\alpha}\right)^{3/2} \end{aligned}$$

which gives:

$$N = \frac{1}{24}\left(\frac{2m}{\alpha\hbar^2}\right)^{3/2} \mu^3 \tag{9.121}$$

This shows that $\mu \propto N^{1/3}$ is determined by the total number of particles in a system.

9.5 Exercises

1. Two Bosons occupy the $n=4$ and $n=3$ states of a 1D infinite potential well of width a. Write the properly normalized wave function for the system.
2. Consider the two electrons confined to move freely within a solid of volume V. They occupy the momentum states $\mathbf{k}$ and $-\mathbf{k}$ and their spin functions are represented by α and β. Use Eqn. (9.15) and write the wave function for the singlet state ($S=0$) and triplet states ($S=1$).
3. 1D infinite potential well described by $V(x)=0$ for $0<x<a$ and $V(x)=\infty$ for $0>x>a$, contains three electrons. If the width of the well is 0.5 nm, calculate the total energy of the system by neglecting the Coulombic interaction between the electrons.
4. Suppose that the Schrödinger equation for a singly ionized helium atom is solved by considering it as a hydrogen-like atom with charge $2e$ at the nucleus. (a) How does the Bohr radius change in the $\psi_N(\mathbf{r})$, where $N(=nlm)$ is a composite quantum number, as compared to that in wave functions of the hydrogen atom? (b) Include the spin parts of wave functions α and β for up and down spins, respectively. Now treat the helium atom having two electrons without any interaction between them and then write down the typical two electron wave function. Do not choose ground state. (c) Show that your wave functions comply with Pauli's exclusion principle.
5. Write down the Hamiltonian for a system of two particles having mass m_1 and m_2. They interact via potential $V(|x|)$ which depends only on relative position $x=x_1-x_2$. Show that the Hamiltonian can be transformed to $H=-\frac{\hbar^2}{2M}\frac{\partial^2}{\partial X^2}-\frac{\hbar^2}{2\mu}\frac{\partial^2}{\partial x^2}+V(x)$, where $X=\frac{m_1x_1+mx_2}{m_1+m_2}$, $M=m_1+m_2$, and $\frac{1}{\mu}=\frac{1}{m_1}+\frac{1}{m_2}$.
6. Consider a pair of electrons which are constrained to move along the x-axis in a spin $S=1$ state. Electrons interact through an attractive potential $V(x_1,x_2)=-V_0$ for $0<|x_1-x_2|<a$ and $V(x_1,x_2)=\infty$ for $0\geq|x_1-x_2|\geq a$. Find the lowest energy E_0 and corresponding eigenstate $\psi_0(x_1,x_2)$ in the case where total momentum is zero.
7. Two non-interacting and indistinguishable particles, each of mass m, are oscillating with frequency ω in a 1D harmonic potential. One of the particles is in the ground state while the other particle is in the first excited state. (a) Construct the antisymmetric spatial part of the wave function of the system and (b) calculate $\langle (x_2-x_1)^2\rangle$.

10

Time-dependent Perturbations and Semi-classical Treatment of Interaction of Field with Matter

In the preceding chapters, we discussed solutions and the applications of the Schrödinger wave equation having time-independent potentials. Schrödinger quantum mechanics assume that operators and the Hamiltonian are independent of time. And the time evolution of a system is taken care of by wave function. The time independence of the Hamiltonian allows to us to factorize the wave function into space- and time-dependent parts. Therefore, in Schrödinger representation we write:

$$i\hbar\frac{\partial\psi(\mathbf{r},t)}{\partial t}=\left[-\frac{\hbar^2}{2m}\nabla^2+V(\mathbf{r})\right]\psi(\mathbf{r},t),\ \text{with}\ \psi(\mathbf{r},t)=\varphi(\mathbf{r})e^{\frac{-iEt}{\hbar}} \tag{10.1}$$

This allows us to obtain an eigenvalue equation:

$$\left[-\frac{\hbar^2}{2m}\nabla^2+V(\mathbf{r})\right]\varphi(\mathbf{r})=E\varphi(\mathbf{r}) \tag{10.2}$$

where E is the energy eigenvalue. The $|\psi(\mathbf{r},t)|^2$, which is interpreted as probability density, is constant in time. In addition to this, expectation values of operators (observables) are also constant in time. Equation (10.2), which involves time-independent potential, is solved exactly or approximately, depending upon the nature of the potential $V(\mathbf{r})$.

10.1 Time-dependent Potentials

The Hamiltonians of a large number of systems depend on time through the time-dependent potentials. When, the Hamiltonian is time-dependent, factorization of the wave function into space- and time-dependent parts is not possible and hence the Schrödinger picture is inadequate to describe the system. In this chapter, we are going to deal with the Hamiltonian that has the form $H(\mathbf{r},t)=H_0(\mathbf{r})+H_1(\mathbf{r},t)$, where $H_0(\mathbf{r})$ is time-independent and it is assumed to be an unperturbed part, whose energy eigenvalues and eigenstates are known exactly. $H_1(\mathbf{r},t)$ is the time-dependent part of the Hamiltonian, and it is treated as a perturbation applied to the system. A perturbation theory can be used if $|H_1|\ll H_0$. Because of the time dependence of $H_1(\mathbf{r},t)$, calculation for stationary energy eigenstates is not expected to be of any importance. $\psi(\mathbf{r},t)$ for a time-dependent Hamiltonian cannot be found in the Schrödinger representation. The interaction picture of quantum mechanics, which utilizes some elements of the Heisenberg picture and some elements of the Schrödinger picture have been used to find $\psi(\mathbf{r},t)$ for the time-dependent Hamiltonian.

However, we are not going to use the interaction picture here. The approach adopted by us is as follows. Let us say H_0 satisfies the eigenvalue equation:

$$H_0|\phi_n\rangle = E_n|\phi_n\rangle \tag{10.3}$$

where E_n and $|\phi_n\rangle$ are the energy eigenvalue and eigenstate of the nth state of the system in the absence of perturbation. Let us also assume that $|\phi_n\rangle$ form a complete set. Unlike time-independent perturbation theory, we here used E_n in place of $E_n^{(0)}$, because here we are not going to determine energy corrections, which are irrelevant in the case of time-dependent perturbations. The eigenstate $|\phi_n\rangle$ can be taken as basis vectors of Hilbert space to write $|\varphi\rangle = \sum_n c_n(0)|\phi_n\rangle$, where $c_n(0)$ is a constant in time. In the absence of $H_1(\mathbf{r},t)$, we are dealing with stationary state problems and hence we have:

$$\psi(\mathbf{r},t) = \sum_n c_n(0)e^{-\frac{iE_nt}{\hbar}}\phi_n(\mathbf{r}) \tag{10.4}$$

Now, let us consider the situation where initially, in the absence of $H_1(\mathbf{r},t)$, only one of the eigenstates, say $|\phi_i\rangle$, is populated. But as time goes on and $H_1(\mathbf{r},t)$ is turned on, states other than $|\phi_i\rangle$ get populated. We are then no longer dealing with stationary state problems and hence the time evolution of $\psi(\mathbf{r},t)$ is not as simple as given by Eqn. (10.4). The time-dependence of $H_1(\mathbf{r},t)$ causes transition to states other than $|\phi_i\rangle$. The basic question we would like to address is: What change in $\psi(\mathbf{r},t)$ is expected on introducing $H_1(\mathbf{r},t)$? Or, how does an arbitrary state ket change as time goes on, where the Hamiltonian is time-dependent? We can expect that $\psi(\mathbf{r},t)$ is still represented in the same form as in Eqn. (10.4), provided c_n is made time-dependent: We therefore write:

$$|\psi(t)\rangle = \sum_n c_n(t)e^{-\frac{iE_nt}{\hbar}}|\phi_n\rangle \tag{10.5}$$

Note that in this procedure, time-dependence of $c_n(t)$ is solely arising due to $H_1(\mathbf{r},t)$ and it must go to $c_n(0)$ whenever $H_1(\mathbf{r},t)=0$. Then the probability of finding a particle in the state $|\phi_n\rangle$ is given by $|c_n(t)|^2$ in place of $|c_n(0)|^2$. The time-dependence of $c_n(t)$ is thus determined from the Schrödinger equation having the Hamiltonian $H(\mathbf{r},t) = H_0(\mathbf{r}) + H_1(\mathbf{r},t)$. We then have:

$$i\hbar\frac{\partial|\psi(t)\rangle}{\partial t} = (H_0 + H_1)|\psi(t)\rangle \tag{10.6}$$

which with the use of Eqn. (10.5) gives:

$$\begin{aligned}
&i\hbar\frac{\partial}{\partial t}\sum_n c_n(t)e^{-\frac{iE_nt}{\hbar}}|\phi_n\rangle = (H_0 + H_1)\sum_n c_n(t)e^{-\frac{iE_nt}{\hbar}}|\phi_n\rangle \\
&\Rightarrow i\hbar\sum_n \dot{c}_n(t)e^{-\frac{iE_nt}{\hbar}}|\phi_n\rangle + \sum_n E_nc_n(t)e^{-\frac{iE_nt}{\hbar}}|\phi_n\rangle \\
&= \sum_n E_nc_n(t)e^{-\frac{iE_nt}{\hbar}}|\phi_n\rangle + H_1(\mathbf{r},t)\sum_n c_n(t)e^{-\frac{iE_nt}{\hbar}}|\phi_n\rangle \\
&\Rightarrow i\hbar\sum_n \dot{c}_n(t)e^{-\frac{iE_nt}{\hbar}}|\phi_n\rangle = H_1(\mathbf{r},t)\sum_n c_n(t)e^{-\frac{iE_nt}{\hbar}}|\phi_n\rangle
\end{aligned} \tag{10.7}$$

On taking the inner product with $\langle \phi_m | e^{\frac{iE_m t}{\hbar}}$ and then defining $\omega_{mn} = \frac{(E_m - E_n)}{\hbar}$, we get:

$$i\hbar \sum_n \dot{c}_n(t) e^{\frac{i(E_m - E_n)t}{\hbar}} \langle \phi_m | \phi_n \rangle = \sum_n c_n(t) e^{\frac{i(E_m - E_n)t}{\hbar}} \langle \phi_m | H_1(\mathbf{r},t) | \phi_n \rangle$$
$$\Rightarrow \dot{c}_m(t) = \frac{1}{i\hbar} \sum_n h_{mn}(t) c_n(t) e^{i\omega_{mn} t} \tag{10.8a}$$

Here, we have used $\langle \phi_m | \phi_n \rangle = \delta_{m,n}$ and defined $h_{mn}(t) = \langle \phi_m | H_1(\mathbf{r},t) | \phi_n \rangle$. We thus obtained a matrix equation:

$$\frac{d}{dt}\begin{bmatrix} c_1 \\ c_2 \\ . \\ . \end{bmatrix} = \frac{1}{i\hbar}\begin{bmatrix} h_{11}e^{i\omega_{11}t} & h_{12}e^{i\omega_{12}t} & . & . \\ h_{21}e^{i\omega_{21}t} & h_{22}e^{i\omega_{22}t} & . & . \\ . & . & . & . \\ . & . & . & . \end{bmatrix}\begin{bmatrix} c_1 \\ c_2 \\ . \\ . \end{bmatrix} \tag{10.8b}$$

which yields coupled equations that are solved to find $c_1(t), c_2(t).....c_n(t)......$ After knowing $c_1(t), c_2(t).....c_n(t).....$, one determines the probability of finding the system in any particular state at a later time. *Note that we have not used any approximation until now and hence all treatments are exact.* Exactly solvable problems with time-dependent potentials are rather rare. In most of the cases, perturbation expansion is used to solve coupled equations. However, there are problems of enormous practical importance such as nuclear magnetic resonance and MASERs, which are exactly solvable.

10.2 Exactly Solvable Time-dependent Two-state Systems

The matrix Eqn. (10.8b) is solved exactly for two-state systems perturbed by a periodic external field. The real physical systems may have more than two states, but for some important cases two of the states are very weakly coupled to other states, and hence two-state analysis becomes relevant. One of the examples is the ammonia MASER.

A two-state system problem with sinusoidal perturbing potential is described as follows:

$$|\psi\rangle = c_1(t) e^{\frac{-iE_1 t}{\hbar}} |\phi_1\rangle + c_2(t) e^{\frac{-iE_2 t}{\hbar}} |\phi_2\rangle \tag{10.9}$$

$$H_0 = E_1 |\phi_1\rangle\langle\phi_1| + E_2 |\phi_2\rangle\langle\phi_2|, \text{with } E_2 > E_1 \tag{10.10a}$$

and:

$$H_1 = \gamma e^{i\omega t} |\phi_1\rangle\langle\phi_2| + \gamma e^{-i\omega t} |\phi_2\rangle\langle\phi_1| \tag{10.10b}$$

where γ and ω are real positive. We thus have time-dependent perturbing potential that connects the two states and the transition between two states takes place. *Note that* Eqn. (10.10a) *is consistent with* Eqn. (10.3). The matrix elements h_{11}, h_{12}, h_{21}, and h_{22} are:

$$h_{11} = \langle\phi_1|H_1|\phi_1\rangle = \gamma e^{i\omega t}\langle\phi_1|\phi_1\rangle\langle\phi_2|\phi_1\rangle + \gamma e^{-i\omega t}\langle\phi_1|\phi_2\rangle\langle\phi_1|\phi_1\rangle$$
$$\Rightarrow h_{11} = \gamma e^{i\omega t}\delta_{21} + \gamma e^{-i\omega t}\delta_{12} = 0 \tag{10.11a}$$

Similarly, we find that $h_{22} = 0$. We next take:

$$h_{12} = \langle\phi_1|H_1|\phi_2\rangle = \gamma e^{i\omega t}\langle\phi_1|\phi_1\rangle\langle\phi_2|\phi_2\rangle + \gamma e^{-i\omega t}\langle\phi_1|\phi_2\rangle\langle\phi_1|\phi_2\rangle$$
$$\Rightarrow h_{12} = \gamma e^{i\omega t} \tag{10.11b}$$

and:

$$h_{21} = \langle\phi_2|H_1|\phi_1\rangle = \gamma e^{i\omega t}\langle\phi_2|\phi_1\rangle\langle\phi_2|\phi_1\rangle + \gamma e^{-i\omega t}\langle\phi_2|\phi_2\rangle\langle\phi_1|\phi_1\rangle$$
$$\Rightarrow h_{21} = \gamma e^{-i\omega t} \tag{10.11c}$$

On substituting Eqns. (10.11) into Eqn. (10.8a), we get:

$$i\hbar\frac{dc_1(t)}{dt} = \gamma e^{i\omega t}e^{i\omega_{12}t}c_2(t) = \gamma e^{i(\omega+\omega_{12})t}c_2(t) \tag{10.12a}$$

and:

$$i\hbar\frac{dc_2(t)}{dt} = \gamma e^{-i\omega t}e^{i\omega_{21}t}c_1(t) = \gamma e^{-i(\omega-\omega_{21})t}c_1(t) \tag{10.12b}$$

where, $\omega_{12} = -\omega_{21} = \frac{(E_1 - E_2)}{\hbar}$. The two first order differential equations are combined to give one second order differential equation:

$$i\hbar\frac{d^2c_2(t)}{dt^2} = \frac{d}{dt}\left(\gamma e^{-i(\omega-\omega_{21})t}c_1(t)\right) = -i(\omega-\omega_{21})\gamma e^{-i(\omega-\omega_{21})t}c_1(t) + \gamma e^{-i(\omega-\omega_{21})t}\frac{dc_1(t)}{dt} \tag{10.13a}$$

which with the use of Eqns. (10.12a) and (10.12b) goes to:

$$\frac{d^2c_2(t)}{dt^2} + i(\omega-\omega_{21})\frac{dc_2(t)}{dt} + \left(\frac{\gamma}{\hbar}\right)^2 c_2(t) = 0 \tag{10.13b}$$

This is a standard second order differential equation. We solve it with the use of a trial solution, $c_2(t) = c_2(0)e^{i\Omega t}$, which satisfies Eqn. (10.13b) if:

$$\left[\Omega^2 + \Omega(\omega-\omega_{21}) - \left(\frac{\gamma}{\hbar}\right)^2\right] = 0 \tag{10.14}$$

whose solution gives:

$$\Omega = -\frac{1}{2}(\omega-\omega_{21}) \pm \left\{\left(\frac{\omega-\omega_{21}}{2}\right)^2 + \left(\frac{\gamma}{\hbar}\right)^2\right\}^{1/2} \tag{10.15}$$

Therefore, a general solution of Eqn. (10.13b) is:

$$c_2(t) = e^{-i\frac{(\omega-\omega_{21})t}{2}}\left[Ae^{i\sqrt{\left(\frac{\omega-\omega_{21}}{2}\right)^2+\left(\frac{\gamma}{\hbar}\right)^2}t} + Be^{-i\sqrt{\left(\frac{\omega-\omega_{21}}{2}\right)^2+\left(\frac{\gamma}{\hbar}\right)^2}t}\right] \tag{10.16}$$

if initially at $t = 0$ only the lower state $|\phi_1\rangle$ is populated so that $c_1(0) = 1$ and $c_2(0) = 0$. This implies $A = -B$ and from Eqn. (10.12b):

$$i\hbar\left.\frac{dc_2(t)}{dt}\right|_{t=0} = \gamma \tag{10.17}$$

Differentiating Eqn. (10.16) with respect to t and then using Eqn. (10.17), we get:

$$A = -\frac{\gamma}{\left\{\hbar^2(\omega-\omega_{21})^2 + 4\gamma^2\right\}^{1/2}} \tag{10.18}$$

The probability that both the states $|\phi_1\rangle$ and $|\phi_2\rangle$ are populated at later time t is then given by:

$$|c_2(t)|^2 = \frac{\gamma^2}{\left\{\hbar^2\left(\frac{\omega-\omega_{21}}{2}\right)^2+\gamma^2\right\}}\sin^2\left\{\left(\sqrt{\left(\frac{\omega-\omega_{21}}{2}\right)^2+\left(\frac{\gamma}{\hbar}\right)^2}\right)t\right\} \tag{10.19}$$

and $|c_1(t)|^2 = 1 - |c_2(t)|^2$.

Equation (10.19) is the famous Rabi's formula. I.I. Rabi is known as the father of molecular beam technologies.

The plot of $|c_2(t)|^2$ as a function of ω at $t = \frac{\hbar\pi}{2\gamma}$ is displayed in Fig. 10.1.

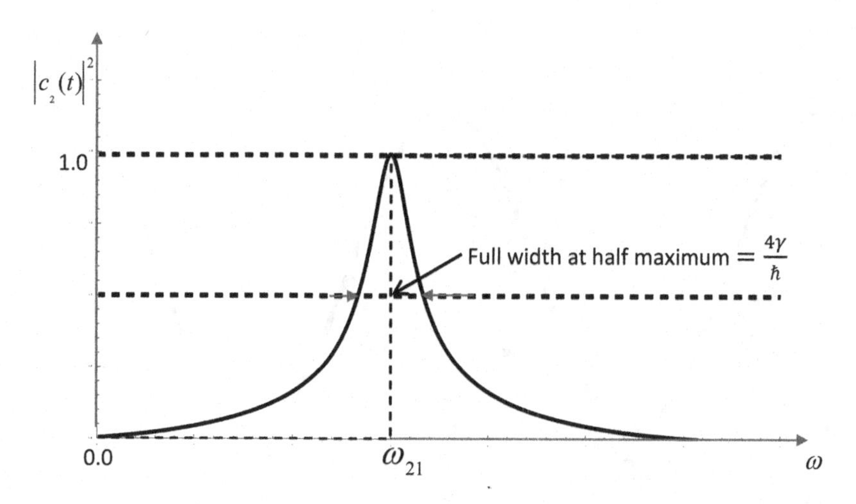

FIGURE 10.1
Plot of $|c_2(t)|^2$ versus ω at $t = \frac{\hbar\pi}{2\gamma}$.

As is seen from Fig. 10.1, a maximum occurs at:

$$\omega = \omega_{21} = \frac{E_2 - E_1}{\hbar} \tag{10.20}$$

which is known as the *Resonance condition*. The full width at half maximum of the curve is $\frac{4\gamma}{\hbar}$. Note that the weaker time-dependent potential (smaller γ) makes the resonance peak narrower.

The $|c_2(t)|^2$ and $|c_1(t)|^2$ are plotted as the function of t at $\omega = \omega_{21}$ (resonance condition) in Fig. 10.2.

As is seen from the figure, both $|c_2(t)|^2$ and $|c_1(t)|^2$ exhibit oscillatory behavior. The behavior of $|c_2(t)|^2$ is the opposite of that of $|c_1(t)|^2$. At $t = 0$, $|c_1(t)|^2 = 1$ and $|c_2(t)|^2 = 0$, which means the initially lower state $|\phi_1\rangle$ is populated and the upper state $|\phi_2\rangle$ is empty. As time grows, the system absorbs the energy from the time-dependent perturbation $H_1(\mathbf{r},t)$ and $|c_1(t)|^2$ decreases while $|c_2(t)|^2$ increases. At $t = \frac{\hbar\pi}{2\gamma}$, $|c_2(t)|^2 = 1$ and $|c_1(t)|^2 = 0$ the upper state is populated and the lower state is empty. Then between $t = \frac{\hbar\pi}{2\gamma}$ and $t = \frac{\hbar\pi}{\gamma}$, $|c_1(t)|^2$ increases and $|c_2(t)|^2$ decreases, which means that the system gives up the excess energy back to the perturbing field during this time interval. This cycle of absorption and emission is repeated infinitely, as is seen from Fig. 10.2. We thus see that time-dependent perturbation is causing transitions from $|\phi_1\rangle$ to $|\phi_2\rangle$ (absorption) and then from $|\phi_2\rangle$ to $|\phi_1\rangle$ (emission) as time increases.

The absorption-emission cycle takes place even if $\omega \neq \omega_{21}$. However, in that case, $|c_2(t)|^2$ no longer reaches to 1 and $|c_1(t)|^2$ does not reduce down to zero. Transition from $|\phi_2\rangle$ to $|\phi_1\rangle$ starts even if the state $|\phi_1\rangle$ is not completely empty.

There are many applications of the general time-dependent two state problem, such as nuclear magnetic resonance, spin-magnetic resonance, MASER (microwave amplification of stimulated emission of radiations), the atomic clock, and optical pumping. *It is amazing*

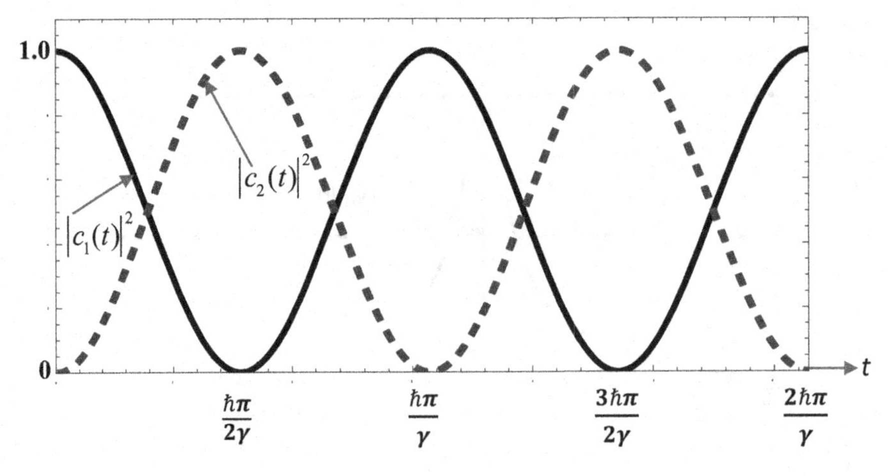

FIGURE 10.2
Plot $|c_1(t)|^2$ and $|c_2(t)|^2$ as a function t at $\omega = \omega_{21}$ (resonance condition).

to know that four Nobel Prizes have been awarded to those who exploited the applications of a time-dependent two state system in some form.

We have seen that an oscillating field can drive a collection of molecules from the ground state to an excited state. In an ammonia MASER, a stream of ammonia molecules travels down with known velocity through a tube of definite length. The tube has an oscillating microwave field, so that all (or almost all) molecules emerging at the other end of the tube are in the first excited state. Then, the excited outgoing molecules will decay on application of a small amount of electromagnetic radiation of the same frequency, generating an intense and coherent radiation because of the shorter period for all to decay.

10.3 Time-dependent Perturbation Theory

Except for a few problems such as two state time-dependent systems, the exact solution to differential equations that determine $c_n(t)$ is not possible. One then attempts a perturbative approach to solve Eqn. (10.7). Similar to the case of time-independent perturbation theory, we include the small dimensionless parameter λ to write the Hamiltonian $H = H_0 + \lambda H_1$ and expand the $c_n(t)$ in terms of λ as follows:

$$c_n(t) = c_n^{(0)}(t) + \lambda c_n^{(1)}(t) + \lambda^2 c_n^{(2)}(t) + \lambda^3 c_n^{(3)}(t) + \ldots\ldots \tag{10.21}$$

Substituting Eqn. (10.21) into Eqn. (10.8a) and replacing H_1 by λH_1 we get:

$$\begin{aligned} &\dot{c}_m^{(0)}(t) + \lambda \dot{c}_m^{(1)}(t) + \lambda^2 \dot{c}_m^{(2)}(t) + \ldots\ldots\ldots \\ &= \frac{1}{i\hbar}\sum_n h_{mn}(t)\left\{\lambda c_n^{(0)}(t) + \lambda^2 c_n^{(1)}(t) + \lambda^3 c_n^{(2)}(t) + + \ldots\ldots\right\} e^{i\omega_{mn}t} \end{aligned} \tag{10.22}$$

To satisfy this, the coefficient of each power of λ must vanish. We therefore have:

$$\frac{dc_m^{(0)}(t)}{dt} = 0 \tag{10.23a}$$

$$\frac{dc_m^{(1)}(t)}{dt} = \frac{1}{i\hbar}\sum_n h_{mn}(t) c_n^{(0)}(t) e^{i\omega_{mn}t} \tag{10.23b}$$

$$\frac{dc_m^{(2)}(t)}{dt} = \frac{1}{i\hbar}\sum_n h_{mn}(t) c_n^{(1)}(t) e^{i\omega_{mn}t} \tag{10.23c}$$

and so on.

Equation (10.23a) implies that $c_m^{(0)}(t)$ is constant in time, which is because of the fact that the zeroth order Hamiltonian H_0 is independent of time.

10.3.1 First Order Perturbation

The first order contribution to $c_m(t)$ is given by Eqn. (10.23b). We obtain:

$$c_m^{(1)}(t) = \frac{1}{i\hbar}\int \sum_n h_{mn}(t') c_n^{(0)}(t') e^{i\omega_{mn}t'}\, dt' \tag{10.24}$$

The majority of problems of practical interest are definable by assuming that the system evolves according to H_0 until $t=0$, and then a time-dependent perturbation is turned on. If the system initially at $t=0$ is in only one unperturbed state $|\phi_n\rangle$ then $c_n^{(0)}$ is non-zero (equal to 1) only for one value of n belonging to $|\phi_n\rangle$, and it is zero for all other values of n. When $t>0$ but very small, such that we still have $c_n^{(0)}(t)\simeq 1$, we then can drop all terms on right side of Eqn. (10.24) except one term. We get:

$$c_m^{(0)}(t)=\delta_{mn} \tag{10.25a}$$

$$c_m^{(1)}(t)=\frac{1}{i\hbar}\int_0^t h_{mn}(t')e^{i\omega_{mn}t'}dt', m\neq n \tag{10.25b}$$

The perturbation $H_1(\mathbf{r},t)$ has introduced transitions to other states. If we confine ourselves only to first order perturbation theory, then the probability that the transition from $|\phi_n\rangle$ *to* $|\phi_m\rangle$ has occurred after time t is given by:

$$|c_m(t)|^2=|c_m^{(1)}(t)|^2. \tag{10.26}$$

To see that the first order perturbation theory gives a reasonably correct answer, notice that if all elements in the column matrix on the right side of Eqn. (10.8b) are zero except one element $c_n(t)$ that is equal to unity, we have the equation:

$$\frac{dc_m(t)}{dt}=\frac{1}{i\hbar}h_{mn}e^{i\omega_{mn}t},\ \text{with}\, m\neq n \tag{10.27}$$

This is equivalent to saying that initially at $t=0$ only the $|\phi_n\rangle$ state is populated, and other states are empty.

The integration of Eqn. (10.27) between $0\rightarrow t$ yields:

$$c_m(t)=\frac{1}{i\hbar}\int_0^t h_{mn}(t')e^{i\omega_{mn}t'}dt' \tag{10.28}$$

which has right hand side identical to that in Eqn. (10.25b), justifying the use of first order perturbation theory.

10.4 Harmonic Perturbation

Let us consider a system initially in the state of $|\phi_n\rangle$ perturbed by a periodic potential that is described by:

$$H_1(\mathbf{r},t)=2H_1(\mathbf{r})\cos(\omega t)=H_1(\mathbf{r})\{e^{i\omega t}+e^{-i\omega t}\} \tag{10.29}$$

which is switched on at $t=0$. An example of this is an atom or a molecule exposed to electromagnetic radiation (harmonic perturbation). The probability amplitude for an atom in the initial state $|\phi_n\rangle$ to be in state $|\phi_m\rangle$ after time t is then obtained from Eqn. (10.25b) with the use of:

$$h_{mn}(t') = H_{mn}\left(e^{i\omega t'} + e^{-i\omega t'}\right), \text{where } H_{mn} = \langle\phi_m|H_1(\mathbf{r})|\phi_n\rangle \tag{10.30}$$

We thus have:

$$\begin{aligned} c_m^{(1)}(t) &= \frac{H_{mn}}{i\hbar}\int_0^t \left(e^{i(\omega+\omega_{mn})t'} + e^{-i(\omega-\omega_{mn})t'}\right)dt' \\ \Rightarrow c_m^{(1)}(t) &= H_{mn}\left[\frac{e^{-i(\omega-\omega_{mn})t} - 1}{\hbar(\omega-\omega_{mn})} - \frac{e^{i(\omega+\omega_{mn})t} - 1}{\hbar(\omega+\omega_{mn})}\right] \end{aligned} \tag{10.31}$$

where the first term on the right-hand side will reach to its maximum value when $(E_m - E_n) \to \hbar\omega$, while the second term will tend to a maximum value for $(E_m - E_n) \to -\hbar\omega$. On application of an harmonic perturbating field, the system either receives energy from the field (absorption) or it transfers the energy to the field (induced or stimulated emission). We thus find that the first term on the right-hand side represents absorption, while the second corresponds to stimulated emission. Since both absorption and stimulated emission cannot take place simultaneously, only one term is to be discussed at a time. We retain the first term for further discussions and obtain:

$$\begin{aligned} c_m^{(1)}(t) &= H_{mn}\frac{e^{-i(\omega-\omega_{mn})t} - 1}{\hbar(\omega-\omega_{mn})} \\ &= \frac{H_{mn}}{\hbar(\omega-\omega_{mn})}e^{\frac{i(\omega_{mn}-\omega)t}{2}}\left[e^{\frac{i(\omega_{mn}-\omega)t}{2}} - e^{\frac{-i(\omega_{mn}-\omega)t}{2}}\right] \\ &= \frac{2iH_{mn}}{\hbar(\omega-\omega_{mn})}e^{\frac{i(\omega_{mn}-\omega)t}{2}}\sin\left\{\frac{(\omega_{mn}-\omega)t}{2}\right\} \end{aligned} \tag{10.32}$$

10.4.1 Transition Probability

If the system was in the $|\phi_n\rangle$ state initially at $t = 0$, then the probability to find the system in the state $|\phi_m\rangle$ after time t is given by:

$$P_{n\to m}(t) = \left|c_m^{(1)}(t)\right|^2 = \frac{|H_{mn}|^2}{\hbar^2\left(\frac{\omega_{mn}-\omega}{2}\right)^2}\sin^2\left\{\frac{(\omega_{mn}-\omega)t}{2}\right\} \tag{10.33}$$

Note that $\frac{\sin\alpha t}{\alpha} \to t$, its maximum value, when $\alpha = \frac{(\omega_{mn}-\omega)}{2} \to 0$ or $(E_m - E_n) \to \hbar\omega$. Thus, $P_{n\to m}(t)$ exhibits a peak of height equal to $\frac{|H_{mn}|^2 t^2}{\hbar^2}$ and width equal to $2\pi/t$ at $\omega = \frac{(E_m - E_n)}{\hbar}$, when plotted as a function of ω. The $P_{n\to m}(t)$ shows more peaks at $\frac{(\omega_{mn}-\omega)}{2}t = \left(n + \frac{1}{2}\right)\pi$ where n is an integer. A plot $P_{n\to m}(t)$ versus $\omega_{mn} - \omega$ is displayed in Fig. 10.3.

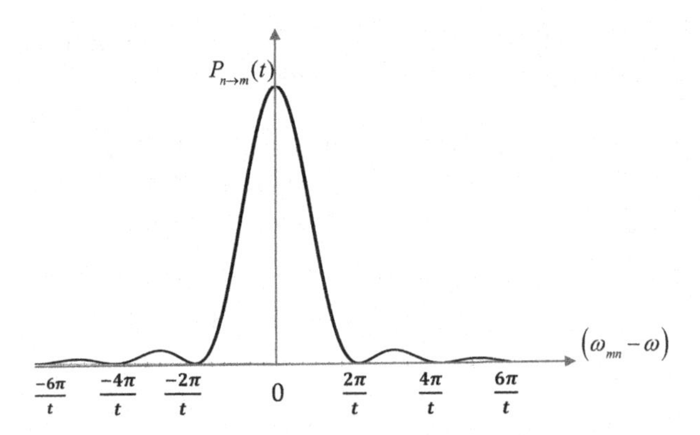

FIGURE 10.3
Plot of $P_{n\to m}(t)$ as a function of $(\omega_{mn}-\omega)$ at fixed t.

A central peak has the largest height, as is seen in Fig. 10.3. As t increases, the height of the central peak enhances as t^2, while its width decreases as $1/t$. For $t\to\infty$, one expects that the function:

$$\frac{\sin^2\alpha t}{\alpha^2}\to\pi t\delta(\alpha) \tag{10.34}$$

10.4.2 Fermi's Golden Rule

We next consider the transitions from one of the initial discrete states to a final state that is part of a continuum, with density of states $\rho(E_m)$. For example consider the case of ionization of an atom in which the initial state may be one of the discrete bound states and the final state includes a free electron momentum eigenstate that is part of a continuum of non-normalizable states. As has been shown above, the probability of transition between discrete states exhibits periodic dependence in time. But, if the final state is part of a continuum, an integral over all final states is required to get a resultant transition probability. A transition rate is associated with such a probability function. *The final transition rate is given by Fermi's Golden Rule.*

An approximation to the sum of transition probabilities over final states is given by the integral:

$$P(t)=\sum_m P_{n\to m}(t)=\int P_{n\to m}(t)\rho(E_m)dE_m \tag{10.35a}$$

which gives:

$$P(t)=\int\frac{H_{mn}^2}{\hbar^2}\frac{\sin^2(\alpha t)}{\alpha^2}\rho(E_m)dE_m \tag{10.35b}$$

As discussed above, when transitions are allowed for a longer time, the width of the central peak becomes very narrow, and therefore a very small energy range is expected to contribute to the integral (Eqn. 10.35b). In this range of energy, $|H_{mn}|^2$ and $\rho(E_m)$ almost remain the same, and they are treated independent of energy in the continuum of states. We thus have:

$$P(t)=\frac{|H_{mn}|^2}{\hbar^2}\rho(E_m)\int\frac{\sin^2(\alpha t)}{\alpha^2}dE_m \tag{10.36}$$

If all final states do not necessarily have the same matrix element, then $|H_{mn}|^2$ must be replaced by an average value of $|H_{mn}|^2$.

Note that:

$$\alpha = \frac{(\omega_{mn} - \omega)}{2} = \frac{E_m - E_n}{2\hbar} - \frac{\omega}{2} \\ \Rightarrow dE_m = 2\hbar d\alpha \tag{10.37}$$

Since the integral is contributed by a very small energy range, our answer will not be changed even if the limit of integration over α is extended to $\pm\infty$. We therefore have:

$$P(t) = \frac{2|H_{mn}|^2}{\hbar}\rho(E_m)\int_{-\infty}^{\infty}\frac{\sin^2(\alpha t)}{\alpha^2}d\alpha = \frac{2\pi}{\hbar}t|H_{mn}|^2\rho(E_m) \tag{10.38}$$

where we made use of:

$$\int_{-\infty}^{\infty}\left(\frac{\sin x}{x}\right)^2 dx = \pi \tag{10.39}$$

Note that the total transition probability is proportional to t for larger values of t. This linearity in t is a consequence of the fact that total transitional probability is proportional to the area under the main peak whose height varies as t^2 and width varies as $1/t$.

In obtaining Eqn. (10.38), we restricted the integral to the central peak only, which yields our answer 90% correct. Inclusion of more peaks from to the left and right of the central peak improves accuracy further.

The probability of transition per unit time $\frac{P(t)}{t}$, also termed the *transition rate* is:

$$W = \frac{P(t)}{t} = \frac{2\pi}{\hbar}|H_{mn}|^2\rho(E_m) \tag{10.40}$$

This equation is known as *Fermi's Golden Rule*. Equation (10.40) shows that total transition probability is independent of time, or $\frac{d}{dt}\sum_m |c_m^{(1)}(t)|^2$ is constant in time, provided that the first order perturbation theory is valid.

Note that for larger t, Eqn. (10.33), with the use of Eqn. (10.34), gives:

$$\begin{aligned} w_{n\to m}(t) &= \frac{P_{n\to m}(t)}{t} = \frac{\pi|H_{mn}|^2}{\hbar^2}\delta(\alpha) \\ &= \frac{\pi|H_{mn}|^2}{\hbar^2}\delta\left(\frac{\omega_{mn}-\omega}{2}\right) \end{aligned} \tag{10.41}$$

which is the rate of transition from state $|\phi_n\rangle$ to $|\phi_m\rangle$. The total transition probability is then obtained from the integration $\int w_{n\to m}\rho(E_m)dE_m = 2\hbar\int w_{n\to m}\rho(E_m)d\alpha$, which yields the same results as are given by Eqn. (10.40). Equation (10.41) also is called Fermi's Golden Rule. It agrees well with experimental results when applied to atomic systems, and it is of great practical importance.

10.5 Constant Perturbation

Let us consider that a constant perturbation is turned on at $t = 0$:

$$H_1(\mathbf{r},t) = \begin{cases} 0 & t < 0 \\ H'(\mathbf{r}) & t \geq 0 \end{cases} \tag{10.42}$$

$H'(\mathbf{r})$ is constant in time (independent of t), but it can be a function of position, spin, and momentum. Now suppose that at $t = 0$ only $|\phi_n\rangle$ is populated. Equations (10.25) now can be written as:

$$c_m^{(0)}(t) = \delta_{mn} \tag{10.43}$$

$$c_m^{(1)}(t) = \frac{H'_{mn}}{i\hbar}\int_0^t e^{i\omega_{mn}t'}dt'$$
$$\Rightarrow c_m^{(1)}(t) = \frac{-iH'_{mn}}{(E_m - E_n)}\left[e^{i\omega_{mn}t} - 1\right] \tag{10.44}$$

where $H'_{mn} = \langle\phi_m|H'(\mathbf{r})|\phi_n\rangle$, with $m \neq n$. The probability to find a system in the $|\phi_m\rangle$ state after time t is:

$$P_{n\to m}(t) = \left|c_m^{(1)}(t)\right|^2 = \frac{4|H'_{mn}|^2}{(E_m - E_n)^2}\sin^2\left(\frac{\omega_{mn}t}{2}\right) \tag{10.45}$$

which depends on $E_m - E_n$ along with time. An interesting study in this case is to see how transition probability varies with E_m, when E_m belongs to a continuum of final states all having nearly the same energy. Fig. 10.4 displays the plot of $\frac{4}{\gamma^2}\sin^2\left(\frac{\gamma t}{2}\right)$ as a function of $\gamma = \frac{E_m - E_n}{\hbar}$ for fixed t, the time interval during which perturbation has been turned on.

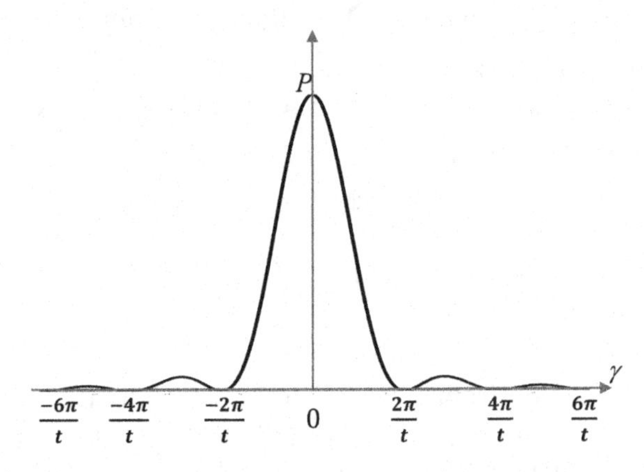

FIGURE 10.4
Plot of $P = \frac{4}{\gamma^2}\sin^2\left(\frac{\gamma t}{2}\right)$ versus γ at fixed t.

The central peak that occurs at $\gamma = 0$ has height t^2 and the width is proportional to $1/t$. For a larger t, the $\left|c_m^{(1)}(t)\right|^2$ is appreciable only for those final states that satisfy:

$$t \sim \frac{2\pi}{|\gamma|} = \frac{2\pi\hbar}{|E_m - E_n|} \tag{10.46}$$

If we say that constant perturbation is turned on for time Δt, then the transition probability is going to be appreciable to only those energy levels, which satisfy the relation $\Delta t \Delta E \sim \hbar$, where ΔE is the mean energy involved in a transition. The shorter Δt results into a larger peak width.

10.5.1 Fermi's Golden Rule

Similar to the case of harmonic perturbation, we can derive Fermi's Golden Rule for the constant a perturbation. With the use of:

$$\lim_{t\to\infty} \frac{1}{t} \frac{\sin^2(\gamma t/2)}{(\gamma^2/4)} = \pi\delta(\gamma/2) = 2\pi\hbar\delta(E_m - E_n) \tag{10.47}$$

we get:

$$P_{n\to m}(t) = 2\pi t \frac{|H'_{mn}|^2}{\hbar}\delta(E_m - E_n) \tag{10.48}$$

and:

$$w_{n\to m} = 2\pi \frac{|H'_{mn}|^2}{\hbar}\delta(E_m - E_n) \tag{10.49}$$

Hence, the total transition rate is:

$$\begin{aligned} W &= \int w_{n\to m}\rho(E_m)dE_m = \frac{2\pi}{\hbar}\int |H'_{mn}|^2 \delta(E_m - E_n)\rho(E_m)dE_m \\ &= \frac{2\pi}{\hbar}|H'_{mn}|^2 \rho(E_m)\Big|_{E_m = E_n} \end{aligned} \tag{10.50}$$

10.6 Semi-classical Treatment of Interaction of a Field with Matter

In this section, we discuss the interaction of electrons in an atom or molecule with a classically defined electromagnetic field. The electric and magnetic fields are derived from classical Maxwell equations, not from quantized electromagnetic radiations.

10.6.1 Absorption and Stimulated Emission

In the presence of an electromagnetic field, the basic Hamiltonian for an electron of mass m_e can be written as:

$$H = \frac{(\mathbf{p} - e\mathbf{A})^2}{2m_e} + e\phi(\mathbf{r}) + V(\mathbf{r}) \tag{10.51}$$

where $\mathbf{A}(\mathbf{r})$ and $\phi(\mathbf{r})$ are vector and scalar potentials, respectively. The $\mathbf{p}$ is momentum and $V(\mathbf{r})$ is the potential energy of the electron. The $\mathbf{A}(\mathbf{r})$ and $\phi(\mathbf{r})$ are determined from Maxwell's equations. In the space, which is free from sources of charge and current (charge density and current density are zero), we have:

$$\begin{aligned} \nabla.\mathbf{D} &= 0, \\ \nabla.\mathbf{B} &= 0, \\ \nabla\times\mathbf{E} &= -\frac{\partial\mathbf{B}}{\partial t}, \\ \nabla\times\mathbf{H} &= \frac{\partial\mathbf{D}}{\partial t} \end{aligned} \tag{10.52}$$

where $\mathbf{D}=\varepsilon_0\mathbf{E}$, $\mathbf{B}=\mu_0\mathbf{H}$ and $\varepsilon_0\mu_0=1/c^2$. Since charge density is zero, $\phi(\mathbf{r})=0$. Also, in the transverse (radiation) gauge, $\nabla.\mathbf{A}=0$ The $\mathbf{E}$ and $\mathbf{B}$ are then determined by:

$$\begin{aligned} \mathbf{E} &= -\frac{\partial\mathbf{A}}{\partial t}, \\ \mathbf{B} &= \nabla\times\mathbf{A} \end{aligned} \tag{10.53}$$

substitution of which into the last Maxwell's equation gives:

$$\begin{aligned} &\frac{1}{\mu_0}\nabla\times(\nabla\times\mathbf{A})=\varepsilon_0\frac{\partial^2\mathbf{A}}{\partial t^2} \\ &\Rightarrow \nabla^2\mathbf{A}-\frac{1}{c^2}\frac{\partial^2\mathbf{A}}{\partial t^2}=0 \end{aligned} \tag{10.54}$$

the solution of which is:

$$\mathbf{A}(\mathbf{r},t)=\mathrm{A}_0\hat{\varepsilon}\left(e^{i(\mathbf{k}.\mathbf{r}-\omega t)}+e^{-i(\mathbf{k}.\mathbf{r}-\omega t)}\right) \tag{10.55a}$$

for monochromatic radiation. If the field is not monochromatic, the frequency spread over a frequency range of $d\omega$, then $A(\mathbf{r},t)$ can be taken as:

$$\mathbf{A}(\mathbf{r},t)=\int\hat{\varepsilon}B_0(\omega)\left(e^{i(\mathbf{k}.\mathbf{r}-\omega t)}+e^{-i(\mathbf{k}.\mathbf{r}-\omega t)}\right)d\omega \tag{10.55b}$$

Both Eqns. (10.55a) and (10.55b) satisfy Eqn. (10.54) with the condition $k^2=\frac{\omega^2}{c^2}$. The $\hat{\varepsilon}$ and $\mathbf{k}$ are unit polarization vector and propagation vector, respectively. Note that A_0 and $B_0(\omega)$ have different dimensions.

Since $\phi(\mathbf{r})=0$ and $\nabla.\mathbf{A}=0$, Eqn. (10.51) simplifies to:

$$H=\frac{p^2}{2m_e}-\frac{e\mathbf{A}.\mathbf{p}}{m_e}+V(\mathbf{r}) \tag{10.56}$$

where we have used $(\mathbf{p}.\mathbf{A}+\mathbf{A}.\mathbf{p})\psi=-i\hbar(\nabla.\mathbf{A})\psi+(\mathbf{A}.\mathbf{p})\psi+(\mathbf{A}.\mathbf{p})\psi=2(\mathbf{A}.\mathbf{p})\psi$ and have dropped $\frac{e^2|A|^2}{2m_e}$ because it is much smaller as compared to other terms. Thus, for an electron in an atom or molecule interacting with an electromagnetic field, we have $H = H_0 + H_1(\mathbf{r},t)$, with:

$$H_0 = \frac{p^2}{2m_e} + V(\mathbf{r}) \tag{10.57a}$$

$$H_1(\mathbf{r},t) = -\frac{e\mathbf{A}(\mathbf{r},t).\mathbf{p}}{m_e} \tag{10.57b}$$

For the monochromatic field:

$$H_1(\mathbf{r},t) = -\frac{eA_0\hat{\varepsilon}.\mathbf{p}}{m_e}\left(e^{i(\mathbf{k}.\mathbf{r}-\omega t)} + e^{-i(\mathbf{k}.\mathbf{r}-\omega t)}\right) \tag{10.58}$$

On comparing it with Eqn. (10.29) and then following the discussions of Section 10.4, we find that the first term on the right-hand side is responsible for absorption, while the second term relates to stimulated emission. As one of the absorption and emission takes place at a time, we here too discuss only the absorption part in detail. In Eqn. (10.33), we now have:

$$H_{mn} = i\frac{e\hbar A_0}{m_e}\langle\phi_m|e^{i\mathbf{k}.\mathbf{r}}(\hat{\varepsilon}.\nabla)|\phi_n\rangle \tag{10.59}$$

Equation (10.41) then yields:

$$\begin{aligned} w_{n\to m} &= \pi\left(\frac{e}{m_e}\right)^2 |A_0|^2 \left|\langle\phi_m|e^{i\mathbf{k}.\mathbf{r}}(\hat{\varepsilon}.\nabla)|\phi_n\rangle\right|^2 \delta\left(\frac{\omega_{mn}-\omega}{2}\right) \\ &= \frac{2\pi}{\hbar}\left(\frac{e}{m_e}\right)^2 |A_0|^2 \left|\langle\phi_m|e^{i\mathbf{k}.\mathbf{r}}(\hat{\varepsilon}.\mathbf{p})|\phi_n\rangle\right|^2 \delta(E_m - E_n - \hbar\omega) \end{aligned} \tag{10.60}$$

The presence of the δ-function can be understood as follows: If $|\phi_m\rangle$ is part of the continuum, we can integrate with $\rho(E_m)$ to get total transition rate. But if $|\phi_m\rangle$ is discrete, its energy is not infinitely sharp, and there may a natural broadening due to the finite life time or due to collisions. In such a case, ω is replaced by $(\omega + i\alpha)$ in Eqns. (10.55a) and (10.58); here α is very small but real, so that perturbation is switched on gradually in past. For such a case, Eqn. (10.31), with the absorption term only, modifies to:

$$\begin{aligned} c_m^{(1)}(t) &= \frac{H_{mn}}{i\hbar}\int_{-\infty}^{t} e^{-i(\omega-\omega_{mn})t'} e^{\alpha t'} dt' \\ &= \frac{H_{mn}}{\hbar}\frac{e^{-i(\omega-\omega_{mn})t} e^{\alpha t}}{\omega - \omega_{mn} + i\alpha} \end{aligned} \tag{10.61}$$

Then:

$$\left|c_m^{(1)}(t)\right|^2 = \frac{|H_{mn}|^2}{\hbar^2}\frac{e^{2\alpha t}}{(\omega-\omega_{mn})^2 + \alpha^2} \tag{10.62}$$

and the rate of change of the transition probability is:

$$\frac{d}{dt}\left|c_m^{(1)}(t)\right|^2 = \frac{|H_{mn}|^2}{\hbar^2}\frac{2\alpha e^{2\alpha t}}{(\omega-\omega_{mn})^2+\alpha^2} \tag{10.63}$$

which for $\alpha \to 0$ gives:

$$w_{n\to m} = \frac{2\pi}{\hbar}|H_{mn}|^2\,\delta(E_m - E_n - \hbar\omega) \tag{10.64}$$

where we used:

$$\lim_{\alpha\to 0}\frac{2\alpha}{(\omega-\omega_{mn})^2+\alpha^2} = 2\pi\delta(\omega-\omega_{mn}) = 2\pi\hbar\delta(E_m - E_n - \hbar\omega) \tag{10.65}$$

The electromagnetic field may not be strictly monochromatic. In that case, the more useful quantity is the absorption cross-section, which is defined as the *energy per unit time absorbed by the atom for transition* $(n \to m)$ *divided by the energy flux of the radiation field*. The energy flux (energy per unit area per unit time) is calculated as:

$$cu = \frac{c}{2}\left(\varepsilon_0|\mathbf{E}|^2 + \frac{1}{\mu_0}|\mathbf{B}|^2\right) = c\varepsilon_0\omega^2|A_0|^2 \tag{10.66}$$

We thus get the absorption cross-section:

$$\begin{aligned}\sigma_{abs} &= \frac{w_{n\to m}\hbar\omega}{cu}\\ &= \frac{2\pi}{c\varepsilon_0\omega}\left(\frac{e}{m_e}\right)^2\left|\langle\phi_m|e^{i\mathbf{k}.\mathbf{r}}(\hat{\varepsilon}.\mathbf{p})|\phi_n\rangle\right|^2\delta(E_m - E_n - \hbar\omega)\end{aligned} \tag{10.67}$$

10.6.2 Electric Dipole Approximation

The electric dipole approximation is based on the fact that the wavelength of the radiation field is much larger than the dimension of an atom or molecule. Therefore, $|\mathbf{k}.\mathbf{r}| = \left|\frac{2\pi}{\lambda}\hat{n}.\mathbf{r}\right| \ll 1$ and the series:

$$e^{i\mathbf{k}.\mathbf{r}} = 1 + \frac{2\pi}{\lambda}i(\hat{n}.\mathbf{r}) - \frac{1}{2}\left\{\frac{2\pi}{\lambda}(\hat{n}.\mathbf{r})\right\}^2 + \ldots.. \tag{10.68}$$

can be approximated by 1. The validity of such an approximation follows from the fact that for transitions in an atom, the energy of field $\hbar\omega$ must be the order of spacing between energy levels in an atom. The average energy difference between two neighboring electronic levels for a hydrogen-like atom can be approximated by $\frac{Ze^2}{4\pi\varepsilon_0 R_a}$, where Ze is charge and R_a is the average atomic radius. We then have:

$$\begin{aligned}\hbar\omega &\sim \frac{Ze^2}{4\pi\varepsilon_0 R_a}\\ \Rightarrow \frac{R_a}{\lambda} &\sim \frac{Ze^2}{2\pi(4\pi\varepsilon_0)\hbar c} \sim \frac{Z}{137}\end{aligned} \tag{10.69}$$

which says that for a lighter atom $\frac{R_a}{\lambda} \ll 1$ and therefore the use of $e^{i\mathbf{k}.\mathbf{r}} \approx 1$ is justified. We therefore have:

$$\langle \phi_m | e^{i\mathbf{k}.\mathbf{r}} (\hat{\varepsilon}.\mathbf{p}) | \phi_n \rangle \simeq \langle \phi_m | (\hat{\varepsilon}.\mathbf{p}) | \phi_n \rangle \tag{10.70}$$

If we take the polarization vector $\hat{\varepsilon}$ along the x-axis and $\hat{n}$ along the z-axis, then $\hat{\varepsilon}.\mathbf{p} = p_x$ and we need to evaluate $\langle \phi_m | p_x | \phi_n \rangle$ only, which is done as follows:

$$\begin{aligned} [x, H_0] &= x\left(\frac{p^2}{2m_e} + V\right) - \left(\frac{p^2}{2m_e} + V\right)x \\ &= x\frac{p^2}{2m_e} - \frac{p^2}{2m_e}x \\ &= \frac{1}{2m_e}\left[x\left(p_x^2 + p_y^2 + p_z^2\right) - \left(p_x^2 + p_y^2 + p_z^2\right)x\right] \\ &= \frac{1}{2m_e}\left(xp_x^2 - p_x^2 x\right) \\ &= \frac{1}{2m_e}\left(xp_x^2 - p_x x p_x + p_x x p_x - p_x^2 x\right) = \frac{i\hbar p_x}{m_e} \end{aligned} \tag{10.71}$$

which implies:

$$p_x = \frac{m_e}{i\hbar}[x, H_0] \tag{10.72}$$

and:

$$\begin{aligned} \langle \phi_m | p_x | \phi_n \rangle &= -\frac{im_e}{\hbar} \langle \phi_m | (xH_0 - H_0 x) | \phi_n \rangle \\ &= \frac{im_e}{\hbar}(E_m - E_n) \langle \phi_m | x | \phi_n \rangle \\ &= im_e \omega_{mn} \langle \phi_m | x | \phi_n \rangle \end{aligned} \tag{10.73}$$

Here we have made use of:

$$\left[H_0 |\phi_m\rangle = E_m |\phi_m\rangle\right]^\dagger \equiv \left[\langle \phi_m | H_0 = E_m \langle \phi_m |\right] \tag{10.74}$$

because E_m is real and H_0 is Hermitian.

From Eqn. (10.60), we now have:

$$w_{n\to m} = \frac{2\pi}{\hbar}\omega_{mn}^2 |A_0|^2 \left|e\langle \phi_m | x | \phi_n \rangle\right|^2 \delta(E_m - E_n - \hbar\omega) \tag{10.75a}$$

which with the use of Eqn. (10.66) becomes:

$$w_{n\to m} = \frac{2\pi}{\varepsilon_0 \hbar}\frac{\omega_{mn}^2}{\omega^2} u(\omega) \left|e\langle \phi_m | x | \phi_n \rangle\right|^2 \delta(E_m - E_n - \hbar\omega) \tag{10.75b}$$

The transition rate for absorption is then obtained as:

$$W = \int w_{n\to m}\, d\omega = \frac{2\pi}{\varepsilon_0 \hbar^2} u(\omega_{mn}) \mu_{mn}^2 \tag{10.76}$$

with $\mu_{mn} = e\langle \phi_m | x | \phi_n \rangle$. Since $e\mathbf{r}$ is the electric dipole moment, this approximation is called the **electric dipole approximation**, where μ_{mn} is known as the *transition dipole moment.* When the wavelength of the radiation field is larger as compared to the size of the atom/molecule, the field sees them as an electric dipole. In the case of when the wavelength of the radiation field is comparable with the size of the atom/molecule, use of the electric dipole approximation is not justified.

Substitution of Eqn. (10.73) into Eqn. (10.67) yields the absorption cross-section in the dipole approximation:

$$\sigma_{abs}(\omega) = \frac{2\pi e^2}{c\varepsilon_0 \omega} \omega_{mn}^2 |\langle \phi_m | x | \phi_n \rangle|^2 \delta(E_m - E_n - \hbar\omega) \tag{10.77a}$$

The absorption cross-section exhibits δ-function-like peaks whenever $\omega \to \omega_{mn}$. If $|\phi_n\rangle$ is ground state, then ω_{mn} is positive. Integration of Eqn. (10.77a) gives:

$$\begin{aligned} \int \sigma_{abs}(\omega) d\omega &= \int \frac{2\pi e^2}{c\varepsilon_0 \omega} \omega_{mn}^2 |\langle \phi_m | x | \phi_n \rangle|^2 \delta(E_m - E_n - \hbar\omega) d\omega \\ &= \frac{2\pi e^2}{c\varepsilon_0 \hbar} \sum_m \omega_{mn} |\langle \phi_m | x | \phi_n \rangle|^2 \end{aligned} \tag{10.77b}$$

In atomic physics:

$$f_{mn} = \frac{2 m_e \omega_{mn}}{\hbar} |\langle \phi_m | x | \phi_n \rangle|^2 \tag{10.78}$$

is defined as **oscillator strength**. The **Thomas-Reiche-Kuhn** sum rule says $\sum_m f_{mn} = 1$. We therefore have:

$$\int \sigma_{abs}(\omega) d\omega = \frac{\pi e^2}{c\varepsilon_0 m_e} \tag{10.79}$$

which is the classical result, suggesting that quantum mechanics yields the correct limiting value.

10.7 Spontaneous Emission and Einstein Coefficients

The semi-classical treatment, which is discussed in the previous section, includes absorption and stimulated emission. However, it is well known that the electrons in atoms or molecules which are in an excited state can make a transition to a lower energy state, even if no field is applied to the atom/molecule. Such an emission is known as spontaneous emission.

Thus, emission of energy by excited atoms should include both spontaneous emission and the stimulated emission (emission triggered by the radiation field). A complete explanation of the entire process of absorption, spontaneous emission, and stimulated emission requires the approach that involves the quantization of electromagnetic field. However, an approach within the semi-classical treatment to explain the phenomenon of absorption, spontaneous emission, and stimulated emission was formulated by Einstein in 1927. He considered a system of an atom/molecule having two states in equilibrium with the radiation field of energy density $u(\omega)$. Let us say that the number of atoms/molecules in the lower energy state $|\phi_n\rangle$ is N_n, and the number of those that are in the upper energy state $|\phi_m\rangle$ is N_m. The probability per unit time (transition rate) that an atom makes a transition from $|\phi_n\rangle$ to $|\phi_m\rangle$ by absorbing energy from the field is proportional to $u(\omega)$. Thus, in the case of an atom,

$$\textit{The transition rate for absorption} = B_{n\to m}u(\omega_{nm}) \tag{10.80}$$

where $B_{n\to m}$ is called Einstein's B-coefficient for induced absorption. Therefore,

$$\textit{The transition rate for the entire system from } |\phi_n\rangle \textit{ to } |\phi_m\rangle = N_n B_{n\to m}u(\omega_{nm}) \tag{10.81}$$

Similarly, if we denote Einstein's B coefficient for stimulated emission by $B_{m\to n}$, then:

$$\textit{The transition rate for the system from } |\phi_m\rangle \text{ to } |\phi_n\rangle = N_m B_{m\to n}u(\omega_{mn}) \tag{10.82}$$

In thermal equilibrium, when there is no transfer of energy between the system and radiation field, we should have:

$$N_n B_{n\to m}u(\omega_{nm}) = N_m B_{m\to n}u(\omega_{mn}) \tag{10.83}$$

Since $u(\omega_{mn}) = u(\omega_{nm})$, Einstein inferred $B_{n\to m} = B_{m\to n}$ from this, which implies that $N_n = N_m$. But this is not in accordance with the Boltzmann distribution law that demands:

$$\frac{N_n}{N_m} = e^{\frac{\hbar\omega_{mn}}{k_B T}} \tag{10.84}$$

where k_B is the Boltzmann constant and T is temperature. Einstein proposed that there should be an additional contribution to emission, called spontaneous emission, that is independent of the radiation field and takes place even in the absence of the field. He wrote the additional contribution to emission as $N_m A_{m\to n}$, where $A_{m\to n}$ is the *Einstein coefficient for spontaneous emission*. Then in thermal equilibrium, we have:

$$N_n B_{n\to m}u(\omega_{mn}) = N_m B_{m\to n}u(\omega_{mn}) + N_m A_{m\to n} \tag{10.85}$$

which with the use of Eqn. (10.84), gives:

$$A_{m\to n} = B_{n\to m}\left(e^{\frac{\hbar\omega_{mn}}{k_B T}} - 1\right)u(\omega_{mn}) \tag{10.86}$$

Planck's radiation law gives:

$$u(\omega_{mn}) = \frac{\hbar\omega_{mn}^3}{\pi^2 c^3}\frac{1}{e^{\frac{\hbar\omega_{mn}}{k_B T}}-1} \tag{10.87}$$

Therefore:

$$A_{m\to n} = B_{n\to m}\frac{\hbar\omega_{mn}^3}{\pi^2 c^3} \tag{10.88}$$

Note that Eqn. (10.76) also describes the transition rate for induced absorption. Then, on equating Eqns. (10.76) and (10.80), we get:

$$\begin{aligned} &\frac{2\pi}{\varepsilon_0\hbar^2}u(\omega_{mn})\mu_{mn}^2 = B_{n\to m}u(\omega_{mn}) \\ &\Rightarrow B_{n\to m} = \frac{2\pi}{\varepsilon_0\hbar^2}\mu_{mn}^2 \end{aligned} \tag{10.89}$$

We considered $\hat{\varepsilon}$ along the x-axis and $\hat{n}$ along the z-axis, in evaluating $\langle\phi_n|\hat{\varepsilon}.\mathbf{p}|\phi_m\rangle$. Since the radiation field is isotropic and it is not restricted to a given direction only, a more generalized treatment would be to replace $|e\langle\phi_n|x|\phi_m\rangle|^2$ by $\frac{1}{3}|e\langle\phi_n|\mathbf{r}|\phi_m\rangle|^2$. We then obtain:

$$B_{m\to n} = \frac{2\pi}{3\varepsilon_0\hbar^2}|e\langle\phi_m|\mathbf{r}|\phi_n\rangle|^2 \tag{10.90a}$$

and:

$$A_{n\to m} = \frac{2}{3\varepsilon_0\pi\hbar c^3}\omega_{mn}^3|e\langle\phi_m|\mathbf{r}|\phi_n\rangle|^2 \tag{10.90b}$$

To find the ratio of the spontaneous emission rate to the stimulated emission rate, we use Eqns. (10.86), (10.87) and (10.90):

$$\frac{\text{Spontaneous Emission rate}}{\text{Stimulated Emission rate}} = \frac{A_{m\to n}}{B_{m\to n}u(\omega_{mn})} = e^{\frac{\hbar\omega_{mn}}{k_B T}}-1 \tag{10.91}$$

which tells us that in thermal equilibrium, the stimulated emission rate will far exceed the spontaneous emission rate if $\hbar\omega_{mn} \ll k_B T$, the thermal energy. The stimulated emission rate will be suppressed when the condition is reversed.

10.8 Dipole Selection Rules

As seen in the previous section, evaluation of the transition dipole moment $\mathbf{M}_{mn} = \langle\phi_m|e\mathbf{r}|\phi_n\rangle$ is the key element to calculate transition rates for absorption, stimulated emission, and spontaneous emission. As is obvious, transitions for which $\mathbf{M}_{mn} \neq 0$ are allowed transitions, and transitions for which $\mathbf{M}_{mn} = 0$ are forbidden transitions. For a hydrogen-like atom, a state of the atom is specified by quantum numbers (n,l,m). Let us say that $|\phi_n\rangle \equiv |n,l,m\rangle$ and $|\phi_{n'}\rangle \equiv |n',l',m'\rangle$. We then have:

$$\begin{aligned}\mathbf{M}_{n'n} &= \langle \phi_{n'}|e\mathbf{r}|\phi_n\rangle \\ &= \hat{i}e\langle n,l,m|x|n',l',m'\rangle + \hat{j}e\langle n,l,m|y|n',l',m'\rangle + \hat{k}e\langle n,l,m|z|n',l',m'\rangle\end{aligned} \tag{10.92}$$

With use of spherical polar coordinates, we write:

$$\langle n,l,m|z|n',l',m'\rangle = \langle n,l,m|r\cos\theta|n',l',m'\rangle \tag{10.93}$$

Using $\psi_{nlm}(r,\theta,\phi) = N_{nlm}R_{nl}(r)P_l^m(\cos\theta)e^{im\phi}$ for the hydrogen-like atom, we find that integration over angle-ϕ to evaluate Eqn. (10.93) has the form $\int_0^{2\pi} e^{i(m'-m)\phi}d\phi$, which is nonzero if and only if $\Delta m = m' - m = 0$. Note that if radiations are plane polarized with the electric field along the z-axis, we need to evaluate only Eqn. (10.93), because other components will be zero. However, if radiation is polarized in the $x-y$ plane, we have to calculate the x and y components too of Eqn. (10.92). A convenient method to calculate x and y components is found by defining:

$$\begin{aligned}&x^{\pm} = x \pm iy = r\sin\theta\cos\phi \pm ir\sin\theta\sin\phi \\ &\Rightarrow x^{\pm} = r\sin\theta e^{\pm i\phi}\end{aligned} \tag{10.94}$$

Then:

$$\langle n,l,m|x^{\pm}|n',l',m'\rangle = f(r,\theta)\int_0^{2\pi} e^{i(m'-m\pm1)\phi}d\phi \tag{10.95}$$

To make it nonzero one requires that:

$$\begin{aligned}&m' - m \pm 1 = 0 \\ &\Rightarrow \Delta m = \pm 1\end{aligned} \tag{10.96}$$

This implies that to have a nonzero value of both $\langle n,l,m|x|n',l',m'\rangle$ and $\langle n,l,m|y|n',l',m'\rangle$, condition (see Eqn. 10.96) is to be satisfied. Hence, for an arbitrary polarization, the general selection rules are:

$$m' - m = \Delta m = 0, \pm 1 \tag{10.97}$$

In case of integration over angle θ, with the use of the orthogonality condition for Legendre polynomials $P_l^m(\cos\theta)$, it is found that all three matrix elements involved in evaluating Eqn. (10.92) are nonzero if and only if:

$$l' - l = \Delta l = \pm 1 \tag{10.98}$$

The radial part of $\langle n,l,m|\mathbf{r}|n',l',m'\rangle$ will have a nonzero value for all allowed values of n,l,n', and l', provided they are in conformity with the other selection rule. The selection rule for electric dipole transitions in a hydrogen-like atom are then given by:

$$\Delta l = \pm 1 \text{ and } \Delta m = 0, \pm 1 \tag{10.99}$$

Note that the perturbation Hamiltonian does not have any spin operators; therefore, spin quantum numbers do not change during dipole transitions $m_s' = m_s$.

10.9 Solved Examples

1. A charged particle with charge q and mass m is bound to the ground state of a 1D harmonic oscillator potential along the x-axis. A time-dependent perturbation, $H_1 = -qxH'\Theta(t)\sin(\omega t)e^{-t/\tau}$, which is spatially uniform, is switched on at $t = 0$. The H' is constant both in space and time. Using first order perturbation theory, calculate the probability of finding the particle in the excited state for $t \gg \tau$.

SOLUTION

Let us say that the system is in the ground state at $t = 0$ and it is found in the excited state $|\phi_m\rangle$ after time t. The probability of finding the system in the excited state at time t is $|c_m^{(1)}(t)|^2$, where:

$$c_m^{(1)}(t) = \frac{1}{i\hbar}\int_0^t h_{m0}(t')e^{i\omega_{m0}t'}dt' \tag{10.100}$$

with $h_{m0}(t') = -qH'\Theta(t')\sin(\omega t')e^{-t'/\tau}\langle\phi_m|x|\phi_0\rangle$. To calculate $\langle\phi_m|x|\phi_0\rangle$, we use Eqn. (3.125c) of Section 3.19 and get:

$$x = \sqrt{\frac{\hbar}{2m\omega_0}}(a + a^\dagger) \tag{10.101}$$

Here ω_0 is the natural frequency of the oscillator. We thus have:

$$\begin{aligned}\langle\phi_m|x|\phi_0\rangle &= \sqrt{\frac{\hbar}{2m\omega_0}}\langle\phi_m|(a + a^\dagger)|\phi_0\rangle \\ &= \sqrt{\frac{\hbar}{2m\omega_0}}\langle\phi_m|\phi_1\rangle \\ &= \sqrt{\frac{\hbar}{2m\omega_0}}\delta_{m,1}\end{aligned} \tag{10.102}$$

where we have made use of $a|\phi_0\rangle = 0$. We then have:

$$h_{m0}(t') = -qH'\sqrt{\frac{\hbar}{2m\omega_0}}\Theta(t')\sin(\omega t')e^{-t'/\tau}\delta_{m,1} \tag{10.103}$$

and:

$$\begin{aligned} c_m^{(1)}(t) &= \frac{iqH'}{\hbar}\sqrt{\frac{\hbar}{2m\omega_0}}\int_0^t \sin(\omega t')e^{-t'/\tau}e^{i\omega_{10}t'}dt' \\ &= \frac{qH'}{2\hbar}\sqrt{\frac{\hbar}{2m\omega_0}}\int_0^t\left(e^{i(\omega+\omega_{10})t'}e^{-t'/\tau} - e^{-i(\omega-\omega_{10})t'}e^{-t'/\tau}\right)dt' \\ &= \frac{qH'}{2\hbar}\sqrt{\frac{\hbar}{2m\omega_0}}\left[\frac{e^{i(\omega+\omega_{10})t}e^{-t/\tau}-1}{i(\omega+\omega_{10})-1/\tau} + \frac{e^{-i(\omega-\omega_{10})t}e^{-t/\tau}-1}{i(\omega-\omega_{10})+1/\tau}\right] \end{aligned} \tag{10.104}$$

where $\omega_{10} = \frac{3}{2}\omega_0 - \frac{1}{2}\omega_0 = \omega_0$. For $t \gg \tau$, $e^{-t/\tau} \to 0$, and we have:

$$\begin{aligned} c_1^{(1)}(t) &= \frac{iqH'}{2\hbar}\sqrt{\frac{\hbar}{2m\omega_0}}\left[\frac{1}{(\omega+\omega_0)+i/\tau} + \frac{1}{(\omega-\omega_0)-i/\tau}\right] \\ &= \frac{iqH'}{2\hbar}\sqrt{\frac{\hbar}{2m\omega_0}}\left[\frac{2\omega}{\left(\omega^2-\omega_0^2\right) - \frac{2i\omega_0}{\tau} + \frac{1}{\tau^2}}\right] \end{aligned} \tag{10.105}$$

The probability of finding the charged particle in the excited state at time t is:

$$\left|c_1^{(1)}(t)\right|^2 = \frac{(qH')^2}{2\hbar m\omega_0}\frac{\omega^2}{\left(\omega^2-\omega_0^2+\frac{1}{\tau^2}\right)^2 + \frac{4\omega_0^2}{\tau^2}} \tag{10.106}$$

2. Consider a hydrogen atom in its first excited state placed in a cavity. At what temperature of a cavity is the ratio of the spontaneous emission transition rate to the stimulated emission transition rate equal to $\frac{3}{2}$ for the transition from the excited state to the ground state of the atom?

SOLUTION

From Eqn. (10.91), we have:

$$\frac{\text{Spontaneous Emission rate}}{\text{Stimulated Emission rate}} = \frac{A_{m\to n}}{B_{m\to n}u(\omega_{mn})} = e^{\frac{\hbar\omega_{mn}}{k_BT}} - 1 \tag{10.107}$$

Therefore, for the transition from the excited to the ground state of the hydrogen atom we obtain:

$$\frac{3}{2} = e^{\frac{\hbar\omega_{21}}{k_B T}} - 1$$
$$\Rightarrow \frac{\hbar\omega_{21}}{k_B T} = \ln(2.5) \tag{10.108}$$
$$\Rightarrow T = \frac{\hbar\omega_{21}}{k_B \ln(2.5)}$$

where:

$\hbar\omega_{21} = E_2 - E_1 = -21.76 \times 10^{-19}\left[\frac{1}{2^2} - 1\right]$ Joules, and Boltzmann constant $k_B = 1.38 \times 10^{-23}$ Joules/K.

Thus:

$$T = \frac{16.32 \times 10^{-19}}{1.38 \times 10^{-23} \times 0.92} = 1.28 \times 10^5 \text{ K} \tag{10.109}$$

3. Consider a spatially uniform time-dependent force $F(t) = \frac{F_0}{\omega^2(t^2 + \tau^2)}$, with $-\infty \le t \le \infty$, along the x-axis. A 1D harmonic oscillator, which is in its ground state at $t = -\infty$ and oscillating with frequency ω_0, is acted upon by this force. Use the first order time-dependent perturbation theory to calculate the probability of finding the oscillator in its first excited state at $t = \infty$.

SOLUTION

The time-dependent perturbation due to the force is:

$$H_1 = -F(t)x = -\frac{F_0 x}{\omega^2(t^2 + \tau^2)} \tag{10.110}$$

For the transition from the ground state at $t = -\infty$ to the first excited state at $t = \infty$, we get from Eqn. (10.25b):

$$c_1^{(1)}(t) = \frac{1}{i\hbar}\int_{-\infty}^{\infty} h_{10}(t) e^{i\omega_{10} t} dt \tag{10.111}$$

where:

$$h_{10}(t) = \langle \phi_1 | H_1 | \phi_0 \rangle = -\frac{F_0}{\omega^2(t^2 + \tau^2)} \langle \phi_1 | x | \phi_0 \rangle \tag{10.112}$$

With the use of Eqn. (10.101), we have:

$$\langle \phi_1 | x | \phi_0 \rangle = \sqrt{\frac{\hbar}{2m\omega_0}} \langle \phi_1 | (a + a^{\dagger}) | \phi_0 \rangle = \sqrt{\frac{\hbar}{2m\omega_0}} \tag{10.113}$$

Here, we used $a|\phi_0\rangle = 0$ and $a^\dagger|\phi_0\rangle = |\phi_1\rangle$. Also note that $\omega_{10} = \left(\frac{3}{2} - \frac{1}{2}\right)\omega_0 = \omega_0$. We then get:

$$c_1^{(1)}(t) = \frac{iF_0}{\hbar\omega^2}\sqrt{\frac{\hbar}{2m\omega_0}}\int_{-\infty}^{\infty}\frac{e^{i\omega_0 t}}{(t^2+\tau^2)}dt$$
$$= \frac{iF_0\tau}{\hbar\omega^2}\sqrt{\frac{\hbar}{2m\omega_0}}\int_{-\infty}^{\infty}\frac{e^{i\omega_0 t}}{(t+i\tau)(t-i\tau)}dt \tag{10.114}$$

Using contour integration, we obtain:

$$c_1^{(1)}(t) = \frac{iF_0}{\hbar\omega^2}\sqrt{\frac{\hbar}{2m\omega_0}}\left(2\pi i\frac{e^{-\omega_0\tau}}{2i\tau}\right) = \frac{\pi i F_0}{\hbar\omega^2\tau}\sqrt{\frac{\hbar}{2m\omega_0}}e^{-\omega_0\tau} \tag{10.115}$$

Therefore, the probability of finding the oscillator in the first excited state at $t = \infty$ is:

$$P_{0\to 1} = \left|c_1^{(1)}\right|^2 = \frac{\pi^2 F_0^2}{2m\hbar\omega_0\omega^4\tau^2}e^{-2\omega_0\tau} \tag{10.116}$$

which is small for $\omega_0\tau \gg 1$.

4. Consider a hydrogen atom in its ground state $|1,0,0\rangle$, which is acted upon by a time-dependent perturbation defined by:

$$H_1 = \begin{cases} 0 & t \le 0 \\ V_0\sqrt{x^2+y^2}\,e^{-t/\tau} & t > 0 \end{cases}.$$

Use the first order perturbation theory to calculate the probability of finding the atom in the 2s state for $t \gg \tau$.

SOLUTION

The time-dependent perturbation is:

$$H_1 = V_0\sqrt{x^2+y^2}\,e^{-t/\tau} \quad \text{for}\, t > 0 \tag{10.117}$$

For the transition from the ground state (at $t \le 0$) to the 2s state (at $t > 0$) and $t \gg \tau$, we have:

$$c_2^{(1)}(t) = \frac{1}{i\hbar}\int_0^t h_{21}(t')e^{i\omega_{21}t'}dt' \tag{10.118}$$

For the transition:

$$\omega_{21} = \frac{1}{\hbar}[E_2 - E_1] = -\frac{\mu e^4}{2(4\pi\varepsilon_0)^2\hbar^3}\left[\frac{1}{2^2} - 1\right] = \frac{3\mu e^4}{8(4\pi\varepsilon_0)^2\hbar^3} \tag{10.119}$$

And:

$$h_{21}(t) = V_0\langle 2,0,0|\sqrt{x^2+y^2}\,|1,0,0\rangle e^{-t/\tau}, \tag{10.120}$$

We thus have:

$$c_2^{(1)}(t) = \frac{V_0}{i\hbar}\langle 2,0,0|\sqrt{x^2+y^2}\,|1,0,0\rangle \int_0^t e^{i\omega_{21}t'}e^{-t'/\tau}dt'$$
$$= \frac{V_0}{i\hbar}\langle 2,0,0|\sqrt{x^2+y^2}\,|1,0,0\rangle\left[\frac{e^{i\omega_{21}t}e^{-t/\tau}-1}{i\omega_{21}-\frac{1}{\tau}}\right] \tag{10.121}$$

For $t \gg \tau$, $e^{-t/\tau}$ is very small and we therefore have:

$$c_2^{(1)}(t) = \frac{V_0}{\hbar}\langle 2,0,0|\sqrt{x^2+y^2}\,|1,0,0\rangle \frac{1}{\left(\omega_{21}+\frac{i}{\tau}\right)} \tag{10.122}$$

To evaluate the matrix element, we use from Section 6.3:

$$\psi_{100}(r) = \frac{1}{a_0}\left(\frac{1}{\pi a_0}\right)^{1/2} e^{-r/a_0} \tag{10.123}$$

$$\psi_{200}(r) = \frac{1}{4a_0}\left(\frac{1}{2\pi a_0}\right)^{1/2}\left(2-\frac{r}{a_0}\right)e^{-r/2a_0}. \tag{10.124}$$

With the use of spherical polar coordinates, we obtain:

$$\langle 2,0,0|\sqrt{x^2+y^2}\,|1,0,0\rangle = \frac{1}{4\sqrt{2}\pi a_0^3}\int_0^\infty r^3 e^{-r/a_0}\left(2-\frac{r}{a_0}\right)e^{-r/2a_0}dr\int_0^\pi \sin^2\theta\, d\theta\int_0^{2\pi} d\phi$$
$$= \frac{\pi a_0}{4\sqrt{2}}\int_0^\infty \left(\frac{r}{a_0}\right)^3 e^{-r/a_0}\left(2-\frac{r}{a_0}\right)e^{-r/2a_0}d(r/a_0) \tag{10.125}$$
$$= \frac{\pi a_0}{4\sqrt{2}}\left[2\frac{3!}{(3/2)^4}-\frac{4!}{(3/2)^5}\right] = -\frac{\pi a_0}{\sqrt{2}}\left(\frac{2}{3}\right)^4$$

Therefore:

$$c_2^{(1)}(t) = -\frac{\pi a_0 V_0}{\hbar\sqrt{2}}\left(\frac{2}{3}\right)^4\frac{1}{(\omega_{21}+i/\tau)} \tag{10.127}$$

The probability of finding the atom in the $2s$ state is:

$$P_{1s\to 2s} = \left|c_2^{(1)}(t)\right|^2 = \frac{1}{2}\left(\frac{\pi V_0 a_0}{\hbar}\right)^2\left(\frac{2}{3}\right)^8\frac{1}{\left(\omega_{21}^2+\frac{1}{\tau^2}\right)} \tag{10.128}$$

5. An electron confined to a cubical box of length L is subjected to the time-dependent perturbation $H_1(z,t)=V_0e^{i(k_z z-\omega t)}$. The perturbation is turned on at $t=0$. Calculate the transition rate, transition probability per unit time, using first order perturbation theory. What is its largest value?

SOLUTION

The transition probability from state $|\psi_{\mathbf{k}}\rangle$ at $t=0$ to state $|\psi_{\mathbf{k}'}\rangle$ at time t is given by: $P_{k\to k'}=\left|c_{k'}^{(1)}(t)\right|^2$, where:

$$c_{k'}^{(1)}(t)=\frac{1}{i\hbar}\int_0^t h_{k'k}(t')e^{i\omega_{k'k}t'}dt' \tag{10.129}$$

The wave function and energy of the electron confined to a cubic box are given by:

$$\psi_{\mathbf{k}}(\mathbf{r})=\frac{1}{L^{3/2}}e^{i\mathbf{k}.\mathbf{r}} \tag{10.130}$$

$$E_k=\frac{\hbar^2k^2}{2m} \tag{10.131}$$

with $\mathbf{k}=\left(\hat{i}n_x+\hat{j}n_y+\hat{k}n_z\right)\frac{\pi}{L}$, where each of n_x, n_y, and n_z takes values 1,2,3....... To evaluate:

$$h_{k'k}(t)=V_0\left\langle\psi_{\mathbf{k}'}\left|e^{ik_z z}\right|\psi_{\mathbf{k}}\right\rangle e^{-\omega t} \tag{10.132}$$

we calculate:

$$\begin{aligned}\left\langle\psi_{\mathbf{k}'}\left|e^{ik_z z}\right|\psi_{\mathbf{k}}\right\rangle&=\frac{1}{L^3}\int_{-\infty}^{\infty}e^{i(k_x-k_x')x}dx\int_{-\infty}^{\infty}e^{i(k_y-k_y')y}dy\int_{-\infty}^{\infty}e^{i(2k_z-k_z')z}dz\\ \Rightarrow\left\langle\psi_{\mathbf{k}'}\left|e^{ik_z z}\right|\psi_{\mathbf{k}}\right\rangle&=\delta_{k_x,k_x'}\delta_{k_y,k_y'}\delta_{2k_z,k_z'}\end{aligned} \tag{10.133}$$

Then:

$$\begin{aligned}c_{k'}^{(1)}(t)&=\frac{V_0}{i\hbar}\int_0^t\delta_{k_x,k_x'}\delta_{k_y,k_y'}\delta_{2k_z,k_z'}e^{-i\omega t'}e^{i\omega_{k'k}t'}dt'\\ &=\frac{V_0}{i\hbar}\int_0^t\delta_{k_x,k_x'}\delta_{k_y,k_y'}\delta_{2k_z,k_z'}e^{-i\omega t'}e^{i\frac{\hbar}{2m}\left(k_x'^2-k_x^2+k_y'^2-k_y^2+k_z'^2-k_z^2\right)t'}dt'\\ &=\frac{V_0}{i\hbar}\int_0^t e^{-i\omega t'}e^{i\frac{3\hbar k_z^2t'}{2m}}dt'\\ \Rightarrow c_{k'}^{(1)}(t)&=\frac{V_0}{\hbar}\frac{e^{i(\alpha-\omega)t}-1}{(\omega-\alpha)}\end{aligned} \tag{10.134}$$

with $\alpha = \dfrac{3\hbar k_z^2}{2m}$. Therefore transition probability per unit time is:

$$\begin{aligned} w_{k\to k'} &= \frac{P_{k\to k'}}{t} \\ &= 2\left(\frac{V_0}{\hbar}\right)^2 \frac{1-\cos\{(\omega-\alpha)t\}}{t(\omega-\alpha)^2} \\ &= \left(\frac{V_0}{\hbar}\right)^2 \frac{\sin^2\{(\omega-\alpha)t/2\}}{\{(\omega-\alpha)t/2\}^2} \end{aligned} \tag{10.135}$$

The largest value of $w_{k\to k'}$ is $\left(\dfrac{V_0}{\hbar}\right)^2$ that occurs for $\omega \to \alpha$.

6. Calculate the Einstein coefficients for the transition $|1,0,0\rangle$ to $|2,1,0\rangle$ in the hydrogen atom.

SOLUTION

Einstein coefficients A and B are given by Eqns. (10.90):

$$A_{2\to 1} = \frac{2e^2}{3\varepsilon_0 \pi\hbar c^3}\omega_{21}^3 \left|\langle 2,1,0|\mathbf{r}|1,0,0\rangle\right|^2 \tag{10.136a}$$

$$B_{2\to 1} = \frac{2\pi e^2}{3\varepsilon_0 \hbar^2}\left|\langle 2,1,0|\mathbf{r}|1,0,0\rangle\right|^2 \tag{10.136b}$$

The ω_{21} is given by Eqn. (10.119) and:

$$\langle 2,1,0|\mathbf{r}|1,0,0\rangle = \hat{i}\langle 2,1,0|x|1,0,0\rangle + \hat{j}\langle 2,1,0|y|1,0,0\rangle + \hat{k}\langle 2,1,0|z|1,0,0\rangle \tag{10.137}$$

Evaluations of $\langle 2,1,0|x|1,0,0\rangle$, $\langle 2,1,0|y|1,0,0\rangle$ and $\langle 2,1,0|z|1,0,0\rangle$, with the use of $|100\rangle$ and $|210\rangle$ defined in Section 6.3, Eqn. (6.38) yield:

$$\langle 2,1,0|x|1,0,0\rangle = \frac{1}{4\sqrt{2}\pi a_0^4}\int_0^\infty r^4 e^{-r/a_0} e^{-r/2a_0}dr \int_0^\pi \sin^2\theta\cos\theta d\theta \int_0^{2\pi}\cos\phi d\phi = 0 \tag{10.138a}$$

$$\langle 2,1,0|y|1,0,0\rangle = \frac{1}{4\sqrt{2}\pi a_0^4}\int_0^\infty r^4 e^{-r/a_0} e^{-r/2a_0}dr \int_0^\pi \sin^2\theta\cos\theta d\theta \int_0^{2\pi}\sin\phi d\phi = 0 \tag{10.138b}$$

and,

$$\begin{aligned} \langle 2,1,0|z|1,0,0\rangle &= \frac{1}{4\sqrt{2}\pi a_0^4}\int_0^\infty r^4 e^{-r/a_0} e^{-r/2a_0}dr \int_0^\pi \sin\theta\cos^2\theta d\theta \int_0^{2\pi} d\phi \\ &= 4\sqrt{2}a_0\left(\frac{2}{3}\right)^5 = 0.745a_0 \end{aligned} \tag{10.138c}$$

Therefore:

$$B_{2\to1} = \frac{3\pi e^2}{2\varepsilon_0\hbar^2}\times(0.745a_0)^2 = \frac{3.33\pi^2}{m_e}a_0 = 1.91\times10^{21}\ \text{meter/kg} \tag{10.139a}$$

and:

$$\begin{aligned}A_{2\to1} &= \frac{2e^2}{3\varepsilon_0\pi\hbar c^3}\omega_{21}^3(0.745a_0)^2 = \frac{1.48\hbar a_0}{m_e c^3}\omega_{21}^3 \\ &= \frac{1.48\hbar a_0}{m_e}\left(\frac{3\mu e^4}{8(4\pi\varepsilon_0)^2\hbar^3 c}\right)^3 \\ &= \frac{1.48\times1.05\times10^{-34}\times.529\times10^{-10}}{9.11\times10^{-31}}\times\left(\frac{16.32}{3\times1.05}\times10^7\right)^3 \\ &= 12.54\times10^8\ /\ \text{sec}\end{aligned} \tag{10.139b}$$

10.10 Exercises

1. A particle of mass m and charge q is oscillating under a 1D harmonic potential. Show that $\sum_{n'}(E_{n'}-E_n)|\langle n'|qx|n\rangle|^2 = \frac{\hbar^2 q^2}{2m}$, where n and n' represent the quantum states. Calculate the absorption cross-section using Eqn. (10.77b). Comment on absorption frequency.
2. A time-dependent perturbation $H_1 = \frac{-eE_0\tau z}{\pi\omega(t^2+\tau^2)}$ is applied to a hydrogen atom. The atom is in its ground state at $t=-\infty$. Use the first order time-dependent perturbation theory to calculate the probability of finding the atom in its first excited state at $t=\infty$.
3. Consider a particle in a 1D infinite potential well of width- a. The particle is in its ground state at time $t=0$. It is subjected to a traveling pulse represented by the time-dependent perturbation $H_1 = A\delta(x-ct)$. Calculate the transition probability of finding the particle in the nth excited state.
4. Take the expression of the transition probability derived in exercise 3 and compute it for the transition from the first excited state to the ground state by taking the mass of the particle equal to the mass of the electron, the width of well 1 nm, and $A = 3.2\times10^{-23}$ Joules $(\text{meter})^3$.
5. The energy levels of an electron confined to a nano-sized particle of radius a are given by $E_n = \frac{\hbar^2\pi^2 n^2}{2m_e a^2}$. The nano-particle is in its first excited state and is placed

in a cavity. At what temperature of the cavity is the ratio of the stimulated emission rate to the spontaneous emission rate equal to $\frac{1}{3}$, for the transition from the excited state to the ground state. Take $a = 4$ nm.

6. Consider a system consisting of a spin-$\frac{1}{2}$ particle, with no orbital angular momentum, placed in a uniform field $B_0\hat{k}$ directed along the z-axis. The Hamiltonian of it is given by $H_0 = -\frac{geB_0}{2m}S_z$. The system is then subjected to a small time-dependent magnetic field rotating in the x-y plane at the angular frequency ω. The perturbative part of the Hamiltonian is $H_1 = \gamma\left(e^{i\omega t}S^- + e^{-i\omega t}S^+\right)$, with $\gamma = -\frac{geB_1}{4m}$. S^- and S^+ are lowering and rising operators. The B_0 and B_1 are constants with $B_1 \ll B_0$. The spin up and spin down states of the system are represented by $|+\rangle$ and $|-\rangle$, respectively. Use the equations presented in Section 5.5.1 and show that $S_z|\pm\rangle = \pm\frac{\hbar}{2}|\pm\rangle$, $S^+|+\rangle = S^-|-\rangle = 0$, $S^+|-\rangle = \hbar|+\rangle$ and $S^-|+\rangle = \hbar|-\rangle$.
7. Consider again the system described in exercise 6 and show that (a) it is an exactly solvable two-state system with $\omega_{21} = \frac{geB_0}{2m}$, and (b) $|c_2(t)|^2$ is given by Eqn. (10.19).

11

Relativistic Quantum Mechanics

The Schrödinger equation describes the quantum mechanics of nonrelativistic particles. Some of the inherent limitations of the Schrödinger equation are the following: (i) The Hamiltonian does not include the spin motion of particle(s); (ii) The Hamiltonian involves the first order derivative with respect to time but the second order derivatives in space coordinates indicating that time and space coordinates are not treated at par; (iii) The Schrödinger formulation of quantum mechanics is not Lorentz–invariant, which requires that the laws of physics must be formulated in such a way that the description should not allow one to differentiate between inertial frames (frames of reference) that are moving relative to each other with a constant uniform velocity.

The Schrödinger equation is applicable when the kinetic energy of the particle is given by $\frac{p^2}{2m_0}$ instead of $\sqrt{(pc)^2+(m_0c^2)^2}$ and its energy is not comparable with its rest mass energy m_0c^2, where **c** is the velocity of light and m_0 is the rest mass of the particle.

If space and time coordinates of an event are represented by $(\mathbf{r},t)$ and $(\mathbf{r}',t')$ in two inertial frames, then the Lorentz–invariance requires the linear transformation $c^2(dt)^2-(d\mathbf{r})^2=c^2(dt')^2-(d\mathbf{r}')^2$, which connects space-time coordinates (x,y,z,t) of one reference frame to that of another (x',y',z',t'). The concept of Minkowski space or 4-vector space is used in the special theory of relativity and relativistic mechanics. Some of the well-known 4-vectors in covariant representation are the position vector $x_\mu \equiv (x_1,x_2,x_3,x_4)$ with $x_4=ict$; linear momentum $p_\mu \equiv (p_1,p_2,p_3,p_4)$ with $p_4=iE/c$ where E is energy; current density $j_\mu=(j_1,j_2,j_3,j_4)$ with $j_4=ic\rho$ where ρ is charge density; and wave vector $k_\mu \equiv (k_1,k_2,k_3,k_4)$ with $k_4=i\omega/c$ where ω is angular frequency.

The first order derivative in the 4-vector space is defined as $\left(\frac{\partial}{c\partial t},\vec{\nabla}\right)$ or $\left(\frac{\partial}{c\partial t},-\vec{\nabla}\right)$. The continuity equation in 4-vector space is:

$$\begin{aligned}
&\frac{\partial\rho}{\partial t}+\vec{\nabla}.\mathbf{J}=0\\
&\Rightarrow \frac{\partial(ic\rho)}{\partial(ict)}+\frac{\partial j_x}{\partial x}+\frac{\partial j_y}{\partial y}+\frac{\partial j_z}{\partial z}=\frac{\partial j_1}{\partial x_1}+\frac{\partial j_2}{\partial x_2}+\frac{\partial j_3}{\partial x_3}+\frac{\partial j_4}{\partial x_4}=0\\
&\Rightarrow \sum_{\nu=1}^{4}\frac{\partial j_\nu}{\partial x_\nu}=0
\end{aligned} \tag{11.1}$$

Here (x,y,z) are represented by (x_1,x_2,x_3) and ict by x_4.

11.1 The Klein-Gordon Equation

Klein-Gordon proposed a relativistic wave equation which is the quantum version of the energy momentum relation. It is of second order in both space and time and manifestly Lorentz-invariant. It relates to the Schrödinger equation as well.

In the nonrelativistic case, we replace $\mathbf{p} \to -i\hbar\nabla$ and $E \to i\hbar\frac{\partial}{\partial t}$ in $\frac{p^2}{2m} = E$ to get the free particle Schrödinger equation $-\frac{\hbar^2}{2m}\nabla^2\psi(\mathbf{r},t) = i\hbar\frac{\partial\psi(\mathbf{r},t)}{\partial t}$. For the relativistic case, the energy and momentum relation is $E^2 = (pc)^2 + (m_0c^2)^2$. Replacing E and $\mathbf{p}$ by operators and then operating both sides on $\psi(\mathbf{r},t)$, one gets:

$$-\hbar^2\frac{\partial^2\psi(\mathbf{r},t)}{\partial t^2} = -\hbar^2c^2\nabla^2\psi(\mathbf{r},t) + (m_0c^2)^2\psi(\mathbf{r},t) \tag{11.2}$$

which is known as the Klein-Gordon equation. One of the solutions of the equation is:

$$\psi(\mathbf{r},t) = \frac{1}{\sqrt{V}}e^{i(\mathbf{p}.\mathbf{r})/\hbar}e^{-iEt/\hbar} \tag{11.3a}$$

where V is volume. It is a second-order equation with respect to both time and space coordinates. Therefore:

$$\psi(\mathbf{r},t) = \frac{1}{\sqrt{V}}e^{i(\mathbf{p}.\mathbf{r})/\hbar}e^{iEt/\hbar} \tag{11.3b}$$

is also an equally acceptable solution. Since $E = \pm c\sqrt{p^2 + (m_0c)^2}$, one of the solutions corresponds to $E > 0$, while the other belongs to $E < 0$. Note that for each value of p, there are negative as well as positive values of E. There is no mechanism to have a transition from positive energy to negative energy. In the presence of some external potential, E and p will be altered to include the effect of field, and then the solutions $\psi(\mathbf{r},t)$ can be expressed as superposition of free particle solutions, provided that the free particle solutions form a complete set. Formation of a complete set does not permit us to discard the negative energy solutions.

11.1.1 Probability Density and Probability Current

Let us rewrite Eqn. (11.2):

$$\frac{\partial^2\psi(\mathbf{r},t)}{\partial t^2} = c^2\nabla^2\psi(\mathbf{r},t) - \left(\frac{m_0c^2}{\hbar}\right)^2\psi(\mathbf{r},t) \tag{11.4a}$$

the complex conjugate of which is:

$$\frac{\partial^2\psi^*(\mathbf{r},t)}{\partial t^2} = c^2\nabla^2\psi^*(\mathbf{r},t) - \left(\frac{m_0c^2}{\hbar}\right)^2\psi^*(\mathbf{r},t) \tag{11.4b}$$

On operating from the left on the first equation by $\psi^*(\mathbf{r},t)$ and then on the second equation by $\psi(\mathbf{r},t)$ and then subtracting the second from the first, we get:

$$\frac{\partial}{\partial t}\left\{\psi^*(\mathbf{r},t)\frac{\partial\psi(\mathbf{r},t)}{\partial t}-\psi(\mathbf{r},t)\frac{\partial\psi^*(\mathbf{r},t)}{\partial t}\right\}=c^2\nabla.\left\{\psi^*(\mathbf{r},t)\nabla\psi(\mathbf{r},t)-\psi(\mathbf{r},t)\nabla\psi^*(\mathbf{r},t)\right\}$$

$$\Rightarrow\frac{\partial}{\partial t}\left[\frac{i\hbar}{2m_0c^2}\left\{\psi^*(\mathbf{r},t)\frac{\partial\psi(\mathbf{r},t)}{\partial t}-\psi(\mathbf{r},t)\frac{\partial\psi^*(\mathbf{r},t)}{\partial t}\right\}\right] \tag{11.5}$$

$$-\left[\frac{i\hbar}{2m_0}\nabla.\left\{\psi^*(\mathbf{r},t)\nabla\psi(\mathbf{r},t)-\psi(\mathbf{r},t)\nabla\psi^*(\mathbf{r},t)\right\}\right]=0$$

Defining probability density P and probability current (flux) $\mathbf{S}$ by:

$$P=\frac{i\hbar}{2m_0c^2}\left\{\psi^*(\mathbf{r},t)\frac{\partial\psi(\mathbf{r},t)}{\partial t}-\psi(\mathbf{r},t)\frac{\partial\psi^*(\mathbf{r},t)}{\partial t}\right\} \tag{11.6a}$$

and:

$$\mathbf{S}=-\frac{i\hbar}{2m_0}\left\{\psi^*(\mathbf{r},t)\nabla\psi(\mathbf{r},t)-\psi(\mathbf{r},t)\nabla\psi^*(\mathbf{r},t)\right\} \tag{11.6b}$$

we get the continuity equation:

$$\frac{\partial P}{\partial t}+\nabla.\mathbf{S}=0 \tag{11.7}$$

In the case of the Schrödinger equation, P is associated with the single particle complex valued wave function so that $Pd^3r=|\psi|^2\,d^3r$ is the probability of finding the particle in volume d^3r. Interpretation of P as probability density requires that $|\psi|^2>0$. Therefore, if relativistic quantum mechanics is to be constructed in analogy with nonrelativistic quantum mechanics, then the relativistic wave function must be able to construct bilinear forms that can be interpreted as probability density and probability current, which satisfy the continuity equation. And, the probability density must be positive. It demands that the P defined by Eqn. (11.6a) must be positive at all values of time t. In addition to this, P should be the fourth component of a 4-vector density and it must transform like $P\to P/\sqrt{1-(v/c)^2}$ to make Pd^3r invariant under Lorentz transformations.

The $\mathbf{S}$ defined by Eqn. (11.6b) is exactly the same as that defined in the case of nonrelativistic quantum mechanics. However, there is difficulty in interpreting P as probability density. The Klein-Gordon equation consists of a second order derivative with respect time. As stated above, both Eqns. (11.3a) and (11.3b) provide equally acceptable solutions for a given situation. Use of Eqn. (11.3a) in Eqn. (11.6a) gives $P=E/Vm_0c^2$, while with the use of Eqn. (11.3b) we get $P=-E/Vm_0c^2$. This implies that P can be positive or negative. Also, the solutions for $E<0$ cannot be omitted because this violates the condition of formation of a complete set of solutions. It then appears that either the interpretation of P defined in Eqn. (11.6a) as probability density is to be abandoned or the Klein-Gordon equation is to be discarded.

To overcome this problem, Pauli and Weisskopf suggested that Eqn. (11.5) can be multiplied throughout with electric charge q and then $\rho = qP$ and $\mathbf{J} = q\mathbf{S}$ can be treated as charge density and current density, respectively. q can take both positive and negative values. The ρ and $\mathbf{J}$ satisfy the continuity equation and ρ can be assigned both positive as well as negative values. Then, the problem of interpretation would be dissolved.

11.1.2 The Klein-Gordon Equation in a Coulombic Field

In the presence of an electromagnetic field, the energy and momentum of a particle are modified according to $E \to E - q\phi$ and $\mathbf{p} \to \mathbf{p} + q\mathbf{A}$, where ϕ and $\mathbf{A}$ are scaler and vector potentials and q is a charge on the particle. We then have:

$$(E - q\phi)^2 = c^2(\mathbf{p} + q\mathbf{A})^2 + \left(m_0 c^2\right)^2 \tag{11.8}$$

When ϕ is the coulombic potential $q\phi = -\dfrac{Ze^2}{4\pi\varepsilon_0 r}$ and a magnetic field is absent $\mathbf{A} = 0$, we get:

$$\begin{aligned} &\left(E + \frac{Ze^2}{4\pi\varepsilon_0 r}\right)^2 \psi(\mathbf{r},t) = c^2 p^2 \psi(\mathbf{r},t) + \left(m_0 c^2\right)^2 \psi(\mathbf{r},t) \\ \Rightarrow &\left(i\hbar\frac{\partial}{\partial t} + \frac{Ze^2}{4\pi\varepsilon_0 r}\right)^2 \psi(\mathbf{r},t) = -\hbar^2 c^2 \nabla^2 \psi(\mathbf{r},t) + \left(m_0 c^2\right)^2 \psi(\mathbf{r},t) \end{aligned} \tag{11.9}$$

Taking $\psi(\mathbf{r},t) = \varphi(\mathbf{r}) e^{-iEt/\hbar}$, we obtain:

$$\left(E + \frac{Ze^2}{4\pi\varepsilon_0 r}\right)^2 \varphi(\mathbf{r}) = -\hbar^2 c^2 \nabla^2 \varphi(\mathbf{r}) + \left(m_0 c^2\right)^2 \varphi(\mathbf{r}) \tag{11.10}$$

Note that E is an operator in Eqn. (11.9), while it is energy eigenvalue in Eqn. (11.10). Writing:

$$\nabla^2 = \frac{1}{r^2}\frac{\partial}{\partial r}\left(r^2 \frac{\partial}{\partial r}\right) + \frac{1}{r^2}\left[\frac{1}{\sin\theta}\frac{\partial}{\partial\theta}\left(\sin\theta\frac{\partial}{\partial\theta}\right) + \frac{1}{\sin^2\phi}\frac{\partial^2}{\partial\phi^2}\right] \tag{11.11a}$$

and then writing (θ,ϕ) dependent part in terms of L^2, with the use of Eqn. (5.36b), we get:

$$\nabla^2 = \frac{1}{r^2}\frac{\partial}{\partial r}\left(r^2\frac{\partial}{\partial r}\right) - \frac{L^2}{r^2\hbar^2} \tag{11.11b}$$

Thus:

$$\left(E + \frac{Ze^2}{4\pi\varepsilon_0 r}\right)^2 \varphi(\mathbf{r}) = -\hbar^2 c^2 \frac{1}{r^2}\left\{\frac{d}{dr}\left(r^2\frac{d}{dr}\right) - \frac{L^2}{\hbar^2}\right\}\varphi(\mathbf{r}) + \left(m_0 c^2\right)^2 \varphi(\mathbf{r}) \tag{11.12}$$

Writing $\varphi(\mathbf{r}) = R_l(r) Y_{lm}(\theta,\phi)$ and then using $L^2 Y_{lm}(\theta,\phi) = l(l+1)\hbar^2 Y_{lm}(\theta,\phi)$, we have:

$$\frac{1}{r^2}\left\{\frac{d}{dr}\left(r^2\frac{d}{dr}\right) - l(l+1)\right\} R_l(r) = \left(\frac{m_0 c}{\hbar}\right)^2 R_l(r) - \frac{1}{\hbar^2 c^2}\left(E + \frac{Ze^2}{4\pi\varepsilon_0 r}\right)^2 R_l(r) \tag{11.13}$$

Taking $\gamma = \dfrac{Ze^2}{4\pi\varepsilon_0 \hbar c}$, we rewrite the equation as:

$$\frac{1}{r^2}\frac{d}{dr}\left(r^2\frac{dR_l}{dr}\right)+\left(\frac{2E\gamma}{c\hbar r}-\frac{l(l+1)-\gamma^2}{r^2}\right)R_l-\left(\frac{m_0^2c^4-E^2}{\hbar^2c^2}\right)R_l=0 \qquad (11.14)$$

Equation (11.14) resembles Eqn. (6.11) and on writing $E = E' + m_0c^2$ and then taking $\left(E' + \dfrac{Ze^2}{4\pi\varepsilon_0 r}\right) \ll m_0c^2$, it reduces to a nonrelativistic equation for the hydrogen atom. Therefore, the well behaved solution of the equation can be found following the procedure described in Section 6.2. After defining:

$$\lambda = \frac{E\gamma}{\sqrt{m_0^2c^4 - E^2}} \text{ and } s(s+1) = l(l+1) - \gamma^2 \qquad (11.15)$$

we find that a well behaved solution exists if:

$$\lambda = \upsilon + s, \text{with } \upsilon = 1, 2, 3, \ldots.. \qquad (11.16)$$

where s should be positive to make $R_l(r)$ finite at $r = 0$. Solving for s yields:

$$s = -\frac{1}{2} \pm \left[\left(l + \frac{1}{2}\right)^2 - \gamma^2\right]^{1/2} \qquad (11.17)$$

Since γ is very small for all practical purposes, positive values of s are possible for $l > 0$ with the plus sign. Thus, we have:

$$\begin{aligned}\frac{E\gamma}{\sqrt{m_0^2c^4 - E^2}} &= \upsilon - \frac{1}{2} + \left[\left(l + \frac{1}{2}\right)^2 - \gamma^2\right]^{1/2} \\ &= \upsilon - \frac{1}{2} + \left(l + \frac{1}{2}\right)\left[1 - \gamma^2 / \left(l + \frac{1}{2}\right)^2\right]^{1/2} \\ &\approx \upsilon + l - \frac{\gamma^2}{2\left(l + \frac{1}{2}\right)}\end{aligned} \qquad (11.18)$$

To obtain E, we take $\upsilon + l = n$ and $n - \dfrac{\gamma^2}{2\left(l + \frac{1}{2}\right)} = \sqrt{C}$:

$$\begin{aligned}&\frac{E^2\gamma^2}{C} = m_0^2c^4 - E^2 \\ &\Rightarrow E^2\left(1 + \frac{\gamma^2}{C}\right) = m_0^2c^4 \\ &\Rightarrow E = m_0c^2\left(1 + \frac{\gamma^2}{C}\right)^{-1/2} \approx m_0c^2\left(1 - \frac{\gamma^2}{2C} + \frac{3\gamma^4}{8C^2}\right)\end{aligned} \qquad (11.19)$$

Expanding with the use of the binomial series yields:

$$\frac{1}{C} = \left[n - \frac{\gamma^2}{2\left(l+\frac{1}{2}\right)}\right]^{-2} \approx \frac{1}{n^2}\left[1 + \frac{\gamma^2}{n\left(l+\frac{1}{2}\right)} + \frac{3}{4n^2}\frac{\gamma^4}{\left(l+\frac{1}{2}\right)^2}\right] \tag{11.20a}$$

and:

$$\frac{1}{C^2} = \left[n - \frac{\gamma^2}{2\left(l+\frac{1}{2}\right)}\right]^{-4} \approx \frac{1}{n^4}\left[1 + \frac{2\gamma^2}{n\left(l+\frac{1}{2}\right)} + \frac{5}{2n^2}\frac{\gamma^4}{\left(l+\frac{1}{2}\right)^2}\right] \tag{11.20b}$$

and then rearranging the terms, we obtain:

$$E = m_0c^2\left[1 - \frac{\gamma^2}{2n^2} - \frac{\gamma^4}{2n^4}\left(\frac{n}{l+\frac{1}{2}} - \frac{3}{4}\right)\right] \tag{11.21}$$

which finally gives:

$$E' = E - m_0c^2 = -\frac{Z^2e^4m_0}{(4\pi\varepsilon_0)^2 n^2\hbar^2} - \left(\frac{Ze^2}{4\pi\varepsilon_0}\right)^4 \frac{m_0}{2\hbar^4n^4c^2}\left(\frac{n}{\left(l+\frac{1}{2}\right)} - \frac{3}{4}\right) \tag{11.22}$$

The first term on the right-hand side is simply nonrelativistic energy and the second is the relativistic corrections to energy.

11.2 The Dirac Equation

The Klein-Gordon equation is quite satisfactory when it is interpreted properly and therefore it had been used to explain several physical phenomena. However, there are valid reasons to discard it for the description of Fermions. Like the Schrödinger equation, it does not accommodate the spin behavior of a Fermion in a natural manner. Also, if we look closely at the problem of interpreting the probability density, we find that it is originated from the fact that the Klein-Gordon equation is second order in time and hence both the solutions given by Eqns. (11.3a) and (11.3b) are to be accepted. This suggests that the problem of interpreting the probability density can be avoided if the relativistic equation involves only the first order time derivative. And then to maintain the parity between time and space coordinates, the equation should have first order derivatives in space coordinates as well. It can then be inferred that a relativistic quantum mechanical equation should incorporate the spin of a particle, and it should have first order derivatives in both time and space coordinates. In 1920, P.A.M. Dirac successfully derived a wave equation starting with the condition that it should be linear in $\frac{\partial}{\partial t}$. With the use of the equation, he had been able to show that derived probability density is always positive. The Dirac equation is considered the only correct relativistic wave equation.

11.2.1 Derivation of the Dirac Equation

In the case of nonrelativistic quantum mechanics, the operator representing kinetic energy of an electron with rest mass m_0 is:

$$T = \frac{p^2}{2m_0} = -\frac{\hbar^2}{2m_0}\nabla^2 \tag{11.23}$$

On replacing **p** by $\sigma.p$, where $\sigma_1 = \begin{pmatrix} 0 & 1 \\ 1 & 0 \end{pmatrix}; \sigma_2 = \begin{pmatrix} 0 & -i \\ i & 0 \end{pmatrix}$ and $\sigma_3 = \begin{pmatrix} 1 & 0 \\ 0 & -1 \end{pmatrix}$ are Pauli matrices, which are discussed in Section 5.5.1, see Eqn. (5.63), we get $T = \frac{(\sigma.\mathbf{p})(\sigma.\mathbf{p})}{2m_0}$. In the solved example 6 of Chapter 5, it was shown that for any two operators **A** and **B**, $(\sigma.\mathbf{A})(\sigma.\mathbf{B}) = \mathbf{A}.\mathbf{B} + i\sigma.(\mathbf{A}\times\mathbf{B})$, which gives $(\sigma.\mathbf{p})(\sigma.\mathbf{p}) = p^2$. This suggests that replacing **p** by $\sigma.\mathbf{p}$ *leaves kinetic energy unchanged.*

Next let us consider the effect of magnetic field on the kinetic energy of an electron. In the presence of a magnetic field, the effective linear momentum of the electron would be $\mathbf{p}' = \mathbf{p} - e\mathbf{A}$, where $-e$ is the charge on the electron and **A** is the vector potential. Again on replacing $\mathbf{p}'$ by $\sigma.\mathbf{p}'$ the kinetic energy operator in the presence of the magnetic field is:

$$T' = \frac{(\sigma.\mathbf{p}')(\sigma.\mathbf{p}')}{2m_0} = \frac{\{\sigma.(\mathbf{p}-e\mathbf{A})\}\{\sigma.(\mathbf{p}-e\mathbf{A})\}}{2m_0} \tag{11.24}$$

which with the use of the identity $(\sigma.\mathbf{A})(\sigma.\mathbf{B}) = \mathbf{A}.\mathbf{B} + i\sigma.(\mathbf{A}\times\mathbf{B})$ goes to:

$$\begin{aligned} T' &= \frac{(\mathbf{p}-e\mathbf{A})^2}{2m_0} + \frac{i}{2m_0}\sigma.(\mathbf{p}-e\mathbf{A})\times(\mathbf{p}-e\mathbf{A}) \\ &= \frac{(\mathbf{p}-e\mathbf{A})^2}{2m_0} - \frac{ie}{2m_0}\sigma.(\mathbf{p}\times\mathbf{A}+\mathbf{A}\times\mathbf{p}) \end{aligned} \tag{11.25}$$

The operation of $\mathbf{p}\times\mathbf{A}+\mathbf{A}\times\mathbf{p}$ on a function gives:

$$\begin{aligned} (\mathbf{p}\times\mathbf{A}+\mathbf{A}\times\mathbf{p})f(\mathbf{r}) &= -i\hbar\{\nabla\times(\mathbf{A}f)+\mathbf{A}\times(\nabla f)\} \\ &= -i\hbar\{(\nabla\times\mathbf{A})f - \mathbf{A}\times(\nabla f)+\mathbf{A}\times(\nabla f)\} \\ &= -i\hbar\mathbf{B}f(\mathbf{r}) \end{aligned} \tag{11.26}$$

where **B** is a magnetic field. We thus have:

$$T' = \frac{(\mathbf{p}-e\mathbf{A})^2}{2m_0} - \frac{e\hbar}{2m_0}\sigma.\mathbf{B} \tag{11.27}$$

The second term on the right-hand side is nothing but the energy of the electron due to the interaction of the spin of the electron with a magnetic field, while the first term is the square of the effective linear momentum $\mathbf{p}'$ divided by $2m_0$. An important point emerging here is that if linear momentum is replaced by the scalar product of $\boldsymbol{\sigma}$ with linear momentum, the *spin behavior of the electron is automatically accommodated* in the kinetic energy operator. This suggests that if we follow a similar procedure, electron spin is incorporated in the relativistic expression for the kinetic energy operator. Starting from the relativistic energy momentum relation for a free electron:

$$\left(\frac{E}{c}\right)^2 = p^2 + (m_0 c)^2$$
$$\Rightarrow \left(\frac{E}{c} - \mathbf{p}\right)\cdot\left(\frac{E}{c} + \mathbf{p}\right) = (m_0 c)^2 \tag{11.28}$$

we then replace $\mathbf{p}$ by $\boldsymbol{\sigma}.\mathbf{p}$ to get:

$$\left(\frac{E}{c} - \boldsymbol{\sigma}.\mathbf{p}\right)\left(\frac{E}{c} + \boldsymbol{\sigma}.\mathbf{p}\right) = (m_0 c)^2 \tag{11.29}$$

where $E = i\hbar\frac{\partial}{\partial t}$ and $\mathbf{p} = -i\hbar\nabla$ are operators. A second order differential equation of a particle then is:

$$\left(i\hbar\frac{\partial}{\partial x_0} + i\hbar\boldsymbol{\sigma}.\nabla\right)\left(i\hbar\frac{\partial}{\partial x_0} - i\hbar\boldsymbol{\sigma}.\nabla\right)\phi = (m_0 c)^2\phi \tag{11.30}$$

where we have used $\frac{\partial}{c\,\partial t} = \frac{\partial}{\partial x_0}$. ϕ is a two component wave function. Our aim is to obtain an equation that is linear in $\frac{\partial}{\partial t}$. The relativistic covariance demands that the wave equation linear in $\frac{\partial}{\partial t}$ should also be linear in ∇. Let us define two component wave functions $\phi_L = \phi$ and

$$\phi_R = \frac{i\hbar}{m_0 c}\left(\frac{\partial}{\partial x_0} - \boldsymbol{\sigma}.\nabla\right)\phi \tag{11.31}$$

The total number of components in the wave has now gone to 4 from 2. We then have:

$$\frac{i\hbar}{m_0 c}\left(\frac{\partial}{\partial x_0} + \boldsymbol{\sigma}.\nabla\right)\phi_R = \phi_L$$
$$\Rightarrow i\hbar\left(\frac{\partial}{\partial x_0} + \boldsymbol{\sigma}.\nabla\right)\phi_R = m_0 c\phi_L \tag{11.32a}$$

and:

$$i\hbar\left(\frac{\partial}{\partial x_0} - \boldsymbol{\sigma}.\nabla\right)\phi_L = m_0 c\phi_R \tag{11.32b}$$

In this manner, a second order Eqn. (11.30) is equivalently written in two first order Eqns. (11.32a) and (11.32b). The subscripts R and L are chosen to represent the fact that for $m_0 \to 0$, ϕ_R and ϕ_L describe a right-handed (spin parallel to momentum direction) and left-handed (spin anti-parallel to momentum direction) state of the particle having spin $-\frac{1}{2}$. On taking the sum and difference of these two equations we obtain:

$$i\hbar\frac{\partial}{\partial x_0}(\phi_R+\phi_L)+i\hbar(\sigma.\nabla)(\phi_R-\phi_L)=m_0c(\phi_R+\phi_L) \tag{11.33a}$$

$$i\hbar\frac{\partial}{\partial x_0}(\phi_R-\phi_L)+i\hbar(\sigma.\nabla)(\phi_R+\phi_L)=-m_0c(\phi_R-\phi_L) \tag{11.33b}$$

Defining $\psi_A=\phi_R+\phi_L$ and $\psi_B=\phi_R-\phi_L$, we get:

$$\begin{aligned} i\hbar\frac{\partial\psi_A}{\partial x_0}-\sigma.\mathbf{p}\psi_B &= m_0c\psi_A; \\ i\hbar\frac{\partial\psi_B}{\partial x_0}-\sigma.\mathbf{p}\psi_A &= -m_0c\psi_B \end{aligned} \tag{11.34}$$

Note that:

$$\sigma.\mathbf{p}=\sigma_1p_1+\sigma_2p_2+\sigma_3p_3=\begin{pmatrix} p_3 & p_1-ip_2 \\ p_1+ip_2 & -p_3 \end{pmatrix} \tag{11.35}$$

suggests that each ψ_A and ψ_B must have two components. Hence $\psi=\begin{pmatrix}\psi_A\\ \psi_B\end{pmatrix}$ has four components. Equation (11.34) can be written as a matrix equation:

$$\left[i\hbar\frac{\partial}{\partial x_0}\begin{pmatrix} I & 0 \\ 0 & I\end{pmatrix}-\begin{pmatrix} 0 & \sigma.\mathbf{p} \\ \sigma.\mathbf{p} & 0\end{pmatrix}\right]\begin{bmatrix}\psi_A\\ \psi_B\end{bmatrix}=m_0c\begin{bmatrix} I & 0 \\ 0 & -I\end{bmatrix}\begin{bmatrix}\psi_A\\ \psi_B\end{bmatrix} \tag{11.36}$$

where $\sigma.\mathbf{p}=\sigma_1p_1+\sigma_2p_2+\sigma_3p_3$ is the sum of three 2×2 matrices. Therefore:

$$\begin{pmatrix} 0 & \sigma.\mathbf{p} \\ \sigma.\mathbf{p} & 0\end{pmatrix}=\alpha_1p_1+\alpha_2p_2+\alpha_3p_3, \quad \text{with} \quad \alpha_i=\begin{pmatrix} 0 & \sigma_i \\ \sigma_i & 0\end{pmatrix} \quad \text{and} \quad i=1,2,3 \tag{11.37}$$

Since σ_1, σ_2, and σ_3 are 2×2 matrices, each of α_1, α_2, and α_3 will be a 4×4 matrix. This also suggests that I in Eqn. (11.36) is a 2×2 unit matrix. Equation (11.36) thus takes the form:

$$i\hbar\frac{\partial\psi}{\partial t}=\left(c\alpha.\mathbf{p}+\beta m_0c^2\right)\psi \tag{11.38}$$

with:

$$\beta=\begin{pmatrix} 1 & 0 & 0 & 0 \\ 0 & 1 & 0 & 0 \\ 0 & 0 & -1 & 0 \\ 0 & 0 & 0 & -1\end{pmatrix} \tag{11.39}$$

and:

$$\psi = \begin{bmatrix} \psi_1 \\ \psi_2 \\ \psi_3 \\ \psi_4 \end{bmatrix} \tag{11.40}$$

It is to be noted here that though ψ has four components, it not a 4-vector because it has nothing to do with the 4-dimensional nature of space-time. It is commonly known as a bispinor or Dirac spinor.

The matrices $\alpha_1, \alpha_2, \alpha_3$ and β anti-commute with each other:

$$\alpha_1\alpha_2 + \alpha_2\alpha_1 = \begin{bmatrix} 0 & \sigma_1 \\ \sigma_1 & 0 \end{bmatrix}\begin{bmatrix} 0 & \sigma_2 \\ \sigma_2 & 0 \end{bmatrix} + \begin{bmatrix} 0 & \sigma_2 \\ \sigma_2 & 0 \end{bmatrix}\begin{bmatrix} 0 & \sigma_1 \\ \sigma_1 & 0 \end{bmatrix}$$
$$= \begin{bmatrix} \sigma_1\sigma_2 + \sigma_2\sigma_1 & 0 \\ 0 & \sigma_1\sigma_2 + \sigma_2\sigma \end{bmatrix} \tag{11.41a}$$

Since $\sigma_1\sigma_2 = i\sigma_3$ and $\sigma_2\sigma_1 = -i\sigma_3$, we get $\alpha_1\alpha_2 + \alpha_2\alpha_1 = 0$. Similar algebra shows that $\alpha_2\alpha_3 + \alpha_3\alpha_2 = 0$; $\alpha_3\alpha_1 + \alpha_1\alpha_3 = 0$, and $\alpha_1\beta + \beta\alpha_1 = 0$. Also, we see that:

$$\alpha_i^2 = \begin{bmatrix} 0 & \sigma_i^2 \\ \sigma_i^2 & 0 \end{bmatrix} = 1 \quad \text{and} \quad \beta^2 = \begin{bmatrix} 1 & 0 \\ 0 & 1 \end{bmatrix} = 1 \tag{11.41b}$$

The properties of matrices can be summarized by $\alpha_i\alpha_j + \alpha_j\alpha_i = 2\delta_{ij}$ and $\alpha_i\beta + \beta\alpha_i = 0$. The α matrices and the β matrix are Hermitian and traceless.

11.2.2 Covariant Form of the Dirac Equation

Equation (11.38) can be written in *covariant form.* On dividing each term with $\hbar c$ and multiplying by β and then using $\beta^2 = 1$, we get:

$$\left[\beta\frac{\partial}{\partial(ict)} - i\beta\alpha.\nabla + \frac{m_0c}{\hbar}\right]\psi(\mathbf{r},t) = 0 \tag{11.42}$$

Writing $ict = x_4$; $\beta = \gamma_4$ and $\gamma_k = -i\beta\alpha_k = \begin{pmatrix} 0 & -i\sigma_k \\ i\sigma_k & 0 \end{pmatrix}$; $k = 1,2,3$, we obtain *the covariant form of the Dirac equation*:

$$\left(\sum_{\mu=1}^{4}\gamma_\mu\frac{\partial}{\partial x_\mu} + \frac{m_0c}{\hbar}\right)\psi = 0 \tag{11.43}$$

which is also expressed as:

$$\sum_{j=1}^{4}\left[\sum_{\mu=1}^{4}(\gamma_\mu)_{ij}\frac{\partial}{\partial x_\mu} + \left(\frac{m_0c}{\hbar}\right)\delta_{ij}\right]\psi_j = 0 \tag{11.44}$$

11.2.3 Probability Density and Probability Current

Multiplying Eqn. (11.38) from the left with the conjugate adjoint of Eqn. (11.40):

$$\psi^{\dagger} = \begin{bmatrix} \psi_1^* & \psi_2^* & \psi_3^* & \psi_4^* \end{bmatrix} \tag{11.45}$$

we get:

$$i\hbar\psi^{\dagger}\frac{\partial\psi}{\partial t} = -i\hbar c\psi^{\dagger}\alpha.\nabla\psi + \psi^{\dagger}\beta m_0c^2\psi \tag{11.46a}$$

We next multiply the conjugate adjoint of Eqn. (11.38) with ψ from the right, to obtain:

$$-i\hbar\frac{\partial\psi^{\dagger}}{\partial t}\psi = i\hbar c\left(\nabla\psi^{\dagger}\right).\alpha^{\dagger}\psi + \psi^{\dagger}\beta^{\dagger}m_0c^2\psi \tag{11.46b}$$

α_i and β are Hermitian and m_0c^2 scalar, subtraction of Eqn. (11.46b) from (11.46a) yields:

$$\begin{aligned} &i\hbar\frac{\partial}{\partial t}\left(\psi^{\dagger}\psi\right) = -i\hbar c\left\{\psi^{\dagger}\alpha.\left(\nabla\psi\right) + \left(\nabla\psi^{\dagger}\right).\alpha\psi\right\} \\ &\Rightarrow \frac{\partial}{\partial t}\left(\psi^{\dagger}\psi\right) + \nabla.\left(c\psi^{\dagger}\alpha\psi\right) = 0 \end{aligned} \tag{11.47}$$

or $\frac{\partial P}{\partial t} + \nabla.\mathbf{S} = 0$, with probability density (P) and probability current ($\mathbf{S}$):

$$P = \psi^{\dagger}\psi \tag{11.48a}$$

and:

$$S_j = c\left(\psi^{\dagger}\alpha_j\psi\right) \tag{11.48b}$$

Thus $P = |\psi_1|^2 + |\psi_2|^2 + |\psi_3|^2 + |\psi_4|^2$, which is contributed by four probable states of an electron, is positive definite. The probability current has three components:

$$\begin{aligned} S_1 &= c\psi^{\dagger}\alpha_1\psi \\ &= c\begin{bmatrix} \psi_1^* & \psi_2^* & \psi_3^* & \psi_4^* \end{bmatrix}\begin{bmatrix} 0 & 0 & 0 & 1 \\ 0 & 0 & 1 & 0 \\ 0 & 1 & 0 & 0 \\ 1 & 0 & 0 & 0 \end{bmatrix}\begin{bmatrix} \psi_1 \\ \psi_2 \\ \psi_3 \\ \psi_4 \end{bmatrix} \\ &= c\begin{bmatrix} \psi_1^* & \psi_2^* & \psi_3^* & \psi_4^* \end{bmatrix}\begin{bmatrix} \psi_4 \\ \psi_3 \\ \psi_2 \\ \psi_1 \end{bmatrix} \\ &= c\left(\psi_1^*\psi_4 + \psi_2^*\psi_3 + \psi_3^*\psi_2 + \psi_4^*\psi_1\right) \end{aligned} \tag{11.49a}$$

$$S_2 = c\psi^{\dagger}\alpha_2\psi$$

$$= c\begin{bmatrix} \psi_1^* & \psi_2^* & \psi_3^* & \psi_4^* \end{bmatrix}\begin{bmatrix} 0 & 0 & 0 & -i \\ 0 & 0 & i & 0 \\ 0 & -i & 0 & 0 \\ i & 0 & 0 & 0 \end{bmatrix}\begin{bmatrix} \psi_1 \\ \psi_2 \\ \psi_3 \\ \psi_4 \end{bmatrix} \tag{11.49b}$$

$$= c\begin{bmatrix} \psi_1^* & \psi_2^* & \psi_3^* & \psi_4^* \end{bmatrix}\begin{bmatrix} -i\psi_4 \\ i\psi_3 \\ -i\psi_2 \\ i\psi_1 \end{bmatrix}$$

$$= -ic\left(\psi_1^*\psi_4 - \psi_2^*\psi_3 + \psi_3^*\psi_2 - \psi_4^*\psi_1\right)$$

and

$$S_3 = c\psi^{\dagger}\alpha_3\psi$$

$$= c\begin{bmatrix} \psi_1^* & \psi_2^* & \psi_3^* & \psi_4^* \end{bmatrix}\begin{bmatrix} 1 & 0 & 0 & 0 \\ 0 & 1 & 0 & 0 \\ 0 & 0 & -1 & 0 \\ 0 & 0 & 0 & -1 \end{bmatrix}\begin{bmatrix} \psi_1 \\ \psi_2 \\ \psi_3 \\ \psi_4 \end{bmatrix} \tag{11.49c}$$

$$= c\begin{bmatrix} \psi_1^* & \psi_2^* & \psi_3^* & \psi_4^* \end{bmatrix}\begin{bmatrix} \psi_1 \\ \psi_2 \\ -\psi_3 \\ -\psi_4 \end{bmatrix}$$

$$= c\left(|\psi_1|^2 + |\psi_2|^2 - |\psi_3|^2 - |\psi_4|^2\right)$$

The continuity equation in terms of 4-vector probability current is:

$$\sum_{\mu=1}^{4}\frac{\partial S_\mu}{\partial x_\mu} = 0, \quad \text{where } S_4 = icP = ic\left(|\psi_1|^2 + |\psi_2|^2 + |\psi_3|^2 + |\psi_4|^2\right) \tag{11.50}$$

Equation (11.38) is the Dirac equation for a free particle. In the presence of potential energy $V(\mathbf{r})$, the Dirac Hamiltonian is:

$$H = c\alpha.\mathbf{p} + \beta m_0 c^2 + V(\mathbf{r}) \tag{11.51}$$

11.3 Free Particle Solutions of the Dirac Equation

For a freely moving particle, we can write:

$$\psi = \begin{pmatrix} \psi_A \\ \psi_B \end{pmatrix} = \frac{1}{\sqrt{V}} \begin{pmatrix} \chi_A(\mathbf{p}) \\ \chi_B(\mathbf{p}) \end{pmatrix} e^{i(\mathbf{p}.\mathbf{r}-Et)/\hbar} \tag{11.52}$$

where V is volume and $\chi_A(\mathbf{p})$ and $\chi_A(\mathbf{p})$ are spinors. Substitution of Eqn. (11.52) into Eqn. (11.38) yields:

$$c\begin{pmatrix} 0 & \sigma.\mathbf{p} \\ \sigma.\mathbf{p} & 0 \end{pmatrix}\begin{pmatrix} \chi_A(\mathbf{p}) \\ \chi_B(\mathbf{p}) \end{pmatrix} + m_0c^2\begin{pmatrix} I & 0 \\ 0 & -I \end{pmatrix}\begin{pmatrix} \chi_A(\mathbf{p}) \\ \chi_B(\mathbf{p}) \end{pmatrix} = E\begin{pmatrix} I & 0 \\ 0 & I \end{pmatrix}\begin{pmatrix} \chi_A(\mathbf{p}) \\ \chi_B(\mathbf{p}) \end{pmatrix} \tag{11.53}$$

which gives two equations:

$$c\sigma.\mathbf{p}\chi_B + m_0c^2\chi_A = E\chi_A \Rightarrow \chi_A = \frac{c\sigma.\mathbf{p}}{\left(E - m_0c^2\right)}\chi_B \tag{11.54a}$$

and:

$$c\sigma.\mathbf{p}\chi_A - m_0c^2\chi_B = E\chi_B \Rightarrow \chi_B = \frac{c\sigma.\mathbf{p}}{\left(E + m_0c^2\right)}\chi_A \tag{11.54b}$$

From Eqn. (11.35) we notice that $\sigma.\mathbf{p}$ is a 2×2 matrix and hence each of χ_A and χ_B should have two components. E takes both positive and negative values. Note that Eqn. (11.54a) cannot be used for $E > 0$, because it leads to an unacceptable solution when $|E| \to m_0c^2$. Similarly, Eqn. (11.54b) is applicable only for $E < 0$. For $E = c\sqrt{p^2 + m_0c^2} > 0$, we take, apart from the normalization constant, $\begin{pmatrix} 1 \\ 0 \end{pmatrix}$ and $\begin{pmatrix} 0 \\ 1 \end{pmatrix}$ for $\chi_A(\mathbf{p})$. The corresponding two components of $\chi_B(\mathbf{p})$, call these $\chi_{B1}(\mathbf{p})$ and $\chi_{B2}(\mathbf{p})$, are:

$$\begin{aligned} \chi_{B1}(\mathbf{p}) &= \frac{c\sigma.\mathbf{p}}{\left(E + m_0c^2\right)}\begin{pmatrix} 1 \\ 0 \end{pmatrix} \\ &= \frac{c}{\left(E + m_0c^2\right)}\begin{pmatrix} p_3 & p_1 - ip_2 \\ p_1 + ip_2 & -p_3 \end{pmatrix}\begin{pmatrix} 1 \\ 0 \end{pmatrix} \\ &= \begin{pmatrix} cp_3 / \left(E + m_0c^2\right) \\ cp^+ / \left(E + m_0c^2\right) \end{pmatrix} \end{aligned} \tag{11.55a}$$

and:

$$\chi_{B2}(\mathbf{p}) = \frac{c\sigma.\mathbf{p}}{\left(E+m_0c^2\right)}\begin{pmatrix} 0 \\ 1 \end{pmatrix}$$

$$= \frac{c}{\left(E+m_0c^2\right)}\begin{pmatrix} p_3 & p_1 - ip_2 \\ p_1 + ip_2 & -p_3 \end{pmatrix}\begin{pmatrix} 0 \\ 1 \end{pmatrix} \tag{11.55b}$$

$$= \begin{pmatrix} cp^- / \left(E+m_0c^2\right) \\ -cp_3 / \left(E+m_0c^2\right) \end{pmatrix}$$

where $p^+ = p_1 + ip_2$ and $p^- = p_1 - ip_2$. Thus, two solutions for $E > 0$ are:

$$\chi^{(1)}(\mathbf{p}) = N\begin{bmatrix} 1 \\ 0 \\ cp_3 / \left(E+m_0c^2\right) \\ cp^+ / \left(E+m_0c^2\right) \end{bmatrix}, \text{ and } \chi^{(2)}(\mathbf{p}) = N\begin{bmatrix} 0 \\ 1 \\ cp^- / \left(E+m_0c^2\right) \\ -cp_3 / \left(E+m_0c^2\right) \end{bmatrix} \tag{11.56a}$$

Here N is the normalization constant that is to be determined later. Following a similar procedure and choosing $\begin{pmatrix} 1 \\ 0 \end{pmatrix}$ and $\begin{pmatrix} 0 \\ 1 \end{pmatrix}$ for $\chi_B(\mathbf{p})$, we obtain another two solutions for the case of $E = -c\sqrt{p^2 + m_0c^2} < 0$:

$$\chi^{(3)}(\mathbf{p}) = N\begin{bmatrix} -cp_3 / \left(|E|+m_0c^2\right) \\ -cp^+ / \left(|E|+m_0c^2\right) \\ 1 \\ 0 \end{bmatrix}, \text{ and } \chi^{(4)}(\mathbf{p}) = N\begin{bmatrix} -cp^- / \left(|E|+m_0c^2\right) \\ cp_3 / \left(|E|+m_0c^2\right) \\ 0 \\ 1 \end{bmatrix} \tag{11.56b}$$

Four values of $\chi^{(l)}(\mathbf{p})$ with $l = 1,2,3,4$ represent four possible states: (i) $E > 0$ and spin up, (ii) $E > 0$ and spin down, (iii) $E < 0$ with spin up, and (iv) $E < 0$ with spin down. Since $\psi(\mathbf{r},t) = \chi^{(l)}(\mathbf{p})e^{\frac{i}{\hbar}(\mathbf{p}.\mathbf{r}-Et)}$ satisfies the free particle Dirac Eqn. (11.38), the spinors $\chi^{(l)}(\mathbf{p})$ satisfy

$$\left(c\alpha.\mathbf{p} + \beta m_0c^2\right)\chi^{(l)}(\mathbf{p}) = E\chi^{(l)}(\mathbf{p}) \tag{11.57}$$

The spinors are orthogonal: $\left(\chi^{(l')}(\mathbf{p})\right)^{\dagger}\chi^{(l)}(\mathbf{p})=0$. To see it, let us take:

$$\begin{aligned}
&\left(\chi^{(1)}(\mathbf{p})\right)^{\dagger}\chi^{(2)}(\mathbf{p})\\
&=N^2\begin{bmatrix}1 & 0 & cp_3/\left(E+m_0c^2\right) & cp^-/\left(E+m_0c^2\right)\end{bmatrix}\begin{bmatrix}0\\1\\cp^-/\left(E+m_0c^2\right)\\-cp_3/\left(E+m_0c^2\right)\end{bmatrix}\\
&=N^2\left[0+0+\frac{c^2p_3p^- - c^2p_3p^-}{\left(E+m_0c^2\right)^2}\right]=0
\end{aligned} \tag{11.58}$$

Similarly, it can be proved for other values of l' and l.

The normalization of wave function requires $\int\psi^{\dagger}\psi d^3r=1$. Two approaches have been adopted to normalize the relativistic wave function:

a. $\left(\chi^{(l)}(\mathbf{p})\right)^{\dagger}\chi^{(l)}(\mathbf{p})=1$, which for $l=1$ reduces to:

$$\begin{aligned}
&\left(\chi^{(1)}(\mathbf{p})\right)^{\dagger}\chi^{(1)}(\mathbf{p})\\
&=N^2\begin{bmatrix}1 & 0 & cp_3/\left(E+m_0c^2\right) & cp^-/\left(E+m_0c^2\right)\end{bmatrix}\begin{bmatrix}1\\0\\cp_3/\left(E+m_0c^2\right)\\cp^+/\left(E+m_0c^2\right)\end{bmatrix}\\
&=N^2\left[1+0+\frac{c^2\left(p_3^2+p_1^2+p_2^2\right)}{\left(E+m_0c^2\right)^2}\right]=N^2\left[1+\frac{c^2p^2}{\left(E+m_0c^2\right)^2}\right]\\
&=N^2\left[\frac{2p^2c^2+2m_0^2c^4+2Em_0c^2}{\left(E+m_0c^2\right)^2}\right]=\frac{2N^2E}{\left(E+m_0c^2\right)}
\end{aligned} \tag{11.59a}$$

This yields:

$$N=\sqrt{\frac{\left(|E|+m_0c^2\right)}{2|E|}} \tag{11.59b}$$

Another method of normalization used in some of the textbooks on quantum mechanics is as follows:

b. $\left(\chi^{(l)}(\mathbf{p})\right)^{\dagger}\chi^{(l)}(\mathbf{p}) = |E|/m_0c^2$, which gives:

$$\frac{2N^2|E|}{\left(|E|+m_0c^2\right)} = \frac{|E|}{m_0c^2}$$
$$\Rightarrow N = \sqrt{\frac{\left(|E|+m_0c^2\right)}{2m_0c^2}} \tag{11.59c}$$

The second normalization method says that $\chi^{\dagger}\chi$ transforms like the zeroth component of a 4-vector, which appears somewhat artificial. However, it has been used in the literature.

In the nonrelativistic limit $|\mathbf{p}| \ll m_0c$, $E \approx m_0c^2$ and $N \to 1$. Therefore, for the nonrelativistic case, we have from Eqns. (11.56) to (11.59):

$$\chi^{(1)} = \begin{pmatrix} 1 \\ 0 \\ 0 \\ 0 \end{pmatrix}; \chi^{(2)} = \begin{pmatrix} 0 \\ 1 \\ 0 \\ 0 \end{pmatrix}; \chi^{(3)} = \begin{pmatrix} 0 \\ 0 \\ 1 \\ 0 \end{pmatrix}; \chi^{(4)} = \begin{pmatrix} 0 \\ 0 \\ 0 \\ 1 \end{pmatrix} \tag{11.60}$$

11.3.1 Positive and Negative Energy Eigenvalues

The two energies $E^{+} = \sqrt{p^2c^2 + m_0^2c^4}$ and $E^{-} = -\sqrt{p^2c^2 + m_0^2c^4}$ with $0 \le |\mathbf{p}| \le \infty$, suggests that $m_0c^2 \le E^{+} \le \infty$ and $-\infty \le E^{-} \le -m_0c^2$. Therefore, there are two regions of energy; one starts from m_0c^2 and extends up to ∞, while the other begins at $-m_0c^2$ and extends down to $-\infty$. Two regions of energy are separated by a forbidden energy gap $E_g = 2m_0c^2$ and no energy states exist in the gap region. Fig. 11.1 illustrates the energy regions and the gap.

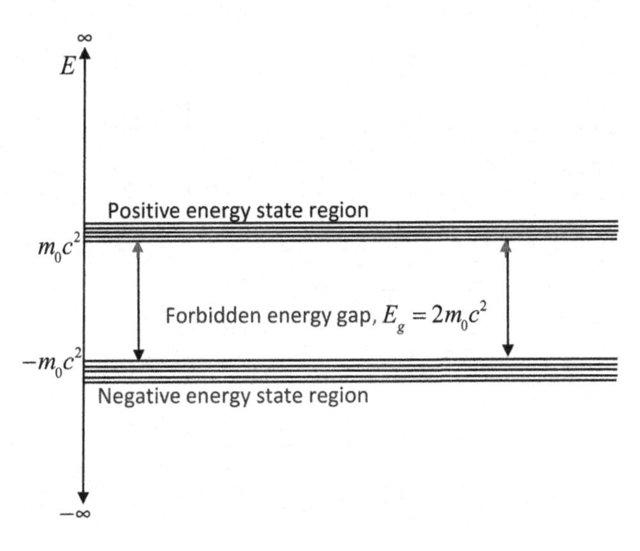

FIGURE 11.1
Energy levels for the free particle Dirac equation.

There is a problem in imaging negative energy states because an electron in a positive energy (bound or free) state should be able to emit a photon and make a transition to a negative energy state. The process could continue giving off an infinite amount of energy, which does not happen in reality. Dirac postulated a solution to this problem, which says that all of the *negative energy* states are completely filled and hence Pauli's exclusion principle does not permit the positive energy electrons to make any transitions to negative energy states. The negative energy states cannot have any *physically observable effect.* It was further postulated that electrons in completely filled negative energy states can make transitions to positive energy states if they pick up energy which is greater than or equal to the forbidden energy gap. Completely filled negative energy states are also referred to as *vacuum states.*

When an electron makes the transition to positive energy states, it leaves behind a vacancy in the negative energy spectrum. The vacancy is termed a *hole,* whose behavior is opposite that of an electron or the particles in negative energy states. A hole has a positive charge and its momentum and spin are in the opposite direction to that of vacuum (negative) states. The transition of an electron from a negative to a positive energy state is the *pair creation,* which produces a positron, a pair of an electron and a hole. The discovery of the positron provided a big support to the theory of the hole. The interpretation of negative energy states is not the hypothetical but has many elements of truth. When the Dirac field is quantized, one no longer needs the idea of an infinite negative energy sea; the electrons and positrons behave as if they were there.

11.4 The Dirac Equation and the Constants of Motion

For an observable A, the Heisenberg equation is:

$$\frac{d\mathrm{A}}{dt} = \frac{i}{\hbar}[\mathrm{A}, H] \tag{11.61}$$

where H is the Hamiltonian. If $[\mathrm{A}, H] = 0$, we say that A is the constant of motion described by the Hamiltonian H. In the case of nonrelativistic quantum mechanics, where H is the Schrödinger Hamiltonian, it was shown that linear momentum $\mathbf{p}$, orbital angular momentum $\mathbf{L}$, and $\mathbf{L}^2$ are the constants of motion. What are the constants of motion when motion is described by the relativistic Dirac equation? Let us answer this.

a. **Linear and Angular Momenta**

We first take the case of linear momentum. The free particle Dirac Hamiltonian $H = c\alpha.\mathbf{p} + \beta m_0 c^2$ gives:

$$[\mathbf{p}, H] = \mathbf{p}H - H\mathbf{p} = \mathbf{p}(c\alpha.\mathbf{p} + \beta m_0 c^2) - (c\alpha.\mathbf{p} + \beta m_0 c^2)\mathbf{p} \tag{11.62}$$

Since β does not depend on space coordinates $\mathbf{p}\beta m_0 c^2 = \beta m_0 c^2 \mathbf{p}$. This yields:

$$[\mathbf{p}, H] = c\left(\hat{i}p_1 + \hat{j}p_2 + \hat{k}p_3\right)\sum_{j=1}^{3}\alpha_j p_j - c\sum_{j=1}^{3}\alpha_j p_j\left(\hat{i}p_1 + \hat{j}p_2 + \hat{k}p_3\right) = 0 \tag{11.63}$$

stating that the linear momentum is a constant of motion even when the motion is described by the relativistic Dirac equation.

We next take $[\mathbf{L},H]=\hat{i}[L_1,H]+\hat{j}[L_2,H]+\hat{k}[L_3,H]$, where:

$$\begin{aligned}[L_1,H]&=L_1H-HL_1\\&=c(L_1\alpha.\mathbf{p}-\alpha.\mathbf{p}L_1)+(L_1\beta-\beta L_1)m_0c^2\\&=c(x_2p_3-x_3p_2)\sum_{j=1}^{3}\alpha_jp_j-c\sum_{j=1}^{3}\alpha_jp_j(x_2p_3-x_3p_2)\\&=c\left[(x_2p_3-x_3p_2)\sum_{j=1}^{3}\alpha_jp_j+i\hbar\left(\sum_{j=1}^{3}\alpha_j\frac{\partial}{\partial x_j}\right)(x_2p_3-x_3p_2)\right]\end{aligned}\tag{11.64}$$

Since β does not depend on space coordinates, we have used $L_1\beta-\beta L_1=0$. Then:

$$\begin{aligned}&\left(-i\hbar\sum_{j=1}^{3}\alpha_j\frac{\partial}{\partial x_j}\right)(x_2p_3-x_3p_2)\\&=-i\hbar\left[\alpha_1\frac{\partial}{\partial x_1}(x_2p_3-x_3p_2)+\alpha_2\frac{\partial}{\partial x_2}(x_2p_3-x_3p_2)+\alpha_3\frac{\partial}{\partial x_3}(x_2p_3-x_3p_2)\right]\\&=-i\hbar\left\{\alpha_1\frac{\partial}{\partial x_1}(x_2p_3-x_3p_2)+\alpha_2p_3+\alpha_2x_2p_3\frac{\partial}{\partial x_2}-\alpha_2x_3p_2\frac{\partial}{\partial x_2}\right\}\\&\quad-i\hbar\left\{\alpha_3x_2p_3\frac{\partial}{\partial x_3}-\alpha_3p_2-\alpha_3x_3p_2\frac{\partial}{\partial x_3}\right\}\\&=\alpha_1p_1(x_2p_3-x_3p_2)-i\hbar\alpha_2p_3+\alpha_2x_2p_3p_2-\alpha_2x_3p_2^2+\alpha_3x_2p_3^2+i\hbar\alpha_3p_2-\alpha_3x_3p_2p_3\\&=\alpha_1p_1(x_2p_3-x_3p_2)+(x_2p_3-x_3p_2)\alpha_2p_2+(x_2p_3-x_3p_2)\alpha_3p_3-i\hbar(\alpha_2p_3-\alpha_3p_2)\\&=(x_2p_3-x_3p_2)\sum_{j=1}^{3}\alpha_jp_j-i\hbar(\alpha_2p_3-\alpha_3p_2)\end{aligned}\tag{11.65a}$$

Therefore:

$$[L_1,H]=i\hbar c(\alpha_2p_3-\alpha_3p_2)\tag{11.66}$$

Adopting a similar procedure we get:

$$[L_2,H]=i\hbar c(\alpha_3p_1-\alpha_1p_3),\ \text{and}\ [L_3,H]=i\hbar c(\alpha_1p_2-\alpha_2p_1)\tag{11.67}$$

Note that we obtain $[L_2,H]$ from $[L_1,H]$ and $[L_3,H]$ from $[L_2,H]$ by changing suffixes in cyclic order. We finally get:

$$\begin{aligned}&[\mathbf{L},H]=i\hbar c\left\{\hat{i}(\alpha_2p_3-\alpha_3p_2)+\hat{j}(\alpha_3p_1-\alpha_1p_3)+\hat{k}(\alpha_1p_2-\alpha_2p_1)\right\}\\&\Rightarrow[\mathbf{L},H]=i\hbar c(\alpha\times\mathbf{p})\end{aligned}\tag{11.68a}$$

which yields:

$$\frac{d\mathbf{L}}{dt} = c(\alpha \times \mathbf{p}) \neq 0 \tag{11.68b}$$

suggesting that **L** is not the constant of motion when we use the relativistic Dirac equation to describe the motion. Now, let us consider an operator:

$$\mathbf{S} = \frac{\hbar}{2}\Sigma = \frac{\hbar}{2}\begin{pmatrix} \sigma & 0 \\ 0 & \sigma \end{pmatrix} \tag{11.69}$$

We then calculate $[\mathbf{S}, H] = \hat{i}[S_1, H] + \hat{j}[S_2, H] + \hat{k}[S_3, H]$. Take:

$$[S_1, H] = S_1 H - H S_1 = \frac{\hbar}{2}(\Sigma_1 H - H\Sigma_1) \tag{11.70a}$$

where:

$$\Sigma_1 H - H\Sigma_1 = c\Sigma_1\left(\sum_{j=1}^{3}\alpha_j p_j + \beta m_0 c\right) - c\left(\sum_{j=1}^{3}\alpha_j p_j + \beta m_0 c\right)\Sigma_1$$

$$= c\left[(\Sigma_1\alpha_1 - \alpha_1\Sigma_1)p_1 + (\Sigma_1\alpha_2 - \alpha_2\Sigma_1)p_2 + (\Sigma_1\alpha_3 - \alpha_3\Sigma_1)p_3 + (\Sigma_1\beta - \alpha_1\beta)m_0 c\right] \tag{11.70b}$$

Now:

$$(i)\quad \Sigma_1\alpha_1 - \alpha_1\Sigma_1 = \begin{pmatrix} \sigma_1 & 0 \\ 0 & \sigma_1 \end{pmatrix}\begin{pmatrix} 0 & \sigma_1 \\ \sigma_1 & 0 \end{pmatrix} - \begin{pmatrix} 0 & \sigma_1 \\ \sigma_1 & 0 \end{pmatrix}\begin{pmatrix} \sigma_1 & 0 \\ 0 & \sigma_1 \end{pmatrix}$$

$$= \begin{pmatrix} 0 & 1-1 \\ 1-1 & 0 \end{pmatrix} = 0;$$

$$(ii)\quad \Sigma_1\alpha_2 - \alpha_2\Sigma_1 = \begin{pmatrix} \sigma_1 & 0 \\ 0 & \sigma_1 \end{pmatrix}\begin{pmatrix} 0 & \sigma_2 \\ \sigma_2 & 0 \end{pmatrix} - \begin{pmatrix} 0 & \sigma_2 \\ \sigma_2 & 0 \end{pmatrix}\begin{pmatrix} \sigma_1 & 0 \\ 0 & \sigma_1 \end{pmatrix}$$

$$= \begin{pmatrix} 0 & \sigma_1\sigma_2 - \sigma_2\sigma_1 \\ \sigma_1\sigma_2 - \sigma_2\sigma_1 & 0 \end{pmatrix} = 2\begin{pmatrix} 0 & \sigma_1\sigma_2 \\ \sigma_1\sigma_2 & 0 \end{pmatrix} = 2i\alpha_3; \tag{11.70c}$$

$$(iii)\quad \Sigma_1\alpha_3 - \alpha_3\Sigma_1 = \begin{pmatrix} \sigma_1 & 0 \\ 0 & \sigma_1 \end{pmatrix}\begin{pmatrix} 0 & \sigma_3 \\ \sigma_3 & 0 \end{pmatrix} - \begin{pmatrix} 0 & \sigma_3 \\ \sigma_3 & 0 \end{pmatrix}\begin{pmatrix} \sigma_1 & 0 \\ 0 & \sigma_1 \end{pmatrix}$$

$$= -2i\alpha_2;$$

$$(iv)\quad \Sigma_1\beta - \beta\Sigma_1 = \begin{pmatrix} \sigma_1 & 0 \\ 0 & \sigma_1 \end{pmatrix}\begin{pmatrix} I & 0 \\ 0 & -I \end{pmatrix} - \begin{pmatrix} I & 0 \\ 0 & -I \end{pmatrix}\begin{pmatrix} \sigma_1 & 0 \\ 0 & \sigma_1 \end{pmatrix}$$

$$= \begin{pmatrix} \sigma_1 - \sigma_1 & 0 \\ 0 & -\sigma_1 + \sigma_1 \end{pmatrix} = 0$$

Therefore:

$$[S_1, H] = -i\hbar c(\alpha_2 p_3 - \alpha_3 p_2) \tag{11.71a}$$

Similarly, we get:

$$[S_2, H] = -i\hbar c(\alpha_3 p_1 - \alpha_1 p_3), \text{and} [S_3, H] = -i\hbar c(\alpha_1 p_2 - \alpha_2 p_1) \tag{11.71b}$$

which yields:

$$[\mathbf{S}, H] = -i\hbar c(\alpha \times \mathbf{p}), \text{and} \frac{d\mathbf{S}}{dt} = -c(\alpha \times \mathbf{p}) \tag{11.72}$$

Combining Eqns. (11.69) and (11.72), we obtain:

$$\frac{d\mathbf{L}}{dt} + \frac{d\mathbf{S}}{dt} = \frac{d\mathbf{J}}{dt} = c(\alpha \times \mathbf{p}) - c(\alpha \times \mathbf{p}) = 0 \tag{11.73}$$

stating that $\mathbf{J} = \mathbf{L} + \mathbf{S}$ is the *constant of motion*. The $\mathbf{S}$ is called spin angular momentum and $\mathbf{J}$ is known as total angular momentum. $[\mathbf{J}, H] = 0$ implies that $[\mathbf{J}^2, H] = 0 = [J_3, H]$.

b. **Operator** K
An operator defined by $K = \beta\Sigma.\mathbf{J} - \frac{\hbar}{2}\beta$ has been found very useful in solving the Dirac equation for the central potential problem. We will show that K too is a constant of motion. Let us take:

$$\begin{aligned}[\beta\Sigma.\mathbf{J}, H] &= \beta\Sigma.\mathbf{J}H - H\beta\Sigma.\mathbf{J} = \beta\Sigma.\mathbf{J}H - \beta H\Sigma.\mathbf{J} - H\beta\Sigma.\mathbf{J} + \beta H\Sigma.\mathbf{J} \\ &= -\beta[H, \Sigma].\mathbf{J} - [H, \beta]\Sigma.\mathbf{J}\end{aligned} \tag{11.74}$$

From Eqn. (11.72), we have $[\Sigma, H] = -2ic(\alpha \times \mathbf{p}) \Rightarrow [H, \Sigma] = 2ic(\alpha \times \mathbf{p})$. And:

$$\begin{aligned}[H, \beta] &= c\alpha.\mathbf{p}\beta - c\beta\alpha.\mathbf{p} + \beta m_0 c^2\beta - \beta m_0 c^2 \beta \\ &= c\left(\sum_{j=1}^{3}(\alpha_j\beta - \beta\alpha_j)p_j\right) \\ &= -2c\beta\alpha.\mathbf{p}\end{aligned} \tag{11.75}$$

Hence:

$$[\beta\Sigma.\mathbf{J}, H] = -2ic\beta(\alpha \times \mathbf{p}).\mathbf{J} + 2c\beta(\alpha.\mathbf{p})(\Sigma.\mathbf{J}) \tag{11.76}$$

The identity $(\sigma.\mathbf{A})(\sigma.\mathbf{B}) = \mathbf{A}.\mathbf{B} + i\sigma.(\mathbf{A} \times \mathbf{B})$ can also be written as $(\Sigma.\mathbf{A})(\Sigma.\mathbf{B}) = \mathbf{A}.\mathbf{B} + i\Sigma.(\mathbf{A} \times \mathbf{B})$. Also, matrices Σ_j can be transformed to α_j though $\alpha_j = -\gamma_5\Sigma_j$, γ_5 is a 4×4 matrix, whose explicit expression is not needed here. To simplify the second term in Eqn. (11.76), we take:

$$(\alpha.\mathbf{p})(\Sigma.\mathbf{J}) = -\gamma_5(\Sigma.\mathbf{p})(\Sigma.\mathbf{J}) = -\gamma_5[\mathbf{p}.\mathbf{J} + i\Sigma.(\mathbf{p} \times \mathbf{J})] \tag{11.77}$$

which yields:

$$\begin{aligned}
[\beta\Sigma.\mathbf{J},H] &= -2ic\beta(\alpha\times\mathbf{p}).\mathbf{J}-2c\beta\gamma_5\left[\mathbf{p}.\mathbf{J}+i\Sigma.(\mathbf{p}\times\mathbf{J})\right] \\
&= -2ic\beta(\alpha\times\mathbf{p}).\mathbf{J}-2c\beta\gamma_5(\mathbf{p}.\mathbf{J})+2ic\beta\alpha.(\mathbf{p}\times\mathbf{J}) \\
&= -2c\beta\gamma_5(\mathbf{p}.\mathbf{J}) \\
&= -2c\beta\gamma_5\mathbf{p}.\left(\mathbf{L}+\frac{\hbar}{2}\Sigma\right) \\
&= -2c\beta\gamma_5\mathbf{p}.\mathbf{L}+\hbar c\beta(\alpha.\mathbf{p}) \\
&= \hbar c\beta(\alpha.\mathbf{p})
\end{aligned} \tag{11.78}$$

where we have used $\mathbf{p}.\mathbf{L}=\mathbf{p}.\mathbf{r}\times\mathbf{p}=-\mathbf{p}\times\mathbf{p}.\mathbf{r}=0$. Then, using $\hbar c\beta(\alpha.\mathbf{p})=-\frac{\hbar}{2}[H,\beta]$, from Eqn. (11.75), we obtain:

$$\begin{aligned}
&[\beta\Sigma.\mathbf{J},H]+\frac{\hbar}{2}[H,\beta]=0 \\
&\Rightarrow \beta\Sigma.\mathbf{J}H-H\beta\Sigma.\mathbf{J}+\frac{\hbar}{2}H\beta-\frac{\hbar}{2}\beta H=0 \\
&\Rightarrow \beta\left(\Sigma.\mathbf{J}-\frac{\hbar}{2}\right)H-H\beta\left(\Sigma.\mathbf{J}-\frac{\hbar}{2}\right)=0 \\
&\Rightarrow \mathrm{K}H-H\mathrm{K}=0
\end{aligned} \tag{11.79}$$

where $\mathrm{K}=\beta\left(\Sigma.\mathbf{J}-\frac{\hbar}{2}\right)=\beta\Sigma.\mathbf{J}-\frac{\hbar\beta}{2}=\beta(\Sigma.\mathbf{L}+\hbar)$. We thus find that K is the constant of motion. $[\mathbf{J},H]=0$ and $[\mathrm{K},H]=0$ imply that $H,\mathrm{K},\mathbf{J}^2$, and J_3 commute with each other.

11.5 Spin Magnetic Moment (the Dirac Electron in an Electromagnetic Field)

We would show that the Dirac equation includes the intrinsic spin magnetic moment of an electron. In the presence of an electromagnetic field, the energy and momentum of a free particle change: $E\rightarrow E-e\phi$ and $\mathbf{p}\rightarrow\mathbf{p}-e\mathbf{A}$, where ϕ and $\mathbf{A}$ are scalar and vector potentials. The free particle Dirac equation in the presence of an electromagnetic field is then given by:

$$\left[E-e\phi-c\alpha.(\mathbf{p}-e\mathbf{A})-\beta m_0c^2\right]\psi=0 \tag{11.80}$$

Here, E and $\mathbf{p}$ are energy and momentum operators. To get the equation that is quadratic in E, we operate both sides from the left with $\left[E-e\phi+c\alpha.(\mathbf{p}-e\mathbf{A})+\beta m_0c^2\right]$ and obtain:

$$\left[E-e\phi+c\alpha.(\mathbf{p}-e\mathbf{A})+\beta m_0c^2\right]\left[E-e\phi-c\alpha.(\mathbf{p}-e\mathbf{A})-\beta m_0c^2\right]\psi=0 \tag{11.81}$$

which expands to:

$$\begin{bmatrix} (E-e\phi)^2 - c(E-e\phi)\alpha.(\mathbf{p}-e\mathbf{A}) - (E-e\phi)\beta m_0c^2 + c\alpha.(\mathbf{p}-e\mathbf{A})(E-e\phi) \\ -c^2\alpha.(\mathbf{p}-e\mathbf{A})\alpha.(\mathbf{p}-e\mathbf{A}) - c\alpha.(\mathbf{p}-e\mathbf{A})\beta m_0c^2 + \beta m_0c^2(E-e\phi) \\ -\beta m_0c^2 c\alpha.(\mathbf{p}-e\mathbf{A}) - m_0^2c^4 \end{bmatrix}\psi = 0 \quad (11.82)$$

which with the use of $\alpha_j\beta + \beta\alpha_j = 0$ and $(E-e\phi)\beta = \beta(E-e\phi)$ simplifies to:

$$\begin{bmatrix} (E-e\phi)^2 - c(E-e\phi)\alpha.(\mathbf{p}-e\mathbf{A}) + c\alpha.(\mathbf{p}-e\mathbf{A})(E-e\phi) - m_0^2c^4 \\ -c^2\alpha.(\mathbf{p}-e\mathbf{A})\alpha.(\mathbf{p}-e\mathbf{A}) \end{bmatrix}\psi = 0 \quad (11.83)$$

The use of the identity $(\alpha.\mathbf{A})(\alpha.\mathbf{B}) = \mathbf{A}.\mathbf{B} + i\Sigma.(\mathbf{A}\times\mathbf{B})$ gives:

$$\begin{aligned} c^2\alpha.(\mathbf{p}-e\mathbf{A})\alpha.(\mathbf{p}-e\mathbf{A})\psi &= c^2\left\{(\mathbf{p}-e\mathbf{A})^2 + i\Sigma.[(\mathbf{p}-e\mathbf{A})\times(\mathbf{p}-e\mathbf{A})]\right\}\psi \\ &= c^2\left\{(\mathbf{p}-e\mathbf{A})^2 - ie\Sigma.\left[\mathbf{A}\times\mathbf{p} + \mathbf{p}\times\mathbf{A}\right]\right\}\psi \\ &= c^2\left\{(\mathbf{p}-e\mathbf{A})^2 - \hbar e\Sigma.\mathbf{B}\right\}\psi \end{aligned} \quad (11.84)$$

where we used $\mathbf{p}\times\mathbf{A}\psi = (\mathbf{p}\times\mathbf{A})\psi - (\mathbf{A}\times\mathbf{p}\psi)$ and $\nabla\times\mathbf{A} = \mathbf{B}$. Here $\mathbf{B}$ is the magnetic field. Since operator E depends on t and $\mathbf{p}$ depends on $\mathbf{r}$ they are interchangeable, and hence $E\mathbf{p}\psi = \mathbf{p}E\psi$. Further, note that $(E\mathbf{A} - \mathbf{A}E)\psi = i\hbar\left(\frac{\partial\mathbf{A}}{\partial t}\right)\psi$ and $(\mathbf{p}\phi - \phi\mathbf{p})\psi = -i\hbar(\nabla\phi)\psi$. On substituting these into Eqn. (11.83), we obtain:

$$\begin{aligned} &\left[(E-e\phi)^2 + ice\hbar\alpha.\left(\frac{\partial\mathbf{A}}{\partial t}\right) + ice\hbar\alpha.(\nabla\phi) - m_0^2c^4 - c^2\left\{(\mathbf{p}-e\mathbf{A})^2 - e\hbar\Sigma.\mathbf{B}\right\}\right]\psi = 0 \\ &\Rightarrow \left[(E-e\phi)^2 - c^2(\mathbf{p}-e\mathbf{A})^2 - m_0^2c^4\right]\psi + e\hbar c^2\Sigma.\mathbf{B}\psi + ice\hbar\alpha.\left\{\nabla\phi + \frac{\partial\mathbf{A}}{\partial t}\right\}\psi = 0 \end{aligned} \quad (11.85)$$

In the Lorentz gauge $\mathbf{E} = -\nabla\phi - \frac{\partial\mathbf{A}}{\partial t}$, where $\mathbf{E}$ is the electric field. Therefore:

$$\left[(E-e\phi)^2 - c^2(\mathbf{p}-e\mathbf{A})^2 - m_0^2c^4\right]\psi + e\hbar c^2\Sigma.\mathbf{B}\psi - ice\hbar\alpha.\mathbf{E}\psi = 0 \quad (11.86)$$

The $\left[(E-e\phi)^2 - c^2(\mathbf{p}-e\mathbf{A})^2 - m_0^2c^4\right]\psi = 0$ is the Klein-Gordon equation in the presence of an electromagnetic field. An additional two terms represent the interaction of electron spin with magnetic and with electric fields, respectively. To understand the physical significance of additional terms, let us take the nonrelativistic limit of Eqn. (11.86). The non-relativistic limit yields the well-known Schrödinger–Pauli equation. In the nonrelativistic limit,

$E' << m_0c^2$ and $e\phi << m_0c^2$, where $E' = E - m_0c^2$. Therefore, $(E - e\phi)^2 - m_0^2c^4 \approx 2m_0c^2(E' - e\phi)$. Then Eqn. (11.86) reduces to:

$$E'\psi = \left[\frac{1}{2m_0}(\mathbf{p} - e\mathbf{A})^2 - \frac{e\hbar}{2m_0}\Sigma.\mathbf{B} + \frac{i\hbar e}{2m_0c}\alpha.\mathbf{E}\right]\psi \tag{11.87}$$

where $-\frac{e\hbar}{2m_0}\Sigma.\mathbf{B}$ is energy due to the interaction of electron spin with a magnetic field. $\frac{i\hbar e}{2m_0c}\alpha.\mathbf{E}$, which yields energy due to electron interaction with the electric field, is very small as compared to other terms. Let us assume that the electron is in a 1D potential well of potential V and width a. Then $|\mathbf{E}| \simeq \mathrm{V}/a$ and $\hbar/a$ is momentum. Therefore $\left|\frac{i\hbar e}{2m_0c}\alpha.\mathbf{E}\right| \approx \frac{\mathrm{v}}{2c}|\alpha|e\mathrm{V} \to 0$, where v is velocity.

Note that: Eqns.(11.87) and (11.27) are in conformity with each other.

11.6 Spin-Orbit Interaction Energy

We would show that spin-orbit interaction energy comes out automatically from the relativistic Dirac equation. The Dirac Hamiltonian for an electron in central potential is given by Eqn. (11.51). On taking $\psi(\mathbf{r},t) = \begin{pmatrix} u_A(\mathbf{r}) \\ u_B(\mathbf{r}) \end{pmatrix} e^{-\frac{i}{\hbar}(\mathbf{p}.\mathbf{r}-Et)}$ equation:

$$\left[c\alpha.\mathbf{p} + \beta m_0c^2 + V(r)\right]\psi(\mathbf{r},t) = i\hbar\frac{\partial\psi(\mathbf{r},t)}{\partial t} \tag{11.88}$$

separates into two equations:

$$c\sigma.\mathbf{p}u_B(\mathbf{r}) + \left(m_0c^2 + V(r)\right)u_A(\mathbf{r}) = Eu_A(\mathbf{r})$$
$$\Rightarrow u_A(\mathbf{r}) = \frac{c\sigma.\mathbf{p}}{\left(E - V - m_0c^2\right)}u_B(\mathbf{r}) \tag{11.89a}$$

and:

$$c\sigma.\mathbf{p}u_A(\mathbf{r}) + \left(V(r) - m_0c^2\right)u_B(\mathbf{r}) = Eu_B(\mathbf{r})$$
$$\Rightarrow u_B(\mathbf{r}) = \frac{c\sigma.\mathbf{p}}{\left(E - V + m_0c^2\right)}u_A(\mathbf{r}) \tag{11.89b}$$

where E is energy eigenvalue. Taking $E = E' + m_0c^2$ and noting that $E' - V = T$ (kinetic energy) $\ll 2m_0c^2$ and $|\sigma.\mathbf{p}| = p$, we get:

$$|u_B(\mathbf{r})| \approx \frac{p}{2m_0c}|u_A(\mathbf{r})| \tag{11.90}$$

which suggests that $|u_B| \ll |u_A|$ and therefore the probability of finding a particle in the state represented by $u_B(\mathbf{r})$ is negligibly small as compared to that denoted by $u_A(\mathbf{r})$. Therefore, for combining Eqns. (11.89a) and (11.89b) into one equation, we should eliminate $u_B(\mathbf{r})$, which gives us:

$$\begin{aligned}
(E'-V)u_A(\mathbf{r}) &= (c\sigma.\mathbf{p})\left\{\frac{c\sigma.\mathbf{p}}{\left(E'-V+2m_0c^2\right)}\right\}u_A(\mathbf{r}) \\
&= \frac{(c\sigma.\mathbf{p})}{2m_0c^2}\left[1+\frac{(E'-V)}{2m_0c^2}\right]^{-1}(c\sigma.\mathbf{p})u_A(\mathbf{r}) \\
&\approx \frac{(c\sigma.\mathbf{p})}{2m_0c^2}\left[1-\frac{(E'-V)}{2m_0c^2}\right](c\sigma.\mathbf{p})u_A(\mathbf{r}) \\
&\approx \frac{(\sigma.\mathbf{p})(\sigma.\mathbf{p})}{2m_0}u_A(\mathbf{r}) - \frac{(\sigma.\mathbf{p})(E'-V)(\sigma.\mathbf{p})}{4m_0^2c^2}u_A(\mathbf{r}) \\
&\approx \frac{p^2}{2m_0}u_A(\mathbf{r}) - \frac{(\sigma.\mathbf{p})(E'-V)(\sigma.\mathbf{p})}{4m_0^2c^2}u_A(\mathbf{r})
\end{aligned} \tag{11.91}$$

where we have used $(\sigma.\mathbf{p})(\sigma.\mathbf{p}) = p^2$. Next, we notice that $\{(\sigma.\mathbf{p})(E'-V)(\sigma.\mathbf{p})\}u_A = E'p^2u_A - \{(\sigma.\mathbf{p})V(\sigma.\mathbf{p})\}u_A$, because E' does not operate on $\mathbf{p}$. Let us take:

$$\begin{aligned}
\{(\sigma.\mathbf{p})V(\sigma.\mathbf{p})\}u_A &= \{(\sigma.\mathbf{p}V)(\sigma.\mathbf{p})\}u_A + V\{(\sigma.\mathbf{p})(\sigma.\mathbf{p})\}u_A \\
&= \{-i\hbar(\nabla V).\mathbf{p} + \hbar\sigma.(\nabla V \times \mathbf{p})\}u_A + Vp^2u_A
\end{aligned} \tag{11.92a}$$

if $V(r)$ is the central potential and it depends only on r, independent of θ and ϕ. Then $\nabla V = \hat{r}\frac{\partial V}{\partial r}$, giving:

$$\begin{aligned}
\{\nabla V \times \mathbf{p}\}u_A &= \left(\hat{r}\frac{\partial V}{\partial r} \times \mathbf{p}\right)u_A \\
&= \left\{\frac{\mathbf{r}\times\mathbf{p}}{r}\left(\frac{\partial V}{\partial r}\right)\right\}u_A \\
&= \frac{1}{r}\left(\frac{\partial V}{\partial r}\right)\mathbf{L}u_A
\end{aligned} \tag{11.92b}$$

and:

$$\{\nabla V.\mathbf{p}\}u_A = -i\hbar\left(\frac{\partial V}{\partial r}\right)\frac{\partial u_A}{\partial r} \tag{11.92c}$$

Hence:

$$\{(\sigma.\mathbf{p})V(\sigma.\mathbf{p})\}u_A = -i\hbar\left(\frac{\partial V}{\partial r}\right)\frac{\partial u_A}{\partial r} + \hbar\frac{\sigma.\mathbf{L}}{r}\left(\frac{\partial V}{\partial r}\right)u_A + Vp^2u_A \tag{11.92d}$$

On substituting this into Eqn. (11.91), we obtain:

$$(E'-V)u_A(\mathbf{r}) \approx \left[\frac{p^2}{2m_0} - \frac{(E'-V)p^2}{4m_0^2c^2} - \frac{\hbar^2}{4m_0^2c^2}\left(\frac{\partial V}{\partial r}\right)\frac{\partial}{\partial r} + \frac{\hbar}{4m_0^2c^2}\frac{\sigma.\mathbf{L}}{r}\left(\frac{\partial V}{\partial r}\right)\right]u_A(\mathbf{r}) \tag{11.93}$$

Using $\mathbf{s} = \frac{\hbar}{2}\sigma$ and $E' - V = \frac{p^2}{2m_0}$ and then rearranging the terms, we get:

$$E'u_A(\mathbf{r}) = \left[\frac{p^2}{2m_0} + V(r) - \frac{p^4}{8m_0^3c^2} - \frac{\hbar^2}{4m_0^2c^2}\left(\frac{dV}{dr}\right)\frac{d}{dr} + \frac{1}{2m_0^2c^2}\frac{\mathbf{s.L}}{r}\left(\frac{dV}{dr}\right)\right]u_A(\mathbf{r}) \tag{11.94}$$

The partial derivative is changed to a full derivative because there is only one derivative. The last term on the right-hand side consists of the scalar product of spin and orbital angular momenta and it is the spin-orbit interaction energy contribution to the Hamiltonian. The third and fourth terms represent the relativistic corrections to kinetic and potential energies, respectively. To see this, let us calculate the limiting value of relativistic kinetic energy:

$$\begin{aligned} E &= \sqrt{p^2c^2 + m_0^2c^4} = m_0c^2\left(1 + \frac{p^2}{m_0^2c^2}\right)^{1/2} \approx m_0c^2\left(1 + \frac{p^2}{2m_0^2c^2} - \frac{p^4}{8m_0^4c^4} + \ldots..\right) \\ \Rightarrow E' &= E - m_0c^2 \approx \frac{p^2}{2m_0} - \frac{p^4}{8m_0^3c^2} \end{aligned} \tag{11.95}$$

which shows that the third term on the right-hand side of Eqn. (11.94) is a relativistic correction to the kinetic energy of the Dirac particle. We thus see that spin-orbit interaction energy appears automatically in the Dirac equation.

11.7 Solution of the Dirac Equation for Central Potential

Our objective here is to solve the energy eigenvalue equation:

$$H\psi(\mathbf{r}) = E\psi(\mathbf{r}) \tag{11.96}$$

where the Dirac Hamiltonian for a particle under central potential is given by Eqn. (11.51). It has also been shown in Section 11.4 that operators K and $\mathbf{J}^2$ commute with the Dirac Hamiltonian. Therefore K, $\mathbf{J}^2$, and H can have simultaneous eigenfunctions. We write:

$$\mathrm{K}\psi(\mathbf{r}) = k\hbar\psi(\mathbf{r}) \tag{11.97a}$$

and:

$$\mathbf{J}^2\psi(\mathbf{r}) = \hbar^2 j(j+1)\psi(\mathbf{r}) \tag{11.97b}$$

where k and j are integers, which are inter-related. To see this, let us take:

$$\begin{aligned} K^2 &= \beta(\Sigma.\mathbf{L}+\hbar)\beta(\Sigma.\mathbf{L}+\hbar) \\ &= (\Sigma.\mathbf{L}+\hbar)^2 \\ &= (\Sigma.\mathbf{L})(\Sigma.\mathbf{L})+2\hbar(\Sigma.\mathbf{L})+\hbar^2 \\ &= \mathbf{L}^2 + i\Sigma.(\mathbf{L}\times\mathbf{L})+2\hbar(\Sigma.\mathbf{L})+\hbar^2 \\ &= \mathbf{L}^2 + i\Sigma.(i\hbar\mathbf{L})+2\hbar(\Sigma.\mathbf{L})+\hbar^2 \\ &= \mathbf{L}^2 + \hbar\Sigma.\mathbf{L}+\hbar^2 \end{aligned} \tag{11.98a}$$

and:

$$\begin{aligned} \mathbf{J}^2 &= (\mathbf{L}+\mathbf{S})^2 \\ &= \left(\mathbf{L}+\frac{\hbar}{2}\Sigma\right)^2 \\ &= \mathbf{L}^2+\frac{\hbar^2}{4}\Sigma^2+\hbar\Sigma.\mathbf{L} \\ &= \mathbf{L}^2+\frac{3\hbar^2}{4}+\hbar\Sigma.\mathbf{L} \end{aligned} \tag{11.98b}$$

Here, we have used $\beta^2 = 1$ and $\Sigma^2 = \Sigma_1^2+\Sigma_2^2+\Sigma_3^2 = 3$. We thus obtain:

$$K^2 = \mathbf{J}^2 + \frac{\hbar^2}{4} \tag{11.99}$$

Therefore, from Eqns. (11.97), we have:

$$\begin{aligned} &K^2\psi = \left(\mathbf{J}^2+\frac{\hbar^2}{4}\right)\psi \\ &\Rightarrow k^2\hbar^2\psi = \left(\hbar^2 j(j+1)+\frac{\hbar^2}{4}\right)\psi \\ &\Rightarrow k = \pm\left(j+\frac{1}{2}\right) \end{aligned} \tag{11.100}$$

Thus k is a nonzero integer that can be positive or negative. The sign of k determines whether spin is parallel $(k<0)$ or antiparallel $(k>0)$ to $\mathbf{J}$, in the relativistic limit. Eqn. (11.97a) in terms of ψ_A and ψ_B is:

$$\left\{\begin{pmatrix} I & 0 \\ 0 & -I \end{pmatrix}\begin{pmatrix} \sigma.\mathbf{L} & 0 \\ 0 & \sigma.\mathbf{L} \end{pmatrix}+\hbar\begin{pmatrix} I & 0 \\ 0 & -I \end{pmatrix}\right\}\begin{pmatrix} \psi_A \\ \psi_B \end{pmatrix} = k\hbar\begin{pmatrix} \psi_A \\ \psi_B \end{pmatrix} \tag{11.101}$$

which implies:

$$\sigma.\mathbf{L}\psi_A = \hbar(k-1)\psi_A \tag{11.102a}$$

and:

$$-\sigma.\mathbf{L}\psi_B = \hbar(k+1)\psi_B \tag{11.102b}$$

Similarly, Eqn. (11.97b) expands to:

$$\left\{\mathbf{L}^2 + \frac{3\hbar^2}{4} + \hbar\begin{pmatrix} \sigma.\mathbf{L} & 0 \\ 0 & \sigma.\mathbf{L} \end{pmatrix}\right\}\begin{pmatrix} \psi_A \\ \psi_B \end{pmatrix} = \hbar^2 j(j+1)\begin{pmatrix} \psi_A \\ \psi_B \end{pmatrix} \tag{11.103}$$

which yields:

$$\begin{aligned} &\left(\mathbf{L}^2 + \frac{3\hbar^2}{4} + \hbar\sigma.\mathbf{L}\right)\psi_A = \hbar^2 j(j+1)\psi_A; \\ &\left(\mathbf{L}^2 + \frac{3\hbar^2}{4} + \hbar\sigma.\mathbf{L}\right)\psi_B = \hbar^2 j(j+1)\psi_B \end{aligned} \tag{11.104}$$

Then, with the use of Eqns. (11.102), we have:

$$\begin{aligned} &\left(\mathbf{L}^2 + \frac{3\hbar^2}{4} + \hbar^2(k-1)\right)\psi_A = \hbar^2 j(j+1)\psi_A \\ &\Rightarrow \mathbf{L}^2\psi_A = \left\{\left(j+\frac{1}{2}\right)^2 - k\right\}\hbar^2\psi_A \end{aligned} \tag{11.105a}$$

and:

$$\begin{aligned} &\left(\mathbf{L}^2 + \frac{3\hbar^2}{4} - \hbar^2(k+1)\right)\psi_B = \hbar^2 j(j+1)\psi_B \\ &\Rightarrow \mathbf{L}^2\psi_B = \left\{\left(j+\frac{1}{2}\right)^2 + k\right\}\hbar^2\psi_B \end{aligned} \tag{11.105b}$$

We thus find that though $\psi(\mathbf{r})$ is not the eigenfunction of $\mathbf{L}^2$ ($\mathbf{L}^2$ does not commute with the Dirac Hamiltonian), its components ψ_A and ψ_B separately are the eigenfunctions of $\mathbf{L}^2$. If we denote the eigenvalues of $\mathbf{L}^2$ by $\hbar^2 l_A(l_A+1)$ in the state of ψ_A and $\hbar^2 l_B(l_B+1)$ for ψ_B, then we get:

$$l_A(l_A+1) = \left(j+\frac{1}{2}\right)^2 - k \text{ and } l_B(l_B+1) = \left(j+\frac{1}{2}\right)^2 + k \tag{11.106}$$

Since $k = \pm\left(j+\frac{1}{2}\right)$, we obtain:

TABLE 11.1

Relation among k, j, l_A and l_B

k	l_A	l_B
$j+\frac{1}{2}$	$j-\frac{1}{2}$	$j+\frac{1}{2}$
$-\left(j+\frac{1}{2}\right)$	$j+\frac{1}{2}$	$j-\frac{1}{2}$

As is obvious, l_A and l_B take values $0,1,2,\ldots$, depending on values of j, and hence these are *orbital quantum numbers*. Also, for a given k-value, the orbital parities of ψ_A and ψ_B are opposite to each other. Let us write:

$$\begin{aligned} \psi_A &= g(r) y_{j,l_A}^{j_3}; \\ \psi_B &= i f(r) y_{j,l_B}^{j_3} \end{aligned} \tag{11.107}$$

where $y_{j,l}^{j_3}$ is a normalized spin-angular momentum eigenfunction (r-independent) of $\mathbf{J}^2, J_3, \mathbf{L}^2$, and $\mathbf{S}^2$. It is a combination of the *Pauli-spinor* with *spherical harmonics* of order l. Mathematically, $y_{j,l}^{j_3}$ for $j = l + \frac{1}{2}$ is:

$$y_{j,l}^{j_3} = \sqrt{\frac{\left(l + j_3 + \frac{1}{2}\right)}{(2l+1)}} Y_{l, j_3 - \frac{1}{2}}(\theta,\phi) \begin{pmatrix} 1 \\ 0 \end{pmatrix} + \sqrt{\frac{\left(l - j_3 + \frac{1}{2}\right)}{(2l+1)}} Y_{l, j_3 + \frac{1}{2}}(\theta,\phi) \begin{pmatrix} 0 \\ 1 \end{pmatrix} \tag{11.108a}$$

where $Y_{l, j_3 \pm \frac{1}{2}}(\theta,\phi)$ are spherical harmonics; $\begin{pmatrix} 1 \\ 0 \end{pmatrix}$ and $\begin{pmatrix} 0 \\ 1 \end{pmatrix}$ are spinors. When $j = l - \frac{1}{2}$, we have:

$$y_{j,l}^{j_3} = -\sqrt{\frac{\left(l - j_3 + \frac{1}{2}\right)}{(2l+1)}} Y_{l, j_3 - \frac{1}{2}}(\theta,\phi) \begin{pmatrix} 1 \\ 0 \end{pmatrix} + \sqrt{\frac{\left(l + j_3 + \frac{1}{2}\right)}{(2l+1)}} Y_{l, j_3 + \frac{1}{2}}(\theta,\phi) \begin{pmatrix} 0 \\ 1 \end{pmatrix} \tag{11.108b}$$

The radial functions $g(r)$ and $f(r)$ depend on quantum number k. The i multiplied to $f(r)$ is taken to keep both $g(r)$ and $f(r)$ real.

We next take:

$$\begin{aligned} (\sigma.\mathbf{p}) &= \frac{(\sigma.\mathbf{r})}{r^2} (\sigma.\mathbf{r})(\sigma.\mathbf{p}) \\ &= \frac{(\sigma.\mathbf{r})}{r^2} \left[\mathbf{r}.\mathbf{p} + i\sigma.(\mathbf{r} \times \mathbf{p}) \right] \\ &= \frac{(\sigma.\mathbf{r})}{r} \left[-i\hbar \frac{\partial}{\partial r} + i \frac{\sigma.\mathbf{L}}{r} \right] \end{aligned} \tag{11.109}$$

The pseudo-scalar operator:

$$\frac{(\sigma.\mathbf{r})}{r} = \frac{\sum_{j=1}^{3}\sigma_j x_j}{r} = \begin{pmatrix} \cos\theta & \sin\theta e^{-i\phi} \\ \sin\theta e^{i\phi} & -\cos\theta \end{pmatrix} \tag{11.110}$$

will operate only on spherical harmonics. Here, we have expressed the Cartesian coordinates in terms of spherical polar coordinates to simplify $\frac{(\sigma.\mathbf{r})}{r}$. As discussed before, $\psi(\mathbf{r})$ is an eigenfunction of operators $\mathbf{J}^2$ and J_z but it is not an eigenfunction of $\mathbf{L}^2$. However, $\psi_A(\mathbf{r})$ and $\psi_B(\mathbf{r})$, which have opposite parities, are eigenfunctions of $\mathbf{L}^2$. Therefore, the new eigenfunction obtained by operating $\frac{(\sigma.\mathbf{r})}{r}$ on $y_{j,l}^{j_3}$ must have the same values of j and j_3 but with *opposite orbital parity*. Thus, $\frac{(\sigma.\mathbf{r})}{r} y_{j,l_A}^{j_3}$ should be equal to $y_{j,l_B}^{j_3}$ multiplied by a phase factor. Since $\frac{(\sigma.\mathbf{r})^2}{r^2} = 1$, the phase convention used in writing Eqns. (11.108) suggests that the multiplying phase factor is –1. We thus have:

$$\frac{(\sigma.\mathbf{r})}{r} y_{j,l_A}^{j_3} = -y_{j,l_B}^{j_3}; \qquad \frac{(\sigma.\mathbf{r})}{r} y_{j,l_B}^{j_3} = -y_{j,l_A}^{j_3} \tag{11.111}$$

Writing Eqn. (11.96) in matrix form:

$$\left\{ c\begin{pmatrix} 0 & \sigma.\mathbf{p} \\ \sigma.\mathbf{p} & 0 \end{pmatrix} + \begin{pmatrix} I & 0 \\ 0 & -I \end{pmatrix} m_0c^2 \right\} \begin{pmatrix} \psi_A \\ \psi_B \end{pmatrix} = (E - V(r)) \begin{pmatrix} I & 0 \\ 0 & I \end{pmatrix} \begin{pmatrix} \psi_A \\ \psi_B \end{pmatrix} \tag{11.112a}$$

we get the following two equations:

$$c(\sigma.\mathbf{p})\psi_B = \left(E - V(r) - m_0c^2\right)\psi_A \tag{11.112b}$$

and:

$$c(\sigma.\mathbf{p})\psi_A = \left(E - V(r) + m_0c^2\right)\psi_B \tag{11.112c}$$

Next, from Eqn. (11.109), we have:

$$c(\sigma.\mathbf{p})\psi_B = c\frac{(\sigma.\mathbf{r})}{r}\left[-i\hbar\frac{\partial\psi_B}{\partial r} + i\frac{(\sigma.\mathbf{L})\psi_B}{r}\right] \tag{11.113a}$$

which with the use of Eqn. (11.102b), gives:

$$c(\sigma.\mathbf{p})\psi_B = c\frac{(\sigma.\mathbf{r})}{r}\left[-i\hbar\frac{\partial\psi_B}{\partial r} - i\hbar\frac{(k+1)\psi_B}{r}\right]$$
$$= c\frac{(\sigma.\mathbf{r})}{r}\left[-i\hbar\frac{\partial}{\partial r} - i\hbar\frac{(k+1)}{r}\right]\left\{if(r)y_{j,l_B}^{j_3}\right\} \tag{11.113b}$$
$$= -c\hbar\left[\frac{\partial}{\partial r} + \frac{(k+1)}{r}\right]f(r)y_{j,l_A}^{j_3}$$

This yields:

$$-c\hbar\left[\frac{\partial}{\partial r} + \frac{(k+1)}{r}\right]f(r)y_{j,l_A}^{j_3} = \left(E - V(r) - m_0c^2\right)g(r)y_{j,l_A}^{j_3}$$
$$\Rightarrow \frac{df}{dr} + \frac{(k+1)f}{r} = -\frac{1}{\hbar c}\left(E - V(r) - m_0c^2\right)g(r) \tag{11.114a}$$

Similarly, we get:

$$\frac{dg}{dr} + \frac{(1-k)g}{r} = \frac{1}{\hbar c}\left(E - V(r) + m_0c^2\right)f(r) \tag{11.114b}$$

On introducing $G(r) = rg(r)$ and $F(r) = rf(r)$, we obtain:

$$\frac{dF(r)}{dr} + \frac{kF(r)}{r} = -\frac{1}{\hbar c}\left(E - V(r) - m_0c^2\right)G(r) \tag{11.115a}$$

and:

$$\frac{dG(r)}{dr} - \frac{kG(r)}{r} = \frac{1}{\hbar c}\left(E - V(r) + m_0c^2\right)F(r) \tag{11.115b}$$

Thus, similar to the case of nonrelativistic quantum mechanics, we have radial equations for central potential. A variety of problems are solved with the use of these radial equations. We here solve these equations for a hydrogen like atom.

11.7.1 Hydrogen-like Atom

The potential energy for a hydrogen-like atom is given by:

$$V(r) = -\frac{Ze^2}{4\pi\varepsilon_0 r} \tag{11.116}$$

Let us introduce:

$$\alpha_1 = \frac{\left(m_0c^2 + E\right)}{\hbar c},\ \alpha_2 = \frac{\left(m_0c^2 - E\right)}{\hbar c},\ \gamma = \frac{Ze^2}{4\pi\varepsilon_0\hbar c},\ \text{and } \rho = \left(\sqrt{\alpha_1\alpha_2}\right)r \tag{11.117}$$

We then have coupled equations:

$$\frac{dF}{\sqrt{\alpha_1\alpha_2}\,dr}+\frac{kF}{\left(\sqrt{\alpha_1\alpha_2}\right)r}=\left[\frac{\left(m_0c^2-E\right)}{\sqrt{\left(m_0^2c^4-E^2\right)}}-\frac{Ze^2}{4\pi\varepsilon_0\hbar c\left(\sqrt{\alpha_1\alpha_2}\right)r}\right]G$$
$$\Rightarrow\left(\frac{d}{d\rho}+\frac{k}{\rho}\right)F=\left(\sqrt{\frac{\alpha_2}{\alpha_1}}-\frac{\gamma}{\rho}\right)G \tag{11.118a}$$

and:

$$\left(\frac{d}{d\rho}-\frac{k}{\rho}\right)G=\left(\sqrt{\frac{\alpha_1}{\alpha_2}}+\frac{\gamma}{\rho}\right)F \tag{11.118b}$$

In the limit $\rho\to\infty$, two equations can be combined to obtain:

$$\frac{d^2F}{d^2\rho}-F=0 \tag{11.119}$$

whose solution is proportional to $e^{-\rho}$. This along with our knowledge of the solution of the Schrödinger equation for the hydrogen atom suggests that F and G can be expressed as:

$$F=e^{-\rho}\rho^s\sum_{l=0}a_l\rho^l \text{ and } G=e^{-\rho}\rho^s\sum_{l=0}b_l\rho^l \tag{11.120}$$

Substitution of Eqn. (11.120) into Eqns. (11.118a) and (11.118b) gives two series in powers of ρ. The coefficient of each power of ρ must separately go to zero to make the series zero. On equating to zero the coefficients of $e^{-\rho}\rho^{s+q-1}$, we get:

$$(s+q+k)a_q-a_{q-1}+\gamma b_q-\sqrt{\frac{\alpha_2}{\alpha_1}}b_{q-1}=0 \tag{11.121a}$$

and:

$$(s+q-k)b_q-b_{q-1}-\gamma a_q-\sqrt{\frac{\alpha_1}{\alpha_2}}a_{q-1}=0 \tag{11.121b}$$

For $q=0$, these equations reduce to:

$$(s+k)a_0+\gamma b_0=0 \text{ and } (s-k)b_0-\gamma a_0=0 \tag{11.122}$$

Note that a_{-1} and b_{-1} do not exist. Since a_0 and b_0 are not zero, the simultaneous solution of Eqn. (11.122) demands that $s^2-k^2+\gamma^2=0$, which implies that $s=\pm\sqrt{k^2-\gamma^2}$. The minimum value that can be assigned to k^2 is 1 and $\gamma^2 \approx \left(\frac{Z}{137}\right)^2$; therefore, in order to avoid $\int |f|^2 r^2 dr \to \infty$ and $\int |g|^2 r^2 dr \to \infty$, at $r \to 0$, s cannot be assigned a negative sign. Hence $s = \sqrt{k^2 - \gamma^2}$. Similar to the case of the nonrelativistic solution for the hydrogen atom, it can be shown here that both F and G increase like e^{ρ} for very large values of ρ. Therefore, the power series given by Eqn. (11.120) must be terminated at some value of l to have the acceptable solutions of Eqns. (11.118). Assuming that the two power series terminate with some power, there should exist an integer n' with property: $a_{n'+1} = b_{n'+1} = 0$, while $a_{n'} \neq 0$ and $b_{n'} \neq 0$. Putting $q = n' + 1$ in Eqns. (11.121), we obtain:

$$a_{n'} = -\sqrt{\frac{\alpha_2}{\alpha_1}} b_{n'} \tag{11.123}$$

Next, multiply Eqn. (11.121a) by α_1 and Eqn. (11.121b) by $\sqrt{\alpha_1\alpha_2}$ and then subtract Eqn. (11.121b) from Eqn. (11.121a) after taking $q = n'$. We obtain:

$$\begin{aligned}
&\left[\alpha_1(s+n'+k)a_{n'} - \alpha_1 a_{n'-1} + \gamma\alpha_1 b_{n'} - \sqrt{\alpha_1\alpha_2}\, b_{n'-1}\right] \\
&-\left[\sqrt{\alpha_1\alpha_2}\,(s+n'-k)b_{n'} - \sqrt{\alpha_1\alpha_2}\, b_{n'-1} - \sqrt{\alpha_1\alpha_2}\,\gamma a_{n'} - \alpha_1 a_{n'-1}\right] \\
&= \alpha_1(s+n'+k)a_{n'} + \gamma\alpha_1 b_{n'} - \sqrt{\alpha_1\alpha_2}\,(s+n'-k)b_{n'} + \sqrt{\alpha_1\alpha_2}\,\gamma a_{n'} = 0
\end{aligned} \tag{11.124}$$

which with the use of Eqn. (11.123) simplifies to:

$$\begin{aligned}
&2\alpha_1(s+n')a_{n'} = -\gamma(\alpha_1 - \alpha_2)b_{n'} = \gamma(\alpha_1 - \alpha_2)\sqrt{\frac{\alpha_1}{\alpha_2}}\, a_{n'} \\
&\Rightarrow 2\sqrt{\alpha_1\alpha_2}\,(s+n') = \gamma(\alpha_1 - \alpha_2) \\
&\Rightarrow \sqrt{\left(m_0^2c^2 - E^2\right)}(s+n') = E\gamma
\end{aligned} \tag{11.125}$$

which yields:

$$E = \frac{m_0c^2}{\left\{1 + \left(\frac{\gamma}{s+n'}\right)^2\right\}^{1/2}} \tag{11.126}$$

Note that E depends on n' and $|k| = j + \frac{1}{2}$. In order to compare the energy eigenvalue obtained here with that obtained from the Schrödinger equation, we approximate Eqn. (11.126) for the case of $(E - m_0c^2) \ll m_0c^2$, the nonrelativistic limit. Let us define $n = n' + |k|$ and then take:

$$s + n' = n' + |k|\left[1 - \frac{\gamma^2}{k^2}\right]^{1/2} \approx n - \frac{\gamma^2}{2|k|} - \frac{\gamma^4}{8|k|k^2} + \ldots. \tag{11.127}$$

Therefore:

$$E = m_0c^2\left\{1+\left(\frac{\gamma}{s+n'}\right)^2\right\}^{-1/2}$$

$$\approx m_0c^2\left\{1-\frac{1}{2}\left(\frac{\gamma}{s+n'}\right)^2+\frac{3}{8}\left(\frac{\gamma}{s+n'}\right)^4+.....\right\}$$

$$\approx m_0c^2\left\{1-\frac{1}{2}\gamma^2\left(n-\frac{\gamma^2}{2|k|}-\frac{\gamma^4}{8|k|k^2}\right)^{-2}+\frac{3}{8}\gamma^4\left(n-\frac{\gamma^2}{2|k|}-\frac{\gamma^4}{8|k|k^2}\right)^{-4}\right\} \tag{11.128}$$

$$\approx m_0c^2\left\{1-\frac{1}{2}\frac{\gamma^2}{n^2}\left(1+\frac{\gamma^2}{n|k|}+\frac{3\gamma^4}{n^2k^2}+\frac{\gamma^4}{4n|k|k^2}\right)+\frac{3}{8}\frac{\gamma^4}{n^4}\left(1+\frac{2\gamma^2}{n|k|}\right)\right\}$$

$$\approx m_0c^2\left[1-\frac{1}{2}\frac{\gamma^2}{n^2}-\frac{1}{2}\frac{\gamma^4}{n^3}\left(\frac{1}{|k|}-\frac{3}{4n}\right)\right]+........$$

Here we have retained the terms up to γ^4. We then have:

$$E' = E - m_0c^2 = -\frac{1}{2}\frac{m_0Z^2e^4}{(4\pi\varepsilon_0)^2n^2\hbar^2}-\frac{m_0}{2n^3c^2}\left(\frac{Ze^2}{4\pi\varepsilon_0\hbar}\right)^4\left(\frac{1}{\left(j+\frac{1}{2}\right)}-\frac{3}{4n}\right)+.... \tag{11.129}$$

The first term on the right-hand side is the nonrelativistic value of energy, while the second term yields the relativistic corrections. The n is the principle quantum number for the nonrelativistic case. Note that the leading correction term is the same as the fine structure splitting term in the Balmer formula. Also, notice that degeneracy in energy level for the given n is partially removed by relativistic effects.

11.8 Solved Examples

1. Show that $\alpha_j (j=1,2,3)$ and β are traceless matrices.

SOLUTION

Consider $tr(\alpha_j)=tr\left(\alpha_i^\dagger\alpha_i\alpha_j\right)$, where we used $\alpha_i^\dagger\alpha_i=\alpha_i^2=1$. The $\alpha_i\alpha_j=-\alpha_j\alpha_i$ gives: $tr(\alpha_j)=tr\left(\alpha_i^\dagger\alpha_i\alpha_j\right)=-tr\left(\alpha_i^\dagger\alpha_j\alpha_i\right)$. Since $tr(ABC)=tr(CAB)$, cyclic invariance, we get:

$$\begin{aligned} &tr(\alpha_j)=-tr\left(\alpha_i\alpha_i^\dagger\alpha_j\right)=-tr(\alpha_j)\\ &\Rightarrow tr(\alpha_j)=0\end{aligned} \tag{11.130}$$

Similarly, it can be proved that $tr(\beta)=0$.

2. Show that $\beta\psi(-\mathbf{r},t)$ is the wave function of the Dirac equation.

SOLUTION

The $\psi(\mathbf{r},t)$ satisfies the Dirac equation:

$$\left(c\alpha.\mathbf{p}+\beta m_0c^2+V(r)\right)\psi(\mathbf{r},t)=i\hbar\frac{\partial\psi(\mathbf{r},t)}{\partial t} \tag{11.131}$$

On replacing $\mathbf{r}$ by $-\mathbf{r}$ and then operating from the left with β, we obtain:

$$\begin{aligned}&\beta\left\{-c\alpha.\mathbf{p}+\beta m_0c^2+V(-r)\right\}\psi(-\mathbf{r},t)=i\hbar\beta\frac{\partial\psi(-\mathbf{r},t)}{\partial t}\\&\Rightarrow\left\{-c\beta\alpha.\mathbf{p}+\beta\beta m_0c^2+\beta V(-r)\right\}\psi(-\mathbf{r},t)=i\hbar\frac{\partial\beta\psi(-\mathbf{r},t)}{\partial t}\end{aligned} \tag{11.132}$$

With the use of $\beta\alpha_j=-\alpha_j\beta$; $\beta V=V\beta$ and assuming that $V(r)=V(-r)$, the spherically symmetric potential, the left-hand side of the equation is rewritten as follows:

$$\begin{aligned}&\left(-c\beta\alpha.\mathbf{p}+\beta\beta m_0c^2+\beta V(-r)\right)\psi(-\mathbf{r},t)\\&=\left(c\alpha.\mathbf{p}\beta+\beta\beta m_0c^2+\beta V(r)\right)\psi(-\mathbf{r},t)\\&=\left(c\alpha.\mathbf{p}+\beta m_0c^2+V(r)\right)\beta\psi(-\mathbf{r},t)\end{aligned} \tag{11.133}$$

Hence:

$$\left(c\alpha.\mathbf{p}+\beta m_0c^2+V(r)\right)\beta\psi(-\mathbf{r},t)=i\hbar\frac{\partial\beta\psi(-\mathbf{r},t)}{\partial t} \tag{11.134}$$

which shows that $\beta\psi(-\mathbf{r},t)$ is the wave function of the Dirac equation.

3. Find the plane wave solution of the 1D Dirac equation:

$$\frac{\partial\psi(x_1,t)}{\partial t}=\left(c\alpha_1p_1+\beta m_0c^2\right)\psi(x_1,t) \tag{11.135}$$

SOLUTION

Let us write the plane wave solution as:

$$\psi(x_1,t)=\chi(p_1)e^{\frac{i}{\hbar}(p_1x_1-Et)} \tag{11.136}$$

Substitution of which into Eqn. (11.135) yields:

$$\left(c\alpha_1p_1+\beta m_0c^2\right)\chi(p_1)=E\chi(p_1) \tag{11.137}$$

which gives:

$$c\sigma_1 p_1 \chi_b = \left(E - m_0c^2\right)\chi_a \tag{11.138a}$$

$$c\sigma_1 p_1 \chi_a = \left(E + m_0c^2\right)\chi_b \tag{11.138b}$$

where we have taken $\chi(p_1) = \begin{pmatrix} \chi_a \\ \chi_b \end{pmatrix}$. Equation (11.138a) can be solved for $E < 0$ while Eqn. (11.138b) is solvable for $E > 0$. For $E > 0$, we take two possible values $\begin{pmatrix} 1 \\ 0 \end{pmatrix}$ and $\begin{pmatrix} 0 \\ 1 \end{pmatrix}$ for χ_a. Then from Eqn. (11.138b), we get:

$$\chi^{(1)}(p_1) = N\begin{bmatrix} 1 \\ 0 \\ 0 \\ cp_1/\left(E + m_0c^2\right) \end{bmatrix} \text{ and } \chi^{(2)}(p_1) = N\begin{bmatrix} 0 \\ 1 \\ cp_1/\left(E + m_0c^2\right) \\ 0 \end{bmatrix} \tag{11.139a}$$

and for $E < 0$, we have from Eqn. (138a):

$$\chi^{(3)}(p_1) = N\begin{bmatrix} 0 \\ -cp_1/\left(|E| + m_0c^2\right) \\ 1 \\ 0 \end{bmatrix} \text{ and } \chi^{(4)}(p_1) = N\begin{bmatrix} -cp_1/\left(|E| + m_0c^2\right) \\ 0 \\ 0 \\ 1 \end{bmatrix} \tag{11.139b}$$

where N is the constant of normalization. The two energies of the particle are: $E^{\pm} = \pm\sqrt{\left(p_1^2c^2 + m_0^2c^4\right)}$.

4. Show that the conservation law for the Dirac probability current $\frac{\partial P}{\partial t} + \nabla.\mathbf{S} = 0$, where $P = \psi^{\dagger}\psi$ and $\mathbf{S} = c\psi^{\dagger}\alpha\psi$, also holds in the presence of an electromagnetic field.

SOLUTION

In the presence of an electromagnetic field, momentum and the Hamiltonian are modified as $\mathbf{p} \to \mathbf{p} - e\mathbf{A}$ and $H \to H + e\phi$, where ϕ and $\mathbf{A}$ are scalar and vector potentials, respectively. We therefore have:

$$i\hbar\frac{\partial\psi}{\partial t} = \left[c\alpha.(-i\hbar\nabla - e\mathbf{A}) + \beta m_0c^2 + e\phi\right]\psi \tag{11.140a}$$

the conjugate adjoint of which is:

$$-i\hbar\frac{\partial\psi^{\dagger}}{\partial t} = i\hbar c\nabla\psi^{\dagger}.\alpha^{\dagger} + \psi^{\dagger}\left[-ce\alpha.\mathbf{A} + \beta m_0c^2 + e\phi\right] \tag{11.140b}$$

Operate on Eqn. (11.140a) from the left with ψ^{+} and on Eqn. (11.140b) with ψ from the right and then subtract Eqn. (11.140b) from Eqn. (11.140a):

$$
\begin{aligned}
i\hbar\left(\psi^{+}\frac{\partial\psi}{\partial t}+\psi\frac{\partial\psi^{+}}{\partial t}\right) &= -i\hbar c\left(\psi^{+}(\alpha.\nabla\psi)+(\alpha.\nabla\psi^{+})\psi\right) \\
&-ce\left(\psi^{+}\alpha.\mathbf{A}\psi-\psi\alpha.\mathbf{A}\psi^{+}\right)+\left(\beta m_0c^2+e\phi\right)\left(\psi^{+}\psi-\psi\psi^{+}\right) \\
&\Rightarrow\frac{\partial}{\partial t}\left(\psi^{+}\psi\right)+c\nabla.\left(\psi^{+}\alpha\psi\right)=0
\end{aligned}
\tag{11.141}
$$

which yields:

$$
\frac{\partial P}{\partial t}+\nabla.\mathbf{S}=0 \tag{11.142}
$$

5. Prove that $\left(\alpha.\mathbf{p}+\beta m_0c\right)^2=\left(\frac{E}{c}\right)^2$.

SOLUTION

$$
\begin{aligned}
\left(\alpha.\mathbf{p}+\beta m_0c\right)^2 &= \left(\alpha_1p_1+\alpha_2p_2+\alpha_3p_3+\beta m_0c\right)^2 \\
&= \left(\alpha_1p_1+\alpha_2p_2+\alpha_3p_3\right)^2+\left(\beta m_0c\right)^2+\left(\alpha_1p_1+\alpha_2p_2+\alpha_3p\right)\beta \\
&\quad+\beta\left(\alpha_1p_1+\alpha_2p_2+\alpha_3p\right) \\
&= \left(\alpha_1p_1+\alpha_2p_2+\alpha_3p_3\right)^2+\beta^2m_0^2c^2+\sum_{j=1}^{3}\left(\alpha_j\beta+\beta\alpha_j\right)p_j
\end{aligned}
\tag{11.143}
$$

Since $\beta^2=1$ and $\alpha_j\beta+\beta\alpha_j=0$, we get:

$$
\begin{aligned}
\left(\alpha.\mathbf{p}+\beta m_0c\right)^2 &= \left(\alpha_1p_1+\alpha_2p_2+\alpha_3p_3\right)^2+m_0^2c^2 \\
&= p^2+(\alpha_1\alpha_2+\alpha_2\alpha_1)p_1p_2+(\alpha_2\alpha_3+\alpha_3\alpha_2)p_2p_3 \\
&\quad+(\alpha_3\alpha_1+\alpha_1\alpha_3)p_3p_1+m_0^2c^2 \\
&= p^2+m_0^2c^2 \\
&= \left(\frac{E}{c}\right)^2
\end{aligned}
\tag{11.144}
$$

Here we used $\alpha_1^2=\alpha_2^2=\alpha_3^2=1$ and $\alpha_i\alpha_j+\alpha_j\alpha_i=0$ when $i\neq j$.

6. Show that an alternative choice for matrices α_j and β in the Dirac equation can be:

$\alpha_j=\begin{pmatrix}-\sigma_j & 0\\ 0 & \sigma_j\end{pmatrix}$ and $\beta=\begin{pmatrix}0 & I\\ I & 0\end{pmatrix}$, where σ_j and I are Pauli matrices and the unit matrix, respectively.

SOLUTION

The necessary conditions on α_j and β matrices is that they should satisfy $\alpha_1^2 = \alpha_2^2 = \alpha_3^2 = 1 = \beta^2$; $\alpha_i\alpha_j + \alpha_j\alpha_i = 0$ when $i \neq j$, and $\alpha_j\beta + \beta\alpha_j = 0$. We see that: $\alpha_j^2 = \begin{pmatrix} -\sigma_j & 0 \\ 0 & \sigma_j \end{pmatrix}\begin{pmatrix} -\sigma_j & 0 \\ 0 & \sigma_j \end{pmatrix} = \begin{pmatrix} \sigma_j^2 & 0 \\ 0 & \sigma_j^2 \end{pmatrix} = 1$ because $\sigma_j^2 = 1$ for any value of j. Also, $\beta^2 = \begin{pmatrix} 0 & I \\ I & 0 \end{pmatrix}\begin{pmatrix} 0 & I \\ I & 0 \end{pmatrix} = \begin{pmatrix} 1 & 0 \\ 0 & 1 \end{pmatrix} = 1.$

We next take:

$$\begin{aligned} \alpha_j\beta + \beta\alpha_j &= \begin{pmatrix} -\sigma_j & 0 \\ 0 & \sigma_j \end{pmatrix}\begin{pmatrix} 0 & I \\ I & 0 \end{pmatrix} + \begin{pmatrix} 0 & I \\ I & 0 \end{pmatrix}\begin{pmatrix} -\sigma_j & 0 \\ 0 & \sigma_j \end{pmatrix} \\ &= \begin{pmatrix} 0 & -\sigma_j \\ \sigma_j & 0 \end{pmatrix} + \begin{pmatrix} 0 & \sigma_j \\ -\sigma_j & 0 \end{pmatrix} = 0 \end{aligned} \tag{11.145}$$

Further:

$$\begin{aligned} \alpha_j\alpha_i + \alpha_i\alpha_j &= \begin{pmatrix} -\sigma_j & 0 \\ 0 & \sigma_j \end{pmatrix}\begin{pmatrix} -\sigma_i & 0 \\ 0 & \sigma_i \end{pmatrix} + \begin{pmatrix} -\sigma_i & 0 \\ 0 & \sigma_i \end{pmatrix}\begin{pmatrix} -\sigma_j & 0 \\ 0 & \sigma_j \end{pmatrix} \\ &= \begin{pmatrix} \sigma_j\sigma_i & 0 \\ 0 & \sigma_j\sigma_i \end{pmatrix} + \begin{pmatrix} \sigma_i\sigma_j & 0 \\ 0 & \sigma_i\sigma_j \end{pmatrix} = \begin{pmatrix} \sigma_j\sigma_i + \sigma_i\sigma_j & 0 \\ 0 & \sigma_j\sigma_i + \sigma_i\sigma_j \end{pmatrix} \end{aligned} \tag{11.146}$$

Since $\sigma_j\sigma_i + \sigma_i\sigma_j = 0$ for $i \neq j$, therefore $\alpha_j\alpha_i + \alpha_i\alpha_j = 0$. We thus find that the given values of α_j and β are the correct alternative choice for the Dirac equation.

7. Show that $\Sigma.\mathbf{p}$ commutes with the free particle Dirac Hamiltonian $H = c\alpha.\mathbf{p} + \beta m_0c^2$.

SOLUTION

The commutator

$$\begin{aligned} [\Sigma.\mathbf{p}, H] &= \Sigma.\mathbf{p}H - H\Sigma.\mathbf{p} \\ &= [\Sigma.\mathbf{p}, c\alpha.\mathbf{p}] + m_0c^2[\Sigma.\mathbf{p}, \beta] \end{aligned} \tag{11.147}$$

Take:

$$\begin{aligned} [\Sigma.\mathbf{p}, c\alpha.\mathbf{p}] &= c\begin{pmatrix} \sigma.\mathbf{p} & 0 \\ 0 & \sigma.\mathbf{p} \end{pmatrix}\begin{pmatrix} 0 & \sigma.\mathbf{p} \\ \sigma.\mathbf{p} & 0 \end{pmatrix} - c\begin{pmatrix} 0 & \sigma.\mathbf{p} \\ \sigma.\mathbf{p} & 0 \end{pmatrix}\begin{pmatrix} \sigma.\mathbf{p} & 0 \\ 0 & \sigma.\mathbf{p} \end{pmatrix} \\ &= c\begin{pmatrix} 0 & p^2 \\ p^2 & 0 \end{pmatrix} - c\begin{pmatrix} 0 & p^2 \\ p^2 & 0 \end{pmatrix} = 0 \end{aligned} \tag{11.148a}$$

and:

$$[\Sigma.\mathbf{p},\beta]=\begin{pmatrix}\sigma.\mathbf{p} & 0\\ 0 & \sigma.\mathbf{p}\end{pmatrix}\begin{pmatrix}I & 0\\ 0 & -I\end{pmatrix}-\begin{pmatrix}I & 0\\ 0 & -I\end{pmatrix}\begin{pmatrix}\sigma.\mathbf{p} & 0\\ 0 & \sigma.\mathbf{p}\end{pmatrix}$$
$$=\begin{pmatrix}\sigma.\mathbf{p} & 0\\ 0 & -\sigma.\mathbf{p}\end{pmatrix}-\begin{pmatrix}\sigma.\mathbf{p} & 0\\ 0 & -\sigma.\mathbf{p}\end{pmatrix}=0 \tag{11.148b}$$

Therefore:

$$[\Sigma.\mathbf{p},H]=0.$$

8. Apply the radial Eqns. (11.115) to the spherically symmetric infinite potential well defined by:

$V(r)=\begin{cases}0 & r<a\\ \infty & r\geq a\end{cases}$ and calculate the radial wave function and determine the energy eigenvalue for the lowest value of the quantum number $|k|$.

SOLUTION

For $r<a$ Eqns. (11.115) reduce to:

$$\frac{dF(r)}{dr}+\frac{kF(r)}{r}=-\frac{1}{\hbar c}\left(E-m_0c^2\right)G(r) \tag{11.149a}$$

and:

$$\frac{dG(r)}{dr}-\frac{kG(r)}{r}=\frac{1}{\hbar c}\left(E+m_0c^2\right)F(r) \tag{11.149b}$$

Combining these equations, we get:

$$\left(\frac{d}{dr}-\frac{k}{r}\right)\left(\frac{dF}{dr}+\frac{kF}{r}\right)=-\frac{1}{(\hbar c)^2}\left(E-m_0c^2\right)\left(E+m_0c^2\right)F$$
$$\Rightarrow\frac{d^2F}{dr^2}-\frac{k(k+1)F}{r^2}+\frac{\left(E^2-m_0^2c^4\right)}{(\hbar c)^2}F=0 \tag{11.150}$$

Taking $\frac{\left(E^2-m_0^2c^4\right)}{(\hbar c)^2}=\lambda^2$ and $\rho=\lambda r$, Eqn. (11.150) goes to:

$$\frac{d^2F}{d\rho^2}-\frac{k(k+1)F}{\rho^2}+F=0 \tag{11.151a}$$

In terms of $f(r)$, we have:

$$\frac{d^2 f}{d\rho^2}+\frac{2}{\rho}\frac{df}{d\rho}+\left(1-\frac{k(k+1)}{\rho^2}\right)f=0 \tag{11.151b}$$

which is a spherical Bessel differential equation, whose solution for $r < a$ is:

$$f(r)=Aj_k(\lambda r) \tag{11.152}$$

The potential is infinite for $r \geq a$, and hence $f(r)$ vanishes at $r = a$. For the lowest value of $|k|$, we obtain:

$$\begin{aligned} &j_1(\lambda a)=0;\\ &\frac{\sin(\lambda a)}{(\lambda a)^2}-\frac{\cos(\lambda a)}{\lambda a}=0;\\ &\Rightarrow \frac{\tan(\lambda a)}{\lambda a}=1 \end{aligned} \tag{11.153}$$

which can be solved to find the energy eigenvalue. For small values of $\lambda a \ll 1$, we have:

$$\tan(\lambda a)\approx \lambda a+\frac{1}{3}(\lambda a)^3+\frac{2}{15}(\lambda a)^5+\ldots\ldots \tag{11.154}$$

substitution of which into Eqn. (11.153) gives:

$$\frac{(\lambda a)^3}{3}\left\{1+\frac{2}{5}(\lambda a)^2\right\}=0 \tag{11.155}$$

This can be satisfied for two cases:

$$E^2-m_0^2c^4=0 \tag{11.156a}$$

and:

$$E^2-m_0^2c^4=-\frac{5}{2}\frac{\hbar^2c^2}{a^2} \tag{11.156b}$$

Equation (11.156a) yields a trivial solution, while Eqn. (11.156b) gives an interesting result. On writing $E=E'+m_0c^2$, and then taking $E' \ll m_0c^2$ we get:

$$E'=-\frac{5}{2}\left(\frac{\hbar^2}{2m_0a^2}\right) \tag{11.157}$$

Note that we get the same answer if we solve the nonrelativistic Schrödinger equation for the spherical symmetric infinite potential well.

11.9 Exercises

1. Show that the Klein-Gordon wave equation can also be derived from the free particle Dirac equation.
2. Show that gamma matrices, defined in Section 11.2.2, satisfy $\gamma_\mu\gamma_\nu + \gamma_\nu\gamma_\mu = 2\delta_{\mu\nu}$ and $\gamma_\mu^\dagger = \gamma_\mu$.
3. Define $\bar{\psi} = \psi^\dagger\gamma_4$ and the four-probability current $S_\mu = c\bar{\psi}\gamma_\mu\psi$. Use the Dirac equation to show that S_μ satisfies Eqn. (11.50) and hence it is known as conserved probability current.
4. For a free particle at rest, the Dirac equation reduces to $\beta m_0 c^2\psi(\mathbf{r},t) = i\hbar\dfrac{\partial\psi(\mathbf{r},t)}{\partial t}$. Find its possible solutions.
5. Show that operators $\mathbf{J}$ and K commute.
6. Show that the matrices obtained from α_j and β through unitary transformations are also the right choice for the Dirac equation. [Hint: calculate $\alpha'_j = U\alpha_j U^\dagger$ and $\beta' = U\beta U^\dagger$ and then show that $\alpha'_j\beta' + \beta'\alpha'_j = 0$, $\alpha'_i\alpha'_j + \alpha'_j\alpha'_i = 2\delta_{ij}$ and $\beta'^2 = 1$].
7. Show that $\dfrac{dx_j}{dt} = c\alpha_j$ $(j = 1,2,3)$, α_j are Dirac matrices.

12

Quantization of Fields and Second Quantization

In this chapter we will discuss the quantization of electromagnetic fields and the second quantization for Bosonic and Fermionic particles, which obey the nonrelativistic Schrödinger equation. Quantization of fields is a transition from a classical concept of physical phenomena to a quantum mechanical understanding of it. It is a procedure to construct quantum field theory starting from classical field theory. Classical fields are converted into operators acting on quantum states of the field theory, after quantization. The concepts of photon, phonon, plasmon, magnon, polaron, etc. have been originated through the quantization of wave fields. The quantization of a wave represented by a wave function, in the Schrödinger picture of nonrelativistic quantum mechanics and Dirac relativistic quantum mechanics, is known as second quantization.

12.1 Quantization of an Electromagnetic Field

After quantization, an electromagnetic field is treated as a system of discretized particles known as photons, which are massless but have definite energy, momentum, and spin. In a space that is free from sources of charge and current, Maxwell's equations are given by Eqn. (10.52). The electric and magnetic fields are then obtained from Eqn. (10.53), within the transverse (radiation) gauge. The vector potential, $\mathbf{A}(\mathbf{r},t)$ satisfies the equation:

$$\nabla^2\mathbf{A} - \frac{1}{c^2}\frac{\partial^2 \mathbf{A}}{\partial t^2} = 0 \tag{12.1}$$

With $\varepsilon_0\mu_0 = 1/c^2$, where ε_0 and μ_0 are permitivity and permeability of free space. The $\mathbf{A}(\mathbf{r},t)$ can be expanded as a Fourier series. This allows us to write:

$$\mathbf{A}(\mathbf{r},t) = \sum_{\mathbf{k}}\sum_{\alpha=1,2}\left[\hat{\varepsilon}_\alpha(\mathbf{k})c_{\mathbf{k}\alpha}(t)u_{\mathbf{k}}(\mathbf{r}) + \hat{\varepsilon}_\alpha(\mathbf{k})c^*_{\mathbf{k}\alpha}(t)u^*_{\mathbf{k}}(\mathbf{r})\right] \tag{12.2}$$

Here, $\hat{\varepsilon}_\alpha(\mathbf{k})$ is a real unit *polarization vector,* whose direction depends on the direction of the propagation vector $\mathbf{k}$ and it takes two mutually perpendicular values $\hat{\varepsilon}_1(\mathbf{k})$ and $\hat{\varepsilon}_2(\mathbf{k})$. The $\hat{\varepsilon}_1(\mathbf{k})$, $\hat{\varepsilon}_2(\mathbf{k})$, and unit vector $\hat{\mathbf{k}}$ are then chosen to form a right-handed set of mutually orthogonal vectors. The second order derivative with respect to time in Eqn. (12.1) permits us to take $c_{\mathbf{k}\alpha}(t) = c_{\mathbf{k}\alpha}(0)e^{-i\omega t}$ and $c^*_{\mathbf{k}\alpha}(t) = c^*_{\mathbf{k}\alpha}(0)e^{i\omega t}$. Then the substitution of Eqn. (12.2) into Eqn. (12.1) gives:

$$\left(\nabla^2 + \frac{\omega^2}{c^2}\right)u_{\mathbf{k}}(\mathbf{r}) = 0 \tag{12.3}$$

the solution for which has to be a sinusoidal function. If the fields are confined to a volume V, $u_{\mathbf{k}}(\mathbf{r}) = \frac{1}{\sqrt{V}} e^{i\mathbf{k}.\mathbf{r}}$ with $k^2 = \frac{\omega^2}{c^2}$ satisfies Eqn. (12.3). The imposition of the periodic boundary condition:

$$u_{\mathbf{k}}(\mathbf{r}) = u_{\mathbf{k}}(\mathbf{r} + \hat{x}L) = u_{\mathbf{k}}(\mathbf{r} + \hat{y}L) = u_{\mathbf{k}}(\mathbf{r} + \hat{z}L) \tag{12.4}$$

demands that:

$$\begin{aligned} & e^{ik_jL} = 1 = e^{2\pi i n_j} \\ & \Rightarrow k_j = \frac{2\pi}{L} n_j, \qquad \text{with} \quad j = 1, 2, 3 \end{aligned} \tag{12.5}$$

The k_1, k_2, and k_3 are components of $\mathbf{k}$ along the x, y, and z-axis, respectively. n_1, n_2, and n_3 are integers and $V = L^3$. Thus, $\mathbf{k}$ takes discretized values. We then have:

$$\mathbf{A}(\mathbf{r},t) = \frac{1}{\sqrt{V}} \sum_{\mathbf{k}} \sum_{\alpha} \hat{\varepsilon}_\alpha(\mathbf{k}) \left\{ c_{\mathbf{k}\alpha}(0) e^{i(\mathbf{k}.\mathbf{r} - \omega t)} + c^*_{\mathbf{k}\alpha}(0) e^{-i(\mathbf{k}.\mathbf{r} - \omega t)} \right\} \tag{12.6}$$

$$\mathbf{E} = -\frac{\partial \mathbf{A}}{\partial t} = -\frac{i}{\sqrt{V}} \sum_{\mathbf{k}} \sum_{\alpha} \omega \left\{ c_{\mathbf{k}\alpha}(0) \hat{\varepsilon}_\alpha(\mathbf{k}) e^{i(\mathbf{k}.\mathbf{r} - \omega t)} - c^*_{\mathbf{k}\alpha}(0) \hat{\varepsilon}_\alpha(\mathbf{k}) e^{-i(\mathbf{k}.\mathbf{r} - \omega t)} \right\} \tag{12.7}$$

and:

$$\mathbf{B} = \nabla \times \mathbf{A} = \frac{i}{\sqrt{V}} \sum_{\mathbf{k}} \sum_{\alpha} \mathbf{k} \times \left\{ c_{\mathbf{k}\alpha}(0) \hat{\varepsilon}_\alpha(\mathbf{k}) e^{i(\mathbf{k}.\mathbf{r} - \omega t)} - c^*_{\mathbf{k}\alpha}(0) \hat{\varepsilon}_\alpha(\mathbf{k}) e^{-i(\mathbf{k}.\mathbf{r} - \omega t)} \right\} \tag{12.8}$$

The Hamiltonian (total energy) of the electromagnetic field is:

$$H = \frac{1}{2} \int d^3r \left(\varepsilon_0 |\mathbf{E}|^2 + \mu_0 |\mathbf{H}|^2 \right) = \frac{\varepsilon_0}{2} \int d^3r \left(|\mathbf{E}|^2 + c^2 |\mathbf{B}|^2 \right) \tag{12.9}$$

We first evaluate:

$$\begin{aligned} \int |\mathbf{E}|^2 d^3r = \frac{1}{V} \sum_{\mathbf{k}} \sum_{\mathbf{k}'} \sum_{\alpha} \sum_{\alpha'} \omega^2 \Big[& \int d^3r c_{\mathbf{k}\alpha}(t) c^*_{\mathbf{k}'\alpha'}(t) \hat{\varepsilon}_\alpha . \hat{\varepsilon}_{\alpha'} e^{i(\mathbf{k}-\mathbf{k}').\mathbf{r}} \\ & - \int d^3r c_{\mathbf{k}\alpha}(t) c_{\mathbf{k}'\alpha'}(t) \hat{\varepsilon}_\alpha . \hat{\varepsilon}_{\alpha'} e^{i(\mathbf{k}+\mathbf{k}').\mathbf{r}} \\ & - \int d^3r c^*_{\mathbf{k}\alpha}(t) c^*_{\mathbf{k}'\alpha'}(t) \hat{\varepsilon}_\alpha . \hat{\varepsilon}_{\alpha'} e^{-i(\mathbf{k}+\mathbf{k}').\mathbf{r}} + \int d^3r c^*_{\mathbf{k}\alpha}(t) c_{\mathbf{k}'\alpha'}(t) \hat{\varepsilon}_\alpha . \hat{\varepsilon}_{\alpha'} e^{i(\mathbf{k}'-\mathbf{k}).\mathbf{r}} \Big] \end{aligned} \tag{12.10}$$

Note that:

$$\frac{1}{V}\hat{\varepsilon}_{\alpha}.\hat{\varepsilon}_{\alpha'}\int e^{\pm i(\mathbf{k}-\mathbf{k}').\mathbf{r}}d^3r = \delta_{\alpha\alpha'}\delta_{k,k'} \tag{12.11}$$

and,

$$\frac{1}{V}\hat{\varepsilon}_{\alpha}.\hat{\varepsilon}_{\alpha'}\int e^{\pm i(\mathbf{k}+\mathbf{k}').\mathbf{r}}d^3r = \delta_{\alpha\alpha'}\delta_{k,-k'} \tag{12.12}$$

In Eqn. (12.10), first and fourth terms do not depend on time, while the second and third terms have sinusoidal behavior with time. In the steady state, the time-averaged contribution to $\int|\mathbf{E}|^2 d^3r$, the contribution from the second and third terms will be zero, because of their sinusoidal nature. We thus obtain:

$$\int|\mathbf{E}|^2 d^3r = 2\sum_{\mathbf{k}}\sum_{\alpha}\omega^2 c^*_{\mathbf{k}\alpha}(t)c_{\mathbf{k}\alpha}(t) = 2\sum_{\mathbf{k}}\sum_{\alpha}\omega^2 c^*_{\mathbf{k}\alpha}(0)c_{\mathbf{k}\alpha}(0) \tag{12.13}$$

We next take:

$$\begin{aligned}\int|\mathbf{B}|^2 d^3r = \frac{1}{V}\sum_{\mathbf{k}}\sum_{\mathbf{k}'}\sum_{\alpha}\sum_{\alpha'}\Big[&\int d^3r c_{\mathbf{k}\alpha}(t)c^*_{\mathbf{k}'\alpha'}(t)(\mathbf{k}\times\hat{\varepsilon}_{\alpha}).(\mathbf{k}'\times\hat{\varepsilon}_{\alpha'})e^{i(\mathbf{k}-\mathbf{k}').\mathbf{r}}\\ &-\int d^3r c_{\mathbf{k}\alpha}(t)c_{\mathbf{k}'\alpha'}(t)(\mathbf{k}\times\hat{\varepsilon}_{\alpha}).(\mathbf{k}'\times\hat{\varepsilon}_{\alpha'})e^{i(\mathbf{k}+\mathbf{k}').\mathbf{r}}\\ &-\int d^3r c^*_{\mathbf{k}\alpha}(t)c^*_{\mathbf{k}'\alpha'}(t)(\mathbf{k}\times\hat{\varepsilon}_{\alpha}).(\mathbf{k}'\times\hat{\varepsilon}_{\alpha'})e^{-i(\mathbf{k}+\mathbf{k}').\mathbf{r}}\\ &+\int d^3r c^*_{\mathbf{k}\alpha}(t)c_{\mathbf{k}'\alpha'}(t)(\mathbf{k}\times\hat{\varepsilon}_{\alpha}).(\mathbf{k}'\times\hat{\varepsilon}_{\alpha'})e^{i(\mathbf{k}'-\mathbf{k}).\mathbf{r}}\Big]\end{aligned} \tag{12.14}$$

Using the vector identity $(\mathbf{A}\times\mathbf{B}).(\mathbf{C}\times\mathbf{D}) = (\mathbf{A.C})(\mathbf{B.D}) - (\mathbf{A.D})(\mathbf{B.C})$, we get:

$$(\mathbf{k}\times\hat{\varepsilon}_{\alpha}).(\mathbf{k}'\times\hat{\varepsilon}_{\alpha'}) = (\mathbf{k.k}')(\hat{\varepsilon}_{\alpha}.\hat{\varepsilon}_{\alpha'}) - (\mathbf{k}.\hat{\varepsilon}_{\alpha'})(\hat{\varepsilon}_{\alpha}.\mathbf{k}') = (\mathbf{k.k}')\delta_{\alpha\alpha'} \tag{12.15}$$

Here, we have utilized the fact that the polarization vector and propagation vectors are perpendicular to each other. Again, the contribution from the second and third is zero. We thus obtain:

$$\int|\mathbf{B}|^2 d^3r = 2\sum_{\mathbf{k}}\sum_{\alpha}k^2 c^*_{\mathbf{k}\alpha}(t)c_{\mathbf{k}\alpha}(t) \tag{12.16}$$

Use of Eqns. (12.13) and (12.16) into Eqn. (12.9) yields:

$$H = 2\varepsilon_0\sum_{\mathbf{k}}\sum_{\alpha}\omega^2 c^*_{\mathbf{k}\alpha}c_{\mathbf{k}\alpha} \tag{12.17}$$

Since $c_{\mathbf{k}\alpha}(t)$ satisfies $\frac{d^2 c_{\mathbf{k}\alpha}}{dt^2} = -\omega^2 c_{\mathbf{k}\alpha}$, Eqn. (12.17) is nothing but the total energy of uncoupled and independent harmonic oscillators. To make it more explicit, let us introduce:

$$Q_{\mathbf{k}\alpha} = \sqrt{\varepsilon_0}\left(c^*_{\mathbf{k}\alpha} + c_{\mathbf{k}\alpha}\right), \text{ and } P_{\mathbf{k}\alpha} = -i\omega\sqrt{\varepsilon_0}\left(c_{\mathbf{k}\alpha} - c^*_{\mathbf{k}\alpha}\right) \tag{12.18}$$

which yields:

$$H = \sum_{\mathbf{k}}\sum_{\alpha}\left(\frac{P^2_{\mathbf{k}\alpha}}{2} + \frac{\omega^2}{2}Q^2_{\mathbf{k}\alpha}\right) \tag{12.19}$$

Thus, the electromagnetic field are treated as a collection of independent harmonic oscillators, where each oscillator is characterized by $(\mathbf{k}, \alpha)$ and has unit mass. Similar to the case of the harmonic oscillator discussed in Section 3.19, we can postulate that $P_{\mathbf{k}\alpha}$ and $Q_{\mathbf{k}\alpha}$ are operators and satisfy:

$$\left[Q_{\mathbf{k}\alpha}, P_{\mathbf{k}'\alpha'}\right] = i\hbar\delta_{\mathbf{k},\mathbf{k}'}\delta_{\alpha,\alpha'} \tag{12.20}$$

and then following the procedure of quantization for an harmonic oscillator described in Section 3.19, we obtain:

$$H = \sum_{\mathbf{k}}\sum_{\alpha}\left(a^\dagger_{\mathbf{k}\alpha}a_{\mathbf{k}\alpha} + \frac{1}{2}\right)\hbar\omega \tag{12.21}$$

where we have defined:

$$a_{\mathbf{k}\alpha} = \frac{1}{\sqrt{2\hbar\omega}}\left(\omega Q_{\mathbf{k}\alpha} + iP_{\mathbf{k}\alpha}\right), \text{ and } a^\dagger_{\mathbf{k}\alpha} = \frac{1}{\sqrt{2\hbar\omega}}\left(\omega Q_{\mathbf{k}\alpha} - iP_{\mathbf{k}\alpha}\right) \tag{12.22}$$

The $a^\dagger_{\mathbf{k}\alpha}$ and $a_{\mathbf{k}\alpha}$ are Bosonic operators and they satisfy:

$$\begin{aligned}&\left[a_{\mathbf{k}\alpha}, a^\dagger_{\mathbf{k}'\alpha'}\right] = \delta_{\mathbf{k},\mathbf{k}'}\delta_{\alpha,\alpha'};\\&\left[a^\dagger_{\mathbf{k}\alpha}, a^\dagger_{\mathbf{k}'\alpha'}\right] = \left[a_{\mathbf{k}\alpha}, a_{\mathbf{k}'\alpha'}\right] = 0\end{aligned} \tag{12.23}$$

The $N_{\mathbf{k}\alpha} = a^\dagger_{\mathbf{k}\alpha}a_{\mathbf{k}\alpha}$ is the number operator for a state identified by $(\mathbf{k}, \alpha)$. As discussed in Section 3.19, application of these operators on a quantum state $|n_{\mathbf{k}\alpha}\rangle$, which consists of $n_{\mathbf{k}\alpha}$- particles (photons), gives:

$$\begin{aligned}&a_{\mathbf{k}\alpha}|n_{\mathbf{k}\alpha}\rangle = \sqrt{n_{\mathbf{k}\alpha}}\,|n_{\mathbf{k}\alpha} - 1\rangle,\\&a^\dagger_{\mathbf{k}\alpha}|n_{\mathbf{k}\alpha}\rangle = \sqrt{n_{\mathbf{k}\alpha} + 1}\,|n_{\mathbf{k}\alpha} + 1\rangle, \text{ and}\\&N_{\mathbf{k}\alpha}|n_{\mathbf{k}\alpha}\rangle = n_{\mathbf{k}\alpha}|n_{\mathbf{k}\alpha}\rangle\end{aligned} \tag{12.24}$$

The quantum state of a system that consists of $n_{\mathbf{k}_1\alpha_1}, n_{\mathbf{k}_2\alpha_2}, n_{\mathbf{k}_3\alpha_3} \dots n_{\mathbf{k}_i\alpha_i}$ in the states $|n_{\mathbf{k}_1\alpha_1}\rangle$, $|n_{\mathbf{k}_2\alpha_2}\rangle$, $|n_{\mathbf{k}_3\alpha_3}\rangle \dots |n_{\mathbf{k}_i\alpha_i}\rangle$, respectively, is:

$$|n_{\mathbf{k}_1\alpha_1}, n_{\mathbf{k}_2\alpha_2}, n_{\mathbf{k}_3\alpha_3}, \dots n_{\mathbf{k}_i\alpha_i}, \dots\dots\rangle = |n_{\mathbf{k}_1\alpha_1}\rangle |n_{\mathbf{k}_2\alpha_2}\rangle |n_{\mathbf{k}_3\alpha_3}\rangle \dots |n_{\mathbf{k}_i\alpha_i}\rangle \dots\dots \quad (12.25)$$

The state:

$$|0\rangle = |0_{\mathbf{k}_1\alpha_1}\rangle |0_{\mathbf{k}_2\alpha_2}\rangle |0_{\mathbf{k}_3\alpha_3}\rangle \dots |0_{\mathbf{k}_i\alpha_i}\rangle \dots\dots \quad (12.26)$$

has zero particles in each and every state $(\mathbf{k},\alpha)$ and it is known as the vacuum state $N_{\mathbf{k}\alpha}|0\rangle = 0$. Operation of $a^{\dagger}_{\mathbf{k}\alpha}$ on $|0\rangle$ will create one photon in the $(\mathbf{k},\alpha)$ state. Similarly, $\frac{1}{\sqrt{2}} a^{\dagger}_{\mathbf{k}\alpha} a^{\dagger}_{\mathbf{k}\alpha}|0\rangle$ represents a normalized eigenvector of the state that has two photons in the $(\mathbf{k},\alpha)$ state, while $a^{\dagger}_{\mathbf{k}'\alpha'} a^{\dagger}_{\mathbf{k}\alpha}|0\rangle$ also is a eigenvector for a two-photon system but they are in two different states. Therefore, Eqn. (12.25) is given by:

$$|\psi\rangle = |n_{\mathbf{k}_1\alpha_1}, n_{\mathbf{k}_2\alpha_2}, n_{\mathbf{k}_3\alpha_3}, \dots n_{\mathbf{k}_i\alpha_i}, \dots\dots\rangle = \prod_{\mathbf{k}_i,\alpha_i} \frac{\left(a^{\dagger}_{\mathbf{k}_i\alpha_i}\right)^{n_{\mathbf{k}_i\alpha_i}}}{\sqrt{(n_{\mathbf{k}_i\alpha_i})!}} |0\rangle \quad (12.27)$$

As is seen $a^{\dagger}_{\mathbf{k}'\alpha'} a^{\dagger}_{\mathbf{k}\alpha}|0\rangle = a^{\dagger}_{\mathbf{k}\alpha} a^{\dagger}_{\mathbf{k}'\alpha'}|0\rangle$, and therefore the wave function ket defined by Eqn. (12.27) is symmetric under the exchange of any two photons. The total energy of the system of photons is obtained by operating H on $|n_{\mathbf{k}_1\alpha_1}, n_{\mathbf{k}_2\alpha_2}, n_{\mathbf{k}_3\alpha_3}, \dots n_{\mathbf{k}_i\alpha_i}, \dots\dots\rangle$:

$$E = \sum_{\mathbf{k}} \sum_{\alpha} \left(n_{\mathbf{k}\alpha} + \frac{1}{2} \right) \hbar\omega, \text{ with } n_{k\alpha} = 0,1,2,\dots.. \quad (12.28)$$

which is discretized.

12.1.1 Field Operators

The Fourier coefficients $c^{*}_{\mathbf{k}\alpha}(t)$ and $c_{\mathbf{k}\alpha}(t)$ in Eqns. (12.6) to (12.8), the classical expressions for an electromagnetic field, must be replaced by corresponding creation and annihilation operators if the canonical variables $P_{\mathbf{k}\alpha}$ and $q_{\mathbf{k}\alpha}$ in Eqn. (12.19) are interpreted as quantum mechanical operators. We obtain:

$$c_{\mathbf{k}\alpha} \to \sqrt{\frac{\hbar}{2\omega\varepsilon_0}} a_{\mathbf{k}\alpha} \text{ and } c^{*}_{\mathbf{k}\alpha} \to \sqrt{\frac{\hbar}{2\omega\varepsilon_0}} a^{\dagger}_{\mathbf{k}\alpha} \quad (12.29)$$

which gives us:

$$\mathbf{A}(\mathbf{r},t) = \frac{1}{\sqrt{V}} \sum_{\mathbf{k}} \sum_{\alpha} \sqrt{\frac{\hbar}{2\omega\varepsilon_0}} \left\{ a_{\mathbf{k}\alpha}(t) \hat{\varepsilon}_{\alpha}(\mathbf{k}) e^{i\mathbf{k}.\mathbf{r}} + a^{\dagger}_{\mathbf{k}\alpha}(t) \hat{\varepsilon}_{\alpha}(\mathbf{k}) e^{-i\mathbf{k}.\mathbf{r}} \right\} \quad (12.30a)$$

$$\mathbf{E} = -\frac{i}{\sqrt{V}} \sum_{\mathbf{k}} \sum_{\alpha} \sqrt{\frac{\hbar\omega}{2\varepsilon_0}} \left\{ a_{\mathbf{k}\alpha}(t) \hat{\varepsilon}_{\alpha}(\mathbf{k}) e^{i\mathbf{k}.\mathbf{r}} - a^{\dagger}_{\mathbf{k}\alpha}(t) \hat{\varepsilon}_{\alpha}(\mathbf{k}) e^{-i\mathbf{k}.\mathbf{r}} \right\} \quad (12.30b)$$

and:

$$\mathbf{B} = \frac{i}{\sqrt{V}} \sum_{\mathbf{k}} \sum_{\alpha} \sqrt{\frac{\hbar}{2\omega\varepsilon_0}} \left\{ a_{\mathbf{k}\alpha}(t) \left[\mathbf{k} \times \hat{\varepsilon}_\alpha(\mathbf{k}) \right] e^{i\mathbf{k}.\mathbf{r}} - a_{\mathbf{k}\alpha}^{\dagger}(t) \left[\mathbf{k} \times \hat{\varepsilon}_\alpha(\mathbf{k}) \right] e^{-i\mathbf{k}.\mathbf{r}} \right\} \tag{12.30c}$$

Note that Eqns. (12.6) to (12.8) describe the classical electromagnetic field, while Eqns. (12.30) are the *field operators or a quantized field*. The field operators act over the state vector. For example operation by $a_{\mathbf{k}\alpha}^{\dagger}(t)\hat{\varepsilon}_\alpha(\mathbf{k})e^{-i\mathbf{k}.\mathbf{r}}$ on the state vector adds a Boson in the state of $(\mathbf{k},\alpha)$ at time t and position $\mathbf{r}$.

The total momentum of the field is given by space integration over the pointing vector:

$$\mathbf{P} = \frac{1}{\mu_0} \int (\mathbf{E} \times \mathbf{B}) d^3 r \tag{12.31}$$

Substituting $\mathbf{E}$ and $\mathbf{B}$ from Eqns. (12.30b) and (12.30c), and then noting the vector products such as $\hat{\varepsilon}_1 \times (\mathbf{k} \times \hat{\varepsilon}_2) = 0$, $\hat{\varepsilon}_1 \times (\mathbf{k} \times \hat{\varepsilon}_1) = \mathbf{k}$ etc., we get:

$$\mathbf{P} = \sum_{\mathbf{k}} \sum_{\alpha} \hbar \mathbf{k} \left(a_{\mathbf{k}\alpha}^{\dagger} a_{\mathbf{k}\alpha} + \frac{1}{2} \right) \tag{12.32}$$

On operating $\mathbf{P}$ on $\left| n_{\mathbf{k}_1\alpha_1}, n_{\mathbf{k}_2\alpha_2}, n_{\mathbf{k}_3\alpha_3}, \ldots n_{\mathbf{k}_i\alpha_i}, \ldots\ldots \right\rangle$, we find that the discretized total momentum of the system is $\sum_{\mathbf{k}} \sum_{\alpha} \hbar \mathbf{k} \left(n_{\mathbf{k}\alpha} + \frac{1}{2} \right)$.

As is seen, $\hbar\omega = \hbar|\mathbf{k}|c$ and $\hbar\mathbf{k}$ are the energy and momentum of a photon, respectively. Then the relation $m_0 c^2 = \left[E^2 - p^2 c^2 \right]^{1/2} = \left[\hbar^2 |\mathbf{k}|^2 c^2 - (\hbar\mathbf{k})^2 c^2 \right]^{1/2} = 0$ proves that the rest mass of the photon is equal to zero. Note that a photon is characterized by both momentum and polarization vector.

12.2 Second Quantization

As has been shown in the previous section, vectors $\mathbf{A}(\mathbf{r},t)$, $\mathbf{E}(\mathbf{r},t)$, and $\mathbf{B}(\mathbf{r},t)$, which satisfy the classical wave equations, are represented by field operators after the quantization. This analogy suggests that a wave function which satisfies a quantum mechanical equation can also be thought of as a field operator. Since the description in which classical observables such as momentum and energy are represented by operators, which is commonly known as *first quantization,* already existed, the representation of wave function by the field operator is termed second quantization. The second quantization is also termed the *occupation number representation* formalism that is used to describe and analyze the quantum theory of many particle systems. The method of second quantization was mostly developed by Vladimir Fock, based on the Dirac field theory of the electromagnetic field.

12.2.1 Second Quantization of the Schrödinger Equation for Bosons

The time-dependent Schrödinger equation is:

$$H\psi(\mathbf{r},t) = i\hbar\frac{\partial\psi(\mathbf{r},t)}{\partial t}, \text{with } H = -\frac{\hbar^2}{2m}\nabla^2 + V(\mathbf{r}) \tag{12.33}$$

which has been derived with the use of $\mathbf{p} = -i\hbar\nabla$ and $E = i\hbar\frac{\partial}{\partial t}$. Therefore, one quantization has already been made, and it is known as the first quantization. Our objective here is to represent the $\psi(\mathbf{r},t)$ as a field operator. The procedure is termed as the second quantization. Our discussions on time-dependent fields, presented in Section 10.1, suggest that for time-dependent fields, the Schrödinger wave function can be expressed as follows:

$$\psi(\mathbf{r},t) = \sum_n a_n(t)u_n(\mathbf{r}) \tag{12.34}$$

where expansion coefficients are time-dependent, and $\int u_l^*(\mathbf{r})u_n(\mathbf{r})d^3r = \langle u_l|u_n\rangle = \delta_{l,n}$. Let us also assume that the basis vectors $|u_n\rangle$ are eigenvectors of H and they form a complete set. We thus have:

$$Hu_n(\mathbf{r}) = E_n u_n(\mathbf{r}) \tag{12.35}$$

On substituting Eqn. (12.34) into (12.33), we get:

$$\begin{aligned} &\sum_n a_n(t)\left[-\frac{\hbar^2}{2m}\nabla^2 + V(\mathbf{r})\right]u_n(\mathbf{r}) = i\hbar\sum_n \frac{\partial a_n(t)}{\partial t}u_n(\mathbf{r}) \\ &\Rightarrow \sum_n \left[a_n(t)E_n - i\hbar\dot{a}_n(t)\right]u_n(\mathbf{r}) = 0 \end{aligned} \tag{12.36}$$

Multiplying from the left with $u_l^*(\mathbf{r})$ and then integrating over space coordinates, we obtain:

$$\dot{a}_l(t) = \frac{1}{i\hbar}a_l(t)E_l \tag{12.37}$$

Let us now assert that a_n are Bosonic operators and they satisfy the commutation relations:

$$\begin{aligned} &\left[a_n, a_l^\dagger\right] = \delta_{n,l} \\ &\left[a_n, a_l\right] = 0 = \left[a_n^\dagger, a_l^\dagger\right] \end{aligned} \tag{12.38}$$

The Hamiltonian in terms of these operators is as follows:

$$\begin{aligned} &\tilde{H} = \int\psi^\dagger(\mathbf{r},t)H\psi(\mathbf{r},t)d^3r = \sum_n\sum_l a_l^\dagger a_n\int u_l^*(\mathbf{r})Hu_n(\mathbf{r})d^3r = \sum_n\sum_l a_l^\dagger a_n E_n\delta_{l,n} \\ &\Rightarrow \tilde{H} = \sum_l a_l^\dagger a_l E_l \end{aligned} \tag{12.39}$$

The $\tilde{H}$ looks similar to the Hamiltonian for the E M field, which is given by Eqn. (12.21). The Heisenberg equation for operator a_n gives:

$$\begin{aligned}
\frac{da_n}{dt} &= \frac{1}{i\hbar}\left[a_n, \tilde{H}\right] = \frac{1}{i\hbar}\left(a_n\tilde{H} - \tilde{H}a_n\right) \\
&= \frac{1}{i\hbar}\left[a_n \sum_l a_l^\dagger a_l E_l - \left(\sum_l a_l^\dagger a_l E_l\right) a_n\right] \\
&= \frac{1}{i\hbar}\sum_l \left(a_n a_l^\dagger a_l - a_l^\dagger a_l a_n\right) E_l \\
&= \frac{1}{i\hbar}\sum_l \left(\delta_{n,l} a_l + a_l^\dagger a_n a_l - a_l^\dagger a_l a_n\right) E_l \\
&= \frac{1}{i\hbar}\sum_l \left(\delta_{n,l} a_l + a_l^\dagger \{a_n a_l - a_l a_n\}\right) E_l
\end{aligned} \tag{12.40}$$

which simplifies to:

$$\dot{a}_n = \frac{1}{i\hbar} a_n E_n \tag{12.41}$$

The Eqns. (12.37) and (12.41) are identical. Hence, our assertion to treat $a_n(t)$ as Bosonic operators is proved to be correct. Note that $\psi(\mathbf{r},t) = \sum_n a_n(t) u_n(\mathbf{r})$, with $a_n(t)$ the Bosonic operator, is a field operator, *no more a wave function*. The Hamiltonian for the Bosonic field is given by $\tilde{H} = \sum_l a_l^\dagger a_l E_l$. Equation (12.24) for Schrödinger Bosonic fields is:

$$\begin{aligned}
a_l |n_l\rangle &= \sqrt{n_l}\,|n_l - 1\rangle, \\
a_l^\dagger |n_l\rangle &= \sqrt{n_l + 1}\,|n_l + 1\rangle, \\
N_l |n_l\rangle &= a_l^\dagger a_l |n_l\rangle = n_l |n_l\rangle
\end{aligned} \tag{12.42}$$

where n_l is a positive integer. The successive application of the annihilation operator will have $a_l|1_l\rangle = |0_l\rangle$ and then $a_l|0_l\rangle = 0$. As discussed in Section 12.1, we can have the situation where all quantum states have zero occupancy; such a state of the system is known as the *vacuum state*, generally represented by $|0\rangle$. Thus, $a_n^\dagger|0\rangle$ has one Boson in the nth state of energy E_n. Similarly, the state vector $a_l^\dagger a_n^\dagger|0\rangle$, which is symmetric with respect to the interchange of $n \leftrightarrow l$, is occupied by two Bosons in two different quantum states, while $\left(a_l^\dagger\right)^2|0\rangle$ also has two Bosons but both in the same state. The state vector:

$$|n_l\rangle = \frac{\left(a_l^\dagger\right)^{n_l}}{\sqrt{n_l!}}|0\rangle \tag{12.43}$$

represents a state that is occupied by n_l Bosons and has energy E_l. Hence, a normalized state vector for the entire system is:

$$|\phi\rangle = \prod_{n_l} |n_l\rangle = \prod_{n_l} \frac{\left(a_l^{\dagger}\right)^{n_l}}{\sqrt{n_l!}}|0\rangle \tag{12.44}$$

The total number of Bosons N and the total energy E of the system is then given by:

$$\begin{aligned} N &= \langle\phi|\sum_l a_l^{\dagger}a_l|\phi\rangle = \sum_l \prod_{n_{l'}} \prod_{n_l} \langle n_{l'}|a_l^{\dagger}a_l|n_l\rangle \\ &= \sum_l n_l \langle\phi|\phi\rangle = n_1 + n_2 + n_3 .... \end{aligned} \tag{12.45a}$$

and:

$$\begin{aligned} E &= \langle\phi|\tilde{H}|\phi\rangle = \prod_{n_{l'}} \prod_{n_l} \sum_l \langle n_{l'}|a_l^{\dagger}a_l E_l|n_l\rangle = \prod_{n_{l'}} \prod_{n_l} \sum_l n_l E_l \langle n_{l'}|n_l\rangle \\ &= \sum_l n_l E_l \langle\phi|\phi\rangle = n_1E_1 + n_2E_2 + n_3E_3 ...... \end{aligned} \tag{12.45b}$$

Note that the ith energy level accommodates n_i particles, each having energy E_i, where i takes values $1, 2, 3, ...$ The n_i is governed by the Bose-Einstein statistics:

$$n_i = \frac{g_i}{e^{(E_i - \mu)/k_B T} - 1} \tag{12.46}$$

where g_i is the number of quantum states that accommodate n_i particles, and k_B is the Boltzmann constant. As is seen from Eqn. (12.46), the number of Bosons per quantum state (n_i / g_i) can take any value between zero to infinity. Therefore, it is quite likely that all particles occupy the lowest energy state at some temperature. In the case of Bosons, a zero energy state exists which has no particle above some temperature, termed as the critical temperature T_c. The thermal energy excites all particles to nonzero energy levels when $k_B T \geq k_B T_c$. However, below T_c some of the Bosons move to the zero energy state and at zero temperature all of them occupy the zero energy state, in the case of an ideal Bosonic gas. This is the well-known phenomenon of *Bose-Einstein condensation,* which takes place in Bosonic gas.

The commutators of the field operators are evaluated as follows:

$$\begin{aligned} \left[\psi(\mathbf{r},t), \psi^{\dagger}(\mathbf{r}',t)\right] &= \sum_l \sum_n a_n(t)u_n(\mathbf{r})a_l^{\dagger}(t)u_l^{*}(\mathbf{r}') - \sum_l \sum_n a_l^{\dagger}(t)u_l^{*}(\mathbf{r}')a_n(t)u_n(\mathbf{r}) \\ &= \sum_l \sum_n \left[a_n(t)a_l^{\dagger}(t) - a_l^{\dagger}(t)a_n(t)\right]u_l^{*}(\mathbf{r}')u_n(\mathbf{r}) \end{aligned} \tag{12.47a}$$

which with the use of Eqn. (12.38) gives:

$$\begin{aligned}\left[\psi(\mathbf{r},t),\psi^{\dagger}(\mathbf{r}',t)\right] &= \sum_{l}\sum_{n}\delta_{n,l}u_l^{*}(\mathbf{r}')u_n(\mathbf{r}) \\ &= \sum_{n}u_n^{*}(\mathbf{r}')u_n(\mathbf{r}) \\ &= \delta(\mathbf{r}-\mathbf{r}')\end{aligned} \tag{12.47b}$$

Similarly, we find that:

$$\left[\psi(\mathbf{r},t),\psi(\mathbf{r}',t)\right] = 0 = \left[\psi^{\dagger}(\mathbf{r},t),\psi^{\dagger}(\mathbf{r}',t)\right] \tag{12.48}$$

12.2.2 Second Quantization of the Schrödinger Equation for Fermions

We again start from Eqn. (12.33). However, we now say that $\psi(\mathbf{r},t)$ represents the wave function for Fermions. Let us write:

$$\psi(\mathbf{r},t) = \sum_{n} b_n(t)u_n(\mathbf{r}) \tag{12.49}$$

Here, again $\langle u_l|u_n\rangle = \delta_{l,n}$ and $H|u_n\rangle = E_n|u_n\rangle$. Adopting the procedure laid down in the previous section, we get:

$$\dot{b}_n(t) = \frac{1}{i\hbar}b_n(t)E_n \tag{12.50}$$

Let us again make an assertion that $b_n(t)$ are operators and they satisfy the anticommutation relations:

$$\begin{aligned}&\left\{b_n, b_l^{\dagger}\right\} = b_n b_l^{\dagger} + b_l^{\dagger}b_n = \delta_{n,l}, \text{ and} \\ &\left\{b_n, b_n\right\} = 0 = \left\{b_n^{\dagger}, b_l^{\dagger}\right\}\end{aligned} \tag{12.51}$$

Next, similar to the case of Eqn. (12.39), we define:

$$\begin{aligned}\tilde{H} &= \int \psi^{\dagger}(\mathbf{r},t)H\psi(\mathbf{r},t)d^3r \\ &= \sum_{n}\sum_{l} b_l^{\dagger}b_n \int u_l^{*}(\mathbf{r})Hu_n(\mathbf{r})d^3r \\ &= \sum_{n}\sum_{l} b_l^{\dagger}b_n E_n \delta_{l,n} \\ &= \sum_{l} b_l^{\dagger}b_l E_l\end{aligned} \tag{12.52}$$

The Heisenberg equation then gives:

$$\begin{aligned}\frac{db_n}{dt} &= \frac{1}{i\hbar}\left[b_n, \tilde{H}\right] = \frac{1}{i\hbar}\left(b_n\tilde{H} - \tilde{H}b_n\right)\\ &= \frac{1}{i\hbar}\left[b_n\sum_l b_l^\dagger b_l E_l - \left(\sum_l b_l^\dagger b_l E_l\right)b_n\right]\\ &= \frac{1}{i\hbar}\sum_l\left(b_n b_l^\dagger b_l - b_l^\dagger b_l b_n\right)E_l = \frac{1}{i\hbar}\sum_l\left(\delta_{n,l}b_l - b_l^\dagger b_n b_l - b_l^\dagger b_l b_n\right)E_l\\ &= \frac{1}{i\hbar}\sum_l\left(\delta_{n,l}b_l - b_l^\dagger\left\{b_n, b_l\right\}\right)E_l\end{aligned} \tag{12.53}$$

Since $\{b_n, b_l\} = 0$, we have:

$$\dot{b}_n = \frac{1}{i\hbar}b_n E_n \tag{12.54}$$

suggesting that our assertions, Eqn. (12.51), are correct. Like a_l and $a_l^\dagger$, the b_l and $b_l^\dagger$ also satisfy:

$$\begin{aligned}b_l|n_l\rangle &= \sqrt{n_l}\,|n_l - 1\rangle,\\ b_l^\dagger|n_l\rangle &= \sqrt{n_l + 1}\,|n_l + 1\rangle,\\ N_l|n_l\rangle &= b_l^\dagger b_l|n_l\rangle = n_l|n_l\rangle\end{aligned} \tag{12.55}$$

This implies:

$$\left(b_l^\dagger b_l\right)^2|n_l\rangle = n_l b_l^\dagger b_l|n_l\rangle = n_l^2|n_l\rangle \tag{12.56}$$

Let us write the left-hand side of this equation as follows:

$$\begin{aligned}\left(b_l^\dagger b_l\right)^2|n_l\rangle &= b_l^\dagger b_l b_l^\dagger b_l|n_l\rangle\\ &= b_l^\dagger\left(1 - b_l^\dagger b_l\right)b_l|n_l\rangle\\ &= b_l^\dagger b_l|n_l\rangle - b_l^\dagger b_l^\dagger b_l b_l|n_l\rangle\end{aligned} \tag{12.57}$$

Since $\{b_l, b_l\} = 2b_l b_l = 0$, $b_l^\dagger b_l^\dagger b_l b_l|n_l\rangle = 0$, therefore:

$$\left(b_l^\dagger b_l\right)^2|n_l\rangle = b_l^\dagger b_l|n_l\rangle = n_l|n_l\rangle \tag{12.58}$$

which along with Eqn. (11.56) implies:

$$\left(n_l^2 - n_l\right)|n_l\rangle = 0 \tag{12.59}$$

If $|n_l\rangle$ is a nonzero state vector then we must have $n_l^2 - n_l = 0$ to fulfill the requirement of Eqn. (12.59). This can be satisfied only for two values, $n_l = 1$ and $n_l = 0$. Therefore, a quantum state should have either one Fermion (filled) or zero Fermions (empty). This is in perfect agreement with Pauli's exclusion principle and the Fermi Dirac statistics.

Next, from Eqn. (12.55) $b_l^\dagger|0_l\rangle = |1_l\rangle$. But $b_l^\dagger b_l^\dagger|0_l\rangle = 0$ because of $\left[b_l^\dagger, b_l^\dagger\right] = 0 \Rightarrow b_l^\dagger b_l^\dagger = 0$, which again suggests that two Fermions cannot occupy the same quantum state. However, we have:

$$b_{l'}^\dagger b_l^\dagger|0\rangle = b_{l'}^\dagger b_l^\dagger|0_1, 0_2, \ldots.0_{l'}\ldots 0_l\ldots.\rangle = |0_1, 0_2, \ldots.1_{l'}\ldots 1_l\ldots.\rangle \neq 0 \tag{12.60a}$$

The $b_{l'}^\dagger b_l^\dagger + b_l^\dagger b_{l'}^\dagger = 0$ implies that $b_{l'}^\dagger b_l^\dagger = -b_l^\dagger b_{l'}^\dagger \neq 0$. We therefore have:

$$b_l^\dagger b_{l'}^\dagger|0\rangle = b_l^\dagger b_{l'}^\dagger|0_1, 0_2, \ldots.0_{l'}\ldots 0_l\ldots.\rangle = -|0_1, 0_2, \ldots.1_l\ldots 1_{l'}\ldots.\rangle \tag{12.60b}$$

This ensures that the wave function ket of a system of Fermions is antisymmetric with respect to the interchange of any two of them.

Therefore, it is possible to construct the state of a system of N-Fermions by successive operation of $b_l^\dagger$ on the vacuum state $|0\rangle$:

$$|\phi\rangle = |n_1, n_2, n_3, \ldots.n_i, \ldots..n_s\rangle = \left(b_1^\dagger\right)^{n_1}\left(b_2^\dagger\right)^{n_2}\left(b_3^\dagger\right)^{n_3}\ldots\ldots\left(b_s^\dagger\right)^{n_s}|0\rangle \tag{12.61}$$

where each of $n_1, n_2, n_3, \ldots\ldots..n_s$ is either 1 or 0. The quantum states that correspond to $n_i = 1$ are filled states and those with $n_i = 0$ are empty states. Note that $|\phi\rangle$ is antisymmetric with respect to the interchange of any two Fermions. Since, one state can hold one Fermion, the quantum states will be filled up to a certain state and then there will be empty states, when the system is in the ground state. The last filled state is termed Fermi-level. This scenario will change if this is not in its ground state.

The total number of Fermions and total energy of the system can be obtained as follows:

$$N = \langle\phi|\sum_l b_l^\dagger b_l|\phi\rangle = n_1 + n_2 + n_3 + \ldots\ldots + n_N + n_{N+1} + \ldots.. + n_s \tag{12.62a}$$

and:

$$E = \langle\phi|\tilde{H}|\phi\rangle = \langle\phi|\sum_l b_l^\dagger b_l E_l|\phi\rangle = n_1E_1 + n_2E_2 + \ldots\ldots. + n_NE_N + n_{N+1}E_{N+1} + \ldots. + n_sE_s \tag{12.62b}$$

If the system consists of N-Fermions, then $n_1 = n_2 = n_3 \ldots\ldots = n_N = 1$ and $n_{N+1} = n_{N+2} \ldots\ldots = n_s = 0$, for the ground state of the system.

Thus, $\psi(\mathbf{r}, t) = \sum_n b_n(t)u_n(\mathbf{r})$ is the Fermionic field operator for Schrödinger fields. With the use of Eqns. (12.51), we find that:

$$\begin{aligned}
\left\{\psi(\mathbf{r}, t), \psi^\dagger(\mathbf{r}', t)\right\} &= \psi(\mathbf{r}, t)\psi^\dagger(\mathbf{r}', t) + \psi^\dagger(\mathbf{r}', t)\psi(\mathbf{r}, t) \\
&= \sum_l\sum_n\left(b_nb_l^\dagger u_n(\mathbf{r})u_l^*(\mathbf{r}') + b_l^\dagger b_n u_l^*(\mathbf{r}')u_n(\mathbf{r})\right) \\
&= \sum_l\sum_n \delta_{n,l}u_l^*(\mathbf{r}')u_n(\mathbf{r}) \\
&= \sum_n u_n^*(\mathbf{r}')u_n(\mathbf{r})
\end{aligned} \tag{12.63}$$

which implies:

$$\left\{\psi(\mathbf{r},t),\psi^{+}(\mathbf{r}',t)\right\}=\delta(\mathbf{r}-\mathbf{r}') \tag{12.64a}$$

Similarly:

$$\left\{\psi^{+}(\mathbf{r},t),\psi^{+}(\mathbf{r}',t)\right\}=0=\left\{\psi(\mathbf{r},t),\psi(\mathbf{r}',t)\right\} \tag{12.64b}$$

12.2.3 Matrix Representation of Fermionic Operators

As stated above, n_l can be assigned one of the two integers 1 and 0. This suggests that the vector $|n_l\rangle$ can be represented, apart from a multiplying constant, by $\begin{pmatrix}1\\0\end{pmatrix}$ or $\begin{pmatrix}0\\1\end{pmatrix}$. Here, it is to be noted that this has nothing to do with the spin of particles. The operators b_l, $b_l^{\dagger}$ and the number operator $N_l=b_l^{\dagger}b_l$ are denoted by 2×2 matrices. The requirements $b_l^{\dagger}|0_l\rangle=|1_l\rangle$, $b_l^{\dagger}b_l^{\dagger}|0_l\rangle=b_l^{\dagger}|1_l\rangle=0$, $b_l|1_l\rangle=|0_l\rangle$, $b_l|0_l\rangle=0$, and $N_l|n_l\rangle=n_l|n_l\rangle$ are fulfilled by the following two choices:

$$\begin{aligned}
&b_l^{\dagger}\rightarrow\begin{pmatrix}0&1\\0&0\end{pmatrix}; && b_l\rightarrow\begin{pmatrix}0&1\\0&0\end{pmatrix};\\
&b_l\rightarrow\begin{pmatrix}0&0\\1&0\end{pmatrix}; && b_l^{\dagger}\rightarrow\begin{pmatrix}0&0\\1&0\end{pmatrix};\\
&N_l=b_l^{\dagger}b_l\rightarrow\begin{pmatrix}1&0\\0&0\end{pmatrix};\quad\text{and} && N_l=b_l^{\dagger}b_l\rightarrow\begin{pmatrix}0&0\\0&1\end{pmatrix};\\
&|0_l\rangle\rightarrow\begin{pmatrix}0\\1\end{pmatrix}; && |0_l\rangle\rightarrow\begin{pmatrix}1\\0\end{pmatrix};\\
&|1_l\rangle\rightarrow\begin{pmatrix}1\\0\end{pmatrix} && |1_l\rangle\rightarrow\begin{pmatrix}0\\1\end{pmatrix}
\end{aligned} \tag{12.65}$$

If v_{α} is the number of occupied states that precede the α-th state, then one chooses:

$$b_l=(-1)^{v_{\alpha}}\begin{pmatrix}0&1\\0&0\end{pmatrix}\text{ and }b_l^{\dagger}=(-1)^{v_{\alpha}}\begin{pmatrix}0&0\\1&0\end{pmatrix}.$$

12.2.4 Number Operator

The number operator for a system of N-particles is given by $N=\int\psi^{+}(\mathbf{r},t)\psi(\mathbf{r},t)d^3r$, which for Bosons is:

$$\begin{aligned}
N&=\int\psi^{+}(\mathbf{r},t)\psi(\mathbf{r},t)d^3r=\sum_{l,l'}a_l^{\dagger}a_{l'}\int u_l^{*}(\mathbf{r})u_{l'}(\mathbf{r})d^3r\\
&=\sum_{l,l'}a_l^{\dagger}a_{l'}\delta_{l,l'}=\sum_{l}a_l^{\dagger}a_l
\end{aligned} \tag{12.66}$$

Similarly, for a system of Fermions, $N = \sum_l b_l^\dagger b_l$. We also see that:

$$\begin{aligned}
\left[N, \tilde{H}\right] &= N\tilde{H} - \tilde{H}N \\
&= \left(\sum_n a_n^\dagger a_n\right)\left(\sum_l a_l^\dagger a_l E_l\right) - \left(\sum_l a_l^\dagger a_l E_l\right)\left(\sum_n a_n^\dagger a_n\right) \\
&= \sum_n \sum_l \left(a_n^\dagger a_n a_l^\dagger a_l - a_l^\dagger a_l a_n^\dagger a_n\right) E_l \\
&= \sum_n \sum_l \left\{a_n^\dagger (a_l^\dagger a_n + \delta_{n,l}) a_l - a_l^\dagger (a_n^\dagger a_l + \delta_{l,n}) a_n\right\} E_l \\
&= \sum_n \sum_l \left(a_n^\dagger \delta_{n,l} a_l - a_l^\dagger \delta_{l,n} a_n\right) E_l = 0
\end{aligned} \tag{12.67a}$$

For a system of N-Fermions;

$$\begin{aligned}
\left[N, \tilde{H}\right] &= N\tilde{H} - \tilde{H}N \\
&= \left(\sum_n b_n^\dagger b_n\right)\left(\sum_l b_l^\dagger b_l E_l\right) - \left(\sum_l b_l^\dagger b_l E_l\right)\left(\sum_n b_n^\dagger b_n\right) \\
&= \sum_n \sum_l \left(b_n^\dagger b_n b_l^\dagger b_l - b_l^\dagger b_l b_n^\dagger b_n\right) E_l
\end{aligned} \tag{12.67b}$$

By interchanging $l \leftrightarrow n$, the first term becomes equal to the second term and hence $\left[N, \tilde{H}\right] = 0$. We thus find that:

$$\frac{dN}{dt} = \frac{1}{i\hbar}\left[N, \tilde{H}\right] = 0 \tag{12.68}$$

stating that the number operator commutes with the Hamiltonian, suggesting that the total number of particles remains constant with time. It does not change during the motion of the system.

12.3 System of Weakly Interacting Bosons

A system such as the ground state of an imperfect Bose gas (superfluid helium) can be treated as system of weakly interacting Bosons, where two particles interact through the potential $v(\mathbf{r}, \mathbf{r}')$. The Bosonic field operator and its conjugate adjoint are: $\psi(\mathbf{r}, t) = \sum_k a_k(t) u_k(\mathbf{r})$ and $\psi^\dagger(\mathbf{r}, t) = \sum_{k'} a_{k'}^\dagger(t) u_{k'}^*(\mathbf{r})$, where operators $a_k(t)$ and $a_k^\dagger(t)$ satisfy the Eqn. (12.38). Also note that:

$$a_k(t) = \int u_k^*(\mathbf{r}) \psi(\mathbf{r}, t) d^3r, \text{ and } a_k^\dagger(t) = \int u_k(\mathbf{r}) \psi^\dagger(\mathbf{r}, t) d^3r \tag{12.69}$$

The number operator and the Hamiltonian operator are:

$$N = \int d^3 r \psi^\dagger(\mathbf{r},t)\psi(\mathbf{r},t) \tag{12.70}$$

and:

$$\tilde{H} = \int d^3 r \psi^\dagger(\mathbf{r},t) H \psi(\mathbf{r},t) \tag{12.71}$$

For weakly interacting particles, we can take:

$$u_k(\mathbf{r}) = \frac{1}{\sqrt{V}} e^{i\mathbf{k}.\mathbf{r}} \tag{12.72}$$

where V is the volume of the system confining the Bosons. Therefore:

$$\begin{aligned}\tilde{H} &= \int d^3 r \psi^\dagger(\mathbf{r},t) H \psi(\mathbf{r},t) \\ &= -\frac{\hbar^2}{2mV}\sum_k \sum_{k'} a_{k'}^\dagger a_k \int d^3 r e^{-i\mathbf{k}'.\mathbf{r}} \nabla^2 e^{i\mathbf{k}.\mathbf{r}} \\ &\quad + \frac{1}{2V^2}\sum_{k_1}\sum_{k_1'}\sum_{k_2}\sum_{k_2'} a_{k_1'}^\dagger a_{k_2'}^\dagger a_{k_1} a_{k_2} \iint d^3 r d^3 r' e^{-i\mathbf{k}_1'.\mathbf{r}} e^{-i\mathbf{k}_2'.\mathbf{r}'} \mathrm{v}(\mathbf{r},\mathbf{r}') e^{i\mathbf{k}_1.\mathbf{r}} e^{-i\mathbf{k}_2.\mathbf{r}'} \\ &= \sum_k a_k^\dagger a_k \varepsilon_k + \frac{1}{2V^2}\sum_{k_1}\sum_{k_1'}\sum_{k_2}\sum_{k_2'} a_{k_1'}^\dagger a_{k_2'}^\dagger a_{k_1} a_{k_2} \iint d^3 r d^3 r' e^{i(\mathbf{k}_1-\mathbf{k}_1').\mathbf{r}} e^{i(\mathbf{k}_2-\mathbf{k}_2').\mathbf{r}'} \mathrm{v}(\mathbf{r},\mathbf{r}')\end{aligned} \tag{12.73}$$

where $\varepsilon_k = \frac{\hbar^2 k^2}{2m}$. The factor of $\frac{1}{2}$ in the second term on the right-hand side arises due to the consideration of two particles in a pair. If the particle at $\mathbf{r}$ makes the transition $|k_1\rangle \to |k_1'\rangle$ and at the same time the other particle at $\mathbf{r}'$ goes $|k_2\rangle \to |k_2'\rangle$ because of interacting potential, then the conservation of momentum requires that:

$$\begin{aligned}&\hbar(\mathbf{k}_1 - \mathbf{k}_1') + \hbar(\mathbf{k}_2 - \mathbf{k}_2') = 0 \\ &\Rightarrow (\mathbf{k}_1 - \mathbf{k}_1') = -(\mathbf{k}_2 - \mathbf{k}_2') = \mathbf{q}\end{aligned} \tag{12.74}$$

Thus:

$$\iint d^3 r d^3 r' e^{i(\mathbf{k}_1-\mathbf{k}_1').\mathbf{r}} e^{i(\mathbf{k}_2-\mathbf{k}')_2.\mathbf{r}'} \mathrm{v}(\mathbf{r}-\mathbf{r}') = \iint d^3 r d^3 r' e^{i\mathbf{q}.(\mathbf{r}-\mathbf{r}')} \mathrm{v}(\mathbf{r}-\mathbf{r}') = V\mathrm{v}(\mathbf{q}) \tag{12.75}$$

Here, we have assumed that the interaction depends on inter-particle distance only. This yields:

$$\tilde{H} = \sum_k a_k^\dagger a_k \varepsilon_k + \frac{1}{2V}\sum_{k_1}\sum_{k_1'}\sum_{k_2}\sum_{k_2'} a_{k_1'}^\dagger a_{k_2'}^\dagger a_{k_1} a_{k_2} \mathrm{v}(\mathbf{q}) \tag{12.76}$$

For a weakly interacting system, $\mathbf{q}$-dependence of $v(\mathbf{q})$ is negligible and it is treated as a constant v_0. Further, Eqn. (12.76) is simplified by noting that $\mathbf{k}_1' = \mathbf{k}_1 - \mathbf{q}$ and $\mathbf{k}_2' = \mathbf{k}_2 + \mathbf{q}$, where momentum transferred to a particle due to inter-particle interaction $\hbar\mathbf{q}$, is very small.

The ground state properties of imperfect Bose gas are studied with the use of Eqn. (12.76) and the field operators. In the ground state at zero temperature, the majority of particles occupy zero momentum state and a very small fraction of these occupies nonzero momentum states, because of particle-particle interaction. At zero temperature, the total number of particles is:

$$\begin{aligned} N &= \langle\phi|\sum_q a_q^\dagger a_q|\phi\rangle \\ &= \langle\phi|a_0^\dagger a_0|\phi\rangle + \langle\phi|\sum_{q\neq 0} a_q^\dagger a_q|\phi\rangle \\ &= N_0 + N_n \end{aligned} \tag{12.77}$$

$N_0(\gg N_n)$ is the number of particles in zero momentum state. Also, $N_0 \gg 1$ allows us to take $\langle\phi|a_0 a_0^\dagger|\phi\rangle = \langle\phi|a_0^\dagger a_0|\phi\rangle - 1 \approx N_0$, suggesting that $a_0^\dagger$ and a_0 can be treated as scalar and we take $a_0^\dagger = a_0 = \sqrt{N_0}$. Then:

$$\sum_{k_1}\sum_{k_1'}\sum_{k_2}\sum_{k_2'} a_{k_1'}^\dagger a_{k_2'}^\dagger a_{k_1} a_{k_2} v(\mathbf{q}) \simeq v_0\left[N^2 + N\sum_{q\neq 0}\left(a_q^\dagger a_{-q}^\dagger + a_q a_{-q} + 2a_q^\dagger a_q\right)\right] \tag{12.78}$$

The Hamiltonian then simplifies to:

$$\tilde{H} = \frac{v_0 N^2}{2V} + \sum_q a_q^\dagger a_q\left(\varepsilon_q + \frac{Nv_0}{V}\right) + \frac{Nv_0}{2V}\sum_{q\neq 0}\left(a_q^\dagger a_{-q}^\dagger + a_q a_{-q}\right) \tag{12.79}$$

which is a non-diagonal Hamiltonian. Bogolubov transformations are used to diagonalize such Hamiltonians. The Hamiltonian Eqn. (12.79) has been used to study the properties of the ground state of superfluid helium, where the presence of small interparticle interaction does not allow all particles to occupy the zero momentum state even at zero temperature.

12.4 Free Electron System

A system of electrons moving freely within the volume V is described by field operators and a Hamiltonian operator. The field operator and its conjugate adjoint are:

$$\psi(\mathbf{r},t) = \frac{1}{\sqrt{V}}\sum_{k\sigma} b_{k\sigma} e^{i\mathbf{k}.\mathbf{r}} \text{ and } \psi^\dagger(\mathbf{r},t) = \frac{1}{\sqrt{V}}\sum_{k'\sigma} b_{k'\sigma}^\dagger e^{i\mathbf{k}'.\mathbf{r}} \tag{12.80}$$

where k represents all relevant quantum numbers and σ represents spin $(\uparrow,\downarrow)$. The Hamiltonian operator that has the kinetic energy of electrons and the electron-electron interaction term is:

$$\tilde{H}=-\frac{\hbar^2}{2m}\int\psi^{\dagger}(\mathbf{r},t)\nabla^2\psi(\mathbf{r},t)d^3r+\iint\psi^{\dagger}(\mathbf{r},t)\psi^{\dagger}(\mathbf{r}',t)\mathrm{v}(|\mathbf{r}-\mathbf{r}'|)\psi(\mathbf{r},t)\psi(\mathbf{r}',t)d^3rd^3r' \quad (12.81)$$

Here $\mathrm{v}(|\mathbf{r}-\mathbf{r}'|)$ is the electron-electron interaction that can be a coulombic repulsive interaction (normal conducting state) or a screened phonon assisted attractive interaction (superconductivity state). Evaluation of the first term gives:

$$T=\sum_{k\sigma}\frac{\hbar^2k}{2m}b^{\dagger}_{k\sigma}b_{k\sigma} \quad (12.82)$$

Evaluation of the second term yields:

$$V_{\text{int}}=\frac{1}{2V^2}\sum_{k_1,k_2,k_1',k_2'}\sum_{\sigma_1,\sigma_2}b^{\dagger}_{k_1'\sigma_1}b^{\dagger}_{k_2'\sigma_2}b_{k_1\sigma_1}b_{k_2\sigma_2}\iint d^3rd^3r'e^{i(\mathbf{k}_1-\mathbf{k}_1').\mathbf{r}}e^{i(\mathbf{k}_2-\mathbf{k}_2').\mathbf{r}'}\mathrm{v}(\mathbf{r},\mathbf{r}') \quad (12.83)$$

The interaction V_{int} is a sum over all processes in which two electrons having momentums $\hbar\mathbf{k}_1$ and $\hbar\mathbf{k}_2$ interact through the potential and a momentum $(\hbar\mathbf{k}_1-\hbar\mathbf{k}_1')=-(\hbar\mathbf{k}_2-\hbar\mathbf{k}_2')=\hbar\mathbf{q}$ is transferred from one to the other, as is seen in Fig. 12.1.

We thus have:

$$\begin{aligned}V_{\text{int}}&=\frac{1}{2V^2}\sum_{k_1,k_2,q}\sum_{\sigma_1,\sigma_2}b^{\dagger}_{\mathbf{k}_1-\mathbf{q},\sigma_1}b^{\dagger}_{\mathbf{k}_2+\mathbf{q},\sigma_2}b_{\mathbf{k}_1\sigma_1}b_{\mathbf{k}_2\sigma_2}\iint d^3rd^3r'e^{i\mathbf{q}.(\mathbf{r}-\mathbf{r}')}\mathrm{v}(|\mathbf{r}-\mathbf{r}'|)\\ \Rightarrow V_{\text{int}}&=\frac{1}{2V}\sum_{k_1,k_2,q}\sum_{\sigma_1,\sigma_2}b^{\dagger}_{\mathbf{k}_1-\mathbf{q},\sigma_1}b^{\dagger}_{\mathbf{k}_2+\mathbf{q},\sigma_2}b_{\mathbf{k}_1\sigma_1}b_{\mathbf{k}_2\sigma_2}\mathrm{v}(\mathbf{q})\end{aligned} \quad (12.84)$$

where, we have used:

$$\mathrm{v}(\mathbf{q})=\frac{1}{V}\iint d^3rd^3r'e^{i\mathbf{q}.(\mathbf{r}-\mathbf{r}')}\mathrm{v}(|\mathbf{r}-\mathbf{r}'|) \quad (12.85)$$

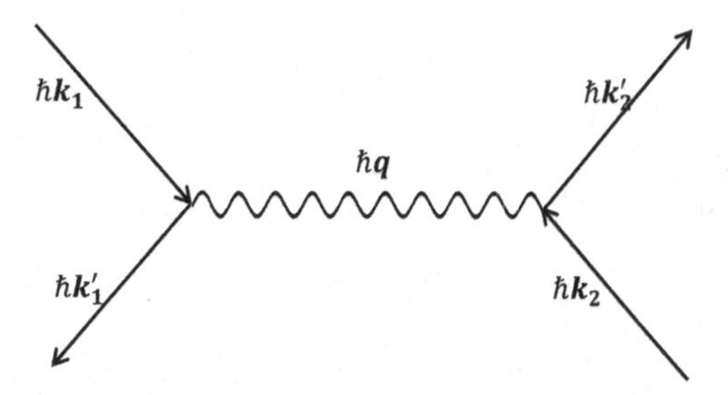

FIGURE 12.1
An electron in the state $|k_1\rangle$ goes to state $|k_1'\rangle$ by emitting a phonon of momentum $\hbar q$, which is absorbed by an electron in state $|k_2\rangle$ to make the transition to state $|k_2'\rangle$.

Therefore:

$$\tilde{H} = \sum_{k\sigma} \frac{\hbar^2 k}{2m} b_{k\sigma}^{\dagger} b_{k\sigma} + \frac{1}{2V} \sum_{k_1,k_2,q\sigma_1,\sigma_2} \sum b_{k_1'\sigma_1}^{\dagger} b_{k_2'\sigma_2}^{\dagger} b_{k_1\sigma_1} b_{k_2\sigma_2} \mathrm{v}(\mathbf{q}) \tag{12.86}$$

For a metal, $\mathrm{v}(\mathbf{q})$ is a screened electron-electron interaction potential. When screening due to both electrons and lattice vibrations (phonons) is taken into account, the screened electron-electron interaction potential is given by:

$$\mathrm{v}(q) = \frac{e^2}{\varepsilon_0 (q^2 + k_0^2)} \left[1 + \frac{\omega^2(q)}{\omega^2 - \omega^2(q)} \right] \tag{12.87}$$

where k_0 is the inverse screening length, $\omega(q)$ is phonon frequency, and $\hbar\omega$ is the difference of energy between the two electronic states. As is seen, effective potential between two electrons is positive (repulsive) for $\omega > \omega(q)$ and it is negative (attractive) when $\omega < \omega(q)$. The positive value of $\mathrm{v}(q)$ causes resistivity and metals exist a normal conducting state. However, a phase transition from the normal conducting state to the superconductivity state occurs in a metal when $\mathrm{v}(\mathbf{q})$ becomes negative. Further details on this topic can be found in specialized textbooks on condensed matter physics.

12.5 Solved Examples

1. Define $\mathbf{C}(\mathbf{r},t) = \left(\frac{i\varepsilon_0}{2\hbar} \right) \mathbf{E}(\mathbf{r},t)$ and then evaluate commutator $\left[\mathbf{A}(\mathbf{r},t), \mathbf{C}^{\dagger}(\mathbf{r}',t) \right]$.

SOLUTION

We have:

$$\mathbf{C}(\mathbf{r},t) = \frac{1}{2\sqrt{V}} \sum_{\mathbf{k}} \sum_{\alpha} \sqrt{\frac{\varepsilon_0 \omega}{2\hbar}} \left\{ a_{\mathbf{k}\alpha}(t) \hat{\varepsilon}_{\alpha}(\mathbf{k}) e^{i\mathbf{k}.\mathbf{r}} - a_{\mathbf{k}\alpha}^{\dagger}(t) \hat{\varepsilon}_{\alpha}(\mathbf{k}) e^{-i\mathbf{k}.\mathbf{r}} \right\} \tag{12.88}$$

Then:

$$\begin{aligned}
&\left[\mathbf{A}(\mathbf{r},t), \mathbf{C}^{\dagger}(\mathbf{r}',t) \right] \\
&= \mathbf{A}(\mathbf{r},t)\mathbf{C}^{\dagger}(\mathbf{r}',t) - \mathbf{C}^{\dagger}(\mathbf{r}',t)\mathbf{A}(\mathbf{r},t) \\
&= \frac{1}{4V} \sum_{\mathbf{k}} \sum_{\alpha} \sum_{\mathbf{k}'} \sum_{\alpha'} \Big[\hat{\varepsilon}_{\alpha} \left\{ a_{\mathbf{k}\alpha}(t) e^{i\mathbf{k}.\mathbf{r}} + a_{\mathbf{k}\alpha}^{\dagger}(t) e^{-i\mathbf{k}.\mathbf{r}} \right\} \hat{\varepsilon}_{\alpha'} \left\{ a_{\mathbf{k}'\alpha'}^{\dagger}(t) e^{-i\mathbf{k}'.\mathbf{r}'} - a_{\mathbf{k}'\alpha'}(t) e^{i\mathbf{k}'.\mathbf{r}'} \right\} \\
&\quad - \hat{\varepsilon}_{\alpha'} \left\{ a_{\mathbf{k}'\alpha'}^{\dagger}(t) e^{-i\mathbf{k}'.\mathbf{r}'} - a_{\mathbf{k}'\alpha'}(t) e^{i\mathbf{k}'.\mathbf{r}'} \right\} \hat{\varepsilon}_{\alpha} \left\{ a_{\mathbf{k}\alpha}(t) e^{i\mathbf{k}.\mathbf{r}} + a_{\mathbf{k}\alpha}^{\dagger}(t) e^{-i\mathbf{k}.\mathbf{r}} \right\} \Big] \\
&= \frac{1}{4V} \sum_{\mathbf{k}} \sum_{\alpha} \sum_{\mathbf{k}'} \sum_{\alpha'} \hat{\varepsilon}_{\alpha} \hat{\varepsilon}_{\alpha'} \Big\{ \left[a_{\mathbf{k}\alpha}, a_{\mathbf{k}'\alpha'}^{\dagger} \right] e^{i\mathbf{k}.\mathbf{r}} e^{-i\mathbf{k}'.\mathbf{r}'} - \left[a_{\mathbf{k}\alpha}, a_{\mathbf{k}'\alpha'} \right] e^{i\mathbf{k}.\mathbf{r}} e^{i\mathbf{k}'.\mathbf{r}'} \\
&\quad + \left[a_{\mathbf{k}\alpha}^{\dagger}, a_{\mathbf{k}'\alpha'}^{\dagger} \right] e^{-i\mathbf{k}.\mathbf{r}} e^{-i\mathbf{k}'.\mathbf{r}'} - \left[a_{\mathbf{k}\alpha}^{\dagger}, a_{\mathbf{k}'\alpha'} \right] e^{-i\mathbf{k}.\mathbf{r}} e^{i\mathbf{k}'.\mathbf{r}'} \Big\}
\end{aligned} \tag{12.89}$$

With the use of $\left[a^{\dagger}_{k\alpha}, a^{\dagger}_{k'\alpha'}\right] = 0 = \left[a_{k\alpha}, a_{k'\alpha'}\right]$, $\left[a_{k\alpha}, a^{\dagger}_{k'\alpha'}\right] = \delta_{k,k'}\delta_{\alpha,\alpha'} = -\left[a^{\dagger}_{k'\alpha'}, a_{k\alpha}\right]$, we get:

$$
\begin{aligned}
\left[\mathbf{A}(\mathbf{r},t), \mathbf{C}^{\dagger}(\mathbf{r}',t)\right] &= \frac{1}{4V}\sum_{\mathbf{k}}\sum_{\alpha}\sum_{\mathbf{k}'}\sum_{\alpha'}\hat{\varepsilon}_{\alpha}\hat{\varepsilon}_{\alpha'}\left[\delta_{\mathbf{k},\mathbf{k}'}\delta_{\alpha\alpha'}e^{i\mathbf{k}.\mathbf{r}}e^{-i\mathbf{k}'.\mathbf{r}'} + \delta_{\mathbf{k}',\mathbf{k}}\delta_{\alpha\alpha'}e^{-i\mathbf{k}.\mathbf{r}}e^{i\mathbf{k}'.\mathbf{r}'}\right] \\
&= \frac{1}{2V}\sum_{\mathbf{k}}\left[e^{i\mathbf{k}.(\mathbf{r}-\mathbf{r}')} + e^{-i\mathbf{k}.(\mathbf{r}-\mathbf{r}')}\right] \\
&= \delta(\mathbf{r}-\mathbf{r}')
\end{aligned}
\tag{12.90}
$$

where we used $\frac{1}{V}\sum_{\mathbf{k}} e^{i\mathbf{k}.\mathbf{x}} = \delta(\mathbf{x})$, $\delta(\mathbf{x}) = \delta(-\mathbf{x})$ and $\left(\hat{\varepsilon}_{\alpha}\right)^2 = \left(\hat{\varepsilon}_1\right)^2 + \left(\hat{\varepsilon}_2\right)^2 = 2$.

2. Consider the Fermionic field operator $\psi^{\dagger}(\mathbf{r}) = \int \frac{d^3k}{(2\pi)^3} b^{\dagger}_{\mathbf{k}} e^{-i\mathbf{k}.\mathbf{r}}$, where $\left\{b_{\mathbf{k}}, b^{\dagger}_{\mathbf{k}'}\right\} = (2\pi)^3\delta(\mathbf{k}-\mathbf{k}')$ show that $\left\{\psi(\mathbf{r}), \psi^{\dagger}(\mathbf{r}')\right\} = \delta(\mathbf{r}-\mathbf{r}')$ and $\psi^{\dagger}(\mathbf{r})\psi^{\dagger}(\mathbf{r}') = 0$.

SOLUTION

We have $\psi(\mathbf{r}) = \int \frac{d^3k}{(2\pi)^3} b_{\mathbf{k}} e^{i\mathbf{k}.\mathbf{r}}$. Thus:

$$
\begin{aligned}
\left\{\psi(\mathbf{r}), \psi^{\dagger}(\mathbf{r}')\right\} &= \psi(\mathbf{r})\psi^{\dagger}(\mathbf{r}') + \psi^{\dagger}(\mathbf{r}')\psi(\mathbf{r}) \\
&= \frac{1}{(2\pi)^6}\iint d^3k d^3k'\left(b_{\mathbf{k}}b^{\dagger}_{\mathbf{k}'}e^{i\mathbf{k}.\mathbf{r}}e^{-i\mathbf{k}'.\mathbf{r}'} + b^{\dagger}_{\mathbf{k}'}b_{\mathbf{k}}e^{-i\mathbf{k}'.\mathbf{r}'}e^{i\mathbf{k}.\mathbf{r}}\right) \\
&= \frac{1}{(2\pi)^6}\iint d^3k d^3k'(2\pi)^3\delta(\mathbf{k}-\mathbf{k}')e^{i\mathbf{k}.\mathbf{r}}e^{-i\mathbf{k}'.\mathbf{r}'} \\
&= \frac{1}{(2\pi)^3}\int e^{i\mathbf{k}.(\mathbf{r}-\mathbf{r}')}d^3k = \delta(\mathbf{r}-\mathbf{r}')
\end{aligned}
\tag{12.91}
$$

and:

$$
\begin{aligned}
\left\{\psi^{\dagger}(\mathbf{r}), \psi^{\dagger}(\mathbf{r}')\right\} &= \psi(\mathbf{r})^{\dagger}\psi^{\dagger}(\mathbf{r}') + \psi^{\dagger}(\mathbf{r}')\psi^{\dagger}(\mathbf{r}) \\
&= \frac{1}{(2\pi)^6}\iint d^3k d^3k'\left(b^{\dagger}_{\mathbf{k}}b^{\dagger}_{\mathbf{k}'}e^{-i\mathbf{k}.\mathbf{r}}e^{-i\mathbf{k}'.\mathbf{r}'} + b^{\dagger}_{\mathbf{k}'}b^{\dagger}_{\mathbf{k}}e^{-i\mathbf{k}'.\mathbf{r}'}e^{-i\mathbf{k}.\mathbf{r}}\right) \\
&= \frac{1}{(2\pi)^6}\iint d^3k d^3k'\left(b^{\dagger}_{\mathbf{k}}b^{\dagger}_{\mathbf{k}'} + b^{\dagger}_{\mathbf{k}'}b^{\dagger}_{\mathbf{k}}\right)e^{-i\mathbf{k}.\mathbf{r}}e^{-i\mathbf{k}'.\mathbf{r}'}
\end{aligned}
\tag{12.92}
$$

Since $\left(b^{\dagger}_{\mathbf{k}}b^{\dagger}_{\mathbf{k}'} + b^{\dagger}_{\mathbf{k}'}b^{\dagger}_{\mathbf{k}}\right) = 0$, we get $\psi(\mathbf{r})^{\dagger}\psi^{\dagger}(\mathbf{r}') + \psi^{\dagger}(\mathbf{r}')\psi^{\dagger}(\mathbf{r}) = 0$, which implies that $\psi^{\dagger}(\mathbf{r})\psi^{\dagger}(\mathbf{r}') = 0$.

3. A quantized scalar field is given by:

$$\phi(r,t)=\sum_k c\sqrt{\frac{\hbar}{2\omega V}}\left[a_k(t)e^{i\mathbf{k}.\mathbf{r}}+a_k^{\dagger}(t)e^{-i\mathbf{k}.\mathbf{r}}\right] \tag{12.93}$$

with:

$$\frac{\omega}{c}=\sqrt{k^2+\left(\frac{m_0c}{\hbar}\right)^2} \text{ and } a_k(t)=a_k(0)e^{-i\omega t} \tag{12.94}$$

a_k and $a_k^{\dagger}$ are Bosonic operators. Show that:

$$\left[\phi(r,t),\pi(r,t)\right]=i\hbar\delta(r-r'), \text{ where } \pi=\frac{1}{c^2}\frac{\partial\phi}{\partial t} \tag{12.95}$$

SOLUTION

Let us calculate:

$$\pi(r,t)=\frac{1}{c^2}\frac{\partial\phi(r,t)}{\partial t}=-\frac{i}{c}\sum_k\sqrt{\frac{\hbar\omega}{2V}}\left[a_k(t)e^{i\mathbf{k}.\mathbf{r}}-a_k^{\dagger}(t)e^{-i\mathbf{k}.\mathbf{r}}\right] \tag{12.96}$$

Then:

$$\begin{aligned}\left[\phi(r,t),\pi(r',t)\right]&=\phi(r,t)\pi(r',t)-\pi(r',t)\phi(r,t)\\&=\sum_k\sum_{k'}\left(\frac{-i\hbar}{2V}\right)\left\{\left[a_k(t)e^{i\mathbf{k}.\mathbf{r}}+a_k^{\dagger}(t)e^{-i\mathbf{k}.\mathbf{r}}\right]\left[a_{k'}(t)e^{i\mathbf{k}'.\mathbf{r}'}-a_{k'}^{\dagger}(t)e^{-i\mathbf{k}'.\mathbf{r}'}\right]\right.\\&\quad\left.-\left[a_{k'}(t)e^{i\mathbf{k}'.\mathbf{r}'}-a_{k'}^{\dagger}(t)e^{-i\mathbf{k}'.\mathbf{r}'}\right]\left[a_k(t)e^{i\mathbf{k}.\mathbf{r}}+a_k^{\dagger}(t)e^{-i\mathbf{k}.\mathbf{r}}\right]\right\}\\&=\left(\frac{-i\hbar}{2V}\right)\sum_k\sum_{k'}\left\{\left[a_k,a_{k'}\right]e^{i\mathbf{k}.\mathbf{r}}e^{i\mathbf{k}'.\mathbf{r}'}-\left[a_k,a_{k'}^{\dagger}\right]e^{i\mathbf{k}.\mathbf{r}}e^{-i\mathbf{k}'.\mathbf{r}'}\right.\\&\quad\left.+\left[a_k^{\dagger},a_{k'}\right]e^{-i\mathbf{k}.\mathbf{r}}e^{i\mathbf{k}'.\mathbf{r}'}-\left[a_k^{\dagger},a_{k'}^{\dagger}\right]e^{-i\mathbf{k}.\mathbf{r}}e^{-i\mathbf{k}'.\mathbf{r}'}\right\}\end{aligned} \tag{12.97}$$

Since $\left[a_k,a_{k'}\right]=0=\left[a_k^{\dagger},a_{k'}^{\dagger}\right]$ and $\left[a_k,a_{k'}^{\dagger}\right]=\delta_{kk'}=-\left[a_k^{\dagger},a_{k'}\right]$, we get:

$$\begin{aligned}\left[\phi(r,t),\pi(r',t)\right]&=\left(\frac{i\hbar}{2V}\right)\sum_k\sum_{k'}\left\{\delta_{k,k'}e^{i\mathbf{k}.\mathbf{r}}e^{-i\mathbf{k}'.\mathbf{r}'}+\delta_{k,k'}e^{-i\mathbf{k}.\mathbf{r}}e^{i\mathbf{k}'.\mathbf{r}'}\right\}\\&=\left(\frac{i\hbar}{2V}\right)\sum_k\left(e^{i\mathbf{k}.(\mathbf{r}-\mathbf{r}')}+e^{-i\mathbf{k}.(\mathbf{r}-\mathbf{r}')}\right)=i\hbar\delta(\mathbf{r}-\mathbf{r}')\end{aligned} \tag{12.98}$$

Here we have used $\frac{1}{V}\sum_k e^{i\mathbf{k}.(\mathbf{r}-\mathbf{r}')}=\delta(\mathbf{r}-\mathbf{r}')$ and $\delta(\mathbf{r}-\mathbf{r}')=\delta(\mathbf{r}'-\mathbf{r})$.

4. Consider the Bosonic field operator $\psi(\mathbf{r}) = \int \frac{d^3p}{(2\pi)^3} a_{\mathbf{p}} e^{i\mathbf{p}.\mathbf{r}}$, where

$$\left[a_{\mathbf{p}}, a_{\mathbf{p}'}^{\dagger}\right] = a_{\mathbf{p}} a_{\mathbf{p}'}^{\dagger} - a_{\mathbf{p}'}^{\dagger} a_{\mathbf{p}} = (2\pi)^3 \delta(\mathbf{p} - \mathbf{p}') \tag{12.99}$$

Show that $\left[\psi(\mathbf{r}), \psi^{\dagger}(\mathbf{r}')\right] = \delta(\mathbf{r} - \mathbf{r}')$ and $\left[\psi^{\dagger}(\mathbf{r}), \psi^{\dagger}(\mathbf{r}')\right] = 0$.

SOLUTION

The conjugate adjoint field operator is:

$$\psi^{\dagger}(\mathbf{r}) = \int \frac{d^3p'}{(2\pi)^3} a_{\mathbf{p}'}^{\dagger} e^{-i\mathbf{p}'.\mathbf{r}} \tag{12.100}$$

Then:

$$\begin{aligned}
\left[\psi(\mathbf{r}), \psi^{\dagger}(\mathbf{r}')\right] &= \psi(\mathbf{r})\psi^{\dagger}(\mathbf{r}') - \psi(\mathbf{r})\psi^{\dagger}(\mathbf{r}') \\
&= \frac{1}{(2\pi)^6} \sum_{\mathbf{p}} \sum_{\mathbf{p}'} \iint d^3p\, d^3p' \left(a_{\mathbf{p}} a_{\mathbf{p}'}^{\dagger} - a_{\mathbf{p}'}^{\dagger} a_{\mathbf{p}}\right) e^{i(\mathbf{p}.\mathbf{r} - \mathbf{p}'.\mathbf{r}')} \\
&= \frac{1}{(2\pi)^6} \sum_{\mathbf{p}} \sum_{\mathbf{p}'} \iint d^3p\, d^3p' (2\pi)^3 \delta_{\mathbf{p},\mathbf{p}'} e^{i(\mathbf{p}.\mathbf{r} - \mathbf{p}'.\mathbf{r}')} \\
&= \frac{1}{(2\pi)^3} \sum_{\mathbf{p}} d^3p\, e^{i\mathbf{p}.(\mathbf{r} - \mathbf{r}')} \\
&= \delta(\mathbf{r} - \mathbf{r}')
\end{aligned} \tag{12.101}$$

Here we used $\left[a_{\mathbf{p}}, a_{\mathbf{p}'}\right] = (2\pi)^3 \delta(\mathbf{p} - \mathbf{p}')$.

Next:

$$\begin{aligned}
\left[\psi^{\dagger}(\mathbf{r}), \psi^{\dagger}(\mathbf{r}')\right] &= \psi(\mathbf{r})^{\dagger} \psi^{\dagger}(\mathbf{r}') - \psi^{\dagger}(\mathbf{r})\psi^{\dagger}(\mathbf{r}') \\
&= \frac{1}{(2\pi)^6} \sum_{\mathbf{p}} \sum_{\mathbf{p}'} \iint d^3p\, d^3p' \left(a_{\mathbf{p}}^{\dagger} a_{\mathbf{p}'}^{\dagger} - a_{\mathbf{p}'}^{\dagger} a_{\mathbf{p}}^{\dagger}\right) e^{i(\mathbf{p}.\mathbf{r} - \mathbf{p}'.\mathbf{r}')}
\end{aligned} \tag{12.102a}$$

Since $\left[a_{\mathbf{p}}^{\dagger}, a_{\mathbf{p}'}^{\dagger}\right] = \left(a_{\mathbf{p}}^{\dagger} a_{\mathbf{p}'}^{\dagger} - a_{\mathbf{p}'}^{\dagger} a_{\mathbf{p}}^{\dagger}\right) = 0$, we have:

$$\left[\psi^{\dagger}(\mathbf{r}), \psi^{\dagger}(\mathbf{r}')\right] = 0 \tag{12.102b}$$

5. Consider the case of a 1D harmonic oscillator and prove that $\left[a, f(a^{\dagger})\right] = \frac{\partial}{\partial a^{\dagger}} f(a^{\dagger})$ and $\left[a^{\dagger}, f(a)\right] = -\frac{\partial}{\partial a} f(a)$. Show that $\left[a, e^{\alpha a^{\dagger}}\right] = \alpha e^{\alpha a^{\dagger}}$ and $e^{-\alpha a^{\dagger}} \alpha e^{-\alpha a^{\dagger}} = \alpha + a$.

SOLUTION

For a 1D harmonic oscillator, a and $a^\dagger$ are defined in terms of the position operator x and the momentum operator p as follows:

$$a = \frac{m\omega x + ip}{\sqrt{2\hbar m\omega}} \text{ and } a^\dagger = \frac{m\omega x - ip}{\sqrt{2\hbar m\omega}} \tag{12.103}$$

Therefore:

$$\begin{aligned}
\left[a, f(a^\dagger)\right]\psi &= af(a^\dagger)\psi - f(a^\dagger)a\psi \\
&= \frac{1}{\sqrt{2\hbar m\omega}}\left[m\omega x f(a^\dagger)\psi + \hbar\frac{\partial}{\partial x}\left\{f(a^\dagger)\psi\right\}\right] - f(a^\dagger)a\psi \\
&= \frac{1}{\sqrt{2\hbar m\omega}}\left[m\omega x f(a^\dagger)\psi + \hbar\frac{\partial f(a^\dagger)}{\partial x}\psi + \hbar f(a^\dagger)\frac{\partial}{\partial x}\psi\right] - f(a^\dagger)a\psi \\
&= \frac{f(a^\dagger)}{\sqrt{2\hbar m\omega}}\left(m\omega x + ip\right)\psi + \frac{\hbar}{\sqrt{2\hbar m\omega}}\frac{\partial f(a^\dagger)}{\partial x}\psi - f(a^\dagger)a\psi \\
&= \frac{\hbar}{\sqrt{2\hbar m\omega}}\frac{\partial f(a^\dagger)}{\partial x}\psi = \frac{\hbar}{\sqrt{2\hbar m\omega}}\frac{\partial f(a^\dagger)}{\partial a^\dagger}\frac{\partial a^\dagger}{\partial x}\psi \\
&= \frac{\partial f(a^\dagger)}{\partial a^\dagger}\psi
\end{aligned} \tag{12.104a}$$

which implies:

$$\left[a, f(a^\dagger)\right] = \frac{\partial}{\partial a} f(a^\dagger) \tag{12.104b}$$

Similarly, we get:

$$\left[a^\dagger, f(a)\right] = -\frac{\partial}{\partial a} f(a) \tag{12.104c}$$

Hence:

$$\begin{aligned}
&\left[a, e^{\alpha a^\dagger}\right] = \frac{\partial}{\partial a^\dagger}\left(e^{\alpha a^\dagger}\right) = \alpha e^{\alpha a^\dagger} \\
&\Rightarrow a e^{\alpha a^\dagger} - e^{\alpha a^\dagger} a = \alpha e^{\alpha a^\dagger}.
\end{aligned}$$

On operating with $e^{-\alpha a^\dagger}$, we obtain:

$$\begin{aligned}
&e^{-\alpha a^\dagger} a e^{\alpha a^\dagger} - e^{-\alpha a^\dagger} e^{\alpha a^\dagger} a = e^{-\alpha a^\dagger}\alpha e^{\alpha a^\dagger} \\
&\Rightarrow e^{-\alpha a^\dagger} a e^{\alpha a^\dagger} = a + \alpha
\end{aligned} \tag{12.105}$$

6. The number operator is defined as $N = \int d^3r\psi^\dagger(\mathbf{r})\psi(\mathbf{r})$, where the Bosonic field operators satisfy $\left[\psi(\mathbf{r}),\psi^\dagger(\mathbf{r}')\right] = \delta(\mathbf{r}-\mathbf{r}')$, $\left[\psi(\mathbf{r}),\psi(\mathbf{r}')\right] = 0$ and $\left[\psi^\dagger(\mathbf{r}),\psi^\dagger(\mathbf{r}')\right] = 0$. Show that $N\psi^\dagger = \psi^\dagger(N+1)$ and $N\psi = \psi(N-1)$.

SOLUTION

Let us take:

$$\begin{aligned}
\left[N,\psi(\mathbf{r})\right] &= N\psi(\mathbf{r}) - \psi(\mathbf{r})N \\
&= \int \psi^\dagger(\mathbf{r}')\psi(\mathbf{r}')d^3r'\psi(\mathbf{r}) - \psi(\mathbf{r})\int \psi^\dagger(\mathbf{r}')\psi(\mathbf{r}')d^3r' \\
&= \int \psi^\dagger(\mathbf{r}')\psi(\mathbf{r}')\psi(\mathbf{r})d^3r' - \int \psi(\mathbf{r})\psi^\dagger(\mathbf{r}')\psi(\mathbf{r}')d^3r'
\end{aligned} \tag{12.106a}$$

$\left[\psi(\mathbf{r}),\psi(\mathbf{r}')\right] = 0$ implies $\psi(\mathbf{r}')\psi(\mathbf{r}) = \psi(\mathbf{r})\psi(\mathbf{r}')$ and hence:

$$\begin{aligned}
\left[N,\psi(\mathbf{r})\right] &= \int \psi^\dagger(\mathbf{r}')\psi(\mathbf{r})\psi(\mathbf{r}')d^3r' - \int \psi(\mathbf{r})\psi^\dagger(\mathbf{r}')\psi(\mathbf{r}')d^3r' \\
&= \int \left[\psi^\dagger(\mathbf{r}')\psi(\mathbf{r}) - \psi(\mathbf{r})\psi^\dagger(\mathbf{r}')\right]\psi(\mathbf{r}')d^3r' \\
&= -\int \delta(\mathbf{r}'-\mathbf{r})\psi(\mathbf{r}')d^3r' \\
&= -\psi(\mathbf{r})
\end{aligned} \tag{12.106b}$$

which means $N\psi(\mathbf{r}) = \psi(\mathbf{r})(N-1)$. We next take:

$$\begin{aligned}
\left[N,\psi^\dagger(\mathbf{r})\right] &= N\psi^\dagger(\mathbf{r}) - \psi^\dagger(\mathbf{r})N \\
&= \int \psi^\dagger(\mathbf{r}')\psi(\mathbf{r}')d^3r'\psi^\dagger(\mathbf{r}) - \psi^\dagger(\mathbf{r})\int \psi^\dagger(\mathbf{r}')\psi(\mathbf{r}')d^3r' \\
&= \int \psi^\dagger(\mathbf{r}')\psi(\mathbf{r}')\psi^\dagger(\mathbf{r})d^3r' - \int \psi^\dagger(\mathbf{r})\psi^\dagger(\mathbf{r}')\psi(\mathbf{r}')d^3r' \\
&= \int \psi^\dagger(\mathbf{r}')\left\{\psi^\dagger(\mathbf{r})\psi(\mathbf{r}') + \delta(\mathbf{r}'-\mathbf{r})\right\}d^3r' - \int \psi^\dagger(\mathbf{r})\psi^\dagger(\mathbf{r}')\psi(\mathbf{r}')d^3r' \\
&= \int \psi^\dagger(\mathbf{r}')\delta(\mathbf{r}'-\mathbf{r})d^3r' \\
&= \psi^\dagger(\mathbf{r})
\end{aligned} \tag{12.107}$$

which implies that $N\psi^\dagger(\mathbf{r}) = \psi^\dagger(\mathbf{r})(N+1)$. This means that N decreases (increases) by one on interchanging its order with $\psi(\mathbf{r})$ $\left(\psi^\dagger(\mathbf{r})\right)$.

12.6 Exercises

1. Show that for a system of Bosons (a) $a_i\left(a_i^\dagger\right)^2 = a_i^\dagger\left(2 + a_i^\dagger a_i\right)$ and (b) $\left[N_i, N_j\right] = 0$ for $i \neq j$. N_i and N_j are number operators.
2. For a system of Bosons, the interaction Hamiltonian is $H_1 = \sum_{i,j,k,l} d_{ijkl} a_i^\dagger a_j^\dagger a_k a_l$. Find the commutator $[H_1, a_i]$.
3. For a system of independent harmonic oscillators, the Hamiltonian is given by $H = \sum_i \varepsilon_i \alpha_i^\dagger \alpha_i$. Determine the equation of motion for the annihilation and creation operators for both the cases of Bosons and Fermions, with the use of the Heisenberg equation.
4. Show that the Bosonic number operator $N = \sum_i a_i^\dagger a_i$ commutes with the Hamiltonian $H = \sum_{j,k} a_j^\dagger t_{jk} a_k + \sum_{j,k,l,m} a_j^\dagger a_k^\dagger v_{jklm} a_m a_l$.
5. Consider a system of non-interacting two particles which can be Bosons or Fermions. The state and the Hamiltonian are defined by $|\alpha, \beta\rangle = a_\alpha^\dagger a_\beta^\dagger |0\rangle$ and $H = \sum_\alpha \varepsilon_\alpha a_\alpha^\dagger a_\alpha$ for both the cases. Find the energy $E_{\alpha\beta}$ of the system.
6. For a system of Fermions, the matrix representation of the annihilation operator b_l, the creation operator $b_l^\dagger$ and the state $|1_l\rangle$ are given as follows:

$$b_l = (-1)^{v_\alpha}\begin{pmatrix} 0 & 1 \\ 0 & 0 \end{pmatrix},\ b_l^\dagger = (-1)^{v_\alpha}\begin{pmatrix} 0 & 0 \\ 1 & 0 \end{pmatrix} \text{ and } |1_l\rangle = \begin{pmatrix} 0 \\ 1 \end{pmatrix}.$$

   Calculate $N_l = b_l^\dagger b_l$ and show that $b_l b_l^\dagger + b_l^\dagger b_l = 1$ and $N_l^2 |1_l\rangle = N_l |1_l\rangle$.
7. For a system of Bosons, $N|n_1, n_2, n_3 ......\rangle = \sum_i N_i |n_1, n_2, n_3 ......\rangle = n|n_1, n_2, n_3 ......\rangle$. Show that:
   a. $Na^\dagger |n_1, n_2, n_3 ......\rangle = (n+1)a^\dagger |n_1, n_2, n_3 ......\rangle$, and
   b. $Na|n_1, n_2, n_3 ......\rangle = (n-1)a|n_1, n_2, n_3 ......\rangle$.
8. A particle of mass m with charge q moves under the influence of a uniform magnetic field $\mathbf{B} = B_0 \hat{k}$. The Hamiltonian of the particle is $H = \frac{1}{2m}(\mathbf{p} - q\mathbf{A})^2$, where $\mathbf{A} = \frac{B_0}{2}\left(-y\hat{i} + x\hat{j}\right)$. Show that the Hamiltonian can be expressed as: $H = \frac{p_z^2}{2m} + \left(a^\dagger a + \frac{1}{2}\right)\hbar\omega$, with $aa^\dagger - a^\dagger a = 1$. What are the values of a and ω?

Annexure A: Useful Formulae

I. Table of Integrals

1. $\int_0^\infty x^n e^{-ax} dx = \frac{n!}{a^{n+1}}, n \geq 0$ and $a > 0$.

2. $\int_0^\infty x^{2n+1} e^{-ax^2} dx = \frac{n!}{2a^{n+1}}, n \geq 0$ and $a > 0$.

3. $\int_0^\infty x^{2n} e^{-ax^2} dx = \frac{(2n-1)!!}{2^{n+1} a^n} \sqrt{\frac{\pi}{a}}$, where $(2l+1)!! = 1.3.5.....(2l+1)$, $0!! = 1$, and $(-1)!! = 1$, $a > 0$.

4. $\int_0^\infty \frac{e^{-ax}}{\sqrt{x}} dx = \sqrt{\frac{\pi}{a}}$.

5. $\int_{-\infty}^\infty \frac{e^{-iqx}}{a^2 + x^2} dx = \frac{\pi}{a} e^{-|aq|}, a \neq 0, q$ is real.

6. Integration by parts:

   $\int_a^b F dG = [FG]_a^b - \int_a^b G dF$, where both F and G are function of x and integration is done from $x = a$ to $x = b$.

II. Series and Expansions

1. Taylor Series:

$$f(x) = \sum_{n=0}^{\infty} \frac{(x - x_0)^n}{n!} \left. \frac{d^n f(x)}{dx^n} \right|_{x=x_0}.$$

2. Exponential Series:

$$e^x = \sum_{k=0}^{\infty} \frac{x^k}{k!}.$$

3. Geometric Series:

$$(1-x)^{-1} = \sum_{k=0}^{\infty} x^k$$

4. Trigonometric series:

$$\sin x = \sum_{k=0}^{\infty} (-1)^k \frac{x^{2k+1}}{(2k+1)!},$$

$$\cos x = \sum_{k=0}^{\infty} (-1)^k \frac{x^{2k}}{(2k)!},$$

$$\tan x = x + \frac{1}{3}x^3 + \frac{2}{15}x^5 + \frac{17}{315}x^7 + \ldots\ldots.$$

5. Binomial Expansion:

$$(a+x)^n = \sum_{k=0}^{n} \frac{n!}{k!(n-k)!} x^k a^{n-k}, \; n > 0$$

$$(1+x)^q = \sum_{k=0}^{q} \frac{q!}{k!(q-k)!} x^k, \; q > 0.$$

III. Basic Functional Relations

1. $e^{ix} = \cos x + i \sin x.$
2. $e^x = \cosh x + \sinh x.$
3. $\sin(x \pm y) = \sin x \cos y \pm \sin y \cos x.$
4. $\cos(x \pm y) = \cos x \cos y \mp \sin x \sin y.$
5. $\tan(x \pm y) = \dfrac{\tan x \pm \tan y}{1 \mp \tan x \tan y}.$

IV. Coordinate Systems

1. $\mathbf{V} = V_x\hat{x} + V_y\hat{y} + V_z\hat{z}$ in Cartesian coordinate system.
2. $\mathbf{V} = V_r\hat{r} + V_\theta\hat{\theta} + V_\phi\hat{\phi}$ in spherical polar coordinate system.
   $(\hat{x}, \hat{y}, \hat{z})$ and $(\hat{r}, \hat{\theta}, \hat{\phi})$ are sets of orthonormal vectors in Cartesian and spherical polar coordinate systems, respectively.
3. $x = r\sin\theta\cos\phi$
   $y = r\sin\theta\sin\phi$
   $z = r\cos\theta$
4. $$\begin{bmatrix} \hat{x} \\ \hat{y} \\ \hat{z} \end{bmatrix} = \begin{bmatrix} \sin\theta\cos\phi & \cos\theta\cos\phi & -\sin\phi \\ \sin\theta\sin\phi & \cos\theta\sin\phi & \cos\phi \\ \cos\theta & -\sin\theta & 0 \end{bmatrix} \begin{bmatrix} \hat{r} \\ \hat{\theta} \\ \hat{\phi} \end{bmatrix}.$$
5. Volume element, $dxdydz = r^2 dr\sin\theta d\theta d\phi$, with $-\infty \le (x, y, z) \le \infty$ and $0 \le r \le \infty$, $0 \le \theta \le \pi$, $0 \le \phi \le 2\pi$.
6. The gradient operator:

$$\nabla = \hat{x}\frac{\partial}{\partial x} + \hat{y}\frac{\partial}{\partial y} + \hat{z}\frac{\partial}{\partial z}$$
$$= \hat{r}\frac{\partial}{\partial r} + \frac{\hat{\theta}}{r}\frac{\partial}{\partial r} + \frac{\hat{\phi}}{r\sin\theta}\frac{\partial}{\partial\phi}.$$

7. The Laplacian operator:

$$\nabla^2 = \frac{\partial^2}{\partial x^2} + \frac{\partial^2}{\partial y^2} + \frac{\partial^2}{\partial z^2}$$
$$= \frac{1}{r^2}\frac{\partial}{\partial r}\left(r^2\frac{\partial}{\partial r}\right) + \frac{1}{r^2\sin\theta}\frac{\partial}{\partial\theta}\left(\sin\theta\frac{\partial}{\partial\theta}\right) + \frac{1}{r^2\sin^2\theta}\frac{\partial^2}{\partial\phi^2}.$$

Annexure B: Dirac Delta Function

The Dirac delta function does not fit into the usual definition of a function that has value at each and every point within its domain. It is a function on the real line, which has infinite value at origin and zero value everywhere else. It is defined by:

$$\delta(x) = \begin{cases} \infty & x = 0 \\ 0 & x \neq 0 \end{cases} \tag{B.1}$$

when origin is taken at $x = 0$ and by:

$$\delta(x-a) = \begin{cases} \infty & x = a \\ 0 & x \neq a \end{cases} \tag{B.2}$$

for change of origin to $x = a$. Dirac delta function is constrained to satisfy:

$$\int_{-\infty}^{\infty} \delta(x)dx = 1 \quad \text{and} \quad \int_{-\infty}^{\infty} \delta(x-a)dx = 1 \tag{B.3}$$

Another defining characteristic of delta function is:

$$\int_{-\infty}^{\infty} f(x)\delta(x)dx = f(0), \text{ and } \int_{-\infty}^{\infty} f(x)\delta(x-a)dx = f(a) \tag{B.4}$$

where $f(x)$ is a well-defined function both at $x = 0$ and $x = a$. The range of integration in Eqn. (B4) need not be from $-\infty$ to ∞, it can be over any domain surrounding the critical point at which the delta function is not zero. Generalization of definition of Dirac delta function to three-dimensional space is made as follows:

$$\delta(\mathbf{r} - a) = \delta(x - a_x)\delta(y - a_y)\delta(z - a_z) \tag{B.5}$$

with the constrain:

$$\int f(\mathbf{r})\delta(\mathbf{r} - a)d\mathbf{r} = f(a) \tag{B.6}$$

where ranges of integration include $\left(a_x, a_y, a_z\right)$.

I. Properties of Delta Function

$$\begin{aligned}
&1.\ x\delta(x)=0\\
&2.\ x\delta(x-a)=a\delta(x-a)\\
&3.\ \delta(x)=\delta(-x)\\
&4.\ f(x)\delta(x-a)=f(a)\delta(x-a)\\
&5.\ \delta(xa)=\frac{1}{a}\delta(x),\ a>0\\
&6.\ \delta(x^2-a^2)=\frac{1}{2|a|}\left[\delta(x-a)+\delta(x+a)\right]\\
&7.\ \delta\left[f(x)\right]=\sum_{i=1}^{n}\frac{\delta(x-x_i)}{|f'(x_i)|},\ \text{with}\ f'(x_i)=\left.\frac{df}{dx}\right|_{x=x_i}\ \text{and}\ f(x_i)=0\\
&8.\ \int\delta(x-a)\delta(x-b)\,dx=\delta(b-a)
\end{aligned} \tag{B.7}$$

II. Representation of Delta Function

Delta function is viewed in various forms. Some of forms are:

1. Limiting value of Rectangle function:

$$R_\sigma(x)=\begin{cases}\dfrac{1}{2\sigma} & -\sigma\le(x-a)\le\sigma\\ 0 & -\sigma>(x-a)>\sigma\end{cases} \tag{B.8a}$$

$$\delta(x-a)=\lim_{\sigma\to0}R_\sigma(x) \tag{B.8b}$$

2. Limiting value of Gaussian function:

$$\delta(x-a)=\lim_{\sigma\to0}\frac{1}{\sigma\sqrt{2\pi}}e^{-(x-a)^2/2\sigma^2},\ \sigma>0 \tag{B.9}$$

3. Limiting value of Lorentzian function:

$$\delta(x)=\lim_{\varepsilon\to0}\frac{1}{\pi}\frac{\varepsilon}{\left(x^2+\varepsilon^2\right)} \tag{B.10}$$

4. Fourier series:

$$\delta(x)=\frac{1}{2\pi}\sum_{n=-\infty}^{\infty}e^{inx}\ \text{on the interval of}\ [-\pi,\pi] \tag{B.11}$$

5. Fourier transformation:

$$\delta(x) = \frac{1}{2\pi}\int_{-\infty}^{\infty} e^{ikx}\,dk \tag{B.12a}$$

A three dimensional generalization of which is:

$$\delta(\mathbf{r}) = \frac{1}{(2\pi)^3}\int e^{i\mathbf{k}.\mathbf{r}}\,d^3k \tag{B.12b}$$

6. Differentiation of Heaviside Step function:

$$\theta(x-a) = \begin{cases} 0 & x < a \\ 1 & x \geq a \end{cases} \tag{B.13a}$$

$$\delta(x-a) = \frac{d}{dx}\theta(x-a) \Leftrightarrow \theta(x-a) = \int_{-\infty}^{x} \delta(t-a)\,dt \tag{B.13b}$$

7. Laplacian:

$$\delta(\mathbf{r}) = \nabla^2\left(-\frac{1}{4\pi|\mathbf{r}|}\right) \tag{B.14}$$

Answers to Exercises

Chapter 1

2. $\nu = 13.14 \times 10^{15}$ Hz and $\lambda = 2.28 \times 10^{-8}$ meter
3. 9.65×10^{8} Hz
4. 2.19×10^{6} meter / sec
6. $v_g = \dfrac{\hbar k}{m}$ and $v_{ph} = v_g / 2$
7. 1.41eV
8. $E = -\dfrac{\hbar^2}{2ma_0^2}$.
9. $N = \dfrac{1}{\sqrt{\pi a_0^3}}$.

Chapter 2

1. 13532 nm
2. 1.71 nm
3. $N = \left(\dfrac{\alpha^3}{\pi}\right)^{3/2}$
4. 0.042
7. $\dfrac{\hbar k}{mr^2}\hat{r}$
9. $V(x) = \dfrac{2x^2\hbar^2}{ma^4}; E = \dfrac{\hbar^2}{ma^2}$
10. (a) $x = \dfrac{1}{\beta}$; (b) $\langle x \rangle = \dfrac{3}{2\beta}; \langle x^2 \rangle = \dfrac{3}{\beta^2}$

    $\langle p \rangle = 0; \langle p^2 \rangle = \beta^2\hbar^2$
11. $E = \dfrac{3\hbar^2}{ma^2}$.

Chapter 3

4. Linearly independent.

5. (a) $|Y\rangle = \begin{bmatrix} 0 \\ -2 \\ -2 \\ 0 \end{bmatrix}$ and (b) $|X\rangle = \begin{bmatrix} 1 \\ 1 \\ 1 \\ 1 \end{bmatrix}$

6. (b) Eigenvalues are ±1.

7. $|A| = 1$

9. When both eigenvectors belong to same eigenvalue, degeneracy.

10. Eigenvalues are; 0,1 and –1. Normalized Eigenvectors are; $\frac{1}{\sqrt{2}}\begin{bmatrix} 1 \\ 0 \\ -1 \end{bmatrix}$, $\begin{bmatrix} 1/2 \\ 1/\sqrt{2} \\ 1/2 \end{bmatrix}$ and $\begin{bmatrix} -1/2 \\ 1/\sqrt{2} \\ -1/2 \end{bmatrix}$. States are nondegenerate.

11. Three eigenvalues of Matrix A are 1, –1 and –1; 2-fold degeneracy. Eigenvalues of matrix B are 2, 2 and –2; 2-fold degeneracy.

13. $|Y\rangle = \frac{1}{\sqrt{2}}\begin{bmatrix} 3 \\ i \end{bmatrix}, \langle X|X\rangle = \langle Y|Y\rangle = 5.$

Chapter 5

1. $\frac{[l(l+1) - m(m-1)]\hbar^2}{2I_1} + \frac{m^2\hbar^2}{2I_2}$

2. Yes, $l = 1$.

5. Yes, $l = 1$ and $m = 0$

6. $[L_x, x] = 0, [L_x, y] = i\hbar z, [L_x, z] = -i\hbar y. [L_x, p_x] = 0, [L_x, p_y] = i\hbar p_z, [L_x, p_z] = -i\hbar p_y$

7. $\mathbf{L}^2 = 2\hbar^2\begin{bmatrix} 1 & 0 & 0 \\ 0 & 1 & 0 \\ 0 & 0 & 1 \end{bmatrix}$, $L_x = \frac{\hbar}{\sqrt{2}}\begin{bmatrix} 0 & 1 & 0 \\ 1 & 0 & 1 \\ 0 & 1 & 0 \end{bmatrix}$, $L_y = \frac{\hbar}{\sqrt{2}}\begin{bmatrix} 0 & -i & 0 \\ i & 0 & -i \\ 0 & i & 0 \end{bmatrix}$, and

$$L_z = \hbar\begin{bmatrix} 1 & 0 & 0 \\ 0 & 0 & 0 \\ 0 & 0 & -1 \end{bmatrix}$$

8. (a) $B = \frac{\sqrt{3}}{2}$, (b) $\psi = \frac{1}{\sqrt{2}}\left(Y_{1,-1} - Y_{11}\right) f(r)$, (c) $\hbar$ and $-\hbar$

9. (a) Probability is $\frac{1}{2}$ for both (b) zero.

Chapter 6

1. $\alpha = \frac{Z}{a_0}$, $A = \left(\frac{Z^3}{\pi a_0^3}\right)^{1/2}$, $E = -\frac{Z^2\hbar^2}{2\mu a_0^2}$, $\langle V \rangle = 2E$ and $\langle T \rangle = -E$.
2. a_0 and $4a_0$
3. $\langle x \rangle = \langle y \rangle = \langle z \rangle = 0$ and $\langle x^2 \rangle = \langle y^2 \rangle = \langle z^2 \rangle = a_0^2$.
4. $\langle p_x \rangle = \langle p_y \rangle = \langle p_z \rangle = 0$ and $\langle p_x^2 \rangle = \langle p_x^2 \rangle = \langle p_x^2 \rangle = \frac{\hbar^2}{3a_0^2}$.
5. $\psi(x,y) = \frac{2}{\sqrt{ab}} \sin\left(\frac{n_1 \pi x}{a}\right) \sin\left(\frac{n_2 \pi y}{b}\right)$ and $E = \frac{\hbar^2\pi^2}{2m}\left(\frac{n_1^2}{a^2} + \frac{n_2^2}{b^2}\right) + V_0$, both $n_1$and n_2 take values; 1,2,3,……
6. $P = \frac{1}{4}$.
7. $E = \left(n_1 + \frac{1}{2}\right)\hbar\omega_1 + \left(n_2 + \frac{1}{2}\right)\hbar\omega_2$, with $\omega_1 = \sqrt{\frac{k_1}{m}}$ and $\omega_2 = \sqrt{\frac{k_2}{m}}$ and

   $\psi_{n_1,n_2}(\rho,\sigma) = \frac{1}{\left(2^{n_1+n_2} n_1! n_2!\right)^{1/2}} \left(\frac{m}{\hbar\pi}\right)^{1/2} (\omega_1\omega_2)^{1/4} e^{-(\rho^2+\sigma^2)/2} H_{n_1}(\rho) H_{n_2}(\sigma)$, with

   $\rho = \sqrt{\frac{m\omega_1}{\hbar}}x$ and $\sigma = \sqrt{\frac{m\omega_2}{\hbar}}x$
8. $A = \frac{1}{4}\left(\frac{1}{6\pi a_0^5}\right)^{1/2}$, probability is 0.053.
9. (a) $E_n = \frac{\hbar^2\pi^2 n^2}{ma^2} + \frac{\hbar^2 k_z^2}{2m}$, (b) $\nu = 8.32 \times 10^{14}$ Hz.

Chapter 7

1. (a) $E_0^{(1)} = \frac{\lambda\hbar}{2m\omega}$, (b) $A = \left(\frac{m\omega}{\pi\hbar}\right)^{1/4}$ and $\beta = \frac{m\omega}{2\hbar}$
2. $E_0^{(1)} = \frac{3}{32}\frac{(\hbar\omega)^2}{mc^2}$

3. $E = E_0^{(0)} + E_0^{(1)} = \dfrac{\hbar^2\pi^2}{2ma_2} + \dfrac{V_0}{2}$

4. $E_0^{(1)} \approx \dfrac{e^2R^2}{6\pi_0 a_0^3}$

5. $E_n^{(1)} = 0$ and $E_n^{(2)} = \dfrac{B^2}{2m\omega^2}$

6. (a) $A = \left(\dfrac{2\alpha}{\pi}\right)^{1/4}$, (b) $\langle T\rangle = \dfrac{\hbar^2\alpha}{2m}$ and $\langle V\rangle = \dfrac{m\omega^2}{8\alpha}$.

7. $E_g = -\dfrac{b^2m}{\pi\hbar^2}$

8. $E_g = \dfrac{1}{\sqrt{2}}\hbar\omega$

9. $E_g = -\dfrac{27}{28}\left(\dfrac{e^2}{4\pi\varepsilon_0 a_0}\right)$

10. $E_n = \left(\dfrac{\hbar^2}{2m}\right)^{1/3}\left[\dfrac{3}{8}(2n+1)qE\pi\right]^{2/3}, n = 0,1,2,....$

Chapter 8

1. $f(k,\theta) = \dfrac{\pi\mu\beta}{q\hbar^2}e^{-qa}$ and $\sigma(k,\theta) = \dfrac{\pi^2\mu^2\beta^2}{q^2\hbar^4}e^{-2qa}$, with $q = 2ksin(\theta/2)$.

2. (a) $\delta_0(k) = \dfrac{\pi\mu\beta}{\hbar^2}$, (b) $\sigma_{tot} = \dfrac{4\pi}{k^2}\left(\dfrac{\pi\mu\beta}{\hbar^2}\right)^2$, (c) $\delta_0 = \dfrac{\pi}{20}$

3. $\left(\nabla^2 + k^2\right)\phi(r) = \dfrac{2m}{\hbar^2}V(r)e^{ikz}$

5. (a) $\dfrac{\lambda^2}{k^2}\sin^2(ka) \ll 1$, (b) $a \ll \dfrac{1}{\lambda}$.

6. $G_0(k,x,x') = \dfrac{i}{2k}e^{ik|x-x'|}$, $\psi(x) = e^{ikx} - \dfrac{im}{\hbar^2k}\displaystyle\int_{-\infty}^{\infty} e^{ik|x-x'|}V(x')\psi(x')dx'$

7. (a) $f(k,k') = \dfrac{-im}{\hbar^2k}\displaystyle\int_{-\infty}^{\infty} e^{-ik'x'}V(x')\psi(x')dx'$; $\sigma(k,k') = |f(k,k')|^2$

8. $\sigma(k) = \dfrac{m^2b^2}{\hbar^4k^2}$ and $R = \dfrac{m^2b^2}{\hbar^4k^2}$. The Born approximation is applicable only for high energy scattering case.

Chapter 9

1. $$\psi(x_1,x_2)=\frac{1}{\sqrt{2}a}\left[\cos\left(\frac{\pi}{a}(4x_1-3x_2)\right)-\cos\left(\frac{\pi}{a}(4x_1+3x_2)\right)+\cos\left(\frac{\pi}{a}(4x_2-3x_1)\right)-\cos\left(\frac{\pi}{a}(4x_2+3x_1)\right)\right]$$

2. $\psi(r_1,r_2)=\frac{1}{2V}\left[e^{i\mathbf{k}.(\mathbf{r}_1-\mathbf{r}_2)}+e^{-i\mathbf{k}.(\mathbf{r}_1-\mathbf{r}_2)}\right]\{\alpha(1)\beta(2)-\beta(1)\alpha(2)\}$ for singlet ($S=0$) state and,

$$\psi(r_1,r_2)=\frac{1}{2V}\left[e^{i\mathbf{k}.(\mathbf{r}_1-\mathbf{r}_2)}-e^{-i\mathbf{k}.(\mathbf{r}_1-\mathbf{r}_2)}\right]\left\{\begin{array}{c}\sqrt{2}\alpha(1)\alpha(2)\\ \alpha(1)\beta(2)+\beta(1)\alpha(2)\\ \sqrt{2}\beta(1)\beta(2)\end{array}\right\}$$ for triplet ($S=1$) state

3. 9.03 eV
4. (a) The Bohr radius in ionized helium atom will be one-half of its value in hydrogen atom.

   (b) $\psi(1,2)=\frac{1}{2}[\psi_N(\mathbf{r}_1)\psi_{N'}(\mathbf{r}_2)+\psi_N(\mathbf{r}_2)\psi_{N'}(\mathbf{r}_1)]\{\alpha(1)\beta(2)-\beta(1)\alpha(2)\}$ for singlet sate ($S=0$) and, $$\psi(1,2)=\frac{1}{2}[\psi_N(\mathbf{r}_1)\psi_{N'}(\mathbf{r}_2)-\psi_N(\mathbf{r}_2)\psi_{N'}(\mathbf{r}_1)]\left\{\begin{array}{c}\sqrt{2}\alpha(1)\alpha(2)\\ \alpha(1)\beta(2)+\beta(1)\alpha(2)\\ \sqrt{2}\beta(1)\beta(2)\end{array}\right\}$$

   for triplet states ($S=1$). Where,

   $$\psi_N(\mathbf{r}_1)=\left[\left(\frac{2}{na_0}\right)^3\frac{(n-l-1)!}{2n\{(n+l)!\}^3}\right]^{1/2}e^{-(r_1/na_0)}\left(\frac{2r_1}{na_0}\right)^l L_{n+l}^{2l+1}\left(\frac{2r_1}{na_0}\right)Y_{lm}(\theta_1,\varphi_1).$$

   The $\psi_{N'}(\mathbf{r}_2)$ is obtained from $\psi_N(\mathbf{r}_1)$ by replacing $r_1\to r_2$, $\theta_1\to\theta_2$, $\phi_1\to\phi_2$, $n\to n'$, $l\to l'$ and $m\to m'$.

   (c) Wave function changes the sign on interchanging 1 and 2 for both the cases of singlet state and triplet states, hence it obeys Pauli's exclusion principle.

6. $E_0=\frac{\hbar^2\pi^2}{ma^2}-V_0$ and $\psi_0(x_1,x_2)=\sqrt{\frac{2}{a}}\sin\left(\frac{\pi}{a}(x_1-x_2)\right)$

7. (a) $\psi(x_1,x_2)=\frac{\alpha^2}{\sqrt{\pi}}(x_2-x_1)e^{-\alpha^2\left(x_1^2+x_2^2\right)/2}$ with $\alpha=\sqrt{\frac{m\omega}{\hbar}}$ and (b) $\left\langle(x_2-x_1)^2\right\rangle=\frac{1}{\alpha^2}$.

Chapter 10

1. $\frac{\pi e^2}{c\varepsilon_0 m}$. Absorption frequency is equal to natural frequency of oscillator.

2. $P_{1s\to 2p} = 0.55\left(\dfrac{eE_0a_0}{\hbar\omega}\right)^2 e^{-2\omega_{21}\tau}$, where $\omega_{21} = \dfrac{3}{8\hbar}\left(\dfrac{e^2}{4\pi\varepsilon_0 a_0}\right)$.

3. $P_{1\to n} = \left(\dfrac{A\gamma}{\hbar ac}\right)^2\left[\dfrac{\alpha^2-\beta^2}{(\alpha^2+\gamma^2)(\beta^2+\gamma^2)}\right]^2$, with

   $\alpha = \dfrac{(n+1)\pi}{a}$; $\beta = \dfrac{(n-1)\pi}{a}$ and $\gamma = \dfrac{\hbar\pi^2}{2ma^2c}(n^2-1)$.

4. $P_{1\to 2} = 0.27$.
5. $T = 590.7$ K

Chapter 11

4. $\begin{bmatrix}1\\0\\0\\0\end{bmatrix} e^{-im_0c^2t/\hbar}$, $\begin{bmatrix}0\\1\\0\\0\end{bmatrix} e^{-im_0c^2t/\hbar}$, $\begin{bmatrix}0\\0\\1\\0\end{bmatrix} e^{im_0c^2t/\hbar}$ and $\begin{bmatrix}0\\0\\0\\1\end{bmatrix} e^{im_0c^2t/\hbar}$.

Chapter 12

3. $a(t) = a(0)e^{-i\varepsilon t/\hbar}$ and $a^\dagger(t) = a^\dagger(0)e^{i\varepsilon t/\hbar}$.
5. $E_{\alpha\beta} = \varepsilon_\alpha + \varepsilon_\beta$.
8. $a = \sqrt{\dfrac{m\omega}{2\hbar}}\left(\dfrac{y}{2} + \dfrac{p_x}{qB_0}\right) + \dfrac{i}{\sqrt{2m\hbar\omega}}\left(p_y - \dfrac{qB_0x}{2}\right)$ and $\omega = \dfrac{qB_0}{m}$.

Bibliography

The existing literature on Quantum mechanics is large and diverse. The bibliography of those books that were referred by the author during the teaching of various courses on quantum mechanics to undergraduates and post-graduates is presented here. Many books on elementary and advanced quantum mechanics have been written and it is very likely that there may have been glaring omissions on part of author. The basic purpose of writing this book has been to provide a textbook on quantum mechanics for undergraduates and post-graduates, and hence no attempt has been made to quote original papers and Advanced level books on quantum mechanics.

Agarwal B K and Hari Prakash, *Quantum Mechanics*, Prentice-Hall of India, New Delhi, 1997.

Aruldhas G, *Quantum Mechanics* (Second Edition), PHI Learning Private Limited, New Delhi, 2009.

Bohm David, *Quantum Theory*, Prentice Hall, New York, 1951.

Fitzpatrick Richard, *Quantum Mechanics*, World Scientific publishing Co. PTE. Ltd., 2015.

Ghatak A K and Lokanathan S, *Quantum Mechanics, Theory and Applications* (Fourth Edition), Macmillan, India Limited, New Delhi, 1999.

Griffiths David J, *Introduction to Quantum Mechanics* (Second Edition), Pearson-Prentice Hall, Upper Saddle River, NJ, 2005.

Kumar Ajit, *Fundamentals of Quantum Mechanics*, Cambridge University Press, Cambridge CB2 8BS, United Kingdom (2018).

Kyriakos Tamvakis, *Problems and Solutions in Quantum Mechanics*, Cambridge University Press, New York, 2005.

Lim Yung-Kuo (editor), *Problems and Solutions on Quantum Mechanics*, World Scientific Publishing Co. PTE. Ltd., Singapore, 1998.

Mathews P M and Venkatesan K, *A Text Book of Quantum Mechanics* (Second Edition), Tata McGraw-Hill, Education Pvt. Ltd., New Delhi, 2010.

Merzbacher E, *Quantum Mechanics* (Second Edition), John Wiley & Sons, New York, 1970.

Neuman John Von, *Mathematical Foundations of Quantum Mechanics*, Princeton University Press, Princeton, 1955.

O'Reilly Eoin, *Quantum Theory of Solids*, Tylor & Francis Group, London and New York, 2002.

Pauling L and Wilson E B, *Introduction to Quantum Mechanics, with Applications to Chemistry*, McGraw-Hill publishing Co., Ltd., London, 1935.

Powell J L and Crasemann B, *Quantum Mechanics*, Dover Publications Inc., 2015; originally published by Addison Wesley Publishing Co. Reading, MA. 1961.

Sakurai J J, *Modern Quantum Mechanics* (Revised Edition), Addison-Wesley Publishing Co., Singapore, PTE. Ltd., Singapore 1999.

Schiff L I, *Quantum Mechanics* (Third Edition), McGraw-Hill, Kogakusha, Tokyo, 1981.

Shankar R, *Principles of Quantum Mechanics* (Second Edition), Plenum Press, New York, 1994.

Yariv Amnon, *An Introduction to Theory and Applications of Quantum Mechanics*, John Wiley & Sons Inc., New York, 1982.

Index

W

Z

Printed in the United States
by Baker & Taylor Publisher Services